Translational Dynamics and Magnetic Resonance

Principles of Pulsed Gradient Spin Echo NMR

Paul T. Callaghan

MacDiarmid Institute for Advanced Materials and Nanotechnology, School of Chemical and Physical Sciences, Victoria University of Wellington, New Zealand

OXFORD

UNIVERSITY PRESS

OXFORD
UNIVERSITY PRESS

Great Clarendon Street, Oxford, OX2 6DP,
United Kingdom

Oxford University Press is a department of the University of Oxford.
It furthers the University's objective of excellence in research, scholarship,
and education by publishing worldwide. Oxford is a registered trade mark of
Oxford University Press in the UK and in certain other countries

First published 2011
First published in paperback 2014

Impression: 1

Published in the United States of America by Oxford University Press
198 Madison Avenue, New York, NY 10016, United States of America

British Library Cataloguing in Publication Data
Data available

ISBN 978–0–19–955698–4 (hbk.)
ISBN 978–0–19–870082–1 (pbk.)

Printed and bound by
Clays Ltd, St Ives plc

For Mavis and Ernest Callaghan

Preface

It is now over 60 years since Hahn pointed out that spin echoes in nuclear magnetic resonance (NMR) could be used to measure molecular translational motion. This facility arises because nuclear spins carry a phase determined by the history of their residence in magnetic fields. If, through our own volition or by consequence of sample structure, magnetic fields can be given some spatial variation, and if the spin-bearing molecules translate, then the spin phases can be made to tell the story of that migration. At its simplest, that story might be 'the molecule undergoes free Brownian motion'. With the advent of new ideas in pulsed magnetic field gradient NMR associated with the work of Stejskal and Tanner in the 1960s, that simple perspective underwent a radical change, and the library of stories was set to grow. However, despite their application by a few specialist practitioners, it was not until magnetic field gradients became widely available in commercial NMR spectrometers in the 1990s that the use of magnetic resonance methods to measure translational dynamics became widespread. Since then the community of NMR users has not only expanded substantially, but NMR is now employed by many more branches of science. Furthermore, with the development of powerful pulse-switching and pulse-shaping capability on modern NMR spectrometers, the variety of experiments available to general users has widened and become more sophisticated.

Many of these changes have arisen because other developments in NMR have led to new insights regarding the way measurements can be made. Principal among these have been magnetic resonance imaging, multi-dimensional NMR, multiple quantum NMR, and low field or one-sided access NMR. A further driver has been the coupling of NMR spectroscopy with rheology and chromatography, where flow is an inherent feature of interest, while another has been the application of NMR to problems of flow and dispersion in chemical engineering.

Unlike modern bio-molecular NMR, the multitude of tricks used to measure molecular translational motion take place without the need for high spectral resolution. While they may be practiced with bio-molecules using the full power of the modern high-resolution, high magnetic field NMR spectrometer, they mostly work, with equal power, at low field and in the absence of spectral discrimination. They naturally lend themselves to that new branch of NMR technology that concerns itself with 'outside the laboratory' applications—in geophysics and petroleum physics, in horticulture, in food technology, in security screening and in environmental monitoring. Furthermore, the diverse physical quantities accessed, the flow and its spatial gradient, the dispersion or diffusion tensors, their anisotropy, their temporal and spatial structure, and their correlation with other physical variables make them transparently useful, with few ambiguities in interpretation. Thus they offer a power and diversity to time domain, low field, and low resolution portable NMR methods, far in excess of that covered by the simple relaxography with which they are usually associated. The translational dynamics of molecules provide a signature for molecular size and shape size, the visco-elasticity of the surrounding fluid medium, their organisation into supramolecular assemblies, their exchange between different sites, their intermittent binding, their confinement by a surrounding matrix or phase boundary, and the topology of that confinement.

This book takes us through the various underlying principles of molecular translational dynamics, outlining the ways in which magnetic resonance, through the use of magnetic field gradients, can reveal those dynamics. The book covers the full range of time and frequency domain methodologies, showing how they can be used. It covers advances in scattering and diffraction methods, multidimensional exchange and correlation experiments, and orientational correlation methods ideal for studying dynamics in anisotropic environments. At the heart of all these new methods resides the ubiquitous spin echo, a phenomenon the discovery of which underpins nearly every major development in magnetic resonance methodology.

Translational Dynamics and Magnetic Resonance is a companion to the author's earlier monograph, *Principles of Nuclear Magnetic Resonance Microscopy*, a book that explains some of the foundations of magnetic resonance, magnetic resonance imaging, and magnetic field gradient methodology. When introducing underlying magnetic resonance concepts in Chapters 3 and 4, the author draws heavily on the earlier text, updating where relevant, sometimes expanding the discussion, sometimes contracting. But the rest of the book contains mostly new material, summarising some of the significant developments in pulsed gradient spin echo NMR from the past two decades, as well as providing primers on thermal energy and diffusion in Chapter 1 and on flow and dispersion in Chapter 2. As with the earlier book, *Translational Dynamics and Magnetic Resonance* is written at a level suitable for a graduate student in physics, chemistry, or engineering, with some background in undergraduate mathematics.

Because of the focus on principles in this book, examples of applications have been selected so as to illustrate key ideas, and no attempt has been made to provide a comprehensive summary of the complete body of experimental work reported in the literature. However, one of the pleasures in writing a book like this is the chance to appreciate the achievements of others who work in the same field of research. I wish to thank the many authors who have given me permission to reproduce examples of their work. I have learned a great deal from so many international colleagues, but I would especially like to acknowledge Bernhard Bluemich, Suzanne Fielding, Ken Packer, Alex Pines, Ed Samulski, Joe Seymour, Siegfied Stapf, and Janez Stepisnik. I am very grateful to my own PhD students, Craig Eccles, Craig Trotter, Peter Daivis, Yang Xia, Andrew Coy, Bertram Manz, Jim Hargreaves, Craig Rofe, Miki Komlosh, Maria Kilfoil, Alexander Khrapitchev, Roger Meder, Ryan Cormier, Rosario Lopez-Gonzalez, Robin Dykstra, Antoine Lutti, Simon Rogers, Kate Washburn, Stefan Hill, Allan Raudsepp, Mark Hunter, Brad Douglass, and Lauren Burcaw, for their beautiful experimental work and for teaching me a great deal. Also assisting my education have been postdoctoral fellows and student visitors Ted Garver, Philip Back, Ross Mair, Ute Skibbe, Melanie Britton, Sarah Codd, Joe Seymour, Elmar Fischer, Song-I Han, William Holmes, Sophie Godefroy, Brett Ryland, Daniel Corbett, David Fairhurst,

Daniel Polders, Ocean Mercier, Petrik Galvosas, Ying Qiao, Andy Jackson, Penny Hubbard, Antje Gottwald, Gui Madelin, Kirk Feindel, and Jen Brown. And we have all benefited so much from group visitors Ken Jeffrey, Milo Shott, Peter Stilbs, Olle Soderman, Noam Kaplan, Ed Samulski, Charles Johnson, Lourdes de Vargas, Yang Xia, Mark Warner, Suzanne Fielding, Istvan Furo, David Ailion, Ralph Colby, Lou Madsen, Thomas Meersman, Galina Pavlovskaya, and Panos Photinos. A segment of the book was written during a period of leave in California, at UC Berkeley, and I am particularly grateful to Alex and Ditsa Pines for their warm hospitality during that time.

I want to thank Gill Sutherland at the Royal Society of New Zealand, and John Spencer, Margaret Brown, Sarah Dadley, David Bibby, and other Victoria University of Wellington colleagues for their support. During the writing of this book, I became diagnosed with cancer. Much was written while I was undertaking courses of chemotherapy, and I could not have possibly completed this project without being supported by a New Zealand Government James Cook Fellowship, administered by the Royal Society of New Zealand. This Fellowship enabled me to conserve my energies and focus on my writing in between my various treatments. Crucially, I was able to take on a remarkably talented assistant, Matthias Meyer. Matthias helped me through the final stages of the book with formatting, references, and diagrams, as well as through valuable comments and insights.

My family have given me great support over so many years; to Jim, Jeanine, and Mary, and to Sue and our children Catherine and Chris, I give my thanks. Finally I express my gratitude to Miang, for her patience and constant encouragement while this book was being written.

Contents

1

Thermal processes and diffusion

1.1 Boltzmann, Einstein, and molecules

During the 19th century an argument raged about the existence of atoms and molecules, and at the centre of that debate was the Austrian physicist Ludwig Boltzmann and his radical idea that temperature was manifest in the restless, random motions of atoms and molecules [1]. Boltzmann proposed that for every independent type of motion, there would be an equipartition of average thermal energy proportional to temperature. Two centuries earlier, Swiss physicist Daniel Bernoulli had used simple Newtonian mechanics to explain gas pressure in terms of particles colliding with the walls. When Boltzmann applied his equipartition idea, taking suitable statistical averages, he not only explained the gas laws, but also the laws of thermodynamics.

It is hardly surprising that he was attacked. After all, the laws of mechanics are reversible. Might that mean, given the mechanical description, that heat could flow from a colder temperature to a hotter temperature under its own volition, something we never observe? Boltzmann, in rebuttal, explained that while such a violation of the second law might be possible, it was extremely improbable, given the huge number of molecules taking part. But in systems with small numbers of molecules, reversal of the law might be seen and, indeed, modern experiments on small systems have revealed precisely that possibility [2]. Today, Boltzmann's statistical mechanics is one of the cornerstones of modern physics, and we honour his work by naming the constant of proportionality for equipartition *Boltzmann's constant*, k_B.

From the standpoint of the 21st century, the idea of a theoretical model that we test indirectly through its predictions seems entirely reasonable. But in the 19th century, theoretical physics was in its infancy, and there were those who found it distasteful to speak of atoms or molecules when no one had seen such entities. That distaste was exemplified by Ernst Mach, another Austrian, after whom we name the speed of sound. To Mach, scientific laws had to be based on what was directly measurable, not on fanciful theory—on the surface a reasonable position that might well resonate with people of common sense. But science is, as Lewis Wolpert has reminded us [3], a means of discovering truths that defy common sense. It is, after all, common sense that the sun revolves around the earth.

The philosophical disagreement was settled in 1905 by another theoretical physicist, Albert Einstein, whose work was to vindicate Boltzmann completely. That work was on Brownian motion [4], the curious random migration of pollen grains observed under the microscope in 1827 by the botanist Robert Brown [5]. Einstein boldly postulated that the pollen grain, large enough to be seen in the optical microscope, was small enough to be buffeted by the random thermal motion of the surrounding water molecules. Einstein used Boltzmann's thermal energy predictions and came up with a rule by which one would be able to estimate Avogadro's number simply by observing how far the pollen grain moved over a given time. That experimental work was completed in 1908 by the French physicist Jean Perrin.[1]

Mach's discomfort with the indirect observation of atoms and molecules is sharply challenged by this present book, in which we describe the measurement of molecular motion, and in particular Brownian motion, via the medium of radio waves emitted from atomic nuclei. Magnetic resonance observations of molecular translational motion draw on quantum mechanics and statistical mechanics in equal parts, reason enough for its fascinating appeal as an experimental technique, and excuse enough for digressions on these beautiful strands of physics as our story unfolds. Of course, both quantum mechanics and statistical mechanics are covered in great detail in hundreds of other texts, and one might ask therefore, what justification could there be in providing introductory material on those subjects in this book? The answer is simply this: a focus on magnetic resonance studies of molecular motion provides a particular context in which non-experts can learn about these other areas of physics. And for the physics experts, the brief descriptions given here provide a compact selection of the key ideas.

1.2 Statistical physics and ensembles

1.2.1 Temperature and entropy

In the case of gas molecules, thermal energy is principally stored as kinetic energy of translation, and in an 'ideal monatomic gas' entirely so.[2] Think of one gas molecule of known kinetic energy, bouncing elastically within a container. The motion, for example the time dependent velocity, is describable by simple classical physics. But for two or more molecules, where collisions are permitted, the description is immensely complicated, while for very large numbers of molecules, only the language of statistics will suffice to describe the distribution of possible velocities that our 'ensemble' of molecules will exhibit. Strangely, the resort to statistics, rather than blurring our ability to describe the behaviour of the gas, will, if the number of molecules is sufficiently large, lead to a high degree of certainty in the description of certain ensemble average properties. The enormous size of Avogadro's number means that even for quite small physical samples, these properties will obey relationships so powerful as to be described as *thermodynamic laws*.

In this section we examine two thermodynamic properties, temperature and entropy, which underpin the process of diffusion. In order to do this, and to simply illustrate the role of statistics, we choose as our example, not gas molecules with kinetic energy, but an example of particles that store energy in an even simpler manner—an example with obvious connection to magnetic resonance. This is the two-state case of quantised spin-$\frac{1}{2}$ particles, either up or down in a magnetic field. Statistical physics

---

[1] Perrin won the Nobel Prize in 1926. Einstein won it in 1921 for another of his 1905 papers, on the photoelectric effect [6].

[2] Even for polyatomic gases, the other storage modes of molecular rotation and vibration are not activated at low temperatures.

rests upon the ideas of systems of *quantum states*, so this example is a particularly helpful one. The use of quantum states as a starting point for an understanding of statistical physics is central to the text by Kittel and Kroemer [7]. Sections 1.2.1 through 1.2.4 paraphrase Kittel and Kroemer's insight, and readers are referred to that remarkable book for further illumination.

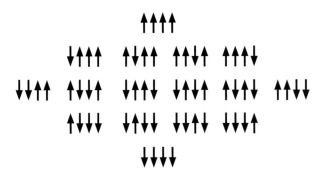

Fig. 1.1 Possible states of four independent spin-$\frac{1}{2}$ particles showing the multiplicities for different numbers of up and down spins.

Let us begin by considering probabilities without reference to energy, easily achieved by having no magnetic field present. As an example, imagine four independent spin particles that may each be either up ($\uparrow$) or down ($\downarrow$). As shown in Fig. 1.1 there is one way that all four spins can be up, four ways one can be down, six ways there can be two up and two down, four ways one can be up, and one way they can all be down. These numbers of possibilities are called *multiplicities*, and they follow a binomial pattern. In particular, if there are N spins with $N_\uparrow$ up and $N_\downarrow$ down, such that the difference between up and down spins is the spin excess, $2x$, then the multiplicity is

$$g\left(N, x\right) = \frac{N!}{N_\uparrow! N_\downarrow!} = \frac{N!}{\left(\frac{1}{2}N + x\right)! \left(\frac{1}{2}N - x\right)!} \tag{1.1}$$

Of course, x ranges between $-N/2$ and $N/2$. Notice that the multiplicity function is peaked about $x = 0$ and that as the total number, N, of spins grows, this peak becomes sharper and sharper in relative terms. Figure 1.2 compares the multiplicity functions of 4, 10, and 100 spins.

Underpinning the present discussion is the idea that all accessible quantum states of a system are equally likely. In the absence of a magnetic field, all spin excess values would be accessible and the probability of finding each value would be given by the multiplicity of states for each configuration. For N spins with spin excess x, that multiplicity is $g\left(N, x\right)$. It is clear that as N grows, the probability distribution associated with the spins becomes more like a certainty that $x = 0$. In fact, the spread in x values grows as $N^{1/2}$, so that the spread in x/N decreases as $N^{-1/2}$. Hence this sharpness is relative to the available range of x between $-N/2$ and $N/2$. This illustration of how, as the numbers of particles in our system increases, near certainty emerges from mere probability, lies at the heart of statistical physics, at the heart of

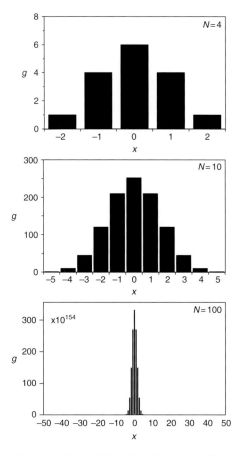

Fig. 1.2 Multiplicities $g(N, x)$ for N values of 5, 10 and 100.

Boltzmann's understanding [8]. The mathematics also simplifies. Note that large N results in another simplification in that one may approximate eqn 1.1 by

$$g(N, x) = g(N, 0) \exp\left(\frac{-2x^2}{N}\right) \tag{1.2}$$

Let us now apply the magnetic field B to the spin system. Each spin has a magnetic moment m such that for spins up along the magnetic field (low energy state) the energy is $-mB$, while for down spins (high energy state) the energy is $+mB$. If the system happened to have spin excess x, then the total system energy would be $U = -2xmB$. The interesting question concerns the rules governing how different systems of spins might exchange energy.

Take two-spin systems, one with N_1 spins and spin excess x_1 and the other with N_2 spins and spin excess x_2. We could say that the total spin excess is $x = x_1 + x_2$. In the presence of the magnetic field, B, the total energy is $-2xmB$. Now, suppose that we bring these systems in thermal contact such that energy can be exchanged.

We will require of course that the total energy be conserved and that in turn means that, although x_1 and x_2 may change as spins exchange energy between the systems, the total value of x must remain constant. In other words, whatever x_1 value results, $x_2 = x - x_1$. Hence the multiplicity function, $g(N, x)$, that defines the combined state of the two systems involves a sum of terms representing each possible value of x_1 and of course fixed total x. Those terms will be products of the individual corresponding multiplicities g_1 and g_2; in other words

$$g(N, x) = \sum_{x_1} g_1(N_1, x_1) g_2(N_2, x - x_1) \tag{1.3}$$

As x_1 varies, one of these products $g_1(N_1, \hat{x}_1) g_2(N_2, x - \hat{x}_1)$ will dominate the summation. This is easily found for the case of the spin system where

$$g(N, x) = g_1(N_1, 0) g_2(N_2, 0) \sum_{x_1} \exp\left(\frac{-2x_1^2}{N_1}\right) \exp\left(\frac{-2(x - x_1)^2}{N_2}\right) \tag{1.4}$$

and the value $\hat{x}_1$ at which $g_1(N, x_1) g_2(N, x - x_1)$ is a maximum is found from setting the derivative, with respect to x_1, of $\exp\left(-2x_1^2/N_1\right) \exp\left(-2(x - x_1)^2/N_2\right)$ to zero, yielding $\hat{x}_1/N_1 = (x - \hat{x}_1)/N_2$. Again, the sharpness of this maximum grows as the number of particles increases and its overwhelming dominance for large N values ensures that this is the 'equilibrium' combined state to which the systems in thermal contact have evolved through the process of exchanging energy. We call this state *thermal equilibrium*. Given that the starting combined multiplicity was just one member of the sum expressed by eqn 1.3, and one which is dominated by a much larger term corresponding to the maximum, it is clear that allowing the spin systems to exchange energy leads to an increased combined multiplicity.

What is the fundamental principle that determines the thermal equilibrium value $\hat{x}_1$, and $\hat{x}_2 = x - \hat{x}_1$? The zeroth law of thermodynamics tells us that it is the equilibration of temperature. But what is temperature in this context? There is an easy way to derive this. We have seen that the system multiplicity in the case of the spin states is related to the number of particles N and the spin excess x. In a magnetic field, that spin excess is simply related to the system energy, U. It is for this reason that energy conservation requires the total spin excess to remain constant in the case of two systems in thermal contact. This idea can be generalised to any system by writing the multiplicity as $g(N, U)$. Then, for two systems in thermal equilibrium,

$$g(N, U) = \sum_{U_1} g_1(N_1, U_1) g_2(N_2, U - U_1) \tag{1.5}$$

Now the largest term in the sum comprising eqn 1.5 is determined by the requirement that $g_1(N_1, U_1) g_2(N_2, U - U_1)$ be a maximum, as illustrated in Fig 1.3. Hence, for an infinitesimal energy exchange dU_1,

$$\frac{d(g_1 g_2)}{dU_1} = \left(\frac{\partial g_1}{\partial U_1}\right)_{N_1} g_2 + g_1 \left(\frac{\partial g_2}{\partial U_1}\right)_{N_2} = 0 \tag{1.6}$$

Then, since by energy conservation $\partial U_2 = -\partial U_1$, dividing by $g_1 g_2$ we have

$$\frac{1}{g_1}\left(\frac{\partial g_1}{\partial U_1}\right)_{N_1} = \frac{1}{g_2}\left(\frac{\partial g_2}{\partial U_2}\right)_{N_2} \tag{1.7}$$

or

$$\left(\frac{\partial \log g_1}{\partial U_1}\right)_{N_1} = \left(\frac{\partial \log g_2}{\partial U_2}\right)_{N_2} \tag{1.8}$$

$\sigma = \log g$ is called the *entropy* of the system. And it is the derivative of the entropy with respect to energy, at fixed numbers of particles, which is equalised between two systems in thermal contact. Thus the *temperature* is defined. By historical convention $\partial \log g / \partial U$ is the inverse temperature and the natural dimension of temperature is energy. Of course, to reconcile the Kelvin units of temperature, a constant is required. Not surprisingly, this is Boltzmann's constant, so that one may define

$$\frac{1}{T} = k_B \left(\frac{\partial \log g}{\partial U}\right)_N \tag{1.9}$$

while for the same historical reason, entropy[3] is defined as $S = k_B \log g$.

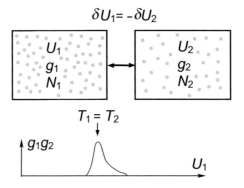

Fig. 1.3 Thermal equilibrium between two systems of N_1 and N_2 particles is attained when the combined multiplicity is a maximum, and is denoted by the temperatures T_1 and T_2 being equal, where $T^{-1} = k_B \frac{\partial \log g}{\partial U}$ at constant N.

The use of spin-$\frac{1}{2}$ particles to illustrate the statistical physics concepts that underpin entropy and temperature is merely a means to an end. Other routes are possible, for example through the more complex consideration of the kinetic energies of gas molecules. That was the path charted by Maxwell and Boltzmann [8, 9]. But the principles are indeed the same. Bring two reservoirs of gas into thermal contact and the energy will redistribute such as to maximise the combined multiplicities of possible states of the gas molecules. One consequence of the increased combined multiplicity when systems are allowed to come into thermal contact is that the energy exchange, known as heat transfer, is associated with a net entropy increase. This idea lies at

---

[3]Given this definition, it might be appropriate to refer to σ as 'informational entropy'.

the heart the second law of thermodynamics, a law which is rooted in the statistical behaviour of immensely large ensembles.[4]

1.2.2   The Boltzmann distribution and partition function

Whether we are considering the system of nuclear spins in a nuclear magnetic resonance (NMR) sample, or the system of molecules of gas in a container, the process of coming to thermal equilibrium involves the exchange of heat with the surrounding environment or *reservoir*. Just as a water reservoir allows small removals or additions without significantly affecting its contents, so the thermal reservoir, by virtue of its much larger heat capacity or internal energy, allows the nuclear spins or the gas molecules to exchange heat without significantly affecting the reservoir temperature. By comparison with that reservoir our system is small. For the gas molecules example, the reservoir might be the laboratory environment surrounding the container, while for the nuclear spins the reservoir is the atomic and molecular environment of the sample, whose thermal motions represent a vastly larger internal energy capacity than the nuclear magnetic energy levels. In NMR we refer to this reservoir as the *lattice*.

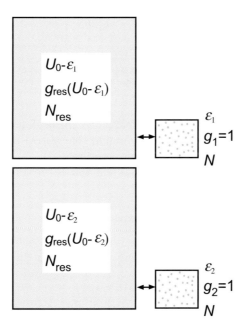

Fig. 1.4 Thermal equilibrium between a reservoir and of a system of N particles at specified states ($g_1 = 1$) of energy ϵ_1 and ϵ_2. The relative probabilities for the two energies are determined by the ratio of reservoir multiplicities.

Suppose we specify a particular quantum state s of a small system in thermal equilibrium with a reservoir at temperature T, as illustrated in Fig 1.4. Since the

---

[4]By implication therefore, when the ensemble becomes sufficiently small, violations of the second law may be seen. This is indeed the case, as may be seen in reference [2].

system state is specified it may be assigned multiplicity 1, and the multiplicity for the combined state reduces to that of the reservoir alone. Assign to the system state the energy ϵ_s, and call the total energy of reservoir plus system U_0. Now ponder the question, how many reservoir states, or in other words, what multiplicity of the reservoir, will permit that system state s? More particularly, what is the relative probability of the system being in thermal equilibrium at T, at two different states $s = 1$ and $s = 2$? We might expect that relative probability to be given by the ratio of those reservoir multiplicities, i.e.

$$\frac{P\left(\epsilon_1\right)}{P\left(\epsilon_2\right)} = \frac{g_{res}\left(U_0 - \epsilon_1\right)}{g_{res}\left(U_0 - \epsilon_2\right)} = \frac{\exp\left(\sigma_{res}\left(U_0 - \epsilon_1\right)\right)}{\exp\left(\sigma_{res}\left(U_0 - \epsilon_2\right)\right)} \tag{1.10}$$

Only reservoir multiplicities are relevant, because the states of the system in thermal equilibrium with that reservoir are specified. Since $\epsilon_s \ll U_0$, by Taylor expansion,

$$\sigma_{res}\left(U_0 - \epsilon_s\right) = \sigma_{res}\left(U_0\right) - \epsilon_s \left(\frac{\partial \sigma_{res}}{\partial U}\right)_N + \dots$$
$$= \sigma_{res}\left(U_0\right) - \frac{\epsilon_s}{k_B T} + \dots \tag{1.11}$$

This leads directly to

$$\frac{P\left(\epsilon_1\right)}{P\left(\epsilon_2\right)} = \frac{\exp\left(\frac{-\epsilon_1}{k_B T}\right)}{\exp\left(\frac{-\epsilon_2}{k_B T}\right)} \tag{1.12}$$

The term $\exp\left(-\epsilon_1/k_B T\right)$ is known as the *Boltzmann factor* for that system's energy state, and gives the relative probability of finding the system at that state when in thermal equilibrium with the reservoir at temperature T.

An extremely useful function, the *partition function*, is the sum over the Boltzmann factors for all possible states of the system,

$$Z\left(T\right) = \sum_s \exp\left(-\epsilon_s/k_B T\right) \tag{1.13}$$

Clearly, the normalised probability for finding our system in the state s will be given by

$$P\left(\epsilon_s\right) = \frac{\exp\left(-\epsilon_s/k_B T\right)}{Z} \tag{1.14}$$

while the average energy of the system will be given by

$$U = \langle \epsilon \rangle = \frac{\sum_s \epsilon_s \exp\left(-\epsilon_s/k_B T\right)}{Z} = k_B T^2 \frac{\partial \log Z}{\partial T} \tag{1.15}$$

1.2.3 Partition function, free energy, and entropy of an ideal monatomic gas

For the ideal gas, where long range interactions between molecules are neglected, the energies are described entirely in terms of particle kinetic energies, $p^2/2m$, where p is the magnitude of the molecular translational momentum vector (p_x, p_y, p_z). For molecules in a container, quantum mechanics dictates that the allowed momenta are restricted by standing de Broglie wave states.[5] If we imagine one cartesian direction (say x) in a cubic container of side length L (see Fig 1.5), the allowed standing waves correspond to integer multiples (n_x) of half wavelengths, such that the allowed momenta in three dimensions are $\frac{h}{2L}(n_x, n_y, n_z)$. The single particle partition function is therefore

$$Z_1 = \sum_{n_x} \sum_{n_y} \sum_{n_z} \exp\left(\frac{-h^2(n_x^2 + n_y^2 + n_z^2)}{8k_B T m L^2}\right) \tag{1.16}$$

When the spacing of the energies is small compared with $k_B T$ the sums can be replaced by integrals as

$$Z_1 = \int_0^\infty \int_0^\infty \int_0^\infty \exp\left(-\alpha^2\left(n_x^2 + n_y^2 + n_z^2\right)\right) dn_x dn_y dn_z \tag{1.17}$$

where $\alpha^2 = h^2/8k_B T m L^2$. It is easy to show that $Z_1 = \left(2\pi k_B T m L^2/h^2\right)^{3/2}$ and, from eqn 1.15, that the single particle average energy is given by $U = \frac{3}{2}k_B T$, equating to $\frac{1}{2}k_B T$, for each of the three 'energy storing modes' or in other words, each of the three degrees of freedom of the single particle motion. The outcome of this simple calculation, namely that there is an average thermal energy of $\frac{1}{2}k_B T$ per energy storing mode, is perfectly general for all systems in thermal equilibrium. This famous result is known as *Boltzmann's equipartition of energy principle*.

For an ideal monatomic gas of N particles in the volume $V = L^3$, the total partition function is given by

$$Z_N = \frac{1}{N!} Z_1^N \tag{1.18}$$

The Z_1^N product of single particle exponential factors gathers up all possible single particle energy sums in the N-particle exponential, while the $1/N!$ term allows for the fact that we need to count each possible N-particle energy sum only once, and not the $N!$ times that each occurs in the product of factors.

The partition function also permits a calculation of the free energy $F = U - TS$ as $F = -k_B T \log(Z)$, and hence the entropy $S = -(\partial F/\partial T)_V$. For the ideal monatomic gas the entropy for N molecules is

$$S = Nk_B\left(\log\left(\frac{n_Q}{n}\right) + \frac{5}{2}\right) \tag{1.19}$$

where n is the number of molecules per unit volume and $n_Q = Z_1/L^3 = \left(2\pi k_B T m/h^2\right)^{3/2}$, the so-called *quantum concentration*

Equation 1.19 is known as the Sackur–Tetrode equation [10, 11]. While it provides an exact expression for the entropy of an ideal gas, it is possible to gain simpler insight in a different way. Take a container of volume V in which we successively introduce N gas molecules. If the effective volume of one molecule is V_m, then the number of

---

[5]The de Broglie wavelength, λ, of a particle of momentum p is h/p, where h is Planck's constant.

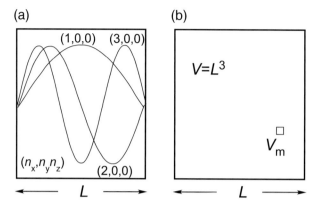

Fig. 1.5 Depictions of allowed states for a gas contained in box of side length L. In (a) the allowed (de Broglie) standing waves are shown, while in (b) a classical picture is presented in terms of the different volumes of size V_m available for each gas molecule.

possible locations for the first molecule is V/V_m, for the second $(V-V_m)/V_m \approx V/V_m$ and so on. Provided that V_m is small, that is $NV_m \ll V$, then the total multiplicity for introducing N molecules is

$$g \approx \left(\frac{V}{V_m}\right)^N \tag{1.20}$$

and so the gas entropy is

$$S \approx Nk_B \log\left(\frac{V}{V_m}\right) \tag{1.21}$$

Both eqns 1.19 and 1.21 tell us that for fixed numbers of molecules N, a volume change δV results in an entropy change $\delta S \sim N\delta V/V$. We will find this result useful in following Einstein's derivation of the self-diffusion coefficient for a Brownian particle.

1.2.4 Ensembles and averages

Statistical physics rests upon the ideas of systems of *quantum states* and *configurations*. Let us return to our system of four spin-$\frac{1}{2}$ particles. The specification 'one spin up and three spins down' represents a particular configuration, one in which the spin excess is −2. But there are four particular quantum state manifestations of that configuration, namely ($\uparrow\downarrow\downarrow\downarrow$), ($\downarrow\uparrow\downarrow\downarrow$), ($\downarrow\downarrow\uparrow\downarrow$), and ($\downarrow\downarrow\downarrow\uparrow$).

The fundamental principle of statistical mechanics is that *all accessible quantum states of a system are equally likely*. If there are W quantum states of a system compatible with all the constraints on energy, volume, and any other parameter defining the physics of the system, then the probability of finding the system in each accessible quantum state is $P = W^{-1}$. To illustrate the idea of accessibility for our spin system, imagine that the spin system had a particular energy available to it when placed in a magnetic field. This in turn would specify the allowed spin excess and hence the allowed configuration. Alternatively, as we saw in the previous section, where two systems of spins are in thermal contact with total spin excess $2x$, the accessible states belonged to configurations x_1 and $x_2 = x - x_1$, and the number of accessible states of all the combined system configurations was $g(N, x) = \sum_{x_1} g_1(N_1, x_1) g_2(N_2, x - x_1)$.

Knowing all the possible quantum states of a system allows us to construct the *ensemble*, the set of all replicas of the system representing each of the accessible quantum states. From the ensemble, we are able to calculate the average value of any physical quantity X as

$$\langle X \rangle = \sum_s P(s) X(s) \tag{1.22}$$

$\langle X \rangle$ is known as the ensemble average of X.[6] Given the fundamental principle, we could carry out the calculation of $\langle X \rangle$ with a sum over configurations, rather than a sum over states. In this case $P(s)$ would be replaced with the configurational probability.

As a first example, consider the case of N spins in the absence of a magnetic field, such that all states are accessible. For N large, the average value of spin excess, $\langle 2x \rangle$, will be given by[7]

$$\langle 2x \rangle = \sum_{-\frac{N}{2}}^{\frac{N}{2}} 2x\, g(N, 0) \exp\left(\frac{-2x^2}{N}\right) = 0 \tag{1.23}$$

A second example concerns the average energy of N spins immersed in a magnetic field B and in thermal equilibrium with a reservoir at temperature T. The possible energies for the two quantum states of each spin are mB and $-mB$ with corresponding probabilities given by the normalised Boltzmann factors, $\exp(\pm mB/k_BT)/(\exp(mB/k_BT)+\exp(-mB/k_BT))$, so that

$$U = N \frac{mB \exp(mB/k_BT) - mB \exp(-mB/k_BT)}{\exp(mB/k_BT) + \exp(-mB/k_BT)} \tag{1.24}$$

Finally, let us use eqn 1.17 to find the average energy of N monatomic gas molecules in thermal equilibrium at temperature T, a result already obtained using the partition function approach. For simplicity we express the particle's momentum in spherical polar coordinates as $p_x = p \sin\theta \cos\phi$, $p_y = p \sin\theta \sin\phi$, $p_z = p \cos\theta$. The single particle Boltzmann factors. are

$$P(\epsilon) = \frac{\exp\left(-p^2/2mk_BT\right)}{4\pi \int\limits_0^\infty p^2 \exp\left(-p^2/2mk_BT\right) dp} \tag{1.25}$$

and the kinetic energy of translation is $\epsilon = p^2/2m$. Integrating by parts, it is simple to show that, as before, the result is $U = \langle p^2/2m \rangle = \frac{3}{2} Nk_BT$.

1.2.5   Fluctuations, ergodicity, and the autocorrelation function

The ideal gas example is instructive for another reason. If we were to follow the individual gas molecules we would find that as a result of random collisions, their momentum

---

[6]Throughout the book the notation $\langle \ldots \rangle$ will be used to represent an ensemble average.

[7]By a similar, but slightly more complex calculation, it may be shown that the fractional root mean-squared spin excess, $\langle 4x^2 \rangle^{1/2}/N = N^{-1/2}$, consistent with the idea that the probability distribution becomes a progressively sharper peak centred at spin excess zero as N increases.

values would fluctuate. Indeed, if we followed a sufficient number of collisions for one molecule, we would find that it would sample all possible values of (p_x, p_y, p_z). Expressed in the language of ensembles, we say that the individual molecules would, over time, sample all possible ensemble states. This particular property is known as *ergodicity*. An ergodic system is one for which the long-time average of a fluctuating variable, $A(t)$, is equal to the instantaneous ensemble average for that variable. In other words

$$\lim_{T \to \infty} \frac{1}{T} \int_0^T A(t)\, dt = \langle A \rangle \tag{1.26}$$

Ergodicity is generally assumed for systems that are *stationary*; that is, when the probabilities associated with various ensemble states do not change with time. . Equation 1.26 will prove particularly useful in evaluating the effect of fluctuations in stationary ensembles.

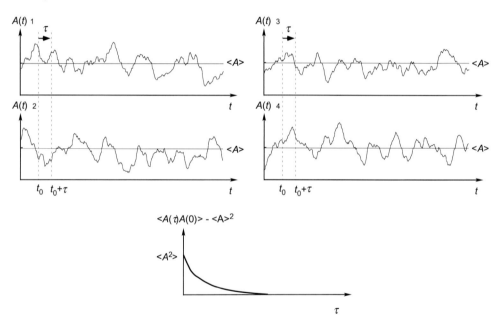

Fig. 1.6 The time-dependence of the stochastic function of time, $A(t)$, for four different members of the ensemble. For an ergodic system the time average is the same as the ensemble average. The lower graph shows the autocorrelation function obtained from the ensemble average of $A(t_0 + \tau)A(t_0)$, identical to $\langle A(\tau)A(0) \rangle$ for a stationary and ergodic system.

A convenient way to deal with a stochastic variable $A(t)$ is to define its behaviour in terms of its autocorrelation function $\langle A(t)A(0) \rangle$ [12, 13]. Figure 1.6 illustrates the noisy behaviour of $A(t)$ along with the well-defined $\langle A(t)A(0) \rangle$. When $t = 0$ the autocorrelation function is simply equal to the mean squared value $\langle A^2 \rangle$. As time increases, $\langle A(t)A(0) \rangle$ decays such that at times much longer than the 'correlation time', $A(t)$ and $A(0)$ are completely uncorrelated so that we may write $\langle A(t) A(0)\rangle = \langle A(t)\rangle\langle A(0)\rangle = \langle A\rangle^2$.

We may define a normalised autocorrelation decay function $g(t)$ as

$$g(t) = \frac{\langle A(t) A(0)\rangle - \langle A\rangle^2}{\langle A^2\rangle - \langle A\rangle^2} \tag{1.27}$$

The correlation time, τ_c, is then defined by the integral

$$\tau_c = \int_0^\infty g(t)\, dt \tag{1.28}$$

Note, for a stationary ensemble, the choice of the time origin is unimportant. Thus $\langle A(\tau) A(0)\rangle = \langle A(t_0 + \tau) A(t_0)\rangle$. In general, $\langle A(t) A(t')\rangle$ depends only on the time difference, $t - t'$.

1.3 Thermal energy and self-diffusion

1.3.1 Fick's law

The process of diffusion may be envisaged as a flux of particles arising from a gradient in concentration, an idea explained by Adolf Fick in 1855 [14]. Given a local concentration of particles $n(\mathbf{r}, t)$, the flux of particles may be written

$$\mathbf{J} = -D\,\nabla n(\mathbf{r}, t) \tag{1.29}$$

where the constant D is known as the *diffusion coefficient*. . In one dimension, the component of flux along the x-axis may be written $J_x = -D\frac{\partial n}{\partial x}$. This relation is sometimes known as Fick's first law and the process is illustrated in Fig 1.7.

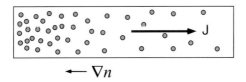

Fig. 1.7 The diffusion flux, $\mathbf{J}$, caused by a concentration gradient ∇n.

Of course, conservation of particles requires that the time rate of change of $n(\mathbf{r}, t)$ is simply related to the local flux divergence, $-\nabla \cdot \mathbf{J} = \frac{\partial n}{\partial t}$. This leads to the Fick's second law, also known as the diffusion equation,

$$\frac{\partial n}{\partial t} = D\nabla^2 n \tag{1.30}$$

Fick's laws were developed to describe the behaviour of solute molecules as a consequence of a non-uniform concentration—the way in which solute molecules will drift from higher to lower concentration so as to equalise concentration gradients. This process is sometimes termed *mutual diffusion* since it requires a countercurrent of solute and solvent particles to maintain the overall mass density. However, the idea may also be applied even to the case of molecules in a liquid comprising a single molecular component, or to the case of solvent molecules in solution or Brownian particles in suspension, where no macroscopic concentration gradients exist. This latter process, first explained by Albert Einstein, is known as *self-diffusion* and it is driven by the random motions associated with thermal energy. Intriguingly, the language of Fick's laws still applies in the case of self-diffusion, and again we take $n(\mathbf{r}, t)$ to be the local probability of finding the solute or solvent molecule of interest. The difference now is that $n(\mathbf{r}, t)$ may be macroscopically uniform, although locally structured. In order to calculate the diffusion coefficient we need to look to the role of thermal fluctuations.

1.3.2   Brownian motion: the Einstein derivation

Einstein's explanation of Brownian motion was based on the idea of the Brownian particles being buffeted by surrounding water molecules, with a net force, K,[8] resulting from the imbalance of exterior collisions [4]. In so reasoning he used the idea that the particles themselves behaved like the molecules in an ideal gas, with pressure $p = k_B T \frac{N}{V}$, where $\frac{N}{V} = n$, the number of Brownian particles per unit volume. Einstein's key step was to consider a small displacement δx of the Brownian particle under the net force, such that the free energy is minimised:

$$\delta F = \delta U - T \delta S = 0 \tag{1.31}$$

The problem was then tackled by calculating δU and δS for a finite volume of particles. Taking a cuboid volume (see Fig 1.8), comprising unit area normal to the x-axis and bounded by $x = 0$ and $x = l$, Einstein expressed the energy change as the total work done by the particles contained within the volume, $-\int_0^l Kn\delta x\, dx$, while, for the entropy change per particle he used the ideal gas result (eqn 1.21), $\delta S = k_B \frac{\delta V}{V}$ or $\delta S = k_B \frac{\partial \delta x}{\partial x}$. Hence

$$\delta S = k_B \int_0^l n \frac{\partial \delta x}{\partial x} dx = -k_B \int_0^l \frac{\partial n}{\partial x} \delta x\, dx \tag{1.32}$$

The free energy relationship yields $Kn = k_B T \frac{\partial n}{\partial x}$ and, noting that the osmotic pressure is $k_B T n$, it emerges that the gradient in osmotic pressure forces provides a balance to the force K.

Next, Einstein considered the balance of mass flow needed to maintain an equilibrium concentration, n. His idea was that the drift of particles caused by K would need to be balanced by a diffusion current in the opposite direction. Using the result that a force K results in a single particle velocity (component along x) K/ζ, where ζ is the friction, he reasoned that the particle current crossing unit area per unit time is nK/ζ, and that this will be counteracted by the self-diffusive flow $-D\frac{\partial n}{\partial x}$, a balance represented by

$$\frac{Kn}{\zeta} - D \frac{\partial n}{\partial x} = 0 \tag{1.33}$$

---

[8]For convenience taken to be the component along one axis, labelled x in the present case.

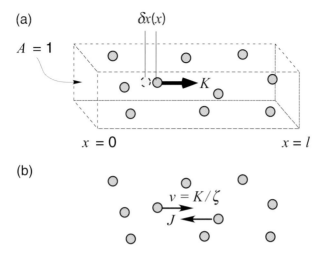

Fig. 1.8 Einstein's diffusion coefficient derivation. (a) The first step, the free energy minimisation with the force on the Brownian particle from the surrounding solvent molecules doing work ($K\delta x$) and resulting in an entropy change as the particle concentration changes; (b) the second step, the balance of the particle drag velocity with the osmotic flux, which maintains the equilibrium concentration.

Using the result $Kn = k_B T \frac{\partial n}{\partial x}$ from eqn 1.31 Einstein found the famous result

$$D = \frac{k_B T}{\zeta} \tag{1.34}$$

Einstein chose for ζ the Stokes drag [15] of a spherical particle of radius R in a medium of viscosity η, $\zeta = 6\pi\eta R$. Hence

$$D = \frac{k_B T}{6\pi\eta R} \tag{1.35}$$

In his remarkable achievement he was not alone. In 1904 an Australian physicist, William Sutherland, using the same reasoning as Einstein, derived the same diffusion relationship and reported the result at a meeting in Dunedin, New Zealand, of the Australian and New Zealand Society for the Advancement of Science [16, 17]. For that reason eqn 1.35 has become known as the Sutherland–Einstein relation.

1.3.3 The probabilistic description

Markov processes and the language of propagators

Einstein's achievement in deriving eqn 1.34 was supplemented by an even more ingenious insight, namely, to ascribe to Fick's laws for the diffusion of molecules in a concentration gradient, an interpretation based on probabilities. This interpretation made possible the description of self-diffusion of molecules or Brownian particles, with a diffusion coefficient as given by eqn 1.34, but in which the average concentration was uniform. Thus the ideas behind Fick's laws were extended to new territory, not only enabling a description of Brownian motion as a stochastic process, but also as one in which the probability densities obeyed differential equations.

Here we lay out some of the basic ideas. For simplicity we will describe the Brownian particle displacement, $x(t)$, over a time t in one dimension. Suppose we were to measure this displacement repeatedly in a set of measurements carried out independently, N times in succession. We would find that we would obtain different results, $x_0(t), x_1(t), x_2(t), \ldots, x_N(t)$. These data represent a sample from a statistical ensemble, and, if N is made sufficiently large, we can find the probabilistic distribution obeyed by the stochastic random variable $x(t)$. This distribution is represented by a probability density $p(x,t)$, where $p(x,t)\,dx$ is the probability that $x(t)$ is in the range $x < x(t) \leq x + dx$. In order to describe the dynamics of Brownian motion we need to know the probabilities associated with the various possible paths of $x(t)$ as time advances. At its simplest, one can define a new probability density [12] $P(x_0, t_0; x_1, t_1)$, where $P(x_0, t_0; x_1, t_1)\,dx_0 dx_1$ is the joint probability that $x(t_0)$ is in the range $x_0 < x(t_0) \leq x_0 + dx_0$, while $x(t_1)$ is in the range $x_1 < x(t_1) \leq x_1 + dx_1$. Clearly, higher order joint probabilities could be described involving combinations $x_0, t_0; x_1, t_1; x_2, t_2$ and so on. However, as we shall see, the second order joint probability will be sufficient for our description of Brownian motion.

This framework enables us to describe a transition probability, the chance that a Brownian particle that certainly starts at x_0 at time t_0 will be found between x_1 and $x_1 + dx_1$ at time t_1; that is:

$$P(x_0, t_0 | x_1, t_1)\,dx_1 = \frac{P(x_0, t_0; x_1, t_1)\,dx_1}{p(x_0, t_0)} \tag{1.36}$$

$P(x_0, t_0 | x_1, t_1)\,dx$ is known as the *conditional probability*. A special feature of Brownian motion is that the probability of finding the particle between x_1 and $x_1 + dx_1$ at time t_1, when it was certainly at x_0 at time t_0, does not depend on where the particle happened to be before time t_0. Such a property is called *Markovian*. For such Markov processes we can calculate the probability of various different paths in a stepwise manner. For example, the joint probability of finding the particle between x_1 and $x_1 + dx_1$ at time t_1 and between x_2 and $x_2 + dx_2$ at time t_2 when it was certainly at x_0 at time t_0 is just

$$P(x_0, t_0 | x_1, t_1; x_2, t_2)\,dx_1 dx_2 = P(x_0, t_0 | x_1, t_1)\,dx_1\ P(x_1, t_1 | x_2, t_2)\,dx_2 \tag{1.37}$$

This enables us to find the conditional probability for any starting and finishing point by integrating over all possible intermediate paths. In other words

$$p(x_0, t_0 | x_2, t_2) = \int p(x_0, t_0 | x_1, t_1)\ p(x_1, t_1 | x_2, t_2)\ dx_1 \tag{1.38}$$

Similarly we may find the probability density $p(x,t)$ by integrating over all possible starting points

$$p(x,t) = \int p(x_0, t_0)\ P(x_0, t_0 | x, t)\ dx_0 \tag{1.39}$$

Because the diffusion process is Markovian, only the time displacement matters, not the absolute time at the origin. This enables us to simplify our notation from $P(x_0, t_0 | x_1, t_1)$ to $P(x_0 | x_1, t)$ where $t = t_1 - t_0$.

Fick's law for probability density and for propagators

Einstein's bold step was to show that the probability density $p(x, t)$ obeyed Fick's laws in the same manner as the more familiar particle concentration function, $n(x, t)$. The details of his idea are based on an analysis of simple Taylor expansions under small increments of time and space. In what follows, we will treat $p(\mathbf{r}, t)$ as a three-dimensional (3-D) probability density. We will see that just as $p(\mathbf{r}, t)$ obeys Fick's Law, so does the conditional probability density, $P(\mathbf{r}|\mathbf{r}', t)$.

Writing $\mathbf{r}$ as our starting coordinate and $\mathbf{r}'$ as the final coordinate,

$$p(\mathbf{r}', t) = \int p(\mathbf{r}, 0) \, P(\mathbf{r}|\mathbf{r}', t) \, d\mathbf{r} \tag{1.40}$$

since $p(\mathbf{r}', t)$ obeys the Fick's law diffusion equation for arbitrary initial conditions $p(\mathbf{r}, 0)$, it is apparent that the conditional probability also obeys the partial differential equation[9]

$$\frac{\partial}{\partial t} P(\mathbf{r}|\mathbf{r}', t) = D\nabla_{\mathbf{r}'}^2 P(\mathbf{r}|\mathbf{r}', t) \tag{1.41}$$

For an fluid of infinite extent, and given the initial condition $P(\mathbf{r}|\mathbf{r}', 0) = \delta(\mathbf{r}' - \mathbf{r})$, the Dirac delta function, the solution to eqn 1.41 is

$$P(\mathbf{r}|\mathbf{r}', t) = (4\pi Dt)^{-\frac{3}{2}} \exp\left(-\frac{(\mathbf{r}' - \mathbf{r})^2}{4Dt}\right) \tag{1.42}$$

Its dependence on time is shown in Fig 1.9. In fact, eqn 1.41 is true only for an isotropic medium, where the diffusion is indeed a simple scalar property. In anisotropic media we shall find it necessary to define a diffusion tensor and rewrite the differential equation for diffusion. This extension is covered in Chapter 2 and elsewhere in this book. The Gaussian nature of the conditional probability for self-diffusion, represented by eqn 1.41, leads to two important results,

$$\langle (x' - x)^2 \rangle = 2Dt \tag{1.43}$$

and

$$\langle (\mathbf{r}' - \mathbf{r})^2 \rangle = 6Dt \tag{1.44}$$

Finite boundaries: the eigenmode solution

Attempting a 'separation of variables' solution to Fick's law (eqn 1.30) we write $p(\mathbf{r}, t) = u(\mathbf{r}) v(t)$ and obtain

$$\frac{D\nabla^2 u(\mathbf{r})}{u(\mathbf{r})} = \frac{\partial v(t)/\partial t}{v(t)} \tag{1.45}$$

and since the left- and right-hand sides are, respectively, functions of $\mathbf{r}$ and t only, both must be a constant $-Dk^2$. This leads to an exponential decay relationship for $v(t)$.[10] The $u(\mathbf{r})$ obey the Helmholtz equation

---

[9] Where more than one spatial coordinate is present, the subscript on the operator is used to refer to the relevant variable. For Markov systems, the operator could act on either $\mathbf{r}$ or $\mathbf{r}'$ and for that reason the subscript is subsequently dropped.

[10] Hence the choice of a negative constant $-Dk^2$. This choice is required to keep $v(t)$ finite as $t \to \infty$.

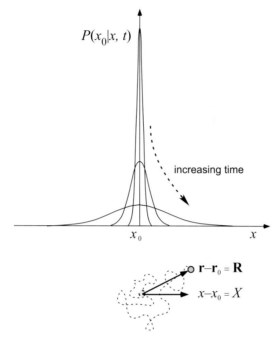

Fig. 1.9 Behaviour of the conditional probability for an ensemble of particles undergoing Brownian motion where the successive Gaussians correspond to successively increasing time.

$$\nabla^2 u\left(\mathbf{r}\right) + k^2 u\left(\mathbf{r}\right) = 0 \tag{1.46}$$

For a finite-sized sample, the $u\left(\mathbf{r}\right)$ must obey some imposed boundary conditions. This constraint fixes the allowed solutions to the eigenmode set $\{u_n\left(\mathbf{r}\right)\}$ with

$$\nabla^2 u_n\left(\mathbf{r}\right) = \lambda_n u_n\left(\mathbf{r}\right) \tag{1.47}$$

where the eigenvalues λ_n are $-k_n^2$. Given real eigenvalues, the $\{u_n\left(\mathbf{r}\right)\}$ form an orthogonal set under $\int_V \ldots d\mathbf{r}$, where V is the volume of the enclosed fluid. Thus the general eigenmode solution may be written as the linear superposition,

$$p\left(\mathbf{r}, t\right) = \sum_n A_n u_n\left(\mathbf{r}\right) \exp\left(-Dk_n^2 t\right) \tag{1.48}$$

with the A_n to be determined from the initial condition.

Suppose that the $u_n\left(\mathbf{r}\right)$ are normalised under $\int_V \ldots d\mathbf{r}$.[11] Then the initial $p\left(\mathbf{r}, t\right)$ may be written

$$p\left(\mathbf{r}, 0\right) = \sum_n A_n u_n\left(\mathbf{r}\right) \tag{1.49}$$

whence $A_m = \int_V p\left(\mathbf{r}', 0\right) u_m^*(\mathbf{r}')d\mathbf{r}'$ and

---

[11]In other words, the $u_n\left(\mathbf{r}\right)$ are an orthonormal set where $\int_V u_m^*(\mathbf{r})u_n(\mathbf{r})d\mathbf{r} = \delta_{mn}$ and we allow that the $u_n(\mathbf{r})$ may be complex.

$$p\left(\mathbf{r},t\right)=\sum_{n}\left(\int_{V}p\left(\mathbf{r}',0\right)u_{n}^{*}\left(\mathbf{r}'\right)d\mathbf{r}'\right)u_{n}\left(\mathbf{r}\right)\exp\left(-Dk_{n}^{2}t\right) \qquad (1.50)$$

In the magnetic resonance examples to be considered in Chapter 6, $p\left(\mathbf{r},t\right)$ will represent a normalised magnetisation density function. For a pore of uniform fluid density, a common situation in magnetic resonance experiments will be that the initial magnetisation across the pore will be uniform so that $p\left(\mathbf{r},0\right)=1/v$ and

$$p\left(\mathbf{r},t\right)=\sum_{n}\frac{\int_{V}u_{n}^{*}\left(\mathbf{r}'\right)d\mathbf{r}'}{V}u_{n}\left(\mathbf{r}\right)\exp\left(-Dk_{n}^{2}t\right) \qquad (1.51)$$

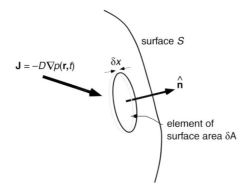

Fig. 1.10 Schematic of surface relaxation involving an element of volume adjacent to δA with layer thickness δx in which spin or particle probability may be 'lost' by some relaxation or permeation process.

The boundary conditions by which the eigenmodes $u_{n}\left(\mathbf{r}\right)$ are determined relate to the degree to which molecules are reflected or absorbed at the walls. This can be easily seen by reference to Fig. 1.10, where we consider an element of volume adjacent to the surface element δA with layer thickness δx. Within this volume element, probability is 'lost' by some process. In the case of magnetic resonance, even under perfectly reflecting wall conditions, collision of molecules with the wall can lead to spin relaxation and hence a loss of spin magnetisation. Alternatively, for a semi-permeable wall, molecules might leak out at the boundary, with none returning. Let us call the loss rate per molecule $1/T$. Then the probability loss rate within the defined volume bounding the surface is $\left(\delta x\,\delta A\,p(\mathbf{r},t)/T\right)\big|_{S}$. Rewriting $\bar{\rho}=\delta x/T$ and allowing that this loss rate in the volume element bounding the surface is fed by the probability current into this volume, we have at the surface $\big|_{S}$,

$$D\hat{\mathbf{n}}\cdot\nabla p\left(\mathbf{r},t\right)+\bar{\rho}\,p\left(\mathbf{r},t\right)=0 \qquad (1.52)$$

Perfectly reflecting walls correspond to $\bar{\rho}=0$, while perfectly permeable or absorbing walls correspond to $\bar{\rho}\to\infty$.

Conditional probability eigenmodes

The simple Gaussian form of eqn 1.42 for the conditional probability applies when diffusion is unrestricted. Where obstacles to diffusion occur, for example due to the bounding surfaces confining liquid molecules interpenetrating a porous medium, that probability will significantly deviate from a Gaussian. While Fick's law still applies within the bulk of the fluid, reflection at the boundaries imparts to the propagator $P(\mathbf{r}|\mathbf{r}', t)$ spatio-temporal properties characteristic of the boundary conditions.

Equation 1.41 may be tackled via the same eigenmode expansion employed for $p(\mathbf{r}, t)$,

$$P(\mathbf{r}|\mathbf{r}', t) = \sum_{n=0}^{\infty} u_n^* (\mathbf{r}) u_n (\mathbf{r}') \exp\left(-Dk_n^2 t\right) \tag{1.53}$$

where, again, the $u_n(\mathbf{r})$ are the orthonormal set of solutions to the Helmholtz equation parametrised by the eigenvalue $-k_n^2$. They are further subject to the identity

$$\delta(\mathbf{r}' - \mathbf{r}) = \sum_{n=0}^{\infty} u_n^* (\mathbf{r}) u_n (\mathbf{r}') \tag{1.54}$$

$P(\mathbf{r}|\mathbf{r}', t)$, thus constructed, satisfies the initial condition

$$P(\mathbf{r}|\mathbf{r}', t{=}0) = \delta(\mathbf{r}' - \mathbf{r}) \tag{1.55}$$

The eigenvalues λ_n depend on the boundary condition. For reflective walls,

$$\hat{\mathbf{n}} \cdot \nabla P(\mathbf{r}|\mathbf{r}', t) = 0 \tag{1.56}$$

where $\hat{\mathbf{n}}$ is the outward surface normal. For the case of permeable or absorbing walls, the boundary condition becomes

$$D\hat{\mathbf{n}} \cdot \nabla P(\mathbf{r}|\mathbf{r}', t) + \bar{\rho} P(\mathbf{r}|\mathbf{r}', t) = 0 \tag{1.57}$$

1.3.4 Relationship of diffusion to velocity autocorrelation function

Equation 1.43 provides a definition for diffusion of the form

$$D = \lim_{t\to\infty} \frac{1}{2} \frac{\partial \langle X^2(t) \rangle}{\partial t} \tag{1.58}$$

where $X(t) = x - x_0 = \int_0^t v(t')\, dt'$, the displacement over time t. This relation leads directly to the result

$$D = \lim_{t\to\infty} \int_0^t \langle v(\tau) v(0) \rangle d\tau \tag{1.59}$$

As an exercise, a detailed proof is given as follows:

$$\langle X^2(t) \rangle = \langle \left(\int_0^t v(t')\, dt' \right)^2 \rangle \tag{1.60}$$

whence[12]

$$\frac{\partial \langle X^2(t) \rangle}{\partial t} = 2 \langle v(t) \int_0^t v(t') \, dt' \rangle$$

$$= 2 \int_0^t \langle v(t) \, v(t') \rangle dt'$$

$$= 2 \int_{-t}^0 \langle v(t) \, v(t+\tau) \rangle d\tau \qquad (1.61)$$

$$= 2 \int_0^t \langle v(t) \, v(t-\tau) \rangle d\tau$$

For a stationary ensemble, the correlation function is independent of starting time t and depends only on the offset τ, and so $\langle v(t) \, v(t-\tau) \rangle = \langle v(0) \, v(-\tau) \rangle$. Using that, and the fact that the correlation function is symmetric with respect to time, eqns 1.61 and 1.58 reduce to eqn 1.59.

1.3.5   The diffusion tensor

The definition of the scalar diffusion coefficient in eqn 1.59 allows for a more general description of a tensor $\underline{\underline{D}}$, the elements of which represent time integrals of the various auto- and cross-correlation functions possible when components of a velocity vector $\mathbf{v}$ are allowed. Namely

$$\underline{\underline{D}} = \lim_{t \to \infty} \int_0^t \langle \mathbf{v}(\tau) \, \mathbf{v}(0) \rangle d\tau \qquad (1.62)$$

with elements:

$$D_{\alpha\beta} = \lim_{t \to \infty} \int_0^t \langle v_\alpha(\tau) \, v_\beta(0) \rangle d\tau \qquad (1.63)$$

In molecular Brownian motion, orthogonal components of velocity are uncorrelated and the diffusion tensor is diagonal. For free diffusion those diagonal elements are equal. By contrast, in an anisotropic medium with axial or biaxial symmetry, such as an oriented liquid crystal or nerve tissue, $\underline{\underline{D}}$ will be diagonal in the principal axis frame, with unequal elements D_{xx}, D_{yy}, and D_{zz}. Of course, in a rotated frame of reference, off-diagonal elements will be apparent even though there exists a similarity transformation to a diagonal representation. As a consequence of the microreversibility of non-equilibrium thermodynamics [18], in the case of self-diffusive motion, $\underline{\underline{D}}$ is symmetric, meaning $\underline{\underline{D}} = \underline{\underline{D}}^T$.

Note, however, that asymmetric and intrinsically non-diagonal properties are possible. For molecules undergoing dispersive flow (see Chapter 2) orthogonal components of velocity may indeed be correlated so that an inherently non-diagonalisable $\underline{\underline{D}}$ matrix is possible.

---

[12]We use the fact that both integration and differentiation are linear operations and that $\langle \ldots \rangle$ is formed from a sum. Hence the ensemble averaging may be interchanged with those operations.

1.3.6   The Smoluchowski equation

Suppose that the particle is subject to an additional potential $U(\mathbf{r})$, with associated force $\mathbf{F} = -\nabla U$. Then this force would result in a velocity $\mathbf{v} = \mathbf{F}/\zeta$. Hence eqn 1.29 needs to be modified to allow for the additional particle flux $n\mathbf{v}$, so that

$$\mathbf{J} = -D\nabla n - \frac{n}{\zeta}\nabla U \tag{1.64}$$

Using the Einstein result, $D = k_B T/\zeta$ and again applying the continuity requirement $\nabla \cdot \mathbf{J} = \partial n/\partial t$, the diffusion equation may now be written

$$\frac{\partial n}{\partial t} = \frac{1}{\zeta}\nabla \cdot [k_B T\,\nabla n\,(\mathbf{r},t) + n\,(\mathbf{r},t)\,\nabla U] \tag{1.65}$$

Equation 1.65 is known as the Smoluchowski equation [19]. Following the same reasoning as in the previous section, we may also write this differential equation in terms of the conditional probability

$$\frac{\partial}{\partial t}P\,(\mathbf{r}_0|\mathbf{r},t) = \frac{1}{\zeta}\nabla \cdot [k_B T\,\nabla P\,(\mathbf{r}_0|\mathbf{r},t) + P\,(\mathbf{r}_0|\mathbf{r},t)\,\nabla U] \tag{1.66}$$

Note that we can use eqn 1.64 to derive the Einstein diffusion coefficient result by simply assuming a thermal equilibrium state in which the flux must vanish. In equilibrium, a time-independent Boltzmann distribution must apply, namely $n\,(\mathbf{r},t) \sim \exp{(-U(\mathbf{r})/k_B T)}$. Setting eqn 1.64 to zero, the result $D = k_B T/\zeta$ follows.

1.3.7   The Langevin equation

A quite different approach to the problem of Brownian motion was demonstrated by Paul Langevin in 1908 [20]. Langevin's starting point was to ascribe to the collisions of neighbouring molecules on the Brownian particle a random fluctuating force, and then attempt a solution to Newton's second law. While Einstein described Brownian motion using a familiar linear partial differential equation, albeit applied to probability densities, Langevin's approach involved a differential equation that was inherently stochastic, and which required special methods for solution.

Again, we start with a simple one-dimensional (1-D) picture of the dynamics, but finish with a generalisation to three dimensions. The fluctuating force we call $F(t)$ while, as previously, we write the particle displacement as $x(t)$. We will, however, allow for the existence of an external force, F_{ext}, noting that we may set this force to zero in the case of simple self-diffusion. Allowing for the frictional drag term, $\zeta\frac{dx}{dt}$, Newton's law gives

$$m\ddot{x} = F_{\text{ext}} - \zeta\dot{x} + F(t) \tag{1.67}$$

where we use a single dot superscript to represent the time derivative d/dt and two dots for d^2/dt^2. Since $F(t)$ is a stochastic random variable, so will be the solution to the 1-D Langevin equation, $x(t)$. Before attempting a solution for the probability distribution

$p(x, t)$, we will attempt a simple analysis based on ensemble averages, $\langle ... \rangle$, and in the case where the external force is zero. Multiplying eqn 1.67 by x we find

$$m x \ddot{x} = -\zeta x \dot{x} + x F(t) \tag{1.68}$$

or

$$m \left[\frac{d}{dt} (x \dot{x}) - \dot{x}^2 \right] = -\zeta x \dot{x} + x F(t) \tag{1.69}$$

We now calculate the ensemble average of eqn 1.69,

$$m \left[\frac{d}{dt} \langle x \dot{x} \rangle - \langle \dot{x}^2 \rangle \right] = -\zeta \langle x \dot{x} \rangle + \langle x F(t) \rangle \tag{1.70}$$

The term $m \langle \dot{x}^2 \rangle$ is twice the ensemble-averaged kinetic energy associated with one of the 'energy storing modes' available for 3-D motion, namely for that associated with motion along the x-axis. By the Boltzmann equipartition theorem, it will be twice $k_B T / 2$. The term $\langle x F(t) \rangle$ is intriguing. There is no reason to expect that $x(t)$ and $F(t)$ will be correlated, although the detailed argument in support of this is delicate [21]. And if uncorrelated we may write $\langle x F(t) \rangle = \langle x(t) \rangle \langle F(t) \rangle$; given that both $\langle x(t) \rangle$ and $\langle F(t) \rangle$ are zero, the term may be neglected.

This leaves us with

$$m \frac{d}{dt} \langle x \dot{x} \rangle - k_B T = -\zeta \langle x \dot{x} \rangle \tag{1.71}$$

and its simple exponential solution

$$\langle x \dot{x} \rangle = C \exp \left(-t/t_\zeta \right) + \frac{k_B T}{\zeta} \tag{1.72}$$

where $t_\zeta = m/\zeta$. This is a curious equation, implying a subtle correlation between x and $\dot{x}$. Of course, at long times, if the particle has diffused along the positive x-axis, then presumably that displacement arises from a predominantly average positive value of $\dot{x}$. Setting $x(0) = x_0 = 0$ at the time origin requires $C = -k_B T / \zeta$, and we may write $x = X$, the displacement from origin.

Of course $\langle X \dot{X} \rangle = \frac{1}{2} \frac{d}{dt} \langle X^2 \rangle$, and so integrating eqn 1.72 from 0 to t results in

$$\langle X^2 \rangle = \frac{2 k_B T}{\zeta} \left[t - t_\zeta \left(1 - \exp \left(-t/t_\zeta \right) \right) \right] \tag{1.73}$$

This leads us to two limiting cases

$$\lim_{t \ll t_\zeta} \langle X^2 \rangle = \frac{k_B T}{m} t^2 \tag{1.74}$$

$$\lim_{t \gg t_\zeta} \langle X^2 \rangle = 2 \frac{k_B T}{\zeta} t \tag{1.75}$$

We can identify the time $t_\zeta = m/\zeta$ as the collision time between molecules. At short times, the motion is velocity-like (ballistic) $(X^2 \sim t^2)$, while beyond the collision time, as shown in Figs 1.11 and 1.12, the motion is diffusive $(X^2 \sim t)$, with the diffusion coefficient given by the Einstein result, $D = k_B T / \zeta$. . This long time limit is where the friction dominates eqn 1.67 and the inertial term on the left-hand side of the equation may be neglected.

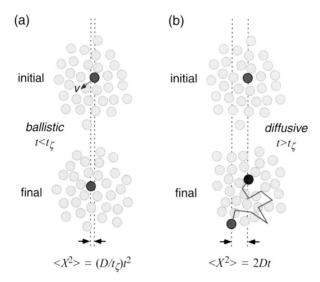

Fig. 1.11 Dense gas of particles with thermal energy, showing the transition from ballistic to diffusive motion for a labelled (black) particle. In (a), the time elapsing between the initial and final state is less than t_ζ and the particle moves with rms velocity $v = \sqrt{2k_B T/m}$ while in (b), where $t > t_\zeta$, the labelled particle suffers many collisions and executes a random walk.

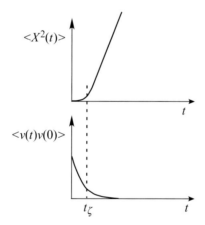

Fig. 1.12 Mean-squared displacement and velocity autocorrelation function showing the transition from ballistic to diffusive motion for colliding molecules at $t = t_\zeta$.

1.3.8 Correlation and the fluctuating force

We now return to an examination of the fluctuating force. In the case of the Langevin force its average value, $\langle F(t) \rangle$, is zero. Its temporal character may be described by the correlation function $\langle F(t)\, F(t') \rangle$. Restricting ourselves to the non-inertial limit, $t \gg t_\zeta$, we may treat the correlation time for fluctuations in $F(t)$ as infinitesimal and so the correlation function may be written

$$\langle F(t)\, F(t') \rangle \sim \delta(t - t')$$
$$= 2\zeta\, k_B T\, \delta(t - t') \tag{1.76}$$

The constant multiplying the Dirac delta function is justified as follows. Setting $F_{\text{ext}} = 0$, $x(0) = x_0$, and neglecting the inertial term in eqn 1.67 we find, on integrating,

$$x(t) - x_0 = \zeta^{-1} \int_0^t F(t')\, dt' \tag{1.77}$$

Assuming that $F(t)$ is a Gaussian variable, then its integral, as represented by eqn 1.77, is also Gaussian,[13] so that we may write the probability distribution of $x(t) - x_0$ as

$$\Psi\left(x(t) - x_0, t\right) = (2\pi B)^{-\frac{1}{2}} \exp\left(-\frac{(x - x_0)^2}{2B}\right) \tag{1.78}$$

where $B = \langle (x(t) - x_0)^2 \rangle$. Evaluating the integral

$$\langle (x(t) - x_0)^2 \rangle = \zeta^{-2} \langle \int_0^t F(t')\, dt' \int_0^t F(t'')\, dt'' \rangle \tag{1.79}$$

yields the Einstein result $\langle (x(t) - x_0)^2 \rangle = 2k_B T t / \zeta$ as required.

Equation 1.78 mirrors the Gaussian form of the conditional probability for free diffusion found earlier, and of course, $\Psi(x - x_0, t)$ obeys the same Fick's law differential equation. Indeed, given that we have specified the start position as x_0, the distribution $\Psi(x - x_0, t)$ must be identical to $P(x_0|x, t)$, the conditional probability that it moved to x after time t if it started at x_0. The distribution on $x(t)$ for a range of start points x_0 may be written

$$p(x(t)) = \int p(x_0)\, P(x_0|x, t)\, dx_0 \tag{1.80}$$

Again, specification of the starting position is equivalent to setting $p(x_0)$ as a delta function, so that $p(x(t))$ becomes identical to the conditional probability.

At this point it is helpful to introduce the probability that a particle displaces by X over a time interval t. This distribution, $\bar{P}(X, t)$, is known in magnetic resonance parlance as an an 'average propagator' and, as in the case of eqn 1.80, is a sum over all possible start positions

$$\bar{P}(X, t) = \int p(x_0)\, P(x_0|x_0 + X, t)\, dx_0. \tag{1.81}$$

The use of the capital letter for X provides a means of reminding us that we are dealing with a displacement from origin, with $\bar{P}(X, t)$ being an ensemble average, $\langle ... \rangle_{x_0}$, over all possible origins x_0. The argument is set out in one dimension, but it can be simply extended to three, where $\mathbf{R}$ becomes the displacement and $\mathbf{r}_0$ the origin vector.

---

[13]This assertion follows from the central limit theorem [22].

1.3.9 Ornstein–Uhlenbeck process

The problem described by eqns 1.74 and 1.75 is known as an 'Ornstein–Uhlenbeck' process or a 'mean-reverting process' [23]. Such problems involve an exponential drift towards a mean behaviour with random fluctuations superposed. In the above problem we see this in the crossover from velocity-like behaviour to diffusive behaviour on a timescale long compared with the particle collision time.

The Gaussian distribution of $\langle F(t) \rangle$ and the linear nature of the Langevin equation for free diffusion means that the probability distributions for displacements are also Gaussian at all times, and not just in the diffusion limit, $t \gg t_\zeta$. Indeed we may write the average propagator for the Brownian Ornstein–Uhlenbeck process as

$$\bar{P}(X,t) = \left(4\pi D \left[t - t_\zeta \left(1 - e^{-t/t_\zeta}\right)\right]\right)^{-\frac{1}{2}} \exp\left(-\frac{X^2}{4D \left[t - t_\zeta \left(1 - e^{-t/t_\zeta}\right)\right]}\right) \quad (1.82)$$

We will revisit these ideas we when come to consider dispersion in the next chapter.

1.3.10 Diffusion in a harmonic potential

There are many examples in nature where diffusing particles are confined in space. A very simple example, again of the Ornstein–Uhlenbeck class, is the case of a diffusion under the simultaneous influence of a harmonic potential. We may write this potential $U(x) = \frac{1}{2}kx^2$, with its associated restoring force $F = -kx$, where k is the force constant.

The Langevin equation for a particle diffusing under such a force is simply

$$m\ddot{x} = -kx - \zeta\dot{x} + F(t) \quad (1.83)$$

In the non-inertial limit, $t \gg t_\zeta$, eqn 1.83 reduces to

$$\zeta\dot{x} = -kx + F(t) \quad (1.84)$$

with solution

$$x(t) = \frac{1}{\zeta} \int_{-\infty}^{t} \exp\left(-\frac{(t-t')}{\tau_k}\right) F(t')\, dt' \quad (1.85)$$

where $\tau_k = \zeta/k$. The correlation function $\langle x(t)\, x(0) \rangle$ may be easily calculated using the relation $\langle F(t')\, F(t'') \rangle = 2\zeta k_B T \delta(t' - t'')$ as

$$\langle x(t)\, x(0) \rangle = \frac{1}{\zeta^2} \int_{-\infty}^{t} \int_{-\infty}^{0} \exp\left(-\frac{(t - t' - t'')}{\tau_k}\right) \langle F(t')\, F(t'') \rangle\, dt'\, dt''$$
$$= \frac{k_B T}{k} \exp\left(-\frac{t}{\tau_k}\right) \quad (1.86)$$

The relative positions of the Brownian particle gradually decorrelate at a rate determined by k/ζ, the ratio of the force constant to the frictional drag.

An expression for $P(x_0|x,t)$ applicable at all times longer than the collision time t_ζ is given in reference [12]. In this non-inertial limit, but for $t \ll \tau_k$, the diffusing particle has insufficient time to feel the effects of the harmonic potential and the conditional probability is the same as for free diffusion

$$P\left(x_0|x,t\right) = \left(4\pi Dt\right)^{-\frac{1}{2}} \exp\left(-\frac{\left(x-x_0\right)^2}{4Dt}\right) \tag{1.87}$$

In the long time limit, $t \gg \tau_k$, the probability distribution on x is time-independent and simply given by the normalised Boltzmann distribution

$$P\left(x\right) = \left(\frac{k}{2\pi k_B T}\right)^{\frac{1}{2}} \exp\left(-\frac{kx^2}{2k_B T}\right) \tag{1.88}$$

The very same function applies independently for the probability distribution of starting positions x_0. Indeed so long as $t \gg \tau_k$, the conditional probability is independent of the starting position of the particle and as well,

$$P\left(x_0|x,t\right) = \left(\frac{k}{2\pi k_B T}\right)^{\frac{1}{2}} \exp\left(-\frac{kx^2}{2k_B T}\right) \tag{1.89}$$

Clearly $P\left(x\right)$ and $P\left(x_0\right)$ obey eqn 1.80. This time-independence of the conditional probability at long time is a feature of restricted diffusion in a bounded system. Note the form of the average propagator in this limit,

$$\bar{P}\left(X,t\right) = \int P\left(x_0\right) P\left(x_0 + X\right) dx_0 \tag{1.90}$$

This integral is the spatial autocorrelation function of the equilibrium density.

References

[1] D. Lindley. *Boltzmann's Atom*. Free Press, New York, 2001.
[2] G. M. Wang, E. M .Sevick, E. Mittag, D.J. Searles, and D.J. Evans. Experimental demonstration of violations of the second law of thermodynamics for small systems and short time scales. *Phys. Rev. Lett.*, 89:050601, 2002.
[3] Lewis Wolpert. In praise of science. In Ralph Levinson and Jeff Thomas, editors, *Science Today*. Routledge, London and New York, 1997.
[4] A. Einstein. Über die von der molekularkinetischen Theorie der Wärme geforderte Bewegung von in ruhenden Flüssigkeiten suspendierten Teilchen (english: On the movement of small particles suspended in a stationary liquid demanded by the molecular-kinetic theory of heat). *Annalen der Physik*, 17:549, 1905.
[5] R. Brown. A brief account of microscopical observations made in the months of June, July and August, 1827, on the particles contained in the pollen of plants; and on the general existence of active molecules in organic and inorganic bodies. *Phil. Mag.*, 4:16, 1829.
[6] A. Einstein. Über einen die Erzeugung und Verwandlung des Lichtes betreffenden heuristischen Gesichtspunkt (english: On a heuristic viewpoint concerning the production and transformation of light.). *Annalen der Physik*, 17:132, 1905.

[7] C. Kittel and H. Kroemer. *Thermal Physics*. W.H. Freeman, New York, 1980.

[8] L. Boltzmann. Über die mechanische Bedeutung des zweiten Hauptsatzes der Wärmetheorie. *Wiener Berichte*, 53:195–220, 1866.

[9] J. C. Maxwell. On the dynamical theory of gases. *Philosophical Transactions of the Royal Society*, 157:49, 1867.

[10] S. Otto. *Lehrbuch der Thermochemie und Thermodynamik (English: A Text Book of Thermo-Chemistry and Thermodynamics)*. BiblioLife (2009), 1912. Original publisher: Springer (1912), English translation: MacMillan (1917).

[11] H. Tetrode. Die chemische Konstante der Gase und das elementare Wirkungsquantum. *Annalen der Physik*, 343(7):434, 1912.

[12] R. Kubo, M. Toda, and N. Hashitsume. *Statistical Physics II: Non-Equilibrium Statistical Mechanics, 2nd Edition*. Springer, New York, 1991.

[13] M. Toda, R. Kubo, and N. Saito. *Statistical Physics I: Equilibrium Statistical Mechanics, 2nd Edition*. Springer, New York, 1992.

[14] A. Fick. Über Diffusion. *Phil. Mag.*, 10:30, 1855.

[15] G. G. Stokes. On the effect of the internal friction of fluids on the motion of pendulums. *Cambridge Philosophical Society Transactions*, 9:8, 1851.

[16] W. Sutherland. The measurement of large molecular masses. *Australasian Association for the Advancement of Science, Report of Meeting*, 10:117, Dunedin, 1904.

[17] W. Sutherland. A dynamical theory of diffusion for nonelectrolytes and the molecular mass of albumin. *Philosophical Magazine*, 6:781, 1905.

[18] L. Onsager. Reciprocal relations in irreversible processes, i. *Phys. Rev.*, 37:405–426, 1931.

[19] M. Smoluchowski. Zur kinetischen Theorie der Brownschen Molekularbewegung und der Suspensionen. *Annalen der Physik*, 21:756, 1906.

[20] P. Langevin. Sur la theorie du mouvement brownien. *C. R. Acad. Sci. Paris*, 146:530, 1908.

[21] A. Manoliu and C. Kittel. Correlation in the Langevin theory of brownian motion. *Am. J. Phys.*, 47:678, 1979.

[22] W. Feller. *An Introduction to Probability Theory and Its Applications*, volume 1. Wiley, 3 edition, 1968.

[23] G. E. Uhlenbeck and L. S. Ornstein. On the theory of Brownian motion. *Phys. Rev.*, 36:823, 1930.

2
Flow and dispersion

In the simplest characterisation of solids, liquids, and gases, our perspective is influenced implicitly by both macroscopic and molecular standpoints, through our descriptors, fluidity and compressibility. Liquids and gases are fluids. They flow continuously when an external force is applied, whereas a solid will merely elastically deform until that external force is balanced by internal stress, returning to its original shape when the stress is removed. These mechanical ideas have their pedigree in Bernoulli, Hooke, and Newton. However, compressibility, significant in gases and vanishingly small in solids and liquids, is an idea rooted in the molecular and atomic model of matter, the former because in liquids and solids we picture molecules in close contact, and the latter because atomic incompressibility ultimately resides in the quantum mechanical principles associated with the Pauli exclusion principle.

Of course such a simplistic subdivision of matter is naive. Polymer networks, solid yet softly so, comprise considerable empty space. They are distinctly compressible. Gels, while dense and elastic, may start to flow once the applied stress exceeds a yield point. Many modern synthetic materials, most food products, and all biological tissues are neither simple solids nor liquids but comprise dispersed solid/liquid phases. Given sufficient time, that which appears solid on a human time scale may well appear fluid. In the song of Deborah [1], 'the mountains shall flow before the Lord'.

We will have occasion to reflect on material complexity in later chapters. Indeed, in the case of soft solids, complex fluids, or condensed matter comprising dispersed liquid/solid phases, nuclear magnetic resonance provides an almost ideal investigative tool, principal amongst the items in the NMR tool kit being the measurement of molecular translational motion. In Chapter 1 we traversed the diffusive motion that arises as a consequence of thermal fluctuations. Here we introduce ideas of flow driven by external stress, and of the stochastic fluctuations that arise when flow is directed through a porous solid matrix. In this latter case, known as dispersive flow, not only will the link with Brownian motion become clear, but we will see that dispersion offers an even richer spectrum of behaviours. We cannot, in one chapter, hope to cover all the essential ideas in fluid mechanics, and readers can find a more complete introduction in a number of texts dedicated to this subject [2–6]. We will, however, highlight a few key concepts, delving in greater depth into areas where NMR measurement of translational motion has the potential for useful insight.

2.1 Flow

2.1.1 Eulerian and Lagrangian descriptions

In describing fluid motion it is possible to adopt two standpoints [4]. In the first, known as the Eulerian perspective, fluid motion is described by assigning to each point in the space of a fluid, a time-dependent velocity $\mathbf{v}_E\,(\mathbf{r},t)$. The set of velocities at different spatial locations represent a velocity field. In the special case where that field is constant with time (steady-state flow), the Eulerian velocity field may be simply written $\mathbf{v}_E\,(\mathbf{r})$. Note that in magnetic resonance imaging experiments we are, in principle, able to 'map the velocity field'—in other words to measure at each location $\mathbf{r}$, $\mathbf{v}_E\,(\mathbf{r})$ in the case of steady-state flow, indexflow!steady state or $\mathbf{v}_E\,(\mathbf{r},t)$ in the case of fluctuating flow, limited as always by some predetermined spatial and temporal resolution.

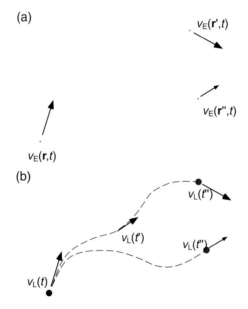

Fig. 2.1 (a) The Eulerian velocity field at time t and three separate locations $\mathbf{r}$, $\mathbf{r}'$, and $\mathbf{r}''$. (b) The Lagrangian velocity trajectories over time t for two particles in the ensemble starting in close proximity at $\mathbf{r}$, each of which has finite probability of passing, at time t'', through positions $\mathbf{r}'$, and $\mathbf{r}''$. Their velocities at time t'' will be identical to $v_E\,(\mathbf{r}')$ and $v_E\,(\mathbf{r}'')$, respectively, if the Eulerian velocity field is steady state.

The second standpoint is known as the Lagrangian perspective and it describes motion in terms of the history of individual fluid particles. For each particle, a time-dependent position $\mathbf{r}_L\,(t)$ or velocity $\mathbf{v}_L\,(t)$ is assigned and, for the fluid as a whole, these vectors represent one member of an ensemble of particle positions or velocity vectors. The members of that ensemble could be labelled in different ways, but one example might be to assign to each particle its position, $\mathbf{r}_0$ when $t = 0$. Hence we could speak of the history of positions or motion of the particle starting at $\mathbf{r}_0$ as $\mathbf{r}_L\,(\mathbf{r}_0,t)$

or $\mathbf{v}_L(\mathbf{r}_0, t)$. Equally we could label each particle, through a component atom, by its nuclear spin magnetisation at some suitable time origin. It is precisely that approach which we may use in magnetic resonance where, even if no imaging gradients are applied, the evolution of individual spin magnetisation may be tracked, albeit as a contribution to some ensemble average taken over the entire fluid sample. For such averaging methods, the Lagrangian perspective is natural. Generally, we will omit the label (for example $\mathbf{r}_0$) and simply note that we must deal with an ensemble of histories $\{\mathbf{r}_L(t)\}$ and $\{\mathbf{v}_L(t)\}$.

Figure 2.1 shows the relationship between the Eulerian and Lagrangian perspectives in the case where the Eulerian field happens to be steady state, and for a particular case where particles beginning in the same vicinity have a finite probability of passing through two different positions in the Eulerian velocity field at a later time t''. In this case there is a simple probabilistic assignment of $v_L(t)$ to $v_E(\mathbf{r})$, a matter which we will analyse in detail later.

The type of flow depicted in Fig. 2.1 is dispersive. Figure 2.2 depicts flow fields in which the flow is, respectively, laminar, turbulent, and dispersive. In laminar flow,

(a)

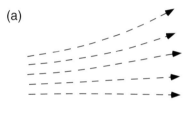

(b)

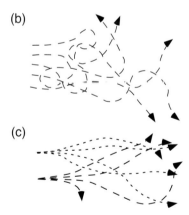

(c)

Fig. 2.2 (a) Steady state laminar flow in which no streamlines cross. (b) Turbulent flow in which the local Eulerian velocities fluctuate with time. (c) Dispersive flow. Particles starting together on the same streamline subsequently follow divergent paths. Dispersive flow can arise from turbulence but may also be associated with laminar flow in a complex porous matrix. The different dashed paths represent subsequent flow histories from different starting points. In steady state flow there are no streamline crossings but trajectories can cross above and below one another in three dimensional flow.

particles starting in nearby proximity follow a time course in which the fluid between them is confined within laminae bounded by the particle streamlines. The laminar flow field is inherently steady state. By contrast, in turbulence the flow field fluctuates in an irregular manner, most commonly due to the effect of fluid inertia at high flow rates where non-linearities start to influence the equations of motion. And, in dispersive flow, while the flow may be steady state, particles that occupy a nearby location at a particular time may subsequently follow very different and diverging paths. Such dispersive flow is characteristic of fluid moving through a porous medium. The non-laminar examples illustrate why the use of a starting coordinate may be a poor descriptor for the Lagrangian ensemble, irrespective of whether the flow field fluctuates or is in steady state.

For two particles that occupy an infinitesimally close location at a particular time, we may, in steady state, use the conditional probability introduced in Chapter 1 to describe the likelihood of different subsequent paths. $P\left(\mathbf{r}|\mathbf{r}',t\right)$ describes the probability that a particle with Eulerian velocity $\mathbf{v}_E\left(\mathbf{r}\right)$ at time zero will have Eulerian velocity $\mathbf{v}_E\left(\mathbf{r}'\right)$ at time t. That conditional probability will have the form of a Dirac delta function for laminar flow. In other words, we may, in the case of laminar flow, predict with certainty where a fluid particle passing through a particular location will end up at any later time. By contrast, for turbulent or dispersive flow the conditional probability will have more complex structure. As we shall see later in section 8.4, $P\left(\mathbf{r}|\mathbf{r}',t\right)$ will provide a valuable tool in linking the Eulerian and Lagrangian descriptions of flow field dynamics.

2.1.2   Substantive derivative and fluid dynamics

In the following sections we will write down some equations of motion for fluid mechanics [7, 8]; equations which have their origin in Newtonian dynamics. That origin raises an interesting issue when we come to describe fluid particle acceleration, since any time derivative of velocity, $\partial\mathbf{v}_E(\mathbf{r},t)/\partial t$, taken in the Eulerian perspective, will not tell us about the acceleration of a particular particle, but instead represents the rate at which the velocity at a particular point is changing as different particles pass through. The derivative we seek, $\partial\mathbf{v}_L(t)/\partial t$, belongs to a single particle and must be calculated in a co-moving frame. Let us consider such a co-moving frame derivative for any quantity $A\left(\mathbf{r},t\right)$, given that the local Eulerian velocity is $\mathbf{v}_E\left(\mathbf{r},t\right)$. We will drop the E subscript in what follows and allow that $\mathbf{v}\left(\mathbf{r},t\right)$ is implicitly Eulerian. At a time later by dt, the fluid particles starting at $\mathbf{r}$ have moved to $\mathbf{r}+dt\,\mathbf{v}$, displacement dx being simply $dt\,v_x$, where v_x is the x component of the velocity, and so on. Hence the change in A to first order may be written [4] via a Taylor expansion as

$$\begin{aligned}
dA &= dt\frac{\partial A}{\partial t} + dx\frac{\partial A}{\partial x} + dy\frac{\partial A}{\partial y} + dz\frac{\partial A}{\partial z} \\
&= dt\frac{\partial A}{\partial t} + dt\,v_x\frac{\partial A}{\partial x} + dt\,v_y\frac{\partial A}{\partial y} + dt\,v_z\frac{\partial A}{\partial z} \qquad (2.1) \\
&= dt\frac{\partial A}{\partial t} + dt\,\mathbf{v}\cdot\nabla A
\end{aligned}$$

For the frame moving with the fluid, the points $\mathbf{r}$ and $\mathbf{r} + \mathbf{v}\,dt$ are of course coincident and so dA/dt is the same as $\partial A/\partial t$ in the co-moving frame. In the stationary frame, dA/dt is called the 'substantive derivative', $\mathcal{D}A/\mathcal{D}t$, and written

$$\frac{\mathcal{D}A}{\mathcal{D}t} = \frac{\partial A}{\partial t} + (\mathbf{v} \cdot \nabla)\,A \tag{2.2}$$

The term $(\mathbf{v} \cdot \nabla)$ is sometimes called the advective or convective component of the substantive derivative $\mathcal{D}/\mathcal{D}t$.

A first example of the use of this substantive derivative concerns the continuity equation, which states that the divergence of mass flow is related to the local change in density by

$$\rho \nabla \cdot \mathbf{v} = -\frac{\mathcal{D}\rho}{\mathcal{D}t} \tag{2.3}$$

The second is the so-called 'momentum equation', which relates the acceleration of the particles to the net force applied

$$-\nabla p + \mathbf{f} = \rho \frac{\mathcal{D}\mathbf{v}}{\mathcal{D}t} \tag{2.4}$$

In this version of Newton's second law, the forces (expressed per unit volume) arise from the pressure gradient and from body forces ($\mathbf{f}$) acting on the fluid, an example of the latter being the gravitational force per unit volume ($\rho\mathbf{g}$). Note again, the velocity $\mathbf{v} = \mathbf{v}\,(\mathbf{r}, t)$ is implicitly Eulerian, and will be so for the rest of this book, unless stated otherwise.

Equations 2.3 and 2.4 are together known as Euler's equations, corresponding respectively to laws of mass and momentum conservation.[1] They apply for the case of inviscid flow, that is when there is no dissipation of energy due to fluid viscosity.

2.1.3 Navier–Stokes, inertia, and the Reynolds number

When viscosity plays a role, the Euler equation must be expanded to read[2]

$$-\nabla \cdot \left(p\,\underline{\underline{\mathbf{I}}}\right) + \nabla \cdot \underline{\underline{\sigma}} + \mathbf{f} - \rho\frac{\mathcal{D}\mathbf{v}}{\mathcal{D}t} \tag{2.5}$$

where the identity matrix, also known as the unit tensor, is

$$\underline{\underline{\mathbf{I}}} = \begin{pmatrix} 1 & 0 & 0 \\ 0 & 1 & 0 \\ 0 & 0 & 1 \end{pmatrix} \tag{2.6}$$

and $\underline{\underline{\sigma}}$ is the deviatoric part of the stress tensor, explained in detail in the next section. Meanwhile, it is sufficient to note that for a liquid with constant dynamic viscosity, η, eqn 2.5 may be written

$$-\nabla p + \eta\,\nabla^2\mathbf{v} + \mathbf{f} = \rho\,\frac{\mathcal{D}\mathbf{v}}{\mathcal{D}t} \tag{2.7}$$

Equation 2.7 is known as the Navier–Stokes equation [7, 8]. For an incompressible fluid we have the additional relation $\nabla \cdot \mathbf{v} = 0$.

---

[1] There is a third Euler equation for energy conservation, which need not concern us here.

[2] Note that we are treating the density, ρ, as constant, as would be appropriate for incompressible fluids, for example liquids.

Note that the convective part of the substantive derivative of velocity, $\rho\left(\mathbf{v}\cdot\nabla\right)\mathbf{v}$, is non-linear in the velocity and, when dominant, may lead to turbulent solutions of the Navier–Stokes equation. By contrast, the remaining terms are linear and, in the absence of convective effects, lead to well-defined solutions. The relative size of the convective and viscous terms may be described by a dimensionless number, obtained by non-dimensionalising eqn 2.7 using a characteristic length L and velocity v so that $\eta\nabla^2\mathbf{v}$ is of order $\eta v/L^2$ while $\rho\left(\mathbf{v}\cdot\nabla\right)\mathbf{v}$ is of order $\rho v^2/L$. This dimensionless number is known as the Reynolds number [9] and is defined by

$$\mathcal{R}_e = \frac{\rho v^2/L}{\eta v/L^2} = \frac{\rho v L}{\eta} = \frac{v L}{\nu} \tag{2.8}$$

where ν is the kinematic viscosity, η/ρ. For low Reynolds numbers the flow in simple geometries will be laminar, with a transition to turbulent flow for $\mathcal{R}_e \gg 1000$.

2.1.4 Stress and strain tensors

The dissipation of energy due to viscous drag, referred to in the previous section, is a consequence of deformational (shape-changing) flow. In simple terms, fluid deformation dissipates energy, while in the case of a solid, deformation causes energy to be stored elastically. In contrast, biological tissue, most foods, and a wide range of modern polymeric materials of interest all have both solid- and liquid-like properties and are consequently termed viscoelastic. Such complex fluids or soft solids are discussed in detail in Section 2.2.

The cause of deformation in materials is applied stress, quite simply the force applied per unit surface area, and the tensorial property of stress arises because of the independent directions associated with the force and the surface normal. The consequential deformation is also a tensor, the indices labelling both the direction of elemental displacements and the direction with respect to which those displacements vary.

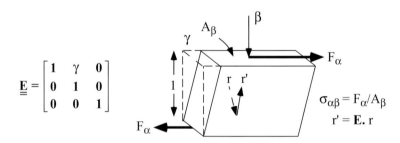

Fig. 2.3 Element of material subject to applied stress. Components of stress tensor $\sigma_{\alpha\beta}$ and deformation gradient tensor $E_{\alpha\beta}$ are shown.

Figure 2.3 shows an element of material subject to an applied stress. We choose to describe an element of the stress tensor $\underline{\sigma}$ by $\sigma_{\alpha\beta}$, where the indices indicate the direction of the force and the relevant surface normal. A corresponding deformation in which a point defined by vector $\mathbf{r}$ is displaced to the position $\mathbf{r}'$ may be defined by the dimensionless strain or deformation gradient tensor

$$E_{\alpha\beta} = \frac{\partial r'_\alpha}{\partial r_\beta} \tag{2.9}$$

where $\mathbf{r}' = \underline{\underline{\mathbf{E}}} \cdot \mathbf{r}$ and α, β may be taken to represent components of the Cartesian axis frame. A shear deformation is given by

$$\underline{\underline{\mathbf{E}}} = \begin{pmatrix} 1 & \gamma & 0 \\ 0 & 1 & 0 \\ 0 & 0 & 1 \end{pmatrix} \tag{2.10}$$

while a volume-conserving uniaxial elongation (along the z-axis) is given by

$$\underline{\underline{\mathbf{E}}} = \begin{pmatrix} \lambda^{-1/2} & 0 & 0 \\ 0 & \lambda^{-1/2} & 0 \\ 0 & 0 & \lambda \end{pmatrix} \tag{2.11}$$

In elastic solids, a shear stress results in a fixed strain and for small strains ($\gamma \ll 1$) the relevant elastic modulus, G, is given by Hooke's law

$$\sigma_{xy} = G\gamma \tag{2.12}$$

In a liquid the deformation under applied stress is continuous and the material flows. Furthermore, the absence of unbalanced torques in a liquid leads to the requirement that the stress tensor be symmetric. Suppose we describe the flow in the liquid by the spatially dependent Eulerian velocity, $\mathbf{v}(\mathbf{r})$. Then the rate of strain tensor $\underline{\underline{\kappa}} = \nabla\mathbf{v}$ is defined by

$$\kappa_{\alpha\beta} = \frac{\partial v_\alpha}{\partial r_\beta} \tag{2.13}$$

For a Newtonian fluid the total stress tensor may be written in terms of the rate of strain tensor via the simple constitutive equation

$$\begin{aligned} T_{\alpha\beta} &= \eta(\kappa_{\alpha\beta} + \kappa_{\beta\alpha}) - pI_{\alpha\beta} \\ &= \sigma_{\alpha\beta} - pI_{\alpha\beta} \end{aligned} \tag{2.14}$$

where η is the constant viscosity, p is the isotropic pressure and $\sigma_{\alpha\beta}$ is the deviatoric part of the stress tensor. Note the sign convention used. External forces bearing on the fluid element and acting normal to the surface are positive when directed inward. By contrast, the intrinsic pressure arising from the fluid element acts outward at the surface normal. In discussing the viscous behaviour of liquids it is customary to neglect the isotropic pressure term in eqn 2.14. However, it is important to note that the force driving the momentum change in the Navier–Stokes equation is related to the gradient in the total stress. Equations 2.5 and 2.7 retain the gradient in both the deviatoric part of the stress, $\underline{\underline{\sigma}}$, as well as in the isotropic pressure term.

$(\kappa_{\alpha\beta} + \kappa_{\beta\alpha})$ is the symmetric rate of strain tensor, sometimes denoted $2\underline{\underline{\mathbf{D}}} = \nabla\mathbf{v} + \nabla\mathbf{v}^T$. An antisymmetric (vorticity) part of the velocity gradient may be defined by $2\underline{\underline{\omega}} = \nabla\mathbf{v} - \nabla\mathbf{v}^T$. One may then conveniently represent eqn 2.14 as

$$\underline{\underline{\sigma}} = 2\eta\underline{\underline{\mathbf{D}}} \tag{2.15}$$

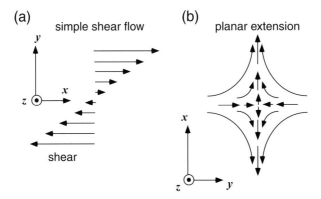

Fig. 2.4 Examples of velocity fields that result in (a) simple shear (across the flow) and (b) planar extension (at the central stagnation point).

As a simple example consider the simple shear case illustrated in Fig. 2.4. The velocity is everywhere directed along x and varies along y. Hence the only non-zero value of $\nabla \mathbf{v}$ is $\dot{\gamma} = \partial v_x / \partial y$. In this case

$$\nabla \mathbf{v} = \begin{pmatrix} 0 & \dot{\gamma} & 0 \\ 0 & 0 & 0 \\ 0 & 0 & 0 \end{pmatrix} \qquad (2.16)$$

and

$$\underline{\underline{\mathbf{D}}} = \frac{1}{2} \begin{pmatrix} 0 & \dot{\gamma} & 0 \\ \dot{\gamma} & 0 & 0 \\ 0 & 0 & 0 \end{pmatrix} \qquad (2.17)$$

and

$$\underline{\underline{\sigma}} = \eta \begin{pmatrix} 0 & \dot{\gamma} & 0 \\ \dot{\gamma} & 0 & 0 \\ 0 & 0 & 0 \end{pmatrix} \qquad (2.18)$$

Clearly this result gives us the Newtonian viscosity law for fluids corresponding to Hooke's law for solids (eqn 2.12),

$$\sigma_{xy} = \eta \dot{\gamma} \qquad (2.19)$$

Also shown in Fig. 2.4 is the case of planar extensional flow. This flow field contains sheared elements but at the centre is a stagnation point at which purely extensional flow results, and where

$$\underline{\underline{\mathbf{D}}} = \begin{pmatrix} \dot{\epsilon} & 0 & 0 \\ 0 & -\dot{\epsilon} & 0 \\ 0 & 0 & 0 \end{pmatrix} \qquad (2.20)$$

2.1.5 Navier–Stokes solutions and lattice Boltzmann

Exact solutions to the Navier–Stokes equation for a Newtonian fluid are possible when the Reynolds number is small. One simple example concerns flow in a long cylindrical pipe of radius a. The steady-state pipe flow is laminar and directed along the axis

(z) of the pipe, with non-slip boundary conditions ($v_z = 0$) at the wall ($r = a$).[3] The natural coordinates are cylindrical polar (r, θ, z) and, by symmetry, the velocity is directed along z such that $v_r = v_\theta = 0$ and v_z is a function of r only. The substantive derivative is exactly zero in this problem, but the solution may be unstable to small symmetry-breaking perturbations if $\mathcal{R}_e$ is large. However, we will examine the low Reynolds number case and hence we may write eqn 2.7, in the absence of body forces, as

$$-\nabla p + \eta \nabla^2 \mathbf{v} = 0 \tag{2.21}$$

or, in cylindrical polars, and consistent with the symmetry outlined above

$$-\frac{\partial p}{\partial z} + \eta \frac{1}{r} \frac{\partial}{\partial r} \left(r \frac{\partial v_z(r)}{\partial r} \right) = 0 \tag{2.22}$$

The solution is clearly

$$v_z(r) = v_0 \left(1 - \frac{r^2}{a^2} \right) \tag{2.23}$$

where

$$v_0 = \frac{a^2}{4\eta} \frac{\partial p}{\partial z} \tag{2.24}$$

This velocity distribution is known as Poiseuille flow [10].[4] Note that v_0 is the

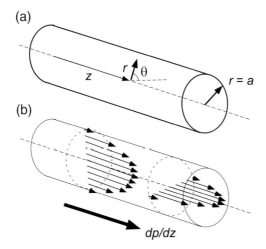

Fig. 2.5 (a) Pipe showing cylindrical polar coordinates and (b) Poiseuille flow profile.

maximum velocity at the pipe centre. It is easy to show that the mean velocity, averaged across the pipe, is $v_0/2$, and that the flow rate $\dot{Q} = \int_0^a 2\pi\, r\, v_z(r)\, dr$ is

---

[3]We shall assume here that the flow to be described is a long way from the entrance to the pipe, and that the pipe diameter is larger than the nanometre scale, where slip effects are apparent even in simple liquids.

[4]This pipe flow velocity profile was discovered independently by G. H. L. Hagen, who did his experiments at the same time as Poiseuille and before the year of Poiseuille's publication. In consequence it is often called the Hagen–Poiseuille law.

$$\dot{Q} = \frac{\pi a^4}{8\eta} \frac{\partial p}{\partial z} \qquad (2.25)$$

Hence, for a known pressure gradient $(\partial p/\partial z)$ down a pipe of radius a, the flow rate provides a simple indication of the viscosity.

The ability to ignore the non-linear effects of convection in the substantive derivative of the Navier–Stokes equation is possible only at low $\mathcal{R}_e$ or for certain regular geometries where symmetry requires $(\mathbf{v} \cdot \nabla)\,\mathbf{v} = 0$. General solutions are a challenge, even for numerical methods, and in complex flows such as the movement of liquids through a packed bed or porous matrix, conventional finite difference or finite element numerical solutions to the Navier–Stokes equation are prohibitively expensive in computer time, even when inertial effects may be neglected. However, an entirely different approach, known as the lattice Boltzmann method [11], provides a particularly efficient route to numerical solution.

Lattice Boltzmann (L-B) relies on the use of a finite lattice of points between which a gas of particles is convected at each time step of a simulation, and at which intersecting particles collide and separate according to rules that can inherently match the requirements of the Navier–Stokes equation. In the case of an incompressible fluid, there is also a requirement that the flow be divergence-less. In a sense, it mirrors a cellular automata approach to physical modelling, but with one important difference. The use of discrete particles in lattice gas automata methods introduces statistical noise as a consequence of the finite number of particles and lattice points. L-B avoids this problem by pre-averaging the lattice gas, replacing the Boolean particle number at each lattice node with an ensemble-averaged density distribution function. In this, the method follows the famous Boltzmann theory for atomistic dynamics.

Boltzmann introduced a probability density $f\,(\mathbf{r}, \mathbf{p}, t)$ in the 'phase space' of particle positions and momenta. The dynamical equation for this one-body distribution function may be written

$$\left(\frac{\partial}{\partial t} + \frac{\mathbf{p}}{m} \cdot \nabla + \mathbf{F} \cdot \nabla_{\mathbf{p}} \right) f\,(\mathbf{r}, \mathbf{p}, t) = C_{12} \qquad (2.26)$$

where the right-hand side represents the interparticle two-body collisions and accounts for gain and loss components as particles stream in or out of the collision point. $\mathbf{F}$ is the external force, equivalent by Newtonian mechanics to $d\mathbf{p}/dt$, while $\nabla_{\mathbf{p}}$ is the vector derivative with respect to the components of momentum. Note the formal similarity of the left-hand side of eqn 2.26 to the substantive derivative of eqn 2.2, the additional momentum derivative term reflecting the influence of the force field. In Boltzmann's treatment the collision term involves two-body distribution functions that are reduced to a simple product of uncorrelated single-body distributions, with local equilibrium defined by a balance of particle gain and loss at each point. Out of that equilibrium comes the famous Maxwell–Boltzmann distribution of eqn 1.25.

The L-B method discretises the Boltzmann equation by dividing space into a set of lattice points (i) and limiting the velocities to a discrete set $\mathbf{v}_i$. The lattice Boltzmann equation is therefore

$$f_i\,(\mathbf{r}_i + \Delta \mathbf{r}, t + \Delta t) - f_i\,(\mathbf{r}_i, t) = C_i \qquad (2.27)$$

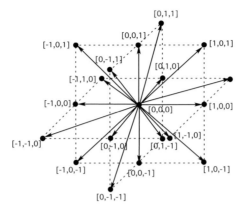

Fig. 2.6 D3Q19 lattice showing the 19 microscopic velocity directions, $\mathbf{v}_i$.

In implementing the lattice Boltzmann equation method for the case of Navier–Stokes flow of a viscous fluid, the collision operator, C_i, takes a form which allows for a single exponential relaxation towards the equilibrium distribution, the relaxation time reflecting the kinematic viscosity of the simulated fluid. One of the most famous forms of the collision operator is that due to Bhatnagar, Gross, and Krook [12], the so-called BGK operator. Note that a number of different lattice geometries are possible, for example the 3-D nineteen velocity (D3Q19) lattice shown in Fig. 2.6. For each lattice node $\mathbf{r}_i$ at time t there are 19 lattice directions labelled by i around the node, as shown in the figure. The local density and velocity of the fluid is then given by $\rho = \sum_i f_i$ and $\mathbf{v} = \rho^{-1} \sum_i f_i \mathbf{v}_i$. A particular advantage of L-B methods is that different boundary conditions are easily accommodated and the code is particularly amenable to parallelisation. A complete description of L-B methods would require a book on its own and readers can find further details in references [11, 13, 14].

2.1.6 Conditional probability and average propagators for flow

The conditional probability for steady-state laminar flow is exceptionally simple. We have certain knowledge as to the subsequent destination $\mathbf{r}'$ of a particle starting at any point $\mathbf{r}$ in the Eulerian flow field, and so we may use a Dirac delta function to describe $P(\mathbf{r}|\mathbf{r}', t)$. While the Navier–Stokes equation ignores Brownian motion, we need to bear in mind that these stochastic molecular excursions will always be superposed on that continuum flow. Generally this may be done by adding the Brownian motion in the co-moving frame.

Let us take a simple example in which all particles in a fluid have common velocity, $\mathbf{V}$. The conditional probability, allowing for flow alone will be

$$P(\mathbf{r}|\mathbf{r}', t) = \delta(\mathbf{r}' - (\mathbf{r} + \mathbf{V}t)) \tag{2.28}$$

The effect of adding the Brownian motion in the co-moving frame is simply to convolve the Brownian and flow probabilities via

$$P\left(\mathbf{r}|\mathbf{r}',t\right) = \delta\left(\mathbf{r}' - (\mathbf{r} + \mathbf{V}t)\right) \otimes \left(4\pi Dt\right)^{-\frac{3}{2}} \exp\left(-\frac{(\mathbf{r}' - \mathbf{r})^2}{4Dt}\right)$$

$$= \left(4\pi Dt\right)^{-3/2} \exp\left(-\frac{(\mathbf{r}' - (\mathbf{r} + \mathbf{V}t))^2}{4Dt}\right)$$

(2.29)

The convolution $\otimes$ is defined by the *faltung* or folding integral

$$f\left(t\right) \otimes g\left(t\right) = \int\limits_{-\infty}^{\infty} f\left(t'\right) g\left(t - t'\right) dt'$$

(2.30)

Note that the conditional probability of eqn 2.29 is, in this special case, independent of the starting position $\mathbf{r}$ and depends only on the displacement $\mathbf{R} = \mathbf{r}' - \mathbf{r}$. Hence, averaging over all starting positions we obtain the average propagator with identical form

$$\bar{P}\left(\mathbf{R},t\right) = \left(4\pi Dt\right)^{-\frac{3}{2}} \exp\left(-\frac{(\mathbf{R} - \mathbf{V}t)^2}{4Dt}\right)$$

(2.31)

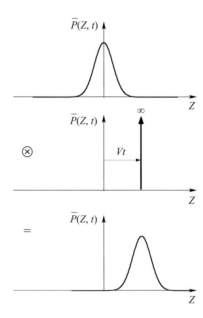

Fig. 2.7 Convolution of propagators for diffusion (top) and flow (middle) to obtain average propagator of eqn 2.31 (bottom). Note that this is a 1-D depiction in which the displacement $\mathbf{R}$ is replaced by Z.

Of course, when the flow field is not uniform, as in the case of pipe flow, for which the velocity varies as a function of radius, the description of the conditional probability is much more complicated when diffusion is included. Here Brownian migration transverse to the flow direction results in particles changing streamlines, moving in effect to positions where the flow velocity differs. Hence, a small change in transverse position due to diffusion may lead to a subsequently wide separation in longitudinal position as the different velocities in adjacent streamlines have the effect of sweeping apart particles that had initially started together. This subtle and complex interplay of diffusion and flow heterogeneity is known as 'Taylor dispersion' [15], and, in the case of pipe flow, is discussed in some detail in Section 2.3.4. In the case of the more complex fluid flow in porous media, Taylor dispersion is one of several mechanisms that cause initially adjacent particles to become separated. These dispersion effects are of immense practical significance and are a major topic of this book. The facility of NMR to provide useful insight regarding dispersion is one of the reasons why NMR of porous media is a major area of international research.

While, for uniform flow, the average propagator, $\bar{P}(\mathbf{R}, t)$, is identical to the conditional probability, in the case of heterogeneous flow, it will differ. As an illustration we calculate the average propagator for pipe flow, choosing for simplicity a pipe of length $L \gg v_0 t$. An an exercise we show each step of the calculation in detail.

$$
\begin{aligned}
\bar{P}(\mathbf{R}, t) &= \int_0^a dr \int_0^{2\pi} r\, d\theta \int_{-L/2}^{L/2} dz\, P(r, \theta, z)\, P(r, \theta, z | r + R, \theta + \Theta, z + Z, t) \\
&= \int_0^a dr \int_0^{2\pi} r\, d\theta \int_{-L/2}^{L/2} dz\, \frac{1}{\pi a^2 L}\, \delta(R)\, \delta(\Theta)\, \delta\left(Z - v_0\left(1 - r^2/a^2\right) t\right) \quad (2.32) \\
&= \int_0^a dr\, \frac{2r}{a^2}\, \delta\left(Z - v_0\left(1 - r^2/a^2\right) t\right)
\end{aligned}
$$

Substituting the variable $\xi = v_0\left(1 - r^2/a^2\right) t$ we rewrite the integral

$$
\begin{aligned}
\bar{P}(Z, t) &= \frac{1}{v_0 t} \int_0^{v_0 t} d\xi\, \delta(Z - \xi) \\
&= \frac{1}{v_0 t} H(Z - v_0 t)
\end{aligned}
\quad (2.33)
$$

where H is the hat function shown in Fig. 2.8. For laminar pipe flow, in the absence of diffusive effects, there is a uniform probability that the particles travel any distance between 0 and the maximum $v_0 t$, or, in other words, an equal probability that the particle velocity lies between 0 and v_0, the peak velocity at the centre of the pipe.

2.2 Non-Newtonian fluids and viscoelasticity

Amongst the many applications of magnetic resonance insights regarding translational motion in fluids is that of rheological characterisation. Rheology concerns the study of various mechanical properties of non-Newtonian fluids, fluids for which a simple linear relation between stress and rate of strain does not hold. Indeed, the simple notion of liquids and solids is never exact, if we deform sufficiently rapidly or sufficiently slowly. For example, even water in the 'liquid' state may exhibit elastic properties if the rate of deformation is sufficiently high, while silicate glass may exhibit viscous flow if the

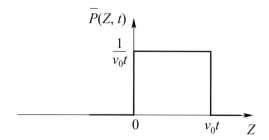

Fig. 2.8 Average propagator for axial displacements in Poiseuille flow.

observation time is sufficiently long. For all fluids the question of timescale is inherent, and in that sense all fluids are 'complex'. However, we generally assign the descriptors 'complex fluids' or 'complex soft matter' to those materials for which the timescales subdividing viscous and elastic behaviour are accessible in mechanical devices.

Complex fluids thus possess timescale complexity and both solid and liquid character. When these materials are subject to flow involving velocity gradients, their physical properties are generally non-linear, often anisotropic, and sometimes spatially heterogeneous. And all these properties have their origin in the molecular organisation and dynamics of each particular system. Hence, while the mechanical description of non-linear rheology is in itself an interesting and sometimes formidable area of engineering mathematics, the linking of these mechanical properties to underlying molecular organisation and dynamics represents one of the great challenges for modern chemical physics, a challenge that has resulted in beautiful new theories concerning the physics of polymers, colloids, and liquid crystals. And of course, this linkage has practical significance in modern materials science, in understanding the nature of biological tissue, the texture of food products, and the way in which fluids may be transported and processed. Rheology is a large field of research and one possessing a number of classic texts [16–25].

The use of magnetic resonance to study rheology is therefore doubly attractive. NMR has the potential to measure, *inter alia*, molecular orientation, intermolecular proximity, molecular re-orientational diffusion, and molecular translational Brownian motion. It is also capable of delivering an image of the local velocity field during the process of fluid deformation. In the case of NMR this information is provided non-invasively and without any requirement for optical transparency. Rheo-NMR is part the of the subject matter of Chapter 10.

2.2.1   Strain fields used in rheology

Effective rheological theories need to be able to handle any flow geometry that may be encountered in practise. However, for rheological characterisation there are a number of easily implemented geometries that allow the experimenter to gain access to relevant parameters of the constitutive behaviour. These generally involve simple cells in which one or more of the containment surfaces move. Figure 2.9 shows two implementations of shear fields using, respectively, cone-and-plate and cylindrical Couette cells, approximating the idealised example shown in Fig. 2.4(a). Figure 2.10 approximates the idealised planar extension example of Fig. 2.4(b).

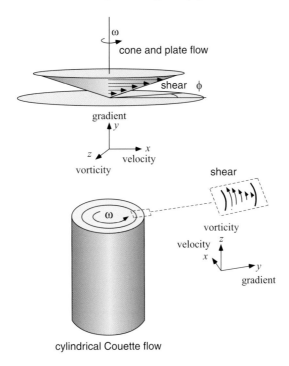

Fig. 2.9 Cone-plate and cylindrical Couette cells used to approximate a simple shear flow.

In simple shear, for which the velocity direction is taken to be x, one has $\kappa_{xy} = \dot{\gamma}$, all other $\kappa_{\alpha\beta}$ zero and

$$v_x = \dot{\gamma}y, \qquad v_y = 0, \qquad v_z = 0 \tag{2.34}$$

The coordinates (x, y, z) define the velocity, gradient, and vorticity axes, respectively.

In simple uniaxial elongational flow, with the axial extension direction taken as z, one has $\kappa_{zz} = \dot{\epsilon} = -2\kappa_{xx} = -2\kappa_{yy}$ and

$$v_x = -\frac{1}{2}\dot{\epsilon}x, \qquad v_y = -\frac{1}{2}\dot{\epsilon}y, \qquad v_z = \dot{\epsilon}z \tag{2.35}$$

This uniaxial extensional deformation can be achieved using a filament stretching rheometer.

By contrast, in planar axial flow, with the extension direction taken as z, $\kappa_{zz} = \dot{\epsilon} = -\kappa_{yy}$ and $\kappa_{xx} = 0$

$$v_x = 0, \qquad v_y = -\dot{\epsilon}y, \qquad v_z = \dot{\epsilon}z \tag{2.36}$$

In shear flow and extensional flow, the respective viscosities are defined by the ratios

$$\eta_S = \frac{\sigma_{xy}}{\dot{\gamma}} \tag{2.37}$$

and

$$\eta_E = \frac{(\sigma_{zz} - \sigma_{yy})}{\dot{\epsilon}} \tag{2.38}$$

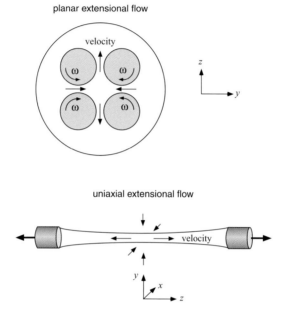

Fig. 2.10 Four roll mill [26] and filament stretch implementations [27] of planar and uniaxial extension, respectively.

The use of the difference of diagonal components of the stress tensor in the latter expression removes the effect of the isotropic term involving the pressure. Using eqn 2.15 it becomes clear that a Newtonian liquid has a uniaxial extensional viscosity $\eta_E = 3\eta_S$ (and planar extensional viscosity $4\eta_S$). The ratio of the extensional and shear viscosities is known as the Trouton ratio (T_r) and may greatly exceed 3 (in the planar case, 4) for viscoelastic fluids. Finally we note that unlike shearing flow, extensional flow cannot be experienced by an element of fluid in steady state in any practically realisable geometry. In consequence extension is always a transient phenomenon.

In the following sections we introduce some basic principles concerning the rheology of non-Newtonian viscoelastic fluids. Most of the discussion will concern shear deformation, although the ideas may be generalised to other deformations such as extensional flow.

2.2.2 Linear viscoelasticity

We start by considering a viscoelastic material deformed 'suddenly' by a small shear strain $\gamma \ll 1$, the strain being accompanied by a sudden stress, which gradually decays as the molecules of the liquid rearrange themselves. Once the stress has decayed to zero the material has permanently deformed, and the elastic energy that was initially stored has dissipated, all 'memory' of the initial deformation being lost. Such a memory effect is characteristic of the material response of all viscoelastic fluids and the associated memory time (by comparison with the suddenness of the applied strain), will determine the extent to which our material can be regarded as solid or liquid.

For sufficiently small strains it is possible to treat the stress–strain response in a linear manner, and in practice the limits of such a linear description can always be found by measuring the extent of strain deformation over which the linear viscoleastic parameters are strain-independent. In the linear description [28] the stress response may be calculated by means of a memory function, $G(t)$, through the relation

$$\sigma(t) = \int_{-\infty}^{t} \dot{\gamma}(t') \, G(t-t') \, dt' \qquad (2.39)$$

$G(t-t')$ gives the stress response of the system at time t when a rate of strain $\dot{\gamma}(t')$ is applied at an earlier time t'. It therefore represents the memory over the interval $(t-t')$. The characteristic time of the memory function, $G(t)$, is given by

$$\tau = \int_{0}^{\infty} \frac{G(t)}{G(0)} \, dt \qquad (2.40)$$

Suppose at time $t' = 0$, a step shear strain γ is established at a rate that is very fast compared with τ (*ie* $\dot{\gamma} \gg \tau^{-1}$). Then $\dot{\gamma}(t')$ can be replaced by the Dirac delta function, $\gamma \, \delta(t')$ and eqn 2.39 reduces to

$$\sigma_{xy}(t) = \gamma \, G(t) \qquad (2.41)$$

Hence we can see that the memory function is identified with the time-dependent shear relaxation modulus, which in turn may be measured directly from the step strain response.

If by contrast the shear strain γ is established at a rate that is very slow compared with τ, then $\dot{\gamma}$ is constant by comparison with $G(t-t')$ and, assuming that linearity holds, we obtain

$$\sigma_{xy}(t) = \dot{\gamma} \int_{-\infty}^{t} G(t-t') \, dt' = \dot{\gamma} \int_{0}^{\infty} G(t) \, dt \qquad (2.42)$$

from which we can identify $\int_{0}^{\infty} G(t) \, dt$ as the Newtonian viscosity, η.

In general, the linear viscoelastic response of a material may be investigated by applying an oscillatory strain, $\gamma_0 \sin(\omega t)$, at small strain amplitudes ($\gamma_0 \ll 1$). Then the stress is given by

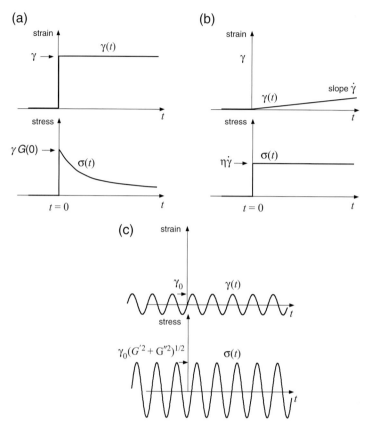

Fig. 2.11 Linear viscoelastic response of stress to (a) small step strain applied more rapidly than characteristic relaxation time τ, (b) steadily increasing strain applied at rate slow compared with τ, and (c) oscillatory strain. The phase shift between the stress and strain is a measure of the relative importance of the elastic and viscous responses.

$$
\begin{aligned}
\sigma_{xy}(t) &= \int_{-\infty}^{t} \gamma_0\, \omega \cos\left(\omega t'\right) G\left(t - t'\right)\, dt' \\
&= \gamma_0\, \omega \int_{0}^{\infty} \cos\left(\omega\left(t - t''\right)\right) G\left(t''\right)\, dt'' \\
&= \gamma_0 \left[\left\{\omega \int_{0}^{\infty} \sin\left(\omega t''\right) G\left(t''\right)\, dt''\right\} \sin\left(\omega t\right) \right. \\
&\qquad \left. + \left\{\omega \int_{0}^{\infty} \cos\left(\omega t''\right) G\left(t''\right)\, dt''\right\} \cos\left(\omega t\right)\right] \\
&= \gamma_0 \left[G' \sin\left(\omega t\right) + G'' \cos\left(\omega t\right)\right]
\end{aligned}
\tag{2.43}
$$

where G' and G'' are, respectively, the frequency-dependent storage and loss moduli and are directly related to the sine and cosine Fourier spectra of the memory function.

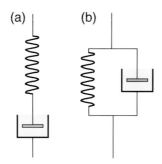

Fig. 2.12 (a) Maxwell and (b) Kelvin–Voigt spring dashpot representations of an 'elastic liquid' and a 'viscous solid', respectively.

Linear viscoelastic properties can be represented by suitable combinations of analogous dissipative and storage elements, such as dashpots and springs. One simple linear picture relevant to a viscoelastic liquid is the series dashpot-spring model of Maxwell shown in Fig. 2.12. The Maxwell memory function is exponential, namely, $G(t) = G_0 \exp(-t/\tau)$, with relaxation time $\tau_M = \eta/G_0$, where η is the viscosity of the dashpot fluid and G_0 is the elastic modulus of the spring. Similarly, viscous soft solids can be modelled using the Kelvin–Voigt model, in which the dashpot and spring are in parallel.

In general, one may model quite complex viscoelastic fluids whose characteristic dynamics cover a wide range of timescales by an appropriate superposition of Maxwell and Voigt elements [28], provided one is restricted to the range of strains for which the response is linear.

2.2.3   Non-linear viscoelasticity

Under sufficiently large strains, the mechanical response of all materials will differ from linear behaviour as a consequence of strain-induced structural rearrangements. For viscoelastic soft solids and liquids, the particular strain at which non-linear effects take over will depend on the details of the molecular structure, but will typically be on the order of unity. The subject of non-linear rheology is too extensive to be covered comprehensively here and readers are referred to a number of texts [17, 19, 22–25]. However, it is helpful to review one facet of non-linear rheology, namely the particularly common case of steady deformation, and to do so for one experimental geometry of interest, that of simple shear.

We start, in a quite general sense, by writing down the deformation over a finite time interval from t' to t as

$$\mathbf{r}(t) = \underline{\underline{\mathbf{E}}}(t, t') \cdot \mathbf{r}(t') \tag{2.44}$$

where the tensor $\underline{\underline{\mathbf{E}}}(t, t')$ describes the deformation history peculiar to the chosen strain geometry. Associated with $\underline{\underline{\mathbf{E}}}(t, t')$ will be a time- and strain-dependent stress in the fluid. Restricting ourselves to steady-state conditions, during which the rate of strain is held constant, results in a great deal of experimental and theoretical simplification. In this limiting case, non-linear viscoelastic behaviour is characterised by 'flow curves', the functions which describe the strain-rate dependencies of the stresses. The flow curve shows the degree to which the structural rearrangements are able to compete with the external rate of deformation. Figure 2.13(a) shows a simple flow curve for a fluid that at low rates of strain behaves in a Newtonian manner, with constant viscosity, but for which the viscosity reduces above a certain characteristic strain rate $\dot{\gamma}_c$. That rate provides an indicator of the time scale, $\tau \sim 1/\dot{\gamma}_c$, for structural rearrangement. The product $\dot{\gamma}_c \tau$ is commonly known as the Weissenberg number for steady shear deformation.

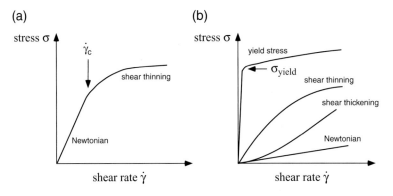

Fig. 2.13 (a) Transition from Newtonian to shear thinning behaviour at critical strain rate $\dot{\gamma}_c$ and (b) shear stress flow curves indicating a range of non-Newtonian behaviours in the non-linear viscosity.

For non-Newtonian viscoelastic liquids, non-linear flow may result not only in shear stress, but also in anisotropic normal stresses, describable by the first and second normal stress differences $(\sigma_{xx} - \sigma_{yy})$ and $(\sigma_{yy} - \sigma_{zz})$. In simple shear the constitutive relations for the fluid are given by

$$\sigma_{xy} = \eta\left(\dot{\gamma}\right)\dot{\gamma}, \quad \left(\sigma_{xx} - \sigma_{yy}\right) = \Psi_1\left(\dot{\gamma}\right)\dot{\gamma}^2, \quad \left(\sigma_{yy} - \sigma_{zz}\right) = \Psi_2\left(\dot{\gamma}\right)\dot{\gamma}^2 \qquad (2.45)$$

where $\eta\left(\dot{\gamma}\right)$ is the shear-rate-dependent viscosity and $\Psi_1\left(\dot{\gamma}\right)$ and $\Psi_2\left(\dot{\gamma}\right)$ the first and second normal stress coefficients.

Figure 2.13(b) shows a family of shear stress flow curves in which a variety of non-Newtonian properties are exhibited, including yield stress, shear-thinning and shear thickening. A wide class of shear-thinning and shear-thickening fluids can be adequately described by a power law constitutive equation of the form

$$\sigma_{xy} = k\dot{\gamma}^n \qquad (2.46)$$

for which $\eta\left(\dot{\gamma}\right) = k\dot{\gamma}^{n-1}$ and a Newtonian liquid has $n = 1$, while shear-thinning and shear-thickening liquids have $n < 1$ and $n > 1$, respectively. Of course such a description tells us nothing about the physical basis of non-linear viscosity.

In an equivalent sense, the rate-dependent non-linear extensional viscosity could be written as $\eta_E\left(\dot{\epsilon}\right)$. However, since extension is always a transient phenomenon this viscosity must inevitably depend on time, t, or equivalently on the total (Hencky) strain $\epsilon = \int_0^t \dot{\epsilon}\left(t'\right) dt'$, so that more correctly we should write the extensional viscosity as the function $\eta_E\left(\dot{\epsilon}, \epsilon\right)$.

2.3 Dispersion

Dispersion is the name we give to the phenomenon in which initially adjacent particles become separated during flow [15, 29]. The spreading and mixing that result from dispersion are of immense practical significance to a wide range of processes, such as oil recovery, ground water remediation, catalysis and the behaviour of packed bed reactors, filtration, chromatography, and biological perfusion. Dispersion is intrinsically governed by stochastic processes and, in this sense, has much in common with Brownian motion. However, whereas Brownian motion is driven by the stochastic fluctuations associated with molecular thermal energy, dispersion is driven by the subtle interplay of advective velocity gradients, molecular diffusion, and boundary layer effects, the latter being of particular importance in the flow of liquids or gases through a porous medium. While the language of dispersion has much in common with that used to describe self-diffusion, the range of phenomena available to be studied is, by comparison, considerably larger. Self-diffusion has unique characteristic lengths and timescales associated with molecular collisions, but in dispersive flow, multiple dynamical processes and structural features impart multiple length and timescales. Remarkably, many of those scales are accessible by NMR, either by magnetic resonance imaging methods or as ensemble averages over all streamlines, the latter obtained from the Lagrangian ensemble in which each molecule is labelled non-invasively by its local precession frequency. The details of how NMR accesses the various details of dispersive processes are left until later chapters. Here we review the essential physics of dispersion.

2.3.1 Stationary random flow and pseudo-diffusion

Before beginning a discussion of dispersion it is helpful to look at two simple examples of flow [30] that result in an isotropic, random distribution of molecular displacements. The first, which might be termed 'stationary random flow', has molecular motions randomly directed in magnitude and/or orientation and which are describable by an Eulerian velocity field, which is time-independent, and an ensemble of Lagrangian velocities, each of which is constant. This is the motion, illustrated in Fig. 2.14(a), which might be associated with laminar flow in an array of randomly directed capillaries in which the local director is fixed. The second, which we label 'pseudo-diffusion', involves molecular velocities that are not only randomly distributed across the ensemble but which fluctuate in time as well. Examples of pseudo-diffusion include turbulence and branched capillary motion, as shown in Fig. 2.14(b). Pseudo-diffusion is characterised by a correlation length l_c and correlation time τ_c.

  The essential difference between these examples lies in the role of fluctuations, and in the time-dependence of mean-squared displacements. In stationary random flow, the absence of fluctuations leads to an inherent reversibility of the dispersion process, either by flow reversal or, as we shall see in the case of magnetic resonance tracer measurements, by reversing the sign of the flow-sensitive NMR parameter. The same reversibility is present in the dispersal caused by the Poiseuille velocity distribution for a single unidirected pipe (provided we neglect the effect of diffusion). Stationary random flow exhibits a quadratic dependence of mean-squared displacements on flow time. At times shorter than the flow time τ_c between branch points, the second flow geometry shown in Fig. 2.14(b) is indistinguishable from stationary random flow. But

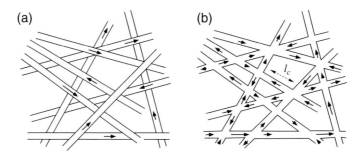

Fig. 2.14 (a) Stationary random flow and (b) pseudodiffusion.

at longer times, the mean-squared displacements become linear in time, akin to the Einstein relation for self-diffusion. This is the pseudo-diffusion regime. Whether the pseudo-diffusion dispersion process is reversible depends of course on the nature of streamline transitions at the branch points. Generally these transitions are at most only partially reversible. The reason for this has to do with the stochastic processes that underlie dispersion in heterogeneous flow fields.

2.3.2 Porous medium characteristics

Porosity and representative elementary volume

A porous medium can be regarded as the interpenetration of two regions in space, one from which fluid is excluded, and one which allows fluid ingress and flow: in the simplest of terms a solid matrix and an inter-connected pore space [31–33]. An illustration of such a medium is shown in Fig. 2.15. The measure that indicates the relative proportions of pore space and matrix is the porosity, ϕ, defined as the pore volume/total volume of medium. Of course it is possible that some pores may be entirely enclosed by the matrix, cut off from the interconnected pore space, but in what follows we shall be concerned only with those pores that contain fluid and therefore belong to the space that permits ingress.

An important characteristic of porous materials is the representative elementary volume (REV), defined as the smallest volume from which one can derive macroscopic properties. Bear [31] presents a nice way of visualising the REV, through the diagram shown in Fig. 2.16. Suppose one starts at some point inside the material and defines a volume ΔV_i surrounding that point. For example, the volume could be a sphere for which our point was the sphere centre. Now calculate a porosity ϕ_i, being the pore volume, ΔV_{pi} within ΔV_i, divided by the total volume ΔV_i. For large volumes the porosity is fairly constant but as the volumes are made smaller, the local porosity fluctuates, as shown in Fig. 2.16. In the limit, as $\Delta V_i \to 0$, ϕ_i will tend to 1 if the point chosen were in the pore space, or 0 if within the matrix. We may then define the REV as being the smallest volume, ΔV_0, beyond which the fluctuations are damped for further volume increases. This simple argument holds for a homogeneous porous medium. For a heterogeneous medium, there may exist longer length scales over which further changes in the mean porosity are observed, even though rapid variations with increasing ΔV_i, are no longer apparent.

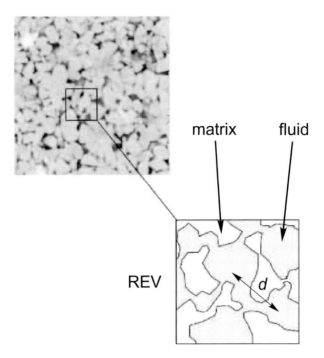

Fig. 2.15 Porous sandstone with schematic representative elementary volume (REV) indicated.

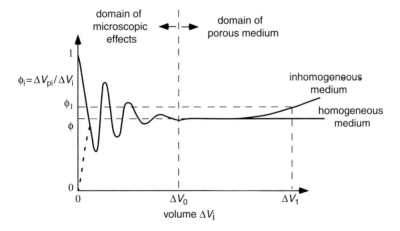

Fig. 2.16 Fluctuations in local porosity as the sampling volume is increased. For a homogeneous medium, there exists a volume ΔV_0, known as the representative elementary volume (REV) beyond which fluctuations are damped out. Heterogeneous materials may exhibit multiple volume and associated length scales. (Adapted from Bear [31].)

Permeability

The permeability, K_p, of the medium gives an indication of the ease with which fluid can be made to pass through the porous matrix, and its definition is due to Darcy in 1856 [34], in a relationship known as Darcy's law. For a fluid of dynamic viscosity η, the permeability is defined by the flow response to a pressure gradient dp/dz as

$$\langle v_z \rangle = -\frac{K_p}{\eta}\frac{dp}{dz} \tag{2.47}$$

where $\langle v_z \rangle$ is the average flow rate along the pressure gradient, as calculated from the volume flow rate across area A of $\dot{Q} = \langle v_z \rangle A$. In most descriptions of fluid flow in porous media, this mean flow rate, $\langle v_z \rangle$, derived from the volume rate, $\dot{Q}$, is known as the 'tube velocity', $\langle v_{\text{tube}} \rangle$. But of course, the fluid motion is confined to the interconnected pore space within the matrix, a space with a volume fraction ϕ. In consequence, an average tube velocity, $\langle v_{\text{tube}} \rangle$, translates to a larger average pore space velocity $\langle v \rangle$ by

$$\langle v \rangle = \frac{\langle v_{\text{tube}} \rangle}{\phi} \tag{2.48}$$

Characteristic lengths, characteristic times, and Péclet number

Inside the matrix the flow velocity fluctuates as molecules stream from pore to pore. The question then arises as to what size or volume is needed in order to calculate a faithful average of the flow. For this purpose we turn to REV, a volume larger than an average pore size (see Fig. 2.15). The length scale associated with the REV will be on the order of the longest correlation length of the pore space structure.

For fluid flow in a porous medium, a volume over which the Eulerian velocity field might be averaged suggests the existence of a suitable time over which the Lagrangian velocity might be averaged in order to obtain the same value of mean flow. Given the wide distribution of velocities present in porous media flow, it is reasonable to assume a wide distribution of correlation times, but there presumably exists a longest time beyond which a molecule has sampled all possible velocities in the Lagrangian ensemble and at which the dispersion may be said to be asymptotic.

A characteristic length shorter than the REV size is the mean pore size or, perhaps more appropriately, pore spacing d. [5] This length determines a fundamental correlation time defining the temporal structure of the velocity field, τ_v, the duration of flow around that characteristic length scale, and written

$$\tau_v = \frac{d}{\langle v \rangle} \tag{2.49}$$

Of course, a second fundamental time, essential to understanding the process of dispersion, is that required to migrate the pore distance by Brownian motion alone, namely $\tau_D \sim d^2/D$. The ratio of τ_D to the velocity correlation time, τ_v, provides a dimensionless number, the Péclet number, $Pe \sim d\langle v \rangle/D$, which characterises the flow dynamics.

[5]The question as to whether pore size or pore space spacing is the more relevant length is a moot point. We prefer to use the definition of pore spacing.

But what is the relevant value of D if we are describing the time to diffuse the mean distance, d, between pores? In defining τ_D, and hence the Péclet number, we need to settle on an appropriate length scale and diffusion coefficient. There are a number of different ways to define Pe for porous media. One convention [35] is

$$Pe = \frac{l\langle v\rangle}{D_0} = \frac{d\langle v_{\text{tube}}\rangle}{D_{\text{eff}}} \tag{2.50}$$

where D_0 is the unrestricted molecular diffusion coefficient. The characteristic dimension l is taken to be the effective pore spacing, defined by $l = \phi d/(1 - \phi)$ and $D_{\text{eff}} = (1 - \phi)D_0$ is an 'effective' or 'reduced' diffusion coefficient. Note that in this definition, Pe can also be regarded as the ratio of d^2/D_{eff}, 'the effective diffusion time to migrate the distance d' to $d/\langle v_{\text{tube}}\rangle$, 'the effective flow time to migrate a distance d'.

In the same way, it is possible to define the dimensionless Reynolds number that compares the magnitude of viscous and inertial forces to which the fluid is subjected, i.e.

$$Re = \frac{l\langle v\rangle}{\nu} = \frac{d\langle v_{\text{tube}}\rangle}{\nu_{\text{eff}}} \tag{2.51}$$

where ν_{eff}, by analogy to D_{eff}, is the effective kinematic viscosity.

The question then arises of what is the effective asymptotic $(t \gg \tau_D)$ diffusion coefficient in the zero flow limit, $Pe = 0$, for molecules of a fluid imbibed in the pore space with self-diffusion coefficient, D_0. Clearly D_{eff}, as defined above, cannot suffice given its contrary dependence on porosity. Furthermore, porosity alone cannot determine the asymptotic diffusion since clearly pore connectivity plays a role. The generally accepted asymptotic result is

$$D^* (Pe = 0) = \frac{D_0}{\phi} \frac{\sigma}{\sigma_0} \tag{2.52}$$

where σ_0 is the electrical conductivity of the fluid and σ the electrical conductivity of the medium when occupied by the fluid. This result is mathematically derived in reference [36].

2.3.3   The dispersion tensor

In Chapter 1 the diffusion process was defined via the Fick's law diffusion equation (eqn 1.41) for the conditional probabilities, $P(\mathbf{r}|\mathbf{r}', t)$. In dispersion theory the same physics applies but with two significant changes. First, we must replace the time derivative by the substantive derivative, and second, we must allow that the dispersion will, in general, be anisotropic and hence represented by a tensor, $\underline{\mathbf{D}}^*$, rather than the scalar diffusion coefficient, D. This anisotropy is an obvious consequence of symmetry-breaking by the flow. Hence

$$\frac{\partial}{\partial t} P(\mathbf{r}|\mathbf{r}', t) + (\mathbf{v} \cdot \nabla) P(\mathbf{r}|\mathbf{r}', t) = \nabla \cdot \left(\underline{\underline{D}}^* \nabla P(\mathbf{r}|\mathbf{r}', t) \right) \tag{2.53}$$

where $\mathbf{v}$ is the Eulerian velocity at $\mathbf{r}$. Solving eqn 2.53, in conjunction with the Navier–Stokes equation, for flow in the complex geometry of a porous matrix is a daunting challenge. However, in the limit of long times, over which the dispersion may be said to be asymptotic, some considerable simplifications arise and indeed the nature of dispersion appears to follow a universal pattern.

The stochastic part of the flow

In the theory of dispersion, the roles of the Eulerian and Lagrangian perspectives are somewhat intertwined. While the Lagrangian approach provides some simplified definitions, when we seek information about spatial correlations a knowledge of the Eulerian velocity field is essential. For the moment, however, we will consider the ensemble of Lagrangian velocities $\{\mathbf{v}_L(t)\}$ [29, 37]. This distribution is characterised by an averaged velocity, $\mathbf{V}_L$, defined by $\mathbf{V}_L = \lim_{t\to\infty}\langle\mathbf{v}_L\rangle$, the ensemble average, $\langle\ldots\rangle$, being taken over the entire velocity distribution.

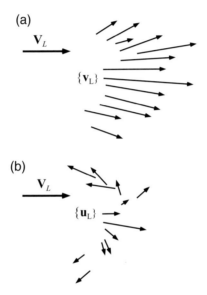

Fig. 2.17 (a) Ensemble of Lagrangian velocities $\{\mathbf{v}_L(t)\}$ with mean flow $\mathbf{V}_L$ and (b) same ensemble but for stochastic part $\{\mathbf{u}_L(t)\}$ (mean flow removed).

The fluctuation in the Lagrangian velocity may be defined by $\mathbf{u}_L(t) = \mathbf{v}_L(t) - \mathbf{V}_L$. It is the behaviour of $\mathbf{u}_L(t)$ and its distribution over the ensemble that contains the details of the dispersion process. For example, fluctuations in $\mathbf{u}_L(t)$ are described via the autocorrelation tensor, $\langle\mathbf{u}_L(\tau)\mathbf{u}_L(0)\rangle$, where the angled brackets refer to the ensemble average over the distribution of streamlines. We shall find the L subscript a nuisance in what follows and so, wherever u is a function of time only, we shall take it to be Lagrangian.

Clearly the averaged flow will be represented by a component of velocity, $V_\parallel$, parallel to the mean flow direction, the mean transverse component being zero. Equally, we may define stochastic components $u_\parallel$ and $u_\perp$. Since these will differ, we will expect the nature of dispersion to be different when measured parallel or perpendicular to the mean flow.

Time-dependent and asymptotic dispersion tensor

Earlier we introduced the symbol $\mathbf{R}$ to define a displacement over some time t. Here we define the displacement $\mathbf{R}_u = \mathbf{R} - \langle \mathbf{R} \rangle$ that arises solely from the stochastic part of the velocity $\mathbf{u}$. We could represent $\mathbf{R}_u$ by the time integral of the velocity $\int_0^t \mathbf{u}\left(t'\right) dt'$. Now consider the component of $\mathbf{u}$ parallel to the mean flow direction, denoting it $u_\parallel$. By direct analogy with the Einstein definition of diffusion[6] a longitudinal dispersion coefficient may be defined by [38]

$$D_\parallel^*\left(t\right) = \frac{1}{2}\frac{d\sigma_\parallel^2\left(t\right)}{dt} \tag{2.54}$$

where $\sigma_\parallel^2\left(t\right) = \langle \left(\int_0^t u_\parallel\left(t'\right) dt'\right)^2 \rangle$.[7] Note that $D_\parallel^*\left(t\right)$ has an implied asymptotic limit

$$D_\parallel^* = \lim_{t\to\infty} \frac{1}{2}\frac{d\sigma_\parallel^2\left(t\right)}{dt} \tag{2.55}$$

Similarly we could define a transverse coefficient $D_\perp^*$.

The implication of anisotropy in the flow is that we will need to reconcile the Einstein description with the full tensorial representation of $\underline{\underline{D}}^*$. This is done by defining the dispersion tensor in a manner similar to that seen in Section 1.3.5. The dispersion tensor is given by [29, 38–40]

$$\underline{\underline{D}}^*\left(t\right) = sym \int_0^t \langle \mathbf{u}\left(\tau\right) \mathbf{u}\left(0\right)\rangle d\tau \tag{2.56}$$

with asymptotic limit

$$\underline{\underline{D}}^* = \lim_{t\to\infty} sym \int_0^t \langle \mathbf{u}\left(\tau\right) \mathbf{u}\left(0\right)\rangle d\tau \tag{2.57}$$

where $sym\left(\underline{\underline{A}}\right) = \frac{1}{2}\left(\underline{\underline{A}} + \underline{\underline{A}}^T\right)$. The tensorial nature of the dispersion is now apparent, since the ensemble average in the integrand of eqn 2.57 includes correlations between different velocity components. To understand how this definition relates to eqn 2.55, we note that the trace of this long timescale 'steady state' dispersion tensor yields a scalar dispersion coefficient that is simply related to the mean-squared displacements, $\sigma^2\left(t\right)$ via

$$\mathrm{Tr}\left(\underline{\underline{D}}^*\right) = \lim_{t\to\infty} \frac{1}{2}\frac{d\sigma^2\left(t\right)}{dt} \tag{2.58}$$

where $\sigma^2\left(t\right) = \langle \left(\int_0^t \mathbf{u}\left(t'\right) dt'\right)^2 \rangle$.

---

[6]Note we here define a time-dependent diffusion coefficient in terms of a time derivative of mean-squared displacement, whereas the Einstein definition is strictly a mean-squared displacement divided by time. The significance of this difference in definition will be apparent in the discussion of asymptotic behaviour in Chapter 6.

[7]This is the fourth different use of the symbol σ after entropy, stress, and electrical conductivity. Context should provide a guide as to meaning.

The proof of eqn 2.58 is as follows. Consider two times, t_1 and t_2, such that $\tau = t_1 - t_2$. Then $\mathbf{R}_u(\tau) = \mathbf{R}_u(t_1) - \mathbf{R}_u(t_2)$ and

$$\mathbf{R}_u \mathbf{R}_u = \langle \mathbf{R}_u(t_1)^2 + \mathbf{R}_u(t_2)^2 - 2\,sym\,\mathbf{R}_u(t_1)\,\mathbf{R}_u(t_2) \rangle \tag{2.59}$$

Differentiation with respect to t_1 and t_2 gives

$$\frac{\partial^2\,\mathbf{R}_u \mathbf{R}_u}{\partial t_1\,\partial t_2} = -2\,sym\,\langle \mathbf{u}(t_1)\,\mathbf{u}(t_2) \rangle \tag{2.60}$$

whence

$$\frac{\partial^2\,\mathbf{R}_u \mathbf{R}_u}{\partial \tau^2} = 2\,sym\,\langle \mathbf{u}(t_1)\,\mathbf{u}(t_2) \rangle \tag{2.61}$$

We will assume that the ensemble is stationary, so that only the difference in time is relevant on the right-hand side of eqn 2.61. This means that $\langle \mathbf{u}(t_1)\,\mathbf{u}(t_2) \rangle$ may be replaced by $\langle \mathbf{u}(\tau)\,\mathbf{u}(0) \rangle$. Thus, on integrating, and noting $\sigma^2 = \mathrm{Tr}\,(\mathbf{R}_u \mathbf{R}_u)$, eqn 2.58 follows from eqn 2.57. Alternatively we may write

$$\underline{\underline{D}}^* = \lim_{t \to \infty} \frac{1}{2} \frac{d\,(\mathbf{R}_u \mathbf{R}_u)}{dt} \tag{2.62}$$

Dispersion mechanisms and scaling

The mechanisms that cause the flow-driven separation of initially adjacent molecules fall into three classes, known respectively as mechanical, diffusive, and holdup dispersion [35, 36, 39, 41]. Mechanical dispersion is due to stochastic variations in velocity induced by flow bifurcations in the advection of the fluid along tortuous paths. Diffusive (Taylor) dispersion [15] arises from molecular diffusion across streamlines. Holdup dispersion arises from boundary layer effects and from the presence of dead-end pores [39]. A simple illustration of these three processes is shown in Fig. 2.18.

The relative importance of these three mechanisms depends partly on flow geometry, but also on the value of Pe. Of course Taylor dispersion is always present, even in simple flows, whereas mechanical and holdup dispersion are a particular consequence of the complexity of flow associated with porous media. Remarkably, when non-dimensionalised against the molecular diffusion coefficient, D_0, and plotted against Péclet number, asymptotic dispersion in porous media follows a universal behaviour as shown in Fig. 2.19 for both $D_\parallel^*/D_0$ and $D_\perp^*/D_0$. Of course, when $Pe \ll 1$, the microscopic Brownian motion dominates and $D^*/D_0 \sim 1$ (see eqn 2.52). For $Pe \gg 1$, D^* obeys an approximate power-law behaviour, $D_\parallel^*/D_0 \sim Pe^\alpha$, where $1 < \alpha < 2$.

The non-dimensionalised mechanical dispersion might be expected to scale as Pe, since the rate of separation of molecules should depend linearly on both the velocity and the pore length scale. Similarly, we expect the non-dimensionalised asymptotic Taylor dispersion, arising from molecular diffusion across streamlines, to scale as Pe^2. The reason is as follows: in the asymptotic limit, the change in streamline velocity experienced by molecules as they sample all the pore space by diffusion transverse to the flow will be on the order of $\langle v \rangle$, while the time taken to traverse the flow by

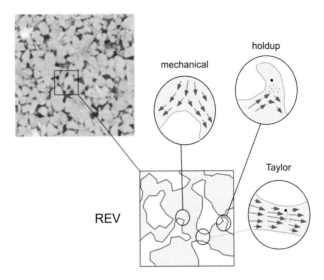

Fig. 2.18 Examples of mechanical, holdup, and Taylor dispersion.

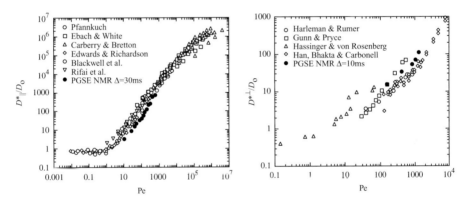

Fig. 2.19 Non-dimensional asymptotic dispersion coefficient vs Péclet number for dispersion longitudinal and transverse to the mean flow direction, taken from references [42, 43].

diffusion will be on the order of τ_D. Hence the mean-squared separation of molecules starting together will be $\langle v \rangle^2 \tau_D^2$, corresponding to a dispersion coefficient

$$
\begin{aligned}
D^*_{\text{Taylor}} &= \frac{\langle v \rangle^2 \tau_D^2}{\tau_D} \\
&= \frac{\langle v \rangle^2 l^2}{D_0}
\end{aligned}
\tag{2.63}
$$

whence $D^*_{\text{Taylor}}/D_0 \sim Pe^2$. Holdup dispersion scales as $Pe \ln Pe$.

A simple example of porous medium dispersion is provided by flow through random bead packs, for which $\alpha \sim 1.2$, gradually reducing with increasing Pe. At the highest numbers for which measurements have been made, around 10^5 to 10^6, the asymptotic longitudinal dispersion scales approximately as Pe, indicating the ultimate dominance of mechanical dispersion.

2.3.4   Taylor dispersion in pipe flow

As an illustration of Taylor dispersion in laminar flow, we here calculate the effect of diffusion across streamlines for Poiseuille flow in a cylindrical pipe of radius a. We will label the direction along the pipe axis z, and try to calculate the mean distance by which particles separate with respect to the mean flow. Of course even in the absence of diffusion perpendicular to the streamlines, shear alone will cause a particle separation. We will be able to identify that part of the dispersion in our result. And, we will always be able to add an additional longitudinal diffusion, D_0, which will be present in the absence of flow.

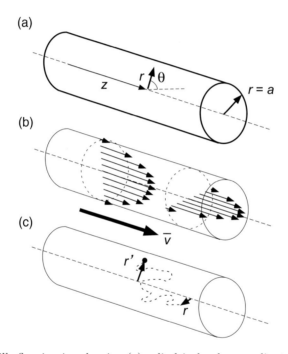

Fig. 2.20 Poiseuille flow in pipe showing (a) cylindrical polar coordinates, (b) velocity profile, and (c) diffusion across streamlines from starting radius r to final radius r'.

We start by writing down the axial displacement from origin at time t as $z(t)$. For Poiseuille flow, the mean axial flow rate, averaged across the pipe, is $\bar{v} = v_0/2$. Noting that $\langle z)t)\rangle = \bar{v}t$, the mean-squared axial displacement with respect to the mean flow is

$$\langle (z(t) - \bar{v}t)^2 \rangle = \langle z(t)^2 \rangle - \bar{v}^2 t^2 \tag{2.64}$$

while

$$\langle z(t)^2 \rangle = \int_0^t \int_0^t \langle v_z(\tau') v_z(\tau) \rangle d\tau d\tau' \tag{2.65}$$

$v_z(\tau)$ being the Lagrangian velocity at time τ. The problem then becomes one of tracking the history of particle velocities. The way we do this is to apply our knowledge of the steady-state Eulerian velocity field, $v_z(\mathbf{r})$, and to use a conditional probability $P(\mathbf{r}|\mathbf{r}',t)$ to determine how the particles move about in that field under diffusion [40]. Of course, given the velocity profile of eqn 2.23, it is clear that only radial displacements will change the velocity, and the problem in cylindrical polar coordinates reduces to a dependence on r only. Noting that the starting probability, $P(r,\tau)$ is independent of time and given by[8] $P(r) = 2/a^2$, we may write

$$\langle z(t)^2\rangle = \int_0^t d\tau \int_0^t d\tau' \int_0^a r\,dr \int_0^a r'\,dr' P(r)\,P(r|r',\tau'-\tau)\,v_z(r)\,v_z(r')\,d\tau d\tau' \tag{2.66}$$

where[9]

$$P(r|r',\tau'-\tau) = \frac{2}{a^2}\left[1 + \sum_n \frac{J_0(\mu_n r/a)\,J_0(\mu_n r'/a)}{\mu_n^2\,J_0(\mu_n)^2}\exp\left(-\frac{D_0\mu_n^2}{a^2}(\tau'-\tau)\right)\right] \tag{2.67}$$

the J_0 being cylindrical Bessel functions with roots μ_n and D_0, the molecular self-diffusion coefficient. Note that eqn 2.67 is appropriate for $\tau' > \tau$ and that part of the integral $\int_0^t d\tau \int_\tau^t d\tau'$. To deal with the complementary case $\tau' < \tau$ we replace $P(r)\,P(r|r',\tau'-\tau)$ by $P(r)\,P(r'|r,\tau-\tau')$. The total integral $\int_0^t d\tau \int_0^t d\tau'$ is a sum of the identical complementary integrals and so

$$\langle z(t)^2\rangle = \left(\frac{2}{a^2}\right)^2 \int_0^t d\tau \int_0^t d\tau' \int_0^a r\,dr \int_0^a r'\,dr'\,v_z(r)\,v_z(r')$$
$$+ 2\left(\frac{2}{a^2}\right)^2 \int_0^t d\tau \int_\tau^t d\tau' \int_0^a r\,dr \int_0^a r'\,dr'\left[v_z(r)\,v_z(r')\right.$$
$$\left.\times \sum_n \frac{J_0(\mu_n r/a)\,J_0(\mu_n r'/a)}{\mu_n^2\,J_0(\mu_n)^2}\exp\left(-\frac{D_0\mu_n^2}{a^2}(\tau'-\tau)\right)\right]$$

The first term in eqn 2.68 reduces to $\bar{v}^2 t^2$ and so the dispersive displacements with respect to the mean flow of eqn 2.64 become

$$\langle(z(t)-\bar{v}t)^2\rangle = 2\int_0^t d\tau \int_\tau^t d\tau' \sum_n \frac{b_n^2}{\mu_n}\exp\left(-\frac{D_0\mu_n^2}{a^2}(\tau'-\tau)\right)$$
$$= 2\frac{a^2}{D_0}\sum_n \frac{b_n^2}{\mu_n^2}t - 2\left(\frac{a^2}{D_0}\right)^2\sum_n \frac{b_n^2}{\mu_n^4}\left[1-\exp\left(-\frac{D_0\mu_n^2 t}{a^2}\right)\right] \tag{2.68}$$

[8]Note that $P(r)$ has been averaged over azimuthal angle and is associated with $r\,dr$ in the integrand.

[9]This result for the conditional probability associated with diffusion of molecules bounded by a cylindrical pipe is explained in Chapter 7.

where the b_n are radial integrals of the Bessel functions over the velocity distributions. Note the limiting cases

$$\lim_{t \to 0} \langle (z(t) - \bar{v}t)^2 \rangle = \sum_n b_n^2 t^2 \tag{2.69}$$

and

$$\lim_{t \to \infty} \langle (z(t) - \bar{v}t)^2 \rangle = 2 \frac{a^2}{D_0} \sum_n \frac{b_n^2}{\mu_n^2} t \tag{2.70}$$

Of course, in the case of both equations, the independent axial diffusion contribution, $2D_0 t$, should be added for a complete description.

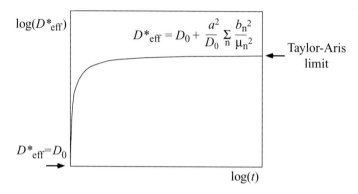

Fig. 2.21 Axial dispersion for flow in a pipe showing increase with time to Taylor–Aris limit.

The short time limit is dominated by the velocity shear across the pipe and contains no contribution from diffusion across streamlines. Accordingly, the quadratic time-dependence indicates that the mean-squared displacements are velocity-like rather than dispersion-like. Note that $\sum_n b_n^2 t^2 = 1/3 \, \bar{u}^2 \, t^2$. As time increases, Taylor dispersion starts to play a role, reaching the limiting case when $t \gg a^2/D_0$. Equation 2.70 defines the Taylor–Aris asymptotic limit [44]

$$D_{\text{Taylor}}^* = D_0 + \frac{a^2}{D_0} \sum_n \frac{b_n^2}{\mu_n^2} \tag{2.71}$$

Figure 2.21 shows the gradual onset of the dispersive limit with increasing time. Intriguingly, as the molecular diffusion rate D_0 increases, the shear contribution to the asymptotic dispersion limit, D^*, decreases. This is an example of motional averaging at work. The faster the molecules diffuse to and fro between the bounding walls of the pipe, the less chance the velocity has to separate them down the flow axis as they cross the streamlines. Note again, given $b_n^2 \sim \bar{u}^2$, that D_{Taylor}^*/D_0 scales as Pe^2.

2.3.5 The velocity autocorrelation function and dispersion spectrum

The asymptotic dispersion tensor defined by eqn 2.57 can be viewed as the zero frequency component in the velocity autocorrelation function spectrum, which may be more generally defined by

$$\underline{\underline{D}}^* (\omega) = \frac{1}{2} sym \int_{-\infty}^{\infty} \langle \mathbf{u} (\tau) \mathbf{u} (0) \rangle \exp (i\omega\tau) \, d\tau \qquad (2.72)$$

where we have taken advantage of the time reversal symmetry of $\langle \mathbf{u} (\tau) \mathbf{u} (0) \rangle$ to define the integral over all time.

The frequency-dependent dispersion tensor defined by eqn 2.72 has a spectral distribution dependent upon the characteristic correlation times for velocity fluctuations, τ_c. Let us for simplicity take a single component of velocity u_z. Noting eqns 1.27 and 1.28 and $\langle u_z \rangle = 0$, the mean correlation time is defined by the relation

$$\tau_c = \int_0^{\infty} \frac{\langle u_z (\tau) \, u_z (0) \rangle}{\langle u_z^2 \rangle} \, d\tau \qquad (2.73)$$

From this definition we derive the zero-frequency amplitude of the dispersion tensor element D_{zz} as,

$$D_{zz} = \langle u_z^2 \rangle \tau_c \qquad (2.74)$$

Finally, note that the definition of dispersion employed could allow for an ensemble in which the velocities, $\{\mathbf{u}\}$, do not vary with time but are simply a static distribution of zero mean. A good example is laminar flow in a straight pipe under conditions of minimal molecular self-diffusion. Such an ensemble will return a dispersion tensor that diverges with observation time, t. We will see that NMR gives us a means of distinguishing such behaviour from fluctuating flows.

2.3.6   Non-local dispersion

The dispersion tensor, as defined by eqn 2.56, has time-dependence, but is independent of spatial coordinates. In the language of fluid mechanics we say that the dispersion tensor is local, in the sense that it does not depend on the behaviour of the velocity field at other places. Indeed, by its very nature it is written in terms of Lagrangian velocities, which are not dependent on space. And, being formed by integrating the velocity autocorrelation function, is only weakly sensitive to its temporal structure, while in the asymptotic limit, $\underline{\underline{D}}^*$ is time-independent.

By comparison with the dispersion tensor, the covariant velocity autocorrelation function (VACF) is more fundamental in understanding fluid dispersion. In fact, we may dig deeper into the VACF and discover an even more fundamental correlation function, which depends not only on temporal displacement but also spatial displacement. Koch and Brady [37] have termed this the *non-local dispersion tensor*. It is the primary quantity describing details of the dispersion process. To understand the non-local dispersion tensor, one needs to link the Eulerian and Lagrangian descriptions. The key tool that enables this linkage is the conditional probability, the device that tells us the chance of finding a particle at two different places at two different times. The way we do this is illustrated in Fig. 2.22. Here, for clarity, we return to explicit use of L and E subscripts.

We will use this approach to rewrite the Lagrangian velocity autocorrelation function, i.e.

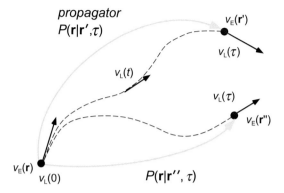

Fig. 2.22 The connection between the Eulerian velocity field and the ensemble of Lagrangian velocities at different times is given by the conditional probability $P(\mathbf{r}|\mathbf{r}',\tau)$.

$$\langle \mathbf{u}_L(\tau)\,\mathbf{u}_L(0)\rangle = \iint P(\mathbf{r})\,\mathbf{u}_E(\mathbf{r},0)\,P(\mathbf{r}|\mathbf{r}',\tau)\,\mathbf{u}_E(\mathbf{r}',\tau)\;d\mathbf{r}\,d\mathbf{r}'$$
$$= \iint P(\mathbf{r})\,\mathbf{u}_E(\mathbf{r},0)\,P(\mathbf{r}|\mathbf{r}+\mathbf{R},\tau)\,\mathbf{u}_E(\mathbf{r}+\mathbf{R},\tau)\;d\mathbf{r}\,d\mathbf{R} \tag{2.75}$$

Equation 2.75 is obtained by writing a starting probability, $P(\mathbf{r})$, of being at position $\mathbf{r}$, and then using a conditional probability to find the subsequent chance to be at $\mathbf{r}'$ (or $\mathbf{r}+\mathbf{R}$), the full Lagrangian ensemble average being formed by integrating the Eulerian velocity product over all starting positions $\mathbf{r}$ and possible displacements $\mathbf{R}$. Note that for steady-state flow, the $\mathbf{u}_E(\mathbf{r},0)$ and $\mathbf{u}_E(\mathbf{r},\tau)$ in the above relations may be exchanged for their time-independent counterparts. The non-local dispersion tensor is then defined as the partial integrand

$$\underline{\underline{D}}^{NL}(\mathbf{R},\tau) = \int P(\mathbf{r})\,\mathbf{u}_E(\mathbf{r},0)\,P(\mathbf{r}|\mathbf{r}+\mathbf{R},\tau)\,\mathbf{u}_E(\mathbf{r}+\mathbf{R},\tau)\;d\mathbf{r} \tag{2.76}$$

This definition leads to the following integrals

$$\langle \mathbf{u}_L(\tau)\,\mathbf{u}_L(0)\rangle = \int \underline{\underline{D}}^{NL}(\mathbf{R},\tau)\;d\mathbf{R} \tag{2.77}$$

and

$$\underline{\underline{D}}^* = \iint \underline{\underline{D}}^{NL}(\mathbf{R},\tau)\;d\mathbf{R}\,d\tau \tag{2.78}$$

The hierarchy of integrations 2.77 and 2.78 provides special insight. Just as the spatial structure of $\underline{\underline{D}}^{NL}(\mathbf{r},\tau)$ is largely lost in the integration to form the velocity autocorrelation function, so in turn the VACF temporal structure is lost when further integration over time is required to obtain the local asymptotic dispersion tensor. Given that perspective, one may regard the non-local dispersion tensor as a primary quantity that retains maximum information. Of course $\underline{\underline{D}}^{NL}(\mathbf{R},\tau)$ is not strictly a dispersion tensor in a dimensional sense. Rather, it is a full spatio-temporal correlation function, nine elements in the symmetric velocity correlation tensor, three dimensions in spatial offset $\mathbf{R}$, and one more dimension in the time offset, τ, sufficiently rich in information that its measurement represents a worthy challenge.

References

[1] Judges 5:4,5.

[2] R. E. Meyer. *Introduction to Mathematical Fluid Dynamics*. Dover, New York, 1982.

[3] D. J. Acheson. *Elementary Fluid Dynamics*. Oxford University Press, Oxford, 1990.

[4] T. E. Faber. *Fluid Dynamics for Physicists*. Cambridge University Press, Cambridge, 1995.

[5] R. A. Granger. *Fluid Mechanics*. Dover, New York, 1995.

[6] F. M. White. *Fluid Mechanics*. McGraw-Hill, New York, 2003.

[7] G. K. Batchelor. *An Introduction to Fluid Dynamics*. Cambridge University Press, Cambridge, 1967.

[8] L. D. Landau and E. M. Lifschitz. *Fluid Mechanics 2nd Edition*. Butterworth-Heinemann, Boston, 1987.

[9] O. Reynolds. An experimental investigation of the circumstances which determine whether the motion of water shall be direct or sinuous, and of the law of resistance in parallel channels. *Philosophical Transactions of the Royal Society*, 174:935, 1883.

[10] J. L. M. Poiseuille. Physiques - recherches experimetales sur le mouvement des liquides dans les tubes de tres petits diametres. *Academie des Sciences, Comptes Rendus*, 111:961 and 1041, 1840.

[11] S. Succi. *The Lattice Boltzmann Equation for Fluid Dynamics and Beyond*. Oxford, New York, 2001.

[12] P. L. Bhatnagar, E. P. Gross, and M. Krook. A model for collision processes in gases. I. Small amplitude processes in charged and neutral one-component systems. *Phys. Rev.*, 94:511, 1954.

[13] D. Wolf-Gladrow. *Automata and Lattice Boltzmann Models*. Springer, New York, 2001.

[14] M. C. Sukop and D. T. Thorne. *Lattice Boltzmann Modeling: An Introduction for Geoscientists and Engineers*. Academic Press, New York, 2007.

[15] G. I. Taylor. Dispersion of soluble matter in solvent flowing slowly through a tube. *Proc. Roy. Soc. A*, 219:186, 1953.

[16] A. S. Lodge. *Elastic Liquids*. Academic Press, New York, 1964.

[17] A. S. Lodge. *Body Tensor Fields in Continuum Mechanics*. Academic Press, New York, 1974.

[18] K. Walters. *Rheometry*. Chapman and Hall, London, 1975.

[19] R. B. Bird, R. C. Armstrong, and O. Hassager. *Dynamics of Polymeric Liquids*. Wiley, New York, 1977.

[20] J. D. Ferry. *Viscoelastic Properties of Polymers*. Wiley, New York, 1980.

[21] R. I. Tanner. *Engineering Rheology*. Oxford, New York, 1985.

[22] M. Doi and S. F. Edwards. *The Theory of Polymer Dynamics*. Oxford, London, 1986.

[23] R. G. Larson. *Constitutive Equations for Polymer Melts and Solutions*. Butterworths, Boston, 1987.

[24] H. A. Barnes, J. J. Hutton, and K. Walters. *An Introduction to Rheology.* Elsevier Amsterdam, 1989.

[25] R.G. Larson. *The Structure and Rheology of Complex Fluids.* Oxford, New York, 1998.

[26] G. I. Taylor. The formation of emulsions in definable fields of flow. *Proc. Roy. Soc. A*, 146:501, 1934.

[27] G. H. McKinley and T. Sridhar. Filament-stretching rheometry of complex fluids. *Annual Review of Fluid Mechanics*, 34:375, 2002.

[28] N. W. Tschoegl. *The Theory of Linear Viscoelastic Behaviour.* Springer, New York, 1980.

[29] J. F. Brady. Dispersion in heterogeneous media. In E. Guyon J. P. Hulin, A. M. Cazabat and F. Carmonas, editors, *Hydrodynamics of Dispersed Media.* Elsevier, New York, 1990.

[30] P. T. Callaghan. *Principles of Nuclear Magnetic Resonance Microscopy.* Oxford University Press, New York, 1991.

[31] J. Bear. *Dynamics of Fluids in Porous Media.* Elsevier, New York, 1972.

[32] F. A. Dullien. *Porous Media.* Academic Press, New York, 1992.

[33] W. Ehlers and J. Bluhm (eds). *Porous Media: Theory, Experiments and Numerical Applications.* Springer, New York, 2002.

[34] H. Dary. *Les Fontaines Publiques de la Ville de Dijon.* Dalmont, Paris, 1856.

[35] M. Quintard and S. Whitaker. Transport in ordered and disordered porous-media – volume-averaged equations, closure problems, and comparison with experiment. *Chem. Eng. Sci.*, 48:2537, 1993.

[36] J. Koplik, S. Redner, and D. Wilkinson. Transport and dispersion in random networks with percolation disorder. *Phys. Rev. A*, 37:2619, 1988.

[37] D. L. Koch and J. F. Brady. A nonlocal description of advection diffusion with application to dispersion in porous-media. *J. Fluid Mech.*, 180:387, 1987.

[38] H. Brenner. Dispersion resulting from flow through spatially periodic porous-media. *Phil. Trans. Roy. Soc. London A*, 297:81, 1980.

[39] J. Salles, J. F. Thovert, R. Delannay, L. Prevors, J. L. Auriault, and P. M. Adler. Taylor dispersion in porous-media – determination of the dispersion tensor. *Phys. Fluids*, 5:2348, 1993.

[40] C. Van Den Broeck. Taylor diffusion revisited. *Physica A*, 168:677, 1990.

[41] O. A. Plumb and S. Whitaker. *Dynamics of Fluids in Hierarchical Porous Media*, chapter Diffusion, Adsorption and Dispersion in Porous Media: Small Scale Averaging and Local Volume Averaging. J. H. Cushman, ed., Academic Press, San Diego, 1990.

[42] J. D. Seymour and P. T. Callaghan. Generalized approach to NMR analysis of flow and dispersion in porous media. *AIChE J.*, 43:2096, 1997.

[43] A. A. Khrapitchev and P. T. Callaghan. Reversible and irreversible dispersion in a porous medium. *Physics of Fluids*, 15:2649, 2003.

[44] R. Aris. On the dispersion of a solute in a fluid flowing through a tube. *Proc. Roy. Soc.*, A 235:67, 1956.

3
Quantum description of nuclear ensembles

When the young Ernest Rutherford arrived at Cambridge University in 1897, his prospective PhD supervisor, John Joseph Thomson, had just discovered the electron, measured its charge and measured its mass. Rutherford had, the previous year, completed a Masters thesis at Canterbury College in New Zealand, developing a new and highly sensitive detector for Hertzian waves. Thomson was impressed with the talented New Zealander but suggested a new direction for his research, the elucidation of the radiation emanating from radium, the element recently extracted by Marie Curie. And so Rutherford discovered the alpha particle, measured its charge, and measured its mass [1]. That alpha particle of his PhD research became a fundamental tool in his subsequent life's work, but perhaps most remarkably in its use to determine atomic structure. At Manchester University in 1911, Ernest Rutherford and his students, Hans Geiger and Ernest Marsden, fired a collimated beam of alpha particles at a thin gold foil. The alpha scattering pattern led Rutherford to propose a model of the atom in which a dense, positively charged nucleus was surrounded by orbiting electrons.

Rutherford's picture of the atom presented a conundrum. Accelerated charges radiate electromagnetic radiation. How could orbiting electrons, with their consequent centripetal acceleration, remain in their orbits if they lost radiative energy? That question was boldly answered by Niels Bohr who was visiting Rutherford's Manchester Laboratory during the period of the famous experiments. In 1913 Bohr simply stated a rule whereby certain orbits of the electron would be allowed to exist without radiative loss, provided that their orbital path was an integral number of electron wavelengths. Given the dependence of the de Broglie wavelength on momentum, Bohr's rule corresponded to a quantisation of electron angular momentum. The quantisation rule fixed the energies of the allowed orbits such that the discrete changes in energy when electrons changed orbits, with energy released as photons of light, corresponded to the discrete nature of the spectral lines of atoms. The Rutherford–Bohr model of atomic structure, while essentially true, was limited in scope. It was, in what is now called 'old quantum theory', the precursor to the revolution in quantum mechanical description that followed Erwin Schrödinger's 1926 explanation of wave mechanics.

Those discoveries of the nucleus and of angular momentum quantisation had, in part, laid the foundations for the discovery of magnetic resonance half a century later. But two more insights were needed. In 1915, using a beautiful experiment in which a demagnetised iron bar suspended on a torsional pendulum was observed to rotate when magnetised along the suspension axis, Einstein and de Haas showed that angular momentum and magnetism were intimately connected [2]. Electrons have a magnetic dipole moment proportional to their angular momentum. Then, in 1923, Wolfgang Pauli interpreted hyperfine structure in the spectra of atoms as evidence for nuclear angular momentum and nuclear magnetism [3]. Nuclear magnetic resonance and electron spin resonance were achieved in the two decades following.

When Rutherford discovered the atomic nucleus he could not have imagined that it might be a window to understanding molecular biology, seeing inside the human body, or revealing how the brain works. NMR has proven an essential tool in physics, it has revolutionised chemistry and biochemistry, it has made astonishing contributions to medicine, and is now making an impact in geophysics, chemical engineering, and food technology. It is even finding applications in new security technologies and in testing fundamental ideas concerning quantum computing.

In the next chapter, the ideas that underpin magnetic resonance are outlined. First, however, we cover some of the essential concepts that govern the quantum behaviour of ensembles of nuclear spins.

3.1 Quantum mechanics and nuclear spin

Nuclear magnetism is inherently quantum mechanical in nature, the fundamental quantity describing the nuclear spin being the angular momentum quantum number I. This fixed integer or half-integer quantity (sometimes called the 'spin') arises from the fundamental symmetry properties of nuclei and characterizes a nucleus in its stable ground state. For example, a proton or ^{13}C nucleus has $I = \frac{1}{2}$ while the deuteron has $I = 1$. Measurement of angular momentum always leads to the projection of one of its vector components along an axis defined by the observation, the most famous example being the Stern–Gerlach experiment [4], in which spin-$\frac{1}{2}$ electronic states were found in 'up' and 'down' projections labelled by $\frac{1}{2}$ and $-\frac{1}{2}$. In general these observational possibilities, defined by the component of angular momentum, m, as measured along some axis z, may be any one of a discrete set of integer or half-integer values in the range $-I$, $-I + 1$, ..., $I - 1$, I. m is also known as the *azimuthal quantum number*. Note that the angular momentum is actually measured in units of $\hbar = h/2\pi$. We will only introduce this unit when it is specifically required.

Despite this underlying quantum behaviour, NMR is often described in semi-classical terms involving the interaction of a vector angular momentum and magnetic dipole moment with a magnetic field. How is this possible? The measurement of nuclear magnetism often involves vast numbers of nuclei, the states of which are statistically distributed in ensembles. When we average across ensembles, macroscopic effects appear continuous. Furthermore, in the case of independent spin-$\frac{1}{2}$ nuclei, the concern of much of this book, all states of the ensemble may be characterised by a simple vector quantity, referred to as the nuclear magnetisation and this vector description further justifies a classical description of the evolution of spin states. But delve a little deeper, allow the nuclear spins to interact with one another, or attempt to describe the time evolution of higher spin quantum number such as $I = 1$, and the underlying quantum mechanics becomes immediately apparent. Understanding those quantum mechanical principles is essential to any understanding of the magnetic resonance phenomenon.

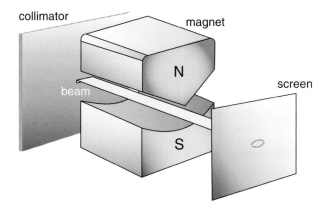

collimator

magnet

N

beam

screen

S

Fig. 3.1 Schematic of Stern–Gerlach apparatus, in which a beam of silver atoms is passed through a magnetic field gradient created by specially shaped magnetic pole pieces. On the screen the beam is found to split into 'up' and 'down' components.

3.1.1 Four key ideas in quantum mechanics

The quantum description of nuclear spin behaviour involves four underlying ideas. The first concerns the concept of a basis set of fundamental states that relate to the observational possibilities of quantum systems. The second relates these basis states to eigenvectors of a linear operator associated with measurement of a physical observable. The third involves a clear description of the measurement process and what that means for the quantum states. And the fourth tells us about dynamics—how quantum states evolve with time. We will deal with each idea in turn.

Before starting it is worth pointing out that, as straightforward as the rules of quantum mechanics may be in mathematical terms, they are deeply counterintuitive from a physical standpoint. Yet quantum mechanics has never failed a single experimental test. The science writer Simon Singh calls quantum mechanics 'The most successful and utterly bizarre theory in the whole of physics' [5]. Neils Bohr said 'Anyone who is not shocked by quantum theory has not understood a single word' [6], while, somewhat modestly, Richard Feynman expressed the mystery by saying 'I think I can safely say that no one understands quantum mechanics' [7].

The quantum states

These quantum states, labelled by α, form the basis of a vector space, $\{|\Psi\rangle\}$, such that any allowed state $|\Psi\rangle$ of a nuclear spin can be written as a linear superposition,

$$|\Psi\rangle = \sum_{\alpha} a_{\alpha}|\alpha\rangle \tag{3.1}$$

where α labels all the elements of the vector space basis set. A significant feature of this (Hilbert) vector space is that the coefficients or 'amplitudes', a_{α}, of the vector space are necessarily complex, causing the amplitudes to have phase δ_{α} as well as magnitude $|a_{\alpha}|$. In particular

$$a_{\alpha} = |a_{\alpha}| \exp\left(i\delta_{\alpha}\right) \tag{3.2}$$

For spins with angular momentum quantum number I, those basis states $|\alpha\rangle$ are labelled by the azimuthal quantum number m and

$$|\Psi\rangle = \sum_m a_m |m\rangle \tag{3.3}$$

where m ranges over all possible values from $-I$ to I. The inherent phase in the amplitudes a_m represents the most important property utilised in the descriptions of nuclear translational motion measurement throughout this book.

Eigenvalue equation and observables

The second idea, indeed the one that underlies the existence of the basis, concerns the so-called eigenvalue equation, in which all measurement possibilities are eigenvalues, α, of a linear operator, A, associated with this measurement—the so-called observable of interest.

$$A|\alpha\rangle = \alpha|\alpha\rangle \tag{3.4}$$

For example, the process of observation of the angular momentum component along the z-axis for a single nucleus in a basis state $|m\rangle$ is described by the eigenvalue equation,

$$I_z|m\rangle = m|m\rangle \tag{3.5}$$

In this example I_z is the 'operator' for angular momentum along the z-axis, while the eigenvalue m is the result of an observation. In general, the complete set, $\{|\alpha\rangle\}$, of eigenvector solutions to the eigenvalue equation provides a basis for describing any state $|\Psi\rangle$ of the system in terms of a superposition of basis states.

A property of all vector spaces concerns the ability to define an 'inner product', a number formed by the ordered pair of any two vectors. The inner product is essentially the projection of one vector onto the other, in our case denoted $\langle\Psi|\Psi'\rangle$, with the property $\langle\Psi|\Psi'\rangle = \langle\Psi'|\Psi\rangle^*$. This product is formed by an ordered multiplication of vectors $\langle\Psi|$ and $|\Psi'\rangle$, known respectively as 'bras' and 'kets'. In that sense, given the existence of the set of ket vectors $\{|\Psi\rangle\}$, the inner product operation defines the set of bra vectors $\{\langle\Psi|\}$.

Physical measurements return real numbers, a consequence of which is that eigenvectors are necessarily orthogonal. By dividing each basis vector by its length, $\langle\alpha|\alpha\rangle$, also known as the 'norm', the vectors are said to be normalised. For such an orthonormal basis set the following relation applies

$$\langle\alpha|\alpha'\rangle = \delta_{\alpha\alpha'} \tag{3.6}$$

where $\delta_{\alpha\alpha'}$ is the Kronecker delta. Since we have a discrete set of orthonormal $|\alpha\rangle$ states, it is helpful to represent these using column vectors. For example, where $I = \frac{1}{2}$ we have the basis

$$|\tfrac{1}{2}\rangle = \begin{bmatrix} 1 \\ 0 \end{bmatrix}, \qquad |-\tfrac{1}{2}\rangle = \begin{bmatrix} 0 \\ 1 \end{bmatrix} \tag{3.7}$$

To fulfil the requirements of the inner product, the bra and ket vectors are each the complex conjugate transpose of the other, $\langle\alpha|$ being the row vector conjugate to column vector $|\alpha\rangle$. So

$$\left\langle \tfrac{1}{2} \right| = \begin{bmatrix} 1 & 0 \end{bmatrix}, \qquad \left\langle -\tfrac{1}{2} \right| = \begin{bmatrix} 0 & 1 \end{bmatrix} \tag{3.8}$$

In this representation, the operator I_z becomes a matrix, namely

$$I_z = \begin{bmatrix} \tfrac{1}{2} & 0 \\ 0 & -\tfrac{1}{2} \end{bmatrix} \tag{3.9}$$

One of the consequences of the fact that the eigenvectors span the space of all possible state vectors is the 'completeness relation' $\sum_\alpha |\alpha\rangle\langle\alpha| = \underline{\underline{1}}$ where $\underline{\underline{1}}$ is the identity operator.

The measurement process

The third idea concerns interpretation of the measurement process. Of course, for a quantum system existing in an eigenstate $|\alpha\rangle$ of an observable A, the result of measurement of A follows directly by calculating $\langle\alpha|A|\alpha\rangle$. For example, a measurement of the I_z observable when the spin-$\tfrac{1}{2}$ system is in the definite 'spin down' state yields

$$\left\langle -\tfrac{1}{2} \right| I_z \left| -\tfrac{1}{2} \right\rangle = \begin{bmatrix} 0 & 1 \end{bmatrix} \begin{bmatrix} \tfrac{1}{2} & 0 \\ 0 & -\tfrac{1}{2} \end{bmatrix} \begin{bmatrix} 0 \\ 1 \end{bmatrix}$$
$$= -\tfrac{1}{2} \tag{3.10}$$

But what if the quantum system is not in an eigenstate of the observable but in some general superposition state $|\Psi\rangle$? This third postulate states that the result of measurement is given by $\langle\Psi|A|\Psi\rangle$. Again, we will illustrate for the case of a nuclear spin in an admixed state given by the superposition, $|\Psi\rangle = \sum_m a_m |m\rangle$. Then the result of a measurement, the so-called 'expectation value', is defined by

$$\langle\Psi|I_z|\Psi\rangle = \sum_{m,m'} a_m^* a_{m'} \langle m'|I_z|m\rangle$$
$$= \sum_{m,m'} a_m^* a_{m'}\, m\, \langle m'|m\rangle \tag{3.11}$$

Because the basis vectors $\{|m\rangle\}$ are orthogonal we have the result

$$\langle\Psi|I_z|\Psi\rangle = \sum_m |a_m|^2\, m \tag{3.12}$$

What is the meaning of this sum? For a large number of identically prepared nuclei it represents a mean of eigenvalue results weighted by the normalised probabilities $|a_m|^2$. For a single nucleus the interpretation is that the measurement will force the superposition state, $|\Psi\rangle$, into one of the eigenstates, $|m\rangle$, returning the result m. This mysterious 'state reduction' is a random process defined by a probability $|a_m|^2$.

Since we chose to represent our states as eigenvalues of I_z, the operator for I_z is diagonal. For measurement of angular momentum about other axes we find that the corresponding operators cannot be diagonal, reflecting the Heisenberg uncertainty principle whereby systems cannot simultaneously exist in definite states of certain pairs of observables. Indeed, the observables I_x, I_y, and I_z are represented by operators that do not commute but follow the famous commutation relationship

$$[I_x, I_y] = I_x I_y - I_y I_x = i I_z \tag{3.13}$$

This relation arises from the fact that successive rotations about different axes do not commute and because the operator for rotation (for example about the z-axis) is related to the operator for angular momentum, I_z, by the rule

$$R_z(\phi) = \exp(i\phi I_z) \tag{3.14}$$

where the exponential function of the operator I_z is interpreted as the usual power series. Equation 3.13 is a fundamental algebra that underpins the behaviour of nuclear spins in magnetic resonance.

Dynamics and the Schrödinger equation

The fourth idea in quantum mechanics concerns dynamics—how quantum systems evolve with time. This is governed by the Schrödinger equation

$$i\hbar \frac{\partial}{\partial t} |\Psi(t)\rangle = \mathcal{H} |\Psi(t)\rangle \tag{3.15}$$

where $\mathcal{H}$ is the Hamiltonian (or energy) operator. Providing we express $\mathcal{H}$ in angular frequency units, we may drop the $\hbar$ in the Schrödinger equation and we will henceforth follow this convention. Note that this formulation (the Schrödinger picture) has the states as functions of time but the operators stationary. $\mathcal{H}$ may have some explicit time-dependence if our quantum system is subjected to some fluctuating disturbance, but if $\mathcal{H}$ is constant, then the Schrödinger equation yields the result

$$|\Psi(t)\rangle = U(t) |\Psi(0)\rangle \tag{3.16}$$

where

$$U(t) = \exp(-i\mathcal{H}t) \tag{3.17}$$

$U(t)$ is known as the evolution operator. Here we demonstrate its action in a simple example. To do so, we take as our Hamiltonian the interaction of the nuclear spin with an external magnetic field. This is known as the Zeeman interaction. Atomic nuclei have a magnetic dipole moment proportional to the angular momentum, the constant of proportionality being known as the gyromagnetic ratio, γ. The interaction energy of a magnetic dipole $\boldsymbol{\mu}$ in a magnetic field $\mathbf{B}_0$ is written classically and quantum mechanically as $-\boldsymbol{\mu} \cdot \mathbf{B}_0$ so that the Hamiltonian operator for the case of $\mathbf{B}_0$ oriented along the z-axis is

$$\mathcal{H} = -\gamma B_0 I_z \tag{3.18}$$

The energy levels associated with the eigenstates of the Zeeman Hamiltonian are shown in Fig. 3.2, the energy separation between adjacent levels being γB_0 for all spin quantum numbers I. Now consider the evolution of a nuclear spin quantum state $|\Psi\rangle$ under the influence of the magnetic field. The evolution operator,

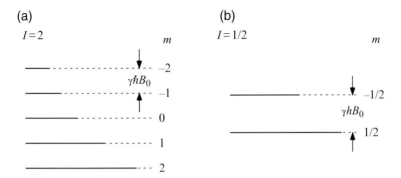

Fig. 3.2 Energy level diagram for spins experiencing a Zeeman interaction. In (a) $I = 2$, while in (b) $I = \frac{1}{2}$. The energy spacing for levels separated by $m = \pm 1$ is $\gamma \hbar B_0$, or in the frequency units adopted here, γB_0. The bold line schematically represents the the relative population in each state for an ensemble of spins in thermal equilibrium.

$U(t) = \exp{(i\gamma B_0 I_z t)}$, is identical to a clockwise rotation of the state about the z-axis by an angle $\gamma B_0 I_z t$. The existence of the field causes all states to precess at the Larmor frequency, ω_0, given by

$$\omega_0 = \gamma B_0 \tag{3.19}$$

For protons ($\gamma = 2.75 \times 10^8$) in a typical superconductive magnet of $B_0 = 9.4\,\text{T}$, the proton precession frequency, in cyclic frequency units, will be $\gamma B_0 / 2\pi = 400\,\text{MHz}$, in the UHF part of the radiofrequency spectrum.

3.1.2 Representation of angular momentum

The starting point for any consideration of angular momentum is the classical orbital angular momentum vector, $\mathbf{L} = \mathbf{r} \times \mathbf{p}$, where $\mathbf{p}$ is a translational momentum vector. By noting that in wave mechanics the operator for momentum is $\mathbf{p} = -i\hbar\nabla$, it may be easily shown that for a wavefunction $\Psi(\mathbf{r})$ whose position vector is rotated by an infinitesimal angle $\underline{\epsilon} = (\epsilon_x, \epsilon_y, \epsilon_z)$ to $\mathbf{r} + \underline{\epsilon} \times \mathbf{r}$, the rotation operator is $\exp((i/\hbar)\,\underline{\epsilon} \cdot \mathbf{L})$. The commutator relations $[L_x, L_y] = i\hbar L_z$ then follow by consideration of geometric non-commutation of classical rotations about orthogonal axes [8], a simple example being $R_x(\epsilon_x)\,R_y(\epsilon_y)R_x(-\epsilon_x)\,R_y(-\epsilon_y) = R_z(-\epsilon_x\epsilon_y)$, as shown in Fig. 3.3. Such orbital angular momentum may be shown to be quantised in units of $\hbar$.

Generalised angular momentum

If, instead of starting from rules for orbital motion, we take as our axiom the algebra of the commutator relations, then we find that the operators that obey this algebra belong to a wider class than orbital angular momentum alone. We have chosen to label this wider class I_x, I_y and I_z and conveniently non-dimensionalise them by dropping the explicit $\hbar$ unit.

Using just the commutator relations alone and choosing I_z as our eigenoperator,[1] we find through simple algebraic manipulation the following rules. First, I_z and

---

[1]Since I_x, I_y and I_z do not commute, only one can be an eigenoperator and hence represented by a diagonal matrix.

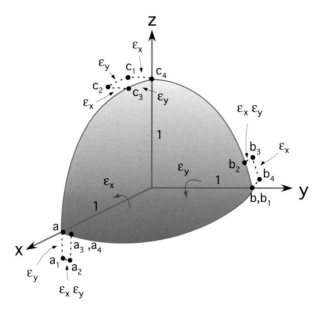

Fig. 3.3 Result of successive infinitesimal rotations on a unit sphere of $R_x(\epsilon_x)$, $R_y(\epsilon_y)$, $R_x(-\epsilon_x)$, and $R_y(-\epsilon_y)$, taking point a to a_1, a_2, etc., and b to b_1, b_2, etc. The final result is $R_z(-\epsilon_x\epsilon_y)$.

$\mathbf{I}^2 = I_x^2 + I_y^2 + I_z^2$ commute. Second, the eigenvalues of I_z range in unit steps between a minimum value $m = -I$, and maximum value $m = I$, where I is integer or half-integer. I is known as the angular momentum or 'spin' quantum number while m is called the azimuthal quantum number. Third, the eigenvalue of $\mathbf{I}^2$ is $I(I+1)$. Because both I_z and $\mathbf{I}^2$ are diagonal operators, we sometimes call this basis the '$\mathbf{I}^2, I_z$' representation. We could label our kets in this basis, $|Im\rangle$, or in shorthand, simply $|m\rangle$. Fourth, the operators $I_+ = I_x + iI_y$ and $I_- = I_x - iI_y$, known as raising and lowering operators, have the effect

$$I_+|m\rangle = \sqrt{I(I+1) - m(m+1)}\,|m+1\rangle \qquad (3.20)$$

and

$$I_-|m\rangle = \sqrt{I(I+1) - m(m-1)}\,|m-1\rangle \qquad (3.21)$$

The simplest non-trivial example of generalised angular momentum is the case of spin $I = \frac{1}{2}$, for which the I_z eigenstates are two-dimensional (2-D), and given by eqns 3.7. The raising and lowering operators have the effect of converting a $\left|-\frac{1}{2}\right\rangle$ state into a $\left|\frac{1}{2}\right\rangle$ state and vice versa according to

$$I_+\left|-\tfrac{1}{2}\right\rangle = \left|\tfrac{1}{2}\right\rangle, \qquad I_-\left|\tfrac{1}{2}\right\rangle = \left|-\tfrac{1}{2}\right\rangle \qquad (3.22)$$

Rotation: active and passive view

Let us write down the rotation operator for spin-$\frac{1}{2}$ in the case of a rotation ϕ about the z-axis. First we need to establish the so-called 'active and passive views'. A frame of reference rotation is termed 'passive' in quantum mechanics, and distinguished from the 'active view' rotation of physical systems. Rotations of coordinate frames about a Cartesian axis, ζ, are represented by the operator, $\exp(i\theta I_\zeta)$, where a positive sign corresponds to an anticlockwise sense. The active- and passive-view operations are identical except for a sign change. For example, as shown in Fig. 3.4, rotating an object by 30° clockwise while keeping the reference frame constant leads to a measurement result identical to that which obtains when the object is held fixed while the reference frame is rotated 30° anticlockwise. Thus physical rotations of spins involve the same operator, $\exp(i\theta I_\zeta)$, but in the opposite sense.

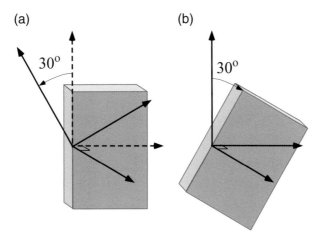

Fig. 3.4 Illustrating the relationship between the active and passive views. Rotating the axes by 30° anticlockwise, as shown in (a) results in the same relationship between axes and object as rotating the object by 30° clockwise, as shown in (b).

We have, for a rotation of axes,

$$R_z(\phi)\left|\tfrac{1}{2}\right\rangle = \begin{bmatrix} \exp\left(i\tfrac{1}{2}\phi\right) & 0 \\ 0 & \exp\left(-i\tfrac{1}{2}\phi\right) \end{bmatrix} \begin{bmatrix} 1 \\ 0 \end{bmatrix} = \begin{bmatrix} \exp\left(i\tfrac{1}{2}\phi\right) \\ 0 \end{bmatrix} \qquad (3.23)$$

and

$$R_z(\phi)\left|-\tfrac{1}{2}\right\rangle = \begin{bmatrix} \exp\left(i\tfrac{1}{2}\phi\right) & 0 \\ 0 & \exp\left(-i\tfrac{1}{2}\phi\right) \end{bmatrix} \begin{bmatrix} 0 \\ 1 \end{bmatrix} = \begin{bmatrix} 0 \\ \exp(-i\tfrac{1}{2}\phi) \end{bmatrix} \qquad (3.24)$$

Note the remarkable result that a rotation of 360° results in the transformation of $\left|\tfrac{1}{2}\right\rangle$ to $-\left|\tfrac{1}{2}\right\rangle$ and $\left|-\tfrac{1}{2}\right\rangle$ to $-\left|-\tfrac{1}{2}\right\rangle$. Spin-$\tfrac{1}{2}$ states, indeed all half-integral angular momentum states, are double-valued on rotation. A 720° rotation is needed to restore the ket to its original state. By contrast, integral angular momentum states (incorporating all orbital angular momentum) are single-valued under rotation. Take the case of spin-1. Here the rotation operator about the 'eigen axis' is

$$R_z(\phi) = \begin{bmatrix} \exp\left(i\phi\right) & 0 & 0 \\ 0 & 1 & 0 \\ 0 & 0 & \exp\left(-i\phi\right) \end{bmatrix} \qquad (3.25)$$

and the operator for a 360° rotation is clearly the identity.

What is spin?

It should now be abundantly clear that 'spin', at least in its half- integral form, has little to do with orbital motion. Spin is really a symmetry property, in the case of spin-$\frac{1}{2}$ associated with an 'upness' or 'downness' of the quantum system, in our case the atomic nucleus. But the fundamental definition of the spin property arises from the phase change observed when the quantum system rotates. Rotation may be of our axes, as we observe the quantum system by rotating our frame of reference around it (the passive view), or it may be of the system itself (active view), as we keep our observational axes fixed and the system physically rotates. The rotation operators are the same in the active and passive views but with opposite sign exponents.

Fig. 3.5 The Möbius strip. Two 360° revolution circuits around the strip are needed to return to origin.

The curious nature of the need for 720° rotations to obtain an identity operator for half-integral spins has often been compared to the symmetry properties of the Möbius strip, where two orbits are needed to return to origin.

General rotations and irreducible representations

Suppose we wish to describe a general reorientation of coordinates. This could be done by specifying a succession of rotations about each of the three Cartesian axes. In fact, while three angles are needed in general, two rotation axes will suffice. These angles (α, β, γ), are known as Euler angles and specify successive rotations α about z, followed by β about y, then γ about z, where the axes remain fixed. The rotation operator for the family of kets $\{|Im\rangle\}$ with angular momentum quantum number I is then the Wigner rotation matrix $D^{(I)}(\alpha, \beta, \gamma)$ [9, 10] where

$$D^{(I)}(\alpha, \beta, \gamma) = \exp\left(-i\gamma I_z\right) \exp\left(-i\beta I_y\right) \exp\left(-i\alpha I_z\right) \tag{3.26}$$

and

$$D^{(I)}|Im\rangle = \sum_{m'} D^{(I)}_{m'm}|Im'\rangle \tag{3.27}$$

$D^{(I)}_{m'm}$ being an element of the Wigner matrix. These matrix elements are well known and may be written in closed form [9, 10].

Equation 3.26 is the perspective appropriate for the active view, where the system itself is rotated. In the passive view, where the coordinates are rotated and the system remains fixed, we require the operator $D^{(I)\dagger}(\alpha, \beta, \gamma)$ and

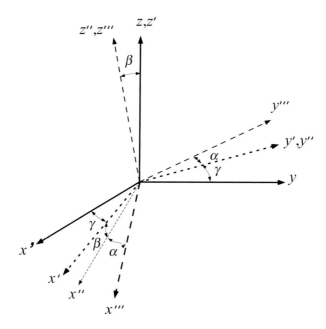

Fig. 3.6 Result of successive passive view Euler angle rotations in which successively rotated axes are used. Here the angle order is reversed from the active view, γ about z, followed by β about y', then α about z''.

$$D^{(I)^\dagger}(\alpha, \beta, \gamma) = \exp\left(i\alpha I_z\right) \exp\left(i\beta I_y\right) \exp\left(i\gamma I_z\right) \qquad (3.28)$$

Note the reversal of order of rotation as a result of the Hermitian conjugation operation, †.

The particular matrix form of $D^{(I)}(\alpha, \beta, \gamma)$ depends on the angular momentum quantum number and hence the chosen basis set. In this sense we say that $D^{(I)}(\alpha, \beta, \gamma)$ is a 'representation' of the rotation for this particular basis. These Wigner matrices will be $2 \times 2, 3 \times 3$ or $(2I + 1) \times (2I + 1)$ for spin-$\frac{1}{2}$ spin-1, or spin-I systems, respectively. Most importantly, these Wigner matrices involve linear superpositions of all the m values of the ket space. And there is no unitary transformation on the kets which can change this fact, that is, to transform $D^{(I)}(\alpha, \beta, \gamma)$ into a block diagonal form involving mixing only within subsets of the m values. In other words, we say that the $D^{(I)}(\alpha, \beta, \gamma)$ are irreducible representations of the full rotation group in 3D.

In the next section we will see that it is possible to combine angular momenta to produce product states, and that these states themselves may be related to new total angular momentum states. Put in the language of group theory, we say that the rotation matrices for the product states may be 'reduced' to a sum of irreducible representations [11].

The product representation and the total angular momentum representation

Suppose that a system is characterised by two independent angular momenta, I_1 and I_2. Here we will be concerned with two nuclear spins, but they might be, for example, the spin and orbital angular momentum of an electron, or the component proton and neutron angular momenta, which combine to form the intrinsic angular momentum of the deuteron. We will only consider the case of two such angular momenta but the argument can be generalised to more.

One important property of angular momentum operators in different spaces is that they commute. They operate in different ket spaces after all. The eigenkets in the '$\mathbf{I}^2, I_z$' representation for I_1 and I_2, respectively, are $|I_1 m_1\rangle$ and $|I_2 m_2\rangle$. We can then define a combined system ket in which the operators in I_1 space have no effect on $|I_2 m_2\rangle$ and vice-versa. Clearly there are $(2I_1 + 1)$ eigenkets $|I_1 m_1\rangle$ and $(2I_2 + 1)$ eigenkets $|I_2 m_2\rangle$. We can write for a general system ket

$$|\psi\rangle_{\text{product}} = \sum_{m_1, m_2} |I_1 m_1\rangle |I_2 m_2\rangle \tag{3.29}$$

To write matrices in the product space, one forms an outer product $\otimes$. By way of illustration, here is the example where $I_1 = 1$, with three- dimensional kets and 3×3, operators and $I_2 = \frac{1}{2}$, where the kets are 2-D and the operators 2×2.

$$|I_1 m_1\rangle |I_2 m_2\rangle = \begin{bmatrix} a_1 \\ a_2 \\ a_3 \end{bmatrix} \otimes \begin{bmatrix} b_1 \\ b_2 \end{bmatrix} = \begin{bmatrix} a_1 \begin{bmatrix} b_1 \\ b_2 \end{bmatrix} \\ a_2 \begin{bmatrix} b_1 \\ b_2 \end{bmatrix} \\ a_3 \begin{bmatrix} b_1 \\ b_2 \end{bmatrix} \end{bmatrix} = \begin{bmatrix} a_1 b_1 \\ a_1 b_2 \\ a_2 b_1 \\ a_2 b_2 \\ a_3 b_1 \\ a_3 b_2 \end{bmatrix} \tag{3.30}$$

The ket $|I_1 m_1\rangle |I_2 m_2\rangle = |1 \ -1\rangle |\frac{1}{2} \frac{1}{2}\rangle$ would correspond to $a_1 = a_2 = 0$, $a_3 = 1$, $b_1 = 1$, $b_2 = 0$.

Operator products are formed in a similar way

$$AB = \begin{bmatrix} A_{11} & A_{12} & A_{13} \\ A_{21} & A_{22} & A_{23} \\ A_{31} & A_{32} & A_{33} \end{bmatrix} \otimes \begin{bmatrix} B_{11} & B_{12} \\ B_{21} & B_{22} \end{bmatrix}$$

$$= \begin{bmatrix} A_{11} \begin{bmatrix} B_{11} & B_{12} \\ B_{21} & B_{22} \end{bmatrix} & A_{12} \begin{bmatrix} B_{11} & B_{12} \\ B_{21} & B_{22} \end{bmatrix} & A_{13} \begin{bmatrix} B_{11} & B_{12} \\ B_{21} & B_{22} \end{bmatrix} \\ A_{21} \begin{bmatrix} B_{11} & B_{12} \\ B_{21} & B_{22} \end{bmatrix} & A_{22} \begin{bmatrix} B_{11} & B_{12} \\ B_{21} & B_{22} \end{bmatrix} & A_{23} \begin{bmatrix} B_{11} & B_{12} \\ B_{21} & B_{22} \end{bmatrix} \\ A_{31} \begin{bmatrix} B_{11} & B_{12} \\ B_{21} & B_{22} \end{bmatrix} & A_{32} \begin{bmatrix} B_{11} & B_{12} \\ B_{21} & B_{22} \end{bmatrix} & A_{33} \begin{bmatrix} B_{11} & B_{12} \\ B_{21} & B_{22} \end{bmatrix} \end{bmatrix} \tag{3.31}$$

Using this rule we can see how to form a matrix to represent the Hamiltonian product operator $\mathbf{I_1} \cdot \mathbf{I_2}$. But what do we do with operators that apparently involve only a single spin, such as I_{1z}? In fact once we commit to the product representation, all operators must be products involving each spin space. What this means for the apparently single-spin operator is that the corresponding operator in the other spin space is just the identity. Hence $I_{1z} = I_{1z} \otimes \underline{1}$ while $I_{2z} = \underline{1} \otimes I_{2z}$, the respective identity operator being of dimension appropriate to each space.

What is the general rotation operator in product space? It is simply the tensor product $D^{(I_1)} \otimes D^{(I_2)}$ [11]. Remarkably it may be shown that

$$D^{(I_1)} \otimes D^{(I_2)} = \sum_{I=|I_1-I_2|}^{I_1+I_2} D^{(I)} \tag{3.32}$$

which means that a unitary transformation U_{cg} exists such that

$$D^{(I_1)} \otimes D^{(I_2)} = U_{cg}^{-1} \begin{bmatrix} D^{(I_1+I_2)} & \underline{0} & \cdots & \underline{0} \\ \underline{0} & D^{(I_1+I_2-1)} & \cdots & \underline{0} \\ \vdots & \vdots & \ddots & \vdots \\ \underline{0} & \underline{0} & \cdots & D^{(|I_1-I_2|)} \end{bmatrix} U_{cg} \tag{3.33}$$

Equation 3.33 shows that the product representation may be reduced to a block diagonal sum of irreducible representations. In other words the combined angular momenta I_1 and I_2 form new total angular momentum states with total angular momentum quantum numbers ranging from $I_1 + I_2$, $I_1 + I_2 - 1$, ..., to $|I_1 - I_2|$. The dimensionality of the product states $(2I_1 + 1)(2I_2 + 1)$ is identical to the sum of total angular momentum dimensions, $\sum_{I=|I_1-I_2|}^{I_1+I_2}(2I+1)$. U_{cg} is known as a Clebsch–Gordon transformation and it connects the product and total angular momentum representations by

$$|\Psi\rangle_{total} = U_{cg}|\Psi\rangle_{product}. \tag{3.34}$$

The matrix elements of U_{cg} are known as Clebsch–Gordon coefficients [12], and are labelled by $\langle I_1 m_1\, I_2 m_2\,|\,Im\rangle$. The relationship between the product states $|I_1 m_1\rangle|I_2 m_2\rangle$ and the total angular momentum states $|Im\rangle$ is given by the Clebsch–Gordon transformation

$$|Im\rangle = \sum_{m_1}\langle I_1 m_1\, I_2 m_2\,|\,Im\rangle\,|I_1 m_1\rangle\,|I_2 m_2\rangle \tag{3.35}$$

with $m_1 + m_2 = m$.

3.2   Spin ensembles and the density matrix

So far the quantum description has been written in terms of the states $|\Psi\rangle$ of individual quantum particles. How do we describe the large ensembles of spins characteristic of macroscopic samples, allowing for the possibility that members of that ensemble may exist in a range of sub-ensembles, each labelled by a different quantum state? A schematic diagram of such a set is shown in Fig. 3.7. Note that the sub-ensemble is defined such that every quantum particle contained within it is in the same quantum state $|\Psi\rangle$. Those states may in fact be superpositions in some chosen basis. However, we start by describing an important set of sub-ensembles for which the best way to represent the respective quantum particles is in terms of eigenstates of energy. This is the case of an ensemble in thermal equilibrium, a common starting point for most experiments!

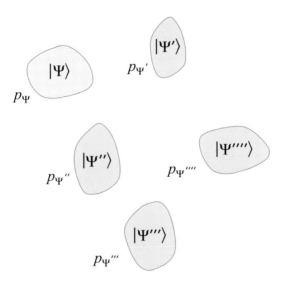

Fig. 3.7 Schematic for an ensemble of sub-ensembles, each a pure state containing quantum systems with common kets $|\Psi\rangle$, $|\Psi'\rangle$, etc., and with probability weightings p_Ψ, $p_{\Psi'}$, etc.

We choose as our example the case of atomic nuclei in thermal equilibrium in a magnetic field. In most condensed matter at physically accessible temperatures, the spacing between adjacent atomic nuclei will exceed the de Broglie wavelength and so Boltzmann statistics will apply. Under the Zeeman Hamiltonian, the Boltzmann distribution will dictate that the probability of occupancy of the quantum state $|\Psi\rangle = |m\rangle$ will be[2]

$$p_m = \frac{\exp\left(-\gamma B_0 m\hbar/k_B T\right)}{\sum_m \exp\left(-\gamma B_0 m\hbar/k_B T\right)} \tag{3.36}$$

These relative populations are shown in Fig. 3.2. Our question concerning the appropriate description is answered by classical statistical mechanics. The result of a measurement A for a statistical ensemble of quantum states $\{|\Psi\rangle\}$ with normalised probabilities p_Ψ is simply the weighted average

$$\overline{\langle A \rangle} = \overline{\langle \Psi|A|\Psi \rangle}$$
$$= \sum_\Psi p_\Psi \langle \Psi|A|\Psi \rangle \tag{3.37}$$

where the bar over a quantity is taken to represent the classical statistical averaging over sub-ensembles while the angular brackets $\langle\ldots\rangle$ represent the quantum mechanical expectation value, the quantum statistical process known as state reduction.

[2]Note the need to reintroduce $\hbar$ so as to write the energy in units of Joules.

3.2.1 Spin-$\frac{1}{2}$ ensembles

Let us illustrate, using the simplest possible non-trivial example of a system in quantum mechanics, the case of a spin with $I = \frac{1}{2}$. In NMR the dominant interaction of a spin with its environment is always via the Zeeman interaction of eqn 3.18. This means that the 'natural' eigenstates are those whose quantum numbers are eigenvalues of I_z, namely $\left|\frac{1}{2}\right\rangle$ and $\left|-\frac{1}{2}\right\rangle$. In general, therefore, we express any state of a particular sub-ensemble in this basis and write

$$|\Psi\rangle = a_{1/2} \begin{bmatrix} 1 \\ 0 \end{bmatrix} + a_{-1/2} \begin{bmatrix} 0 \\ 1 \end{bmatrix} \tag{3.38}$$

Given the operator for I_z of eqn 3.9

$$\overline{\langle\Psi|I_z|\Psi\rangle} = \overline{\left(a^*_{1/2}\, [1\ 0] + a^*_{-1/2}\, [0\ 1]\right) \begin{bmatrix} \frac{1}{2} & 0 \\ 0 & -\frac{1}{2} \end{bmatrix} \left(a_{1/2} \begin{bmatrix} 1 \\ 0 \end{bmatrix} + a_{-1/2} \begin{bmatrix} 0 \\ 1 \end{bmatrix}\right)}$$

$$= \frac{1}{2}\left(\overline{|a_{1/2}|^2} - \overline{|a_{-1/2}|^2}\right) \tag{3.39}$$

The respective classical and quantum statistical averages are both playing a role, the $\overline{|a_{1/2}|^2}$ and $\overline{|a_{-1/2}|^2}$ representing the respective probabilities, for each sub-ensemble, that the act of measurement forces the particles into either the spin-up or spin-down quantum states, while the bar averages these quantum probabilities between each sub-ensemble.

Equation 3.39 may be interpreted by saying that the ensemble averaged expectation value of I_z is determined by the difference in population between the upper and lower energy levels. This difference is said to describe the polarisation of the ensemble. A population difference will arise in thermal equilibrium according to the Boltzmann probability factor. In thermal equilibrium the two levels, separated by $\hbar\gamma B_0$, will have populations

$$\overline{|a_{\pm 1/2}|^2} = \frac{\exp\left(\pm\hbar\gamma B_0/2k_B T\right)}{\exp\left(-\hbar\gamma B_0/2k_B T\right) + \exp\left(\hbar\gamma B_0/2k_B T\right)} \tag{3.40}$$

At room temperature, in a typical laboratory magnet, the energy difference $\hbar\gamma B_0$ is over five orders of magnitude smaller than the Boltzmann energy $k_B T$. In consequence the expression for the populations may be written, with a high degree of accuracy, as

$$\overline{|a_{\pm 1/2}|^2} = \frac{1}{2}\left(1 \pm \frac{1}{2}\hbar\gamma B_0 k_B T\right) \tag{3.41}$$

and in general, for any I, may be written

$$\overline{|a_m|^2} = \frac{1}{2I + 1}\left(1 + m\hbar\gamma B_0/k_B T\right) \tag{3.42}$$

Measurement of the x- or y-components of angular momentum provide interesting examples by way of contrast. This particular observable provides the signal in the NMR experiment. For $I = \frac{1}{2}$, I_x is the off-diagonal operator

$$I_x = \frac{1}{2} \begin{bmatrix} 0 & 1 \\ 1 & 0 \end{bmatrix} \qquad (3.43)$$

so that

$$\overline{\langle \Psi | I_x | \Psi \rangle} = \frac{1}{2} \left[\overline{a_{1/2}^* a_{-1/2}} + \overline{a_{1/2} a_{-1/2}^*} \right] \qquad (3.44)$$

while

$$I_y = \frac{1}{2} \begin{bmatrix} 0 & -i \\ i & 0 \end{bmatrix} \qquad (3.45)$$

and

$$\overline{\langle \Psi | I_y | \Psi \rangle} = \frac{-i}{2} \left[\overline{a_{1/2}^* a_{-1/2}} - \overline{a_{1/2} a_{-1/2}^*} \right] \qquad (3.46)$$

The terms in the brackets are said to describe the degree of 'single-quantum coherence' of the ensemble. This average is quite different from that represented by eqn 3.39 and reflects the degree of phase coherence between the $\left| \frac{1}{2} \right\rangle$ and $\left| -\frac{1}{2} \right\rangle$ states. This coherence is apparent when we write the amplitude terms in Argand form (see eqn 3.2)

$$\begin{aligned} \overline{\langle \Psi | I_x | \Psi \rangle} &= \tfrac{1}{2} \overline{|a_{1/2}||a_{-1/2}| \left(\exp\left(-i \left(\delta_{1/2} - \delta_{-1/2} \right) \right) + \exp\left(i \left(\delta_{1/2} - \delta_{-1/2} \right) \right) \right)} \\ &= \tfrac{1}{2} \overline{|a_{1/2}||a_{-1/2}| \cos\left(\delta_{1/2} - \delta_{-1/2} \right)} \end{aligned} \qquad (3.47)$$

and

$$\begin{aligned} \overline{\langle \Psi | I_y | \Psi \rangle} &= -\tfrac{i}{2} \overline{|a_{1/2}||a_{-1/2}| \left(\exp\left(-i \left(\delta_{1/2} - \delta_{-1/2} \right) \right) - \exp\left(i \left(\delta_{1/2} - \delta_{-1/2} \right) \right) \right)} \\ &= -\tfrac{1}{2} \overline{|a_{1/2}||a_{-1/2}| \sin\left(\delta_{1/2} - \delta_{-1/2} \right)} \end{aligned} \qquad (3.48)$$

$|a_{1/2}||a_{1/2}|$ is always positive. It is the non-random nature of $\overline{\exp(i \left(\delta_{1/2} - \delta_{-1/2} \right))}$, the ensemble average of phase differences between the quantum states, which renders the quantum coherence non-zero.

In the case of spins in thermal equilibrium in a magnetic field directed along z, the phenomena represented by eqns 3.39, 3.44, and 3.46 have a simple interpretation that is illustrated in Fig. 3.8. First, the thermal equilibrium state is such that when the spins are expressed in terms of energy eigenstates, the phase differences between the quantum states are random, leading to the terms $\overline{a_{1/2}^* a_{-1/2}}$ being zero and no transverse magnetisation existing. Second, the thermal equilibrium ensemble has a slightly higher population, $\overline{|a_{1/2}|^2}$, in the lower energy state so that a net positive angular momentum z-component (i.e. longitudinal magnetisation) exists. Indeed, as we will see in the next chapter, when the spin-$\frac{1}{2}$ ensemble is disturbed from its thermal equilibrium state, it is still only the 'spin excess' $\overline{|a_{1/2}|^2} - \overline{|a_{-1/2}|^2}$ that is visible in any experiment.

Of course, once such a disturbance occurs, and we create coherent superposition states with $\overline{a_{1/2}^* a_{-1/2}}$ non zero, then the simple picture given by Fig. 3.8 is of no help. For properly handling the effects of both polarisation and coherence in spin ensembles, we will need a new type of picture and a new type of quantum mechanical operator, one that enables the calculation of statistical averages of quantum mechanical operators with extraordinary ease. This operator we call the density matrix.

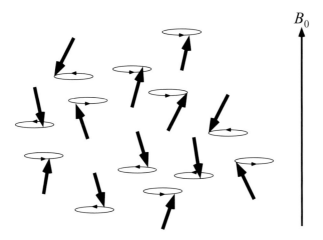

B_0

Fig. 3.8 Schematic for an ensemble of spin-$\frac{1}{2}$ particles immersed in a magnetic field B_0 in which the precessing phases are represented by angular positions in an Argand plane. There is a spin excess of lower energy 'up' states meaning $\overline{|a_{1/2}|^2} - \overline{|a_{-1/2}|^2}$ is non-zero, while phase incoherence leads to $\overline{a_{1/2}^* a_{-1/2}} = 0$.

3.2.2 Density matrix properties

The density matrix ρ, is defined by

$$\rho = \overline{|\Psi\rangle\langle\Psi|} = \sum_{\Psi} p_\Psi |\Psi\rangle\langle\Psi| \tag{3.49}$$

Suppose ρ is written as a matrix in the chosen $\{|m\rangle\}$ angular momentum representation. Then the matrix elements of ρ are

$$\langle m|\rho|m'\rangle = \sum_{\Psi} p_\Psi \langle m|\Psi\rangle\langle\Psi|m'\rangle \tag{3.50}$$
$$= \overline{a_m a_{m'}^*}$$

where, to remind the reader, the bar above a quantity is a shorthand notation for the ensemble average sum $\sum_\Psi p_\Psi$. The usefulness of ρ resides in the fact that the ensemble averaged expectation value for a measurement with operator A is given by

$$\overline{\langle\Psi|A|\Psi\rangle} = \text{Tr}\,(A\rho) \tag{3.51}$$

where the right-hand side represents the trace or diagonal sum over the matrix product $A\rho$. It can easily be seen that eqn 3.51 gives results identical to eqns 3.39, 3.44, and 3.46 for the previously chosen examples. However, let us demonstrate the relation $\overline{\langle A\rangle} = \text{Tr}\,(A\rho)$ by working in a general representation of eigenstates $\{|\alpha\rangle\}$.

$$\mathrm{Tr}\,(A\rho) = \sum_{\alpha} \left\langle \alpha \left| A \sum_{\Psi} p_{\Psi} |\Psi\rangle\langle\Psi| \right| \alpha \right\rangle$$

$$= \sum_{\Psi} p_{\Psi} \sum_{\alpha} \langle\Psi|\alpha\rangle\langle\alpha|A|\Psi\rangle \qquad (3.52)$$

$$= \sum_{\Psi} p_{\Psi} \langle\Psi|A|\Psi\rangle$$

where the completeness relation has been used in the final step. This is precisely the result, eqn 3.37, that we required from a simple consideration of statistics.

3.2.3   Evolution and the quantum Liouville equation

Some properties of ρ should be noted. First it contains all the information we need to calculate the value of any observable quantity of interest. Second, the evolution of ρ with time may be simply deduced from the Schrödinger equation and may be written

$$i\frac{\partial}{\partial t}\rho = [\mathcal{H}, \rho] \qquad (3.53)$$

This relation is known as the Liouville equation [13] and yields for the case of constant Hamiltonian $\mathcal{H}$,

$$\rho(t) = U(t)\rho(0)\,U^{-1}(t) \qquad (3.54)$$

and where the Hamiltonian is time-varying, the overall evolution may be broken down into a succession of sequential incremental evolution steps involving time intervals sufficiently short that $\mathcal{H}(t)$ is taken as constant for that increment.

3.2.4   Pure states, mixed states and quantum coherence

Finally, there are some rules about the elements of ρ. The diagonal elements are probabilities and so must be real numbers between 0 and 1, while the sum of the diagonal elements, $\mathrm{Tr}\,(\rho)$, is just the sum of state probabilities and therefore equals unity. The density matrix is Hermitian, that is to say $\rho^{\dagger} = \rho$, an obvious consequence of the fact that $\langle\alpha|\rho|\alpha'\rangle = \overline{a_{\alpha}a_{\alpha'}^{*}}$. A special case arises when the respective quantum states of all the sub-ensembles are identical, in other words when $p_{\Psi} = 1$ for some particular common Ψ. In that case the ensemble state is deemed 'pure' so that both $\mathrm{Tr}\,(\rho) = 1$ and $\mathrm{Tr}\,(\rho^2) = 1$. This second equality is a consequence of the density matrix for the pure state ensemble being simply $|\Psi\rangle\langle\Psi|$.

Two instructive examples, in the case of spin-$\frac{1}{2}$ are

$$\rho = \begin{bmatrix} \frac{1}{2} & 0 \\ 0 & \frac{1}{2} \end{bmatrix} \qquad (3.55)$$

and

$$\rho = \begin{bmatrix} \frac{1}{2} & \frac{1}{2} \\ \frac{1}{2} & \frac{1}{2} \end{bmatrix} \qquad (3.56)$$

Both are Hermitian and satisfy the requirements for the diagonal elements of density matrices. Both return a zero value for $\overline{\langle I_z \rangle}$. But the first is a mixed state corresponding to equal numbers of spins in up and down states. The second is a pure state consisting of a single sub-ensemble in which the quantum particles are in the same superposition $|\Psi\rangle = \frac{1}{\sqrt{2}}|\frac{1}{2}\rangle + \frac{1}{\sqrt{2}}|-\frac{1}{2}\rangle$. Multiplying this state by a common phase factor $\exp(i\delta)$ would yield the same density matrix, and hence the same result for any measurements. In this sense the absolute phase of a quantum state is irrelevant. It is the relative phase between basis states that determines the off-diagonal elements, $a_m a_{m'}^*$ of the density matrix, and this is measurable as the degree of quantum coherence.

We are now in a position to see how the polarisations, $\overline{|a_\alpha|^2}$, and coherences, $\overline{a_\alpha a_{\alpha'}^*}$ of the superposition states determine the density matrix elements. For spin-$\frac{1}{2}$

$$\rho = \begin{bmatrix} \overline{|a_{1/2}|^2} & \overline{a_{1/2}a_{-1/2}^*} \\ \overline{a_{1/2}^* a_{-1/2}} & \overline{|a_{-1/2}|^2} \end{bmatrix} \tag{3.57}$$

Non-zero off-diagonal elements of the density matrix inform us about the degree of quantum coherence in the ensemble. To have such a coherence, superposition states must be present. To be non-zero, there must exist phase coherence between the amplitudes of those superposition states. We can easily see that the existence of quantum coherence in the case of spin-$\frac{1}{2}$ corresponds to non-zero expectation values for I_x and I_y.

Using eqns 3.49 and 3.51 and the matrices for the operators I_x, I_y, and I_z, it is straightforward to show that for $I = 1/2$

$$\rho = \begin{bmatrix} \frac{1}{2} + \overline{\langle I_z \rangle} & \overline{\langle I_x \rangle} - i\overline{\langle I_y \rangle} \\ \overline{\langle I_x \rangle} + i\overline{\langle I_y \rangle} & \frac{1}{2} - \overline{\langle I_z \rangle} \end{bmatrix} \tag{3.58}$$

The requirements $\mathrm{Tr}(\rho) = 1$ and $\rho^\dagger = \rho$ mean that for spin-$\frac{1}{2}$ only three real numbers are needed to specify all the four matrix elements of ρ and therefore the results of any measurement. In other words, all states of the ensemble for independent spin-$\frac{1}{2}$ nuclei may be described by specifying the components of the vector $\left(\overline{\langle I_x \rangle}\mathbf{i} + \overline{\langle I_y \rangle}\mathbf{j} + \overline{\langle I_z \rangle}\mathbf{k}\right)$. In macroscopic terms this is equivalent to specifying the magnetisation vector, $N\gamma\hbar\left(\overline{\langle I_x \rangle}\mathbf{i} + \overline{\langle I_y \rangle}\mathbf{j} + \overline{\langle I_z \rangle}\mathbf{k}\right)$, where N is the number of spins per unit volume.

3.3 Tensor bases for the density matrix

3.3.1 Liouville space for spin-$\frac{1}{2}$

The set of spin-$\frac{1}{2}$ operators $\{I_{\alpha=x,y,z}\}$, taken together with $I_{\alpha=0} = \frac{1}{2}\underline{\underline{1}}$, satisfy all the requirements of a vector space in which the inner product is defined by $\mathrm{Tr}(I_\alpha I_\beta)$. Indeed that space provides a density matrix basis in which the basis vectors are orthogonal as $\mathrm{Tr}(I_\alpha I_\beta) = \frac{1}{2}\delta_{\alpha\beta}$. In this manner any density matrix may be written as the linear superposition

$$\rho = \rho_0 \frac{1}{2}\underline{\underline{1}} + \rho_x I_x + \rho_y I_y + \rho_z I_z \tag{3.59}$$

where $\rho_0 = 1$ and the ρ_α are real. Table 3.1 shows this Cartesian spin operator density matrix basis. Note that the I_z operator is diagonal and therefore gives information about the population terms $\overline{|a_{1/2}|^2}$, and $\overline{|a_{-1/2}|^2}$, whereas I_x and and I_y are off-diagonal operators and tell us about the coherences, $\overline{a_{1/2}a^*_{-1/2}}$. The I_0 operator gives no information, as we shall see.

Since the set $\{I_{\alpha=0,x,y,z}\}$ span all possible Hermitian operators for spin-$\frac{1}{2}$ we can in fact write any quantum mechanical operator in this basis

$$A = A_0 \frac{1}{2}\underline{1} + A_x I_x + A_y I_y + A_z I_z \tag{3.60}$$

and the ensemble averaged expectation value by

$$\overline{\langle A \rangle} = \mathrm{Tr}\,(\rho A) = \frac{1}{2} \sum_{\alpha=0,x,y,z} A_\alpha \rho_\alpha \tag{3.61}$$

Note how this simple vector basis allows for a very simple description of the Liouville equation. Equation 3.53 may be written

$$i \sum_\alpha \dot{\rho}_\alpha = \sum_{\alpha,\beta} \rho_\alpha \mathcal{H}_\beta \left[I_\beta, I_\alpha \right] \tag{3.62}$$

We can take advantage of the fact the the the identity matrix commutes with all spin operators, and that $\dot{\rho}_0 = 0$, and restrict eqn 3.62 to the terms $\alpha = x, y, z$. This constancy of the identity term in the density matrix and its lack of influence in dynamics is such that it can generally be ignored. Its only role is to ensure $\mathrm{Tr}\,(\rho) = 1$.

Of course, the commutation relationships for angular momentum operators mean that the right-hand side of eqn 3.62 is a simple linear superposition of the I_α. Using the orthogonality of the I_α under the trace operation, we find a simple set of coupled first-order differential equations for the density matrix coefficients, ρ_α. This enables us to solve for the dynamics in terms of a simple precessional motion. For example, if the Hamiltonian is the Zeeman operator, $-\omega_0 I_z$, then the solution is

$$\rho_0(t) = 1$$
$$\rho_x(t) = \rho_x(0)\cos(\omega_0 t) + \rho_y(0)\sin(\omega_0 t)$$
$$\rho_y(t) = \rho_y(0)\cos(\omega_0 t) - \rho_x(0)\sin(\omega_0 t)$$
$$\rho_z(t) = \rho_z(0)$$

This picture of evolution in terms of a simple precession between two components of the density matrix make for powerful insight when considering the behaviour of nuclear spins in magnetic resonance experiments.

The choice of a Cartesian basis for the spin operators is arbitrary and is analogous to the description of light in terms of linear polarisation. The equivalent of a circular polarisation standpoint would be to use the basis set $\{I_z, I_+, I_-\}$, as shown in Table 3.2. As we will see in the next sections, this basis has the advantage of special transformation properties under rotation.

Table 3.1 Cartesian operator density matrix basis for $I=1/2$

Angular momentum operator basis	$\lvert m\rangle\langle m'\rvert$ Ket–bra basis	Name
$\frac{1}{2}\underline{\underline{1}}$	$\frac{1}{2}\begin{bmatrix}1&0\\0&1\end{bmatrix}$	Identity
I_x	$\frac{1}{2}\begin{bmatrix}0&1\\1&0\end{bmatrix}$	1-quantum coherence
I_y	$\frac{1}{2}\begin{bmatrix}0&-i\\i&0\end{bmatrix}$	1-quantum coherence
I_z	$\frac{1}{2}\begin{bmatrix}1&0\\0&-1\end{bmatrix}$	z-polarisation

Table 3.2 Spherical tensor operator basis for $I=1/2$

Tensor component	Angular momentum operator basis	$\lvert m\rangle\langle m'\rvert$ Ket–bra basis	Name
T_{00}	$\underline{\underline{1}}$	$\begin{bmatrix}1&0\\0&1\end{bmatrix}$	Identity
T_{10}	I_z	$\begin{bmatrix}1&0\\0&-1\end{bmatrix}$	z-polarisation
T_{11}	$-\frac{1}{\sqrt{2}}I_+$	$\frac{1}{\sqrt{2}}\begin{bmatrix}0&-1\\0&0\end{bmatrix}$	1-quantum coherence
$T_{1\,-1}$	$\frac{1}{\sqrt{2}}I_-$	$\frac{1}{\sqrt{2}}\begin{bmatrix}0&0\\1&0\end{bmatrix}$	1-quantum coherence

3.3.2 Liouville space for $I > \frac{1}{2}$

These ideas can be extended to higher dimensional systems such as higher angular momenta ($I > 1/2$) or multiple spin problems, where couplings between the spins necessitate a description in terms of multiple spin operators. The Cartesian basis has a natural extension to the case of multiple spin systems. Take, for example, the case of two spins, i and j, which may be coupled via an interaction involving both spin operators (examples are the dipolar and spin–spin scalar interactions covered in Section 3.4). A basis for the density matrix exists, involving products of angular momentum operators, I_{ix}, I_{iy}, I_{iz} and I_{jx}, I_{jy}, I_{jz}. Such a product operator basis is immensely powerful in describing the evolution of spin systems in nuclear magnetic resonance [14, 15].

For the moment we focus on single-spin operators, but for $I > \frac{1}{2}$, for which there exist more than two basis states. A nuclear spin system with three or more basis states and energy levels can no longer be described by a density matrix expressed in

Table 3.3 Cartesian operator density matrix basis for $I=1$

Angular momentum operator basis	$\|m\rangle\langle m'\|$ Ket–bra basis	Name
$\underline{\underline{1}}$	$\begin{bmatrix} 1\,0\,0 \\ 0\,1\,0 \\ 0\,0\,1 \end{bmatrix}$	Identity
I_x	$\frac{1}{\sqrt{2}} \begin{bmatrix} 0\,1\,0 \\ 1\,0\,1 \\ 0\,1\,0 \end{bmatrix}$	1-quantum coherence
I_y	$\frac{1}{\sqrt{2}} \begin{bmatrix} 0 & -i & 0 \\ i & 0 & -i \\ 0 & i & 0 \end{bmatrix}$	1-quantum coherence
I_z	$\begin{bmatrix} 1\,0\,0 \\ 0\,0\,0 \\ 0\,0\,{-1} \end{bmatrix}$	z-polarisation
$\frac{1}{\sqrt{6}} \left(3I_z^2 - 2\underline{\underline{1}}\right)$	$\frac{1}{\sqrt{6}} \begin{bmatrix} 1 & 0 & 0 \\ 0 & -2 & 0 \\ 0 & 0 & 1 \end{bmatrix}$	Quadrupolar order
$\frac{1}{\sqrt{2}} \left(I_z I_y + I_y I_z\right)$	$\frac{1}{2} \begin{bmatrix} 0 & -i & 0 \\ i & 0 & i \\ 0 & -i & 0 \end{bmatrix}$	Antiphase 1-quantum coherence
$\frac{1}{\sqrt{2}} \left(I_z I_x + I_x I_z\right)$	$\frac{1}{2} \begin{bmatrix} 0 & 1 & 0 \\ 1 & 0 & -1 \\ 0 & -1 & 0 \end{bmatrix}$	Antiphase 1-quantum coherence
$\frac{1}{\sqrt{2}} \left(I_x^2 - I_y^2\right)$	$\frac{1}{\sqrt{2}} \begin{bmatrix} 0\,0\,1 \\ 0\,0\,0 \\ 1\,0\,0 \end{bmatrix}$	2-quantum coherence
$\frac{1}{\sqrt{2}} \left(I_x I_y + I_y I_x\right)$	$\frac{1}{\sqrt{2}} \begin{bmatrix} 0 & 0 & i \\ 0 & 0 & 0 \\ -i & 0 & 0 \end{bmatrix}$	2-quantum coherence

a basis of the four operators comprising I_x, I_y, I_z and the identity matrix. What is required is a basis of the appropriate tensor rank. To illustrate, consider the case of $I = 1$ where the kets have a three-dimensional basis $\{|m\rangle\} = \{|1\rangle, |0\rangle, |-1\rangle\}$. Here eight matrices, in addition to the identity operator, are required.[3] We can generate the required Liouville basis by taking products of the spin operators, labelling these products T_α. Again, we will require the T_α, and T_β to be orthogonal under the trace operation. Table 3.3 shows this Cartesian Liouville basis for the case $I = 1$.

---

[3] In general for an n- dimensional representation, the number of independent operators and independent coefficients is $n^2 - 1$, meaning eight for spin-1, 15 for spin-$\frac{3}{2}$ and so on.

Table 3.4 Cartesian product operator basis for coupled $I=1/2$ nuclei (first part)

Angular momentum operator basis	$\lvert m\rangle\langle m'\rvert$ Ket–bra basis	Name
$\frac{1}{2}\underline{\underline{1}}$	$\frac{1}{2}\begin{bmatrix} 1\,0\,0\,0 \\ 0\,1\,0\,0 \\ 0\,0\,1\,0 \\ 0\,0\,0\,1 \end{bmatrix}$	identity
$I_{ix},\,I_{jx}$	$\frac{1}{2}\begin{bmatrix} 0\,0\,1\,0 \\ 0\,0\,0\,1 \\ 1\,0\,0\,0 \\ 0\,1\,0\,0 \end{bmatrix},\ \frac{1}{2}\begin{bmatrix} 0\,1\,0\,0 \\ 1\,0\,0\,0 \\ 0\,0\,0\,1 \\ 0\,0\,1\,0 \end{bmatrix}$	1-quantum coherence
$I_{iy},\,I_{jy}$	$\frac{1}{2}\begin{bmatrix} 0\,0\,-i\,0 \\ 0\,0\,0\,-i \\ i\,0\,0\,0 \\ 0\,i\,0\,0 \end{bmatrix},\ \frac{1}{2}\begin{bmatrix} 0\,-i\,0\,0 \\ -i\,0\,0\,0 \\ 0\,0\,0\,i \\ 0\,0\,i\,0 \end{bmatrix}$	1-quantum coherence
$I_{iz},\,I_{jz}$	$\frac{1}{2}\begin{bmatrix} 1\,0\,0\,0 \\ 0\,1\,0\,0 \\ 0\,0\,-1\,0 \\ 0\,0\,0\,-1 \end{bmatrix},\ \frac{1}{2}\begin{bmatrix} 1\,0\,0\,0 \\ 0\,-1\,0\,0 \\ 0\,0\,1\,0 \\ 0\,0\,0\,-1 \end{bmatrix}$	z-polarisation

3.3.3 Product operator Liouville space for two coupled spin-$\frac{1}{2}$ nuclei

Suppose we have two spins, i and j, coupled by some interaction that requires the system to be treated as a combined quantum system. The product operator basis of Section 3.1.2 is particularly effective when the spins' interactions are described by Hamiltonian terms diagonal in this basis.

Note that the 4×4 matrix required for the two-spin system implies 16 independent basis vectors in Liouville space, so that 16 product operators are needed, including terms such as zero-quantum coherence states, for which no counterpart exists in a single-spin $I > 1/2$ system. Now the density matrix is

$$
\begin{aligned}
\rho = \rho_0 \frac{1}{2}\underline{\underline{1}} &+ \rho_{ix}I_{ix} + \rho_{jx}I_{jx} + \rho_{iy}I_{iy} + \rho_{jy}I_{jy} + \rho_{iz}I_{iz} + \rho_{jz}I_{jz} \\
&+ \rho_{aix}2I_{ix}I_{jz} + \rho_{ajx}2I_{iz}I_{jx} + \rho_{aiy}2I_{iy}I_{jz} + \rho_{ajy}2I_{iz}I_{jy} \\
&+ \rho_{izjz}2I_{iz}I_{jz} + \rho_{ZQCx}(2I_{ix}I_{jx} + 2I_{iy}I_{jy}) + \rho_{ZQCy}(2I_{iy}I_{jx} - 2I_{ix}I_{jy}) \\
&+ \rho_{2QCx}(2I_{ix}I_{jx} - 2I_{iy}I_{jy}) + \rho_{2QCy}(2I_{ix}I_{jx} - 2I_{iy}I_{jy})
\end{aligned}
\tag{3.63}
$$

Table 3.5 Cartesian product operator basis for coupled $I=1/2$ nuclei (continued)

Angular momentum operator basis	$\lvert m\rangle\langle m'\rvert$ Ket-bra basis	Name
$2I_{iz}I_{jz}$	$\frac{1}{2}\begin{bmatrix} 1 & 0 & 0 & 0 \\ 0 & -1 & 0 & 0 \\ 0 & 0 & -1 & 0 \\ 0 & 0 & 0 & 1 \end{bmatrix}$	Longitudinal 2-spin order
$2I_{ix}I_{jz},\ 2I_{iz}I_{jx}$	$\frac{1}{2}\begin{bmatrix} 0 & 0 & 1 & 0 \\ 0 & 0 & 0 & -1 \\ 1 & 0 & 0 & 0 \\ 0 & -1 & 0 & 0 \end{bmatrix},\ \frac{1}{2}\begin{bmatrix} 0 & 1 & 0 & 0 \\ 1 & 0 & 0 & 0 \\ 0 & 0 & 0 & -1 \\ 0 & 0 & -1 & 0 \end{bmatrix}$	Antiphase 1-quantum coherence
$2I_{iy}I_{jz},\ 2I_{iz}I_{jy}$	$\frac{1}{2}\begin{bmatrix} 0 & 0 & -i & 0 \\ 0 & 0 & 0 & i \\ i & 0 & 0 & 0 \\ 0 & -i & 0 & 0 \end{bmatrix},\ \frac{1}{2}\begin{bmatrix} 0 & -i & 0 & 0 \\ i & 0 & 0 & 0 \\ 0 & 0 & 0 & i \\ 0 & 0 & -i & 0 \end{bmatrix}$	Antiphase 1-quantum coherence
$2I_{ix}I_{jx} + 2I_{iy}I_{jy},$ $2I_{iy}I_{jx} - 2I_{ix}I_{jy}$	$\begin{bmatrix} 0 & 0 & 0 & 0 \\ 0 & 0 & 1 & 0 \\ 0 & 1 & 0 & 0 \\ 0 & 0 & 0 & 0 \end{bmatrix},\ \begin{bmatrix} 0 & 0 & 0 & 0 \\ 0 & 0 & -i & 0 \\ 0 & i & 0 & 0 \\ 0 & 0 & 0 & 0 \end{bmatrix}$	0-quantum coherence $ZQC_x,\ ZQC_y$
$2I_{ix}I_{jx} - 2I_{iy}I_{jy},$ $2I_{ix}I_{jx} - 2I_{iy}I_{jy}$	$\begin{bmatrix} 0 & 0 & 0 & 1 \\ 0 & 0 & 0 & 0 \\ 0 & 0 & 0 & 0 \\ 1 & 0 & 0 & 0 \end{bmatrix},\ \begin{bmatrix} 0 & 0 & 0 & -i \\ 0 & 0 & 0 & 0 \\ 0 & 0 & 0 & 0 \\ i & 0 & 0 & 0 \end{bmatrix}$	2-quantum coherence $2QC_x,\ 2QC_y$

Tables 3.4 and 3.5 list these product operators along with their matrix representation in the product representation of I_{1z} and I_{2z} eigenstates. These operators are very convenient to work with when the Hamiltonian involves terms that act as simple rotations in Cartesian space (for example, the effect of resonant radiofrequency pulses in NMR) of terms that are diagonal in the I_{1z} and I_{2z} operators, for example the Zeeman interaction and the scalar coupling[4] $2\pi J I_{iz}I_{jz}$.

In Chapter 4 we will meet specific examples of such terms, and we will also introduce a pictorial representation of these operators due to Sørensen *et al.* [15], which is both extensively used and which provides a helpful visualisation tool.

---

[4]See later in Section 3.4.

3.3.4 Spherical tensors

The Cartesian basis has some merit in the description of magnetic resonance, where the Hamiltonian is generally expressed in terms of Cartesian spin operators. Hence the evolution of the density matrix under the Liouville equation (eqn 3.53) is simply determined by the operator algebra, eqn 3.13, especially for the case of rotations induced by radiofrequency pulses in the Cartesian axis frame. And, as we shall see, the observable quantities in magnetic resonance are the transverse magnetisations, I_x and I_y. For these reasons product combinations of Cartesian angular momentum operators are very convenient for the description of magnetic resonance phenomena. However, for higher-order spin systems, their transformation properties under rotation are somewhat complicated, though straightforward to calculate. Indeed, when considering rotations, a natural basis to use are spherical tensors, for which a rotation characterised by the Euler angles (α, β, γ) is given by the Wigner rotation matrix $D^{(k)}(\alpha, \beta, \gamma)$, the same transformation property as for the $(2k+1)$ spherical harmonics of rank k. That has particular relevance when dealing with spin states, since the rank L $Y_{LM}(\theta, \phi)$ are the eigenfunctions of the orbital angular momentum operators $\mathbf{L}^2$ and L_z as [9–11]

$$\begin{aligned}
\mathbf{L}^2 Y_{LM}(\theta, \phi) &= L\,(L+1)\,Y_{LM}(\theta, \phi) \\
L_z Y_{LM}(\theta, \phi) &= M\,Y_{LM}(\theta, \phi)
\end{aligned} \tag{3.64}$$

where L and M are the angular momentum quantum number and azimuthal quantum number, respectively.

Specifically, the rotation properties of the spherical tensor, T_{kq}, under the Euler rotation is

$$D^{(k)}\,T_{kq}\,D^{(k)^{-1}} = \sum_{q'} D^{(k)}_{q'q}\,T_{kq'} \tag{3.65}$$

where the $D^{(k)}_{q'q}$ are the matrix elements of $D^{(k)}$. In group theoretical terms, eqn 3.65 describes the general transformation properties of an irreducible tensor, T_{kq}, of rank k and order q. Note in particular that a rotation may alter the order but not the rank of the tensor. Consider the example of orbital angular momentum $L = 1$. The three spatial eigenfunctions $Y_{LM}(\theta, \phi)$ are given by $Y_{10}(\theta, \phi)$, $Y_{11}(\theta, \phi)$, and $Y_{1-1}(\theta, \phi)$. The Wigner rotation matrix $D^{(1)}(\alpha, \beta, \gamma)$, which describes the transformation between these states under the Euler rotation, is the same matrix used to describe the rotation of $I = 1$ kets for which $m = 1, 0, -1$. In general, for spin quantum number I for which the $(2I + 1)$ basis kets are $|Im\rangle$, the rotation is given by eqn 3.27.

When we are dealing with transformations of operators in spin space, as opposed to transforming the kets themselves, it is important to distinguish the rank of the operator from the rank of the ket space. For example, the angular momentum operators I_x, I_y, and I_z transform as rank-1 tensors, whether they belong to $I = 1/2$, $I = 1$, or $I > 1$. However, whereas for spin-$\frac{1}{2}$ these rank-1 tensor operators (along with the identity) are all that are needed to form a basis for any density matrix, or indeed any observable operator such as the Hamiltonian, for higher spin ensembles, higher rank operators will be needed. Note also that half-integer rank is possible in ket space, in which case the matrix is double valued under rotation, a rotation by 2π being the negative identity

Table 3.6 Spherical tensor operator basis for $I = 1$

Tensor component	Angular momentum operator basis	$\lvert m\rangle\langle m'\rvert$ Ket–bra basis	Name
T_{00}	$\underline{\underline{1}}$	$\begin{bmatrix} 1\,0\,0 \\ 0\,1\,0 \\ 0\,0\,1 \end{bmatrix}$	Identity
T_{10}	I_z	$\begin{bmatrix} 1\,0\,0 \\ 0\,0\,0 \\ 0\,0\,{-1} \end{bmatrix}$	z-polarisation
T_{11}	$-\frac{1}{\sqrt{2}}I_+$	$\begin{bmatrix} 0\,{-1}\,0 \\ 0\,0\,{-1} \\ 0\,0\,0 \end{bmatrix}$	1-quantum coherence
$T_{1\,-1}$	$\frac{1}{\sqrt{2}}I_-$	$\begin{bmatrix} 0\,0\,0 \\ 1\,0\,0 \\ 0\,1\,0 \end{bmatrix}$	1-quantum coherence
T_{20}	$\frac{1}{\sqrt{6}}\left(3I_z^2 - 2\underline{\underline{1}}\right)$	$\frac{1}{\sqrt{6}}\begin{bmatrix} 1\,0\,0 \\ 0\,{-2}\,0 \\ 0\,0\,1 \end{bmatrix}$	Quadrupolar order
T_{21}	$-\frac{1}{2}\left(I_zI_+ + I_+I_z\right)$	$-\frac{1}{\sqrt{2}}\begin{bmatrix} 0\,1\,0 \\ 0\,0\,{-1} \\ 0\,0\,0 \end{bmatrix}$	Antiphase 1-quantum coherence
$T_{2\,-1}$	$\frac{1}{2}\left(I_zI_- + I_-I_z\right)$	$\frac{1}{\sqrt{2}}\begin{bmatrix} 0\,0\,0 \\ 1\,0\,0 \\ 0\,{-1}\,0 \end{bmatrix}$	Antiphase 1-quantum coherence
T_{22}	$\frac{1}{2}I_+^2$	$\begin{bmatrix} 0\,0\,1 \\ 0\,0\,0 \\ 0\,0\,0 \end{bmatrix}$	2-quantum coherence
$T_{2\,-2}$	$\frac{1}{2}I_-^2$	$\begin{bmatrix} 0\,0\,0 \\ 0\,0\,0 \\ 1\,0\,0 \end{bmatrix}$	2-quantum coherence

and a rotation by 4π the positive identity matrix. By contrast, in the transformation of operators, only integer rank is permitted.

The lowest rank for operators in spin space is the identity operator, T_{00}. Those for $k = 1$ are I_z, and the operators formed by linear combinations of I_x and I_y, namely, $-I_+/\sqrt{2}$ and $I_-/\sqrt{2}$. All higher rank tensors can be formed from appropriate products of these. The rule for forming a spherical tensor $T_{KQ}(\mathbf{A}_1)$, by making products of lower rank tensors, follows that for combining angular momenta, namely

$$T_{KQ}(\mathbf{A}_1, \mathbf{A}_2) = \sum_{q_1} \langle k_1 q_1 k_2 q_2 | KQ\rangle \, T_{k_1 q_1}(\mathbf{A}_1)\, T_{k_2 q_2}(\mathbf{A}_2) \qquad (3.66)$$

where $q_2 = Q - q_1$, the $\langle k_1 q_1 k_2 q_2 | KQ \rangle$ are Clebsch–Gordon coefficients, and $\mathbf{A}_1$ and $\mathbf{A}_2$ are all the other variables upon which the tensors depend. For single-spin operators, $\mathbf{A}_1$ and $\mathbf{A}_2$ are identical. The same Clebsch–Gordon rules for angular momentum addition apply in determining the rank of the tensors. In other words $k = k_1 + k_2, k_1 + k_2 - 1, \ldots, |k_1 - k_2|$ and, as stated above, $q_2 = Q - q_1$. Suppose $k_1 = k_2 = 1$. Then Clebsch–Gordon superpositions of products generate the nine tensors $T_{2\pm2}$, $T_{2\pm1}$, T_{20}, $T_{1\pm1}$, T_{10}, and T_{00}.[5]

Table 3.6 shows the standard spherical tensor basis set for spin-1 [16]. In this form they are orthogonal but not normalised. Unit spherical tensors may be defined by [17]

$$\hat{T}_{KQ} = \frac{1}{K!} \left[\frac{(2K+1)(2I-K)!2K(2K)!}{(2K+I+1)!} \right]^{1/2} T_{KQ} \tag{3.67}$$

with the orthonormal relationship

$$\mathrm{Tr}\left(\hat{T}_{KQ} \hat{T}_{K'Q'}^{\dagger} \right) = \delta_{KK'} \delta_{QQ'} \tag{3.68}$$

the † symbol representing the Hermitian conjugate operation.

Any density matrix may then be written

$$\rho = \sum_{K,Q} \rho_Q^K \, \hat{T}_{KQ} \tag{3.69}$$

where the coefficients, ρ_Q^K are referred to as Fano statistical tensors [9, 10] and are given by

$$\rho_Q^K = \mathrm{Tr}\left(\hat{T}_{KQ}^{\dagger} \rho \right) \tag{3.70}$$

Most importantly when considering the evolution of the density matrix under the effect of a Hamiltonian operator expressed in the spherical tensor basis, we have the following commutation relations, true for all spin quantum number I,

$$[T_{10}, T_{KQ}] = Q \, T_{KQ} \tag{3.71}$$

and

$$[T_{1\pm1}, T_{KQ}] = \mp \frac{1}{\sqrt{2}} \left[(K \mp Q)(K \pm Q + 1) \right]^{1/2} T_{KQ\pm1} \tag{3.72}$$

These relations are useful in determining how our spin system will evolve in the presence of the Zeeman Hamiltonian ($\sim T_{10}$) or, as we shall soon see, radiofrequency pulses ($\sim T_{1\pm1}$). For spin-1, a Hamiltonian of higher rank is possible, namely the electric quadrupole interaction ($\sim T_{20}$). Table 3.7 shows the relevant commutators for spin-1 [18].

Suppose we consider some Hamiltonian $\mathcal{H}$ that may be written in terms of our T_{KQ} basis. Now consider some initial state of the time-dependent density matrix $\rho(t)$, which comprises a single Fano component so that $\rho(0) = \rho_{Q_0}^{K_0} T_{K_0 Q_0}$. Then evolution under $\mathcal{H}$ will be determined by the commutator $[\mathcal{H}, T_{K_0 Q_0}]$. Suppose that commutator

---

[5]For example, $T_{11} = T_{10} T_{11} - T_{11} T_{10}$.

Table 3.7 Commutation relations for $I = 1$ spherical tensor operators.

	T_{10}	T_{11}	T_{1-1}	T_{20}	T_{21}	T_{2-1}	T_{22}	T_{2-2}
T_{10}	0	T_{11}	$-T_{1-1}$	0	T_{21}	$-T_{2-1}$	$2T_{22}$	$-2T_{2-2}$
T_{11}	$-T_{11}$	0	$-T_{10}$	$-\sqrt{3}T_{21}$	$-\sqrt{2}T_{22}$	$-\sqrt{3}T_{20}$	0	$-\sqrt{2}T_{2-1}$
T_{1-1}	T_{1-1}	T_{11}	0	$\sqrt{3}T_{2-1}$	$\sqrt{3}T_{20}$	$\sqrt{3}T_{22}$	$\sqrt{2}T_{21}$	0
T_{20}	0	$\sqrt{3}T_{21}$	$-\sqrt{3}T_{2-1}$	0	$\frac{\sqrt{3}}{2}T_{11}$	$-\frac{\sqrt{3}}{2}T_{1-1}$	0	0
T_{21}	$-T_{21}$	$\sqrt{2}T_{22}$	$-\sqrt{3}T_{20}$	$-\frac{\sqrt{3}}{2}T_{11}$	0	$\frac{1}{2}T_{10}$	0	$\frac{1}{\sqrt{2}}T_{1-1}$
T_{2-1}	T_{2-1}	$\sqrt{3}T_{20}$	$-\sqrt{3}T_{22}$	$\frac{\sqrt{3}}{2}T_{1-1}$	$-\frac{1}{2}T_{10}$	0	$-\frac{1}{\sqrt{2}}T_{11}$	0
T_{22}	$-2T_{22}$	0	$-\sqrt{2}T_{21}$	0	0	$\frac{1}{\sqrt{2}}T_{11}$	0	T_{10}
T_{2-2}	$2T_{2-2}$	$\sqrt{2}T_{2-1}$	0	0	$-\frac{1}{\sqrt{2}}T_{1-1}$	0	$-T_{10}$	0

results in a unique spherical tensor component $T_{K_1Q_1}$. Table 3.7 suggests that this is most likely if our Hamiltonian is also a pure spherical tensor. Now suppose that the commutator $[\mathcal{H}, T_{K_1Q_1}]$ returns $T_{K_0Q_0}$, again a common occurrence in Table 3.7. In that case our evolution problem reduces to a pair of coupled equations and the effect of $\mathcal{H}$ is simply a precession of our density matrix between the states $T_{K_0Q_0}$ and $T_{K_1Q_1}$. It turns out that a wide class of physical phenomena in NMR is so describable.

Spherical tensors for coupled spins

In eqn 3.66 we saw how to construct spherical tensors of two spins, $T_{KQ}(\mathbf{A}_1, \mathbf{A}_2)$ from appropriate tensor products, at that point applying this general relation to single-spin operators where $\mathbf{A}_1 = \mathbf{A}_2$. Here we apply the same relation to spherical tensors for coupled spin systems, where we deal with products of operators, $T_{k_1q_1}(i)$ and $T_{k_2q_2}(j)$, corresponding to different spin spaces, i and j.

$$T_{KQ}(i,j) = \sum_{q_1}\langle k_1q_1k_2q_2|KQ\rangle \, T_{k_1q_1}(i) \, T_{k_2q_2}(j) \tag{3.73}$$

where again, $q_2 = Q - q_1$ and the $T_{KQ}(i,j)$ transform under rotations according to the Wigner matrices $D^{(K)}(\alpha, \beta, \gamma)$.

For two spin-$\frac{1}{2}$ nuclei, the operators $T_{k_1q_1}(i)$ are of rank-1 or, in the case of a single-spin operator where either the i or j spin is represented by the identity, of rank-0. Using the Clebsch–Gordon coefficients for combining spin-1, and spin-0, we obtain the $T_{KQ}(i,j)$ tensors shown in Table 3.8, using the notation $T_{KQ}(R_iR_j)$ to indicate whether we are dealing with two-spin or one-spin operators, R being the rank. [6] Clearly the distinction of two sets of single-spin operators for i and j will make for a larger set than Table 3.6, where there were nine spin tensors. Furthermore, because the products involve operators in different spaces, the commutation relationships for angular momenta no longer apply, as in the case of single-spin operators and this leads to cross product terms with no counterpart in single-spin systems. Finally there exists a scalar product which, unlike the case of single-spin operators, is quite distinct

---

[6]Here we use the abbreviation 0-QC, 1-QC and 2-QC to refer to zero-, single- and double-quantum coherence respectively.

Table 3.8 Spherical tensors for two coupled spin-$\frac{1}{2}$ nuclei

Type	$T_{KQ}(R_i R_j)$	Operator product	Name
Identity	$T_{00}(00)$	$\frac{1}{2}\underline{1} \otimes \underline{1}$	
Scalar product	$T_{00}(11)$	$-\frac{2}{\sqrt{3}}\mathbf{I}_i \cdot \mathbf{I}_j$	Spin–spin coupling
Spin I_i	$T_{10}(10)$	$I_{iz} \otimes \underline{1}$	z-polzn
	$T_{1\pm1}(10)$	$\mp\frac{1}{\sqrt{2}}I_{i\pm} \otimes \underline{1}$	1-QC
Spin I_j	$T_{10}(01)$	$\underline{1} \otimes I_{jz}$	z-polzn
	$T_{1\pm1}(01)$	$\mp\frac{1}{\sqrt{2}}\underline{1} \otimes I_{j\pm}$	1-QC
Cross product	$T_{10}(11)$	$\frac{1}{\sqrt{2}}[I_{i-}I_{j+} - I_{i+}I_{j-}]$	0-QC
	$T_{1\pm1}(11)$	$[I_{iz}I_{j\pm} - I_{i\pm}I_{jz}]$	
Second rank	$T_{20}(11)$	$(\frac{2}{3})^{1/2}[3I_{iz}I_{jz} - \mathbf{I}_i \cdot \mathbf{I}_j]$	Longit 2 spin order
	$T_{2\pm1}(11)$	$\mp[I_{iz}I_{j\pm} + I_{i\pm}I_{jz}]$	Antiphase 1-QC
	$T_{2\pm2}(11)$	$I_{i\pm}I_{j\pm}$	2-QC

from the identity. These factors lead to 16 linearly independent two-spin tensors in Table 3.8 [19–21].

Definition of multiple-quantum coherence

The idea of quantum coherence derives from the non-zero ensemble average of the product of eigenstate amplitudes $\overline{a_m^* a_{m'}}$ that appears off the density matrix diagonal. The ensemble phase coherence between these state amplitudes is brought about by the quanta of radiofrequency energy that couple these states in a transition. Such a non-zero density matrix element corresponds to a coherent superposition of the two eigenstates $|m\rangle$ and $|m'\rangle$ [22].

$$|\psi_{mm'}\rangle = a_m|m\rangle + a_{m'}|m'\rangle \qquad (3.74)$$

This non-equilibrium state contributes non-zero matrix density matrix elements $\rho_{mm'} = |m\rangle\langle m'|$ and $\rho_{m'm} = |m'\rangle\langle m|$. Just as each eigenstate is characterised by magnetic quantum number m, so each coherence $\rho_{mm'}$ is characterised by a magnetic quantum number difference $p = m - m'$. p is known as the coherence order [22].

Subsequently, under the free precession of a Zeeman Hamiltonian, the coupled states precess synchronously. For example, in a single quantum transition, where m and m' differ by one, $\overline{a_m^* a_{m\pm1}}$ oscillates at the Larmor frequency and represents a precessing magnetic dipole detectable using the receiver coil tuned to that Larmor frequency. Where $\Delta m \neq \pm1$, the quantum coherence is not directly detectable in this manner.

In practice, the creation of various orders of multiple-quantum coherence is not performed by direct absorption of multiple radiofrequency (RF) field quanta, but by using a combination of RF pulses and the various interaction terms in the spin

Hamiltonian. These terms allow specific evolutionary migrations of the density matrix elements depending on the spin character of each term. In particular, spin–spin scalar coupling, dipolar interactions, and quadrupole interactions allow the generation of various higher-order coherences beyond $\Delta m = \pm 1$. In turn, the various pathways of evolution provide information to the spectroscopist, pathways which act as a signature for these interactions. Hence, pathway selectivity can provide a filtering process that can be used to identify spins of interest.

The definition of multiple-quantum coherence depends on the chosen basis. Suppose we consider product operators in a basis of I_z, I_+, and I_- operators. Clearly the degree or 'order' p of quantum coherence is also given by the number of raising operators minus the number of lowering operators. In that case we say that a p-quantum coherence corresponds to a density matrix with non-vanishing elements only in the pth off-diagonal, and we could write our density matrix in such a basis as $\rho = \sum_p \rho_p$. These are apparent in Table 3.3. Under the influence of a Zeeman Hamiltonian ωI_z, the p-quantum coherence oscillates at frequency $p\omega$. In general we may write the effect of a rotation about the z-axis by

$$\exp\left(-i\phi I_z\right)\rho_p \exp\left(i\phi I_z\right) = \rho_p \exp\left(-ip\phi\right) \tag{3.75}$$

When spherical tensors, T_{kq}, are used to represent the density matrix, the degree of multiple-quantum coherence is determined by the order q, as apparent in Table 3.6. A p-quantum coherence corresponds to a $T_{k\pm p}$ spherical tensor operator, the maximum order being limited by the tensor rank k. For example, a double-quantum coherence could be represented by $T_{22}+T_{2-2}$ and a single-quantum coherence by $I_x = -\frac{1}{\sqrt{2}}(T_{11} - T_{1-1})$ or by the antiphase single-quantum coherence represented by $T_{21} - T_{2-1}$. For pure orders of defined sign, T_{22} and T_{2-2} represent $p = 2$ and $p = -2$, respectively.

For a system of K coupled spins-$1/2$, p may take values from $-K$ to K. In this case we replace the single-spin operator I_z of eqn 3.75 by $\sum_{i=1}^{K} I_{iz}$. Note that in the case of coupled spins additional orders of coherence are possible. In particular, in Table 3.8 note the cross product terms $T_{10}(11)$ and $T_{11}(1 \pm 1)$, which have no equivalent in single-spin systems. In particular, the term $T_{10}(11)$ reduces to I_z for a single spin while $T_{11}(11)$ is identical to I_+. But in the case of a two-spin system, $T_{10}(11)$ is a unique state of the density matrix known as a zero-quantum coherence. We return to this peculiar state of coupled spin systems in Chapter 4.

3.4 The spin Hamiltonian

Atomic nuclei interact with their atomic and molecular environment through various multipole orders of electromagnetic interaction: electric monopole (charge), magnetic dipole, and electric quadrupole. All nuclei possess charge, the source of Coulombic binding of electrons in the atom. Most possess spin and hence a magnetic dipole, and thereby are sensitive to magnetic fields, whether externally applied, or from surrounding atomic electrons or neighbouring nuclear dipoles. Those nuclei with spin greater than $\frac{1}{2}$ will possess an electric quadrupole moment and so will be sensitive to gradients in electric fields arising from surrounding molecular orbitals. Only the dipole and quadrupole operators involve the nuclear spin and hence only these interactions are apparent in the nuclear spin precession observed in NMR.

Of course the spin electrons that surround atomic nuclei in condensed matter or in gases possess far greater magnetic character than any nearby nuclei. But in diamagnetic materials, the subject of most of our interest, those electrons are spin-paired in accordance with the Pauli principle, and so their magnetic dipole fields vanish. For molecules with unpaired electrons, the electron–nuclear hyperfine interaction plays an important role in the nuclear spin Hamiltonian, contributing a term $A\mathbf{I}\cdot\mathbf{S}$, where S is the electron spin operator. Hyperfine interactions are typically measured in megahertz and may be comparable with the nuclear Zeeman interaction with a large laboratory magnetic field.

3.4.1 The i-spin Hamiltonian

Here we will focus on diamagnetic materials, for which the hyperfine interaction is absent. We can therefore write the spin Hamiltonian for a single nucleus labelled i as

$$\mathcal{H}_i = \mathcal{H}_{\text{Zeeman}} + \mathcal{H}_{\text{CS}} + \mathcal{H}_{\text{scalar}} + \mathcal{H}_{\text{Q}} + \mathcal{H}_{\text{D}}$$
$$= -\gamma B_0 I_{iz} - \gamma \mathbf{I}_i \cdot \underline{\underline{\delta}} \cdot \mathbf{B}_0 + \sum_j 2\pi J_{ij}\,\mathbf{I}_i \cdot \mathbf{I}_j + \mathbf{I}_i \cdot \underline{\underline{\mathbf{Q}}} \cdot \mathbf{I}_i + \sum_j \mathbf{I}_i \cdot \underline{\underline{\mathbf{D}}}_{ij} \cdot \mathbf{I}_j \quad (3.76)$$

The Zeeman interaction is already familiar to us. The remaining terms are, respectively, the chemical shift, the internuclear spin–spin scalar coupling, the electric quadrupole interaction, and the internuclear dipolar interaction. The chemical shift and the spin–spin coupling give spectral information containing a chemical fingerprint for the molecule hosting the nuclear spin of interest. The nuclear quadrupole and internuclear dipolar terms provide a signature for molecular orientation or structural ordering. Each term involves a tensor product of spin and field operators, and in a different context each would be worthy of a complete discussion. But what makes them of interest to us here is the possibility of using these effects to obtain molecular specificity in translational motion measurements. For that reason our discussion is limited. In the next chapter we return to the spin Hamiltonian and consider its effect in NMR experiments, where the additional effect of RF pulses makes for a rich diversity of evolution schemes. In the present chapter we provide a brief outline of these interactions. Further description can be found elsewhere in a number of excellent texts [23–26].

Equation 3.76 is written, for convenience, in terms of products of vector spin operators and Cartesian tensor spatial operators $\underline{\underline{\delta}}$, $\underline{\underline{\mathbf{Q}}}$, and $\underline{\underline{\mathbf{D}}}$. In practice we seldom need to use these full expressions. In laboratory superconducting magnets the Zeeman interaction strength is on the order of thousands of megahertz, much greater than the remaining terms. As a consequence, the Zeeman term defines the zeroth order quantum mechanical basis set in terms of I_z operators. All the remaining terms are at least three orders of magnitude smaller, and so act as perturbations in the zeroth order frame. For the purpose of this book, we need mostly consider only first-order perturbations, corresponding to diagonal (secular) terms of the respective Hamiltonians in the I_z frame, which we write with subscript zero. However, the off-diagonal terms will be of importance in their ability to cause transitions between quantum states, thus contributing to T_1 and T_2 relaxation.

Note that where the full spin Hamiltonian expressions are needed, for example where the Zeeman field is weak or in the case of strong J-couplings where the difference

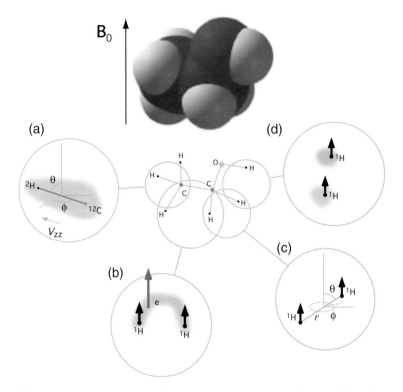

Fig. 3.9 Schematic representation of various interactions in the spin Hamiltonian for an ethanol molecule immersed in a magnetic field, $\mathbf{B}_0$. (a) The electric quadrupole interaction resulting from the electric field gradient, V_{zz} associated with the molecular orbital around a $C-^2H$ bond. (b) The scalar spin–spin interaction mediated by the molecular orbital electrons. (c) The inter-nuclear through-space dipolar interaction. (d) The chemical shift associated with differing diamagetic shielding of the nuclei by electrons in parts of the molecular orbital that differ chemically.

between the chemical shifts is small, it can be convenient to use a spherical tensor representation of the Hamiltonian. This approach is discussed in Section 3.3.4.

3.4.2 The Hamiltonian in terms of tensor products

The Hamiltonian for a spin system is a scalar and, as such, must transform under rotations as a rank-0 tensor. That means that the product of the spin tensor and the spatial tensor that comprise the Hamiltonian operator must be proportional to T_{00} [11]. A special case of eqn 3.66 arises when the two tensors of the same rank $(k = k_1 = k_2)$ are combined to produce a tensor of zero rank and order T_{00}. In this case the Clebsch–Gordon coefficient is just $(-1)^{(k-q)}(2k+1)^{-1/2}$ and

$$T_{00} = (-1)^k (2k+1)^{-1/2} \sum_k (-1)^q T_{kq} T_{k-q} \qquad (3.77)$$

Table 3.9 Spherical harmonics to second rank

Y_{KQ}	
Y_{00}	$\sqrt{\frac{1}{4\pi}}$
Y_{10}	$\sqrt{\frac{3}{4\pi}}\cos\theta$
$Y_{1\pm1}$	$\mp\sqrt{\frac{3}{8\pi}}\sin\theta\exp\left(\pm i\phi\right)$
Y_{20}	$\sqrt{\frac{5}{16\pi}}\left(3\cos^2\theta-1\right)$
$Y_{2\pm1}$	$\mp\sqrt{\frac{15}{8\pi}}\sin\theta\cos\theta\exp\left(\pm i\phi\right)$
$Y_{2\pm2}$	$\mp\sqrt{\frac{15}{32\pi}}\sin^2\theta\exp\left(\pm i2\phi\right)$

Table 3.10 Spin Hamiltonian terms in spherical tensor form, along with secular parts. The quadrupole interaction is taken to be symmetric ($\eta = 0$).

name	spherical tensor form
Zeeman (secular)	$-\gamma B_0 T_{10}$
Scalar coupling (full)	$-2\pi J(3\pi)^{1/2}T_{00}(11)Y_{00}$
Scalar coupling (secular)	$2\pi J\frac{1}{3}\left[T_{20}(11)-\frac{1}{\sqrt{2}}T_{00}(11)\right]$
Quadrupole (full)	$\frac{3eV_{zz}Q}{4I(2I-1)\hbar}(24\pi/5)^{1/2}\sum_q(-1)^q T_{2q}Y_{2-q}(\theta,\phi)$
Quadrupole (secular)	$\frac{3eV_{zz}Q}{4I(2I-1)\hbar}(24\pi/5)^{1/2}T_{20}Y_{20}(\theta,\phi)$
Dipolar (full)	$\frac{\mu_0\gamma^2\hbar}{4\pi r_{ij}^3}(24\pi/5)^{1/2}\sum_q(-1)^q T_{2q}(11)Y_{2-q}(\theta,\phi)$
Dipolar (secular)	$\frac{\mu_0\gamma^2\hbar}{4\pi r_{ij}^3}(24\pi/5)^{1/2}T_{20}(11)Y_{20}(\theta,\phi)$

Since $(-1)^k(2k+1)^{-1/2}$ is a constant for a given spin or spatial tensor, we may write

$$\mathcal{H} = const\sum_q(-1)^q T_{kq}(\mathbf{I})Y_{k-q}(\theta,\phi) \qquad (3.78)$$

where the spatial tensor is represented by a spherical harmonic and $T_{kq}(\mathbf{I})$ is a spin tensor for single (T_{kq}) or coupled ($T_{kq}(11)$) spins, as given respectively in Tables 3.6 and 3.8. All the terms in eqn 3.76 are scalars of this form. For example, the Zeeman Hamiltonian is proportional to $\sum_q(-1)^q T_{1q}Y_{1-q}(\theta,\phi)$, where the angles refer to the orientation of the magnetic field with respect to the representation z-axis.

Because the Zeeman Hamiltonian generally dominates in magnetic resonance, we define, by convention, the representation z-axis as the field direction, making $\theta = 0$ and hence retaining only the $T_{10}Y_{10}(0,\phi) = T_{10}$ As explained in Section 3.4, in calculating the evolution of the spin system we retain only the secular parts of the remaining spin Hamiltonian terms.

3.4.3   Precession diagrams for $I = 1/2$ and $I = 1$

What is the starting point for the density matrix of nuclear spins? Most often we will find that it is the state of thermal equilibrium in a magnetic field. In the next section we look at the thermal equilibrium state in some detail. Having considered that, we will then be in a position to describe the magnetic resonance phenomenon using the following scheme. First we must define the spin Hamiltonian that represents all possible energies of interaction, the dominant term in most cases being the Zeeman interaction due to the laboratory magnetic field used to polarise the spin system, a polarisation direction we label as the z-axis. We then represent the state of the spin system by the density matrix, generally taking as our starting point the thermal equilibrium condition. We will find that equilibrium is often describable by $\rho(0) \sim I_z = T_{10}$. Next we allow the spin system to evolve in the presence of the spin Hamiltonian, using the Liouville equation to predict the outcome. We will see that the starting point for that evolution will be the generation of a non-equilibrium state of the spin system by means of a resonant radiofrequency pulse. Then, at the detection stage of the experiment, we directly measure the precession of macroscopic spin magnetisation. We will see that this corresponds to a measurement of I_x and I_y or $T_{1\pm1}$ components of the density matrix.

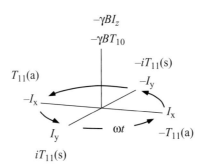

Fig. 3.10 Rotating frame precession diagrams for states of the density matrix in the presence of a Zeeman interaction. The density matrix states and the Hamiltonian are labelled in both Cartesian and spherical tensor form. The precession frequency is $\omega = \gamma B$.

To follow the behaviour of the spin system from thermal equilibrium to ultimate detection, we need only monitor the progress of the density matrix under successive evolution. Understanding this process is often aided by simple precession diagrams, as shown in Fig. 3.10 in which the evolution of I_x and I_y under an I_z Zeeman Hamiltonians is shown. Note that we may cyclically commute this diagram to obtain the evolution of I_y, and I_z under an I_x Hamiltonian and so on. Of course a purely Zeeman-like evolution is all that is possible for $I = 1/2$, but Fig. 3.10 applies equally to higher spins where the Hamiltonian is pure Zeeman. Note that we may also represent the terms in Fig. 3.10 using spherical tensor operators, where $I_z = T_{10}$, $I_x = -T_{11}(a)$, and $I_y = iT_{11}(s)$, the symmetric and antisymmetric combinations being defined by

$$T_{11}(s) = 2^{-1/2} \left(T_{11} + T_{1-1}\right)$$
$$T_{11}(a) = 2^{-1/2} \left(T_{11} - T_{1-1}\right)$$

(3.79)

For higher spin systems, $I \geq 1$, rank 2, and higher tensors may play a role in addition to the rank 1 tensors of Fig. 3.10. In Fig. 3.10 some relevant precessions for $I = 1$ are shown [17]. In each case the vertical axis represents the tensorial nature of the Hamiltonian term, while the two transverse axes represent two density matrix tensor states that interchange in a precessional motion as a result of the commutator relationships returning a simple pair of coupled equations. Note that these refer to the behaviour of independent nuclei experiencing magnetic or quadrupole interactions. Later we will look at the appropriate description of the density matrix for spin systems where special Hamiltonian terms result in a coupling of spins.

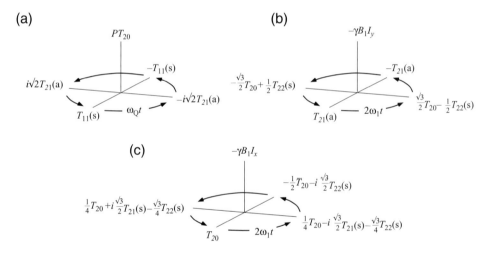

Fig. 3.11 Precession of density matrix states in the spherical tensor formalism applicable for $I = 1$. (s) and (a) refer to symmetric and antisymmetric combinations of tensor components. (a) shows the evolution of I_y ($T_{11}(s)$) under the quadrupole interaction, PT_{20}. The precession frequency is $\sqrt{3/2}P$. (b) shows the evolution of $T_{21}(a)$ into the quadrupole polarisation state, T_{20} under I_y. Note that the Larmor precession frequency is doubled for this second rank tensor. (c) shows the evolution of T_{20} under I_x. Note that this is no longer a simple precession between two states with components given by $\cos(2\omega_1 t)$ and $\sin(2\omega_1 t)$, respectively, but a transformation between T_{20}, T_{21}, and T_{22} states according to the irreducible representation of the rotation group $D^{(2)}$.

The question arises as to whether in practice spherical tensors or Cartesian tensors made of products of Cartesian spin operators, I_α, are easiest to manage in the description of spin evolution. Of course, the spherical tensors do provide a clean definition of higher-order coherences. For example, note the identity of T_{22} with double-quantum coherence in Table 3.6 by comparison with Table 3.3, where superpositions of product operators are needed. But in considering density matrix evolution, the advantage is not so obvious. When considering the effect of higher order spin tensor terms in the Hamiltonian, the scalar coupling, the quadrupole interaction, and the dipolar interaction, the spherical tensor formalism allows us to use the power of irreducible tensor representations to calculate relevant commutators and precessions. But the evolutions of higher-order spherical tensors under both RF pulses and other spin Hamiltonian terms are often complex, as seen in Fig. 3.11. And in considering the effect of the RF field, predicting the behaviour of tensors made from products of Cartesian spin operators is undoubtedly much easier, since the RF field ideally engenders a simple rotation of the I_α.

3.4.4   Spherical tensor precession for coupled spin-$\frac{1}{2}$

To write the two-spin magnetisation in spherical tensor form we note

$$
\begin{aligned}
I_{ix} + I_{jx} &= -2^{-1/2}\left[T_{11}(10) - T_{1-1}(10)\right] - 2^{-1/2}\left[T_{11}(01) - T_{1-1}(01)\right] \\
&= -T_{11}(10, a) - T_{11}(01, a)
\end{aligned}
\tag{3.80}
$$

and

$$
\begin{aligned}
I_{iy} + I_{jy} &= -2^{-1/2}i\left[T_{11}(10) + T_{1-1}(10)\right] - 2^{-1/2}\left[T_{11}(01) + T_{1-1}(01)\right] \\
&= iT_{11}(10, s) + iT_{11}(01, s)
\end{aligned}
\tag{3.81}
$$

where the $T_{11}(01, a)$ and $T_{11}(01, s)$ are antisymmetric and symmetric combinations as defined in eqn 3.79. Suppose the spin Hamiltonian is that of an AB spin system, namely

$$
\begin{aligned}
H &= -\Delta\omega\left[I_{iz} + I_{jz}\right] - \tfrac{1}{2}\delta\left[I_{iz} - I_{jz}\right] + 2\pi J_{ij}\mathbf{I}_i \cdot \mathbf{I}_j \\
&= -\Delta\omega\left[T_{10}(10) + T_{10}(01)\right] - \tfrac{1}{2}\delta\left[(T_{10}(10) - T_{10}(01)\right] - 2\pi J_{ij}\tfrac{2}{\sqrt{3}}T_{00}(11)
\end{aligned}
\tag{3.82}
$$

where $\Delta\omega$ is the average chemical shift in the rotating frame and δ is the chemical shift difference between the i and j spin.

Figure 3.12 shows a rare example of a particularly simple precession of a two-spin tensor term, the double-quantum coherence $T_{22}(11)$ in the presence of the scalar spin–spin interaction and with a chemical shift frequency difference between the i and j spins [19]. Only the mean Larmor frequency plays a role and we find that the precession rate is twice the Larmor frequency. This same frequency doubling behaviour is found for single-spin double-quantum coherence formed as a result of the quadrupole interaction.

3.5   The thermal equilibrium density matrix

3.5.1   The Boltzmann form of the density matrix

What is the meaning of thermal equilibrium? Consider two systems each containing an ensemble of quantum particles, each being in thermal equilibrium with the same 'heat

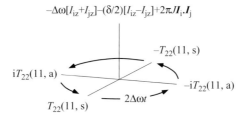

Fig. 3.12 Precession of double-quantum coherence represented by symmetric and antisymmetric combinations of $T_{22}(11)$ under the influence of the general two spin scalar coupling and chemical shift Hamiltonian of eqn 3.82. Only the average frequency appears and the double-quantum coherence precesses at twice the Larmor frequency.

bath' or 'thermal reservoir' at temperature T. If one of the systems is removed from thermal contact with the reservoir, but kept in thermal isolation, then its physical properties will remain unaltered with time. That means that its density matrix does not change, and given our rule for density matrix evolution, the Liouville eqn 3.53, we have [27]

$$[\mathcal{H}, \rho] = 0 \tag{3.83}$$

where $\mathcal{H}$ is the Hamiltonian for this particular system. Hence the thermal equilibrium density matrix must be a simple function of the Hamiltonian, for example a power series.

Now suppose we were to decouple both the two original systems from the reservoir such that they remain coupled together but otherwise isolated. Then, since each was in thermal equilibrium at the same temperature T, each will remain with its physical properties unaltered with time, and the combined system density matrix will also commute with the combined system Hamiltonian. Of course, that Hamiltonian is the sum of each, while the combined system density matrix is an outer product of each individual ρ. Hence we may deduce that the density matrix must be of the form

$$\rho_{eq} = \alpha \exp\left(-\beta \mathcal{H}\right) \tag{3.84}$$

where α is a normalisation constant and β measures the property that each system shares with the reservoir, namely, the temperature T. Following the same arguments outlined in Chapter 1, we may show that β is the familiar Boltzmann factor $1/k_B T$. In thermal equilibrium, the density matrix ρ for an ensemble of nuclear spins obeying Boltzmann statistics may be written

$$\rho_{eq} = \frac{\exp\left(-\mathcal{H}/k_B T\right)}{\mathrm{Tr}\left(\exp\left(-\mathcal{H}/k_B T\right)\right)} \tag{3.85}$$

3.5.2 Nuclear spins in thermal equilibrium—the high temperature approximation

Without writing down a complete Hamiltonian at this point, we will list some of the terms and their relative strengths expressed in frequency units. First and foremost is the Zeeman interaction of the nuclear spins with the laboratory polarizing magnetic field, typically hundreds of megahertz for protons in superconducting magnets. Next is the interaction with laboratory radiofrequency field, at around $100\,\text{kHz}$. Smaller still are the internuclear dipole interactions and electric quadrupole interactions, of a few tens of kilohertz, the chemical shifts caused by electron shielding at parts per million of the magnetic field, and finally the electron mediated spin–spin couplings, typically a few tens of hertz. For other stable nuclei, all with smaller gyromagnetic ratios than the proton, the Zeeman term is weaker but equally so are the remaining interactions.[7] Even for the dominant Zeeman interaction, the ratio $\mathcal{H}/k_B T$ in eqn 3.85 is around 10^{-5} for protons at room temperature. Sub-milliKelvin temperatures are required to Zeeman-polarise nuclear spins in laboratory magnetic field. To a very good approximation therefore, and excepting some very rare cases,[8] eqn 3.85 may be expanded in a power series in which only the leading terms, linear in $\mathcal{H}/k_B T$, are retained, so that

$$\rho_{eq} \approx \frac{\underline{\underline{1}} - \mathcal{H}/k_B T}{\text{Tr}\left(\underline{\underline{1}} - \mathcal{H}/k_B T\right)} \tag{3.86}$$

Allowing that the Zeeman Hamiltonian, $\mathcal{H}_{\text{Zeeman}} = -\gamma B_0 I_z$, dominates and noting $\text{Tr}\left(I_z\right) = 0$, the high temperature approximation (HTA) results,

$$\rho_{eq} \approx \frac{1}{2I+1}\left(\underline{\underline{1}} + \frac{\gamma \hbar B_0 I_z}{k_B T}\right) \tag{3.87}$$

the factor $2I + 1$ being the trace of $\underline{\underline{1}}$.

Three immediate consequences arise from eqn 3.87. First, we can now calculate the thermal equilibrium angular momentum and the thermal equilibrium magnetisation for N spins per unit volume immersed in a magnetic field, B_0, as

$$\begin{aligned}\langle I_z \rangle_{eq} &= \text{Tr}\left(I_z \rho_0\right) \\ &= \frac{1}{2I+1}\frac{\text{Tr}\left(\gamma B_0 \hbar I_z^2\right)}{k_B T} \\ &= \frac{\gamma B_0 \hbar I(I+1)}{3k_B T}\end{aligned} \tag{3.88}$$

and

$$\begin{aligned}M_{eq} &= N\gamma\hbar\,\text{Tr}\left(I_z \rho_{eq}\right) \\ &= \frac{N\gamma^2 B_0 \hbar^2 I(I+1)}{3k_B T}\end{aligned} \tag{3.89}$$

where we have used the identity

$$\text{Tr}\left(I_x^2\right) = \text{Tr}\left(I_y^2\right) = \text{Tr}\left(I_z^2\right) = \frac{1}{3}(2I+1)I(I+1) \tag{3.90}$$

Second, we can see that evolution under the dominant Zeeman Hamiltonian leaves ρ_{eq} unchanged, since the evolution operator arising from a static magnetic field,

---

[7]There are exceptions but these will not concern us here. For example, $I > 1/2$ nuclei in the solid state may have electric quadrupole interactions of tens of megahertz.

[8]An example being the case of long-range intermolecular dipolar interactions for which higher-order terms in the density matrix play a role, as discussed in Chapter 11.

$U(t) = \exp(i\gamma B_0 I_z t)$ clearly commutes with ρ_0. This means that we will need to disturb the spins from equilibrium if we are to observe their precession. Finally, we note that since $\underline{1}$ commutes with any spin Hamiltonian whatever, the only part of the thermal equilibrium state of the density matrix in the HTA that is subject to experimental manipulation is the part proportional to I_z. For that reason, and to within an arbitrary constant, we may generally write our starting condition as

$$\rho_{eq} \sim I_z \tag{3.91}$$

or more precisely

$$\rho_{eq} \approx \frac{1}{2I+1} \frac{\gamma B_0 \hbar}{k_B T} I_z \tag{3.92}$$

3.5.3 Higher terms in the expansion—breakdown of the high temperature approximation

Note that we have used a single-spin picture when calculating the equilibrium magnetisation. In other words, we have described ρ in terms of single-spin operators. A naive extension in the case of many spins would be to write,

$$\rho_{eq} \sim \sum_i I_{iz} \tag{3.93}$$

In most NMR experiments such a starting density matrix serves us well. However, on closer inspection a more precise way of writing an N-spin density matrix is

$$\begin{aligned}\rho_{eq} &= \rho_{1eq} \otimes \rho_{2eq} \otimes \rho_{3eq} \cdots \otimes \rho_{Neq} \\ &= (2I+1)^{-N} \Pi_i \left(\underline{1} + \beta\gamma B_0 \hbar I_{iz}\right)\end{aligned} \tag{3.94}$$

Equation 3.94 involves N one-spin operators of the form $\beta\gamma B_0 \hbar I_{iz}$, $N^2/2$ two-spin operators $(\beta\gamma B_0 \hbar)^2 I_{iz} I_{jz}$, and higher-order spin operators in succession. W.S. Warren [28, 29] has pointed out that despite $\beta\gamma B_0 \hbar$ being typically small ($\sim 10^{-5}$ for protons), the large value of N ensures that the higher-order terms in ρ_{eq} do not vanish whether they become observable or not.

The magnetic resonance experiment starts by rotating all I_{iz} operators in the density matrix to I_{ix} or I_{iy}, so that $(\beta\gamma B_0 \hbar)^2 I_{iz} I_{jz}$ converts to $(\beta\gamma B_0 \hbar)^2 I_{ix} I_{jx}$, for example. As we shall see in the next chapter, such bilinear spin operator terms in the density matrix are not directly observable in NMR experiments, the observable in NMR being $\sum_i (I_{ix} + iI_{iy})$. For these higher-order terms to become observable, there needs to exist in the spin Hamiltonian, terms that can subsequently convert the bilinear $(\beta\gamma B_0 \hbar)^2 I_{ix} I_{jx}$ density matrix term back to simple magnetisation. Such terms must themselves be bilinear in the spin operators and these need to exist for all $N^2/2$ pairs. One candidate is the sum of long range dipolar couplings between distant spins. We will examine the effect of these intermolecular dipolar interactions in more detail in Chapter 11.

3.5.4 A closer look at thermal equilibrium

Our requirements for the density matrix in thermal equilibrium were simply that it be diagonal in the energy representation, and that its diagonal elements represent the thermal equilibrium Boltzmann populations. For thermal equilibrium in a static magnetic field, the energy eigenstates are labelled by the azimuthal quantum number m, basis state amplitudes by $a_m = |a_m| \exp(i\delta_m)$, diagonal density matrix elements by $\overline{|a_m|^2}$, and off-diagonal by $\overline{a_m^* a_{m'}}$. Thermal equilibrium therefore requires a random phase distribution over the δ_m [25, 30].

Of course, any measurement of energy while in thermal equilibrium will, via the state reduction process, force all spins into their energy eigenstates, such that, for each spin, $|a_m|$ is non-zero for only one value of m. Our density matrix is unchanged, but the ensemble now occupies a specific subset of equilibrium possibilities. This new state has more than a lack of phase coherence between a_m and $a_{m'}$. Quite simply, no superposition states exist once the Zeeman thermal equilibrium energy is measured.

Is there any way we could distinguish between these two versions of thermal equilibrium? Certainly there is no ensemble measurement which could separate them, since both have identical density matrices. Only by selecting individual spins from the ensemble and independently measuring their properties could we make such a distinction. So what is the correct description of thermal equilibrium? To understand this we need to appreciate the role of the additional fluctuating interactions that permit the thermal equilibrium state to be obtained by exchange of energy between the spins and the environment. These interactions provide the spin–lattice relaxation process and to be effective they must induce transitions between the energy eigenstates $|m\rangle$ of the Zeeman spin system. That requires both that the interactions have non-zero matrix elements between the energy eigenstates and that they contain fluctuation frequencies matching the energy differences. But if they perform this task, then they can also generate superposition states from eigenstates. The point is that they do so with random phase, keeping any off-diagonal elements of the density matrix zero. Hence, even if we were to start with an ensemble in which every spin was in an energy eigenstate such that the requirement for the thermal equilibrium density matrix was fulfilled, the fluctuating interactions would very soon induce transitions in which randomly phased superposition states existed, but in which the diagonal elements of the density matrix were preserved. The true thermal equilibrium state is this most general manifestation.

References

[1] J. Campbell. *Rutherford, Scientist Supreme*. AAS, Christchurch, 1999.

[2] A. Einstein and W. J. de Haas. Experimenteller Nachweis der Ampereschen Molekularstörme. (English: Experimental proof of Ampére's molecular currents). *Deutsche Physikalische Gesellschaft*, 17:152, 1915.

[3] W. Pauli. Zur Frage der theoretischen Deutung der Satelliten einiger Spektrallinien und ihrer Beeinflussung durch magnetische Felder (English: The question of the theoretical meaning of the satellite of some spectralline and their impact on the magnetic fields). *Naturwissenschaften*, 12:741, 1924.

[4] W. Gerlach and O. Stern. Der experimentelle Nachweis der Richtungsquantelung im Magnetfeld (English: The experimental evidence of direction quantistion in the magnetic field). *Zeitschrift fur Physik*, 9:349, 1922.

[5] Simon Singh. *The Big Bang: The Origin of the Universe*. Fourth Estate, New York, 2005.

[6] Niels Bohr. *The Philosophical Writings of Niels Bohr.* Ox Bow Press, Woodbridge CT, 1987.

[7] R. P. Feynman. *The Character of Physical Law.* Modern Library, New York, 1994.

[8] R. P. Feynman, R. Leighton, and M. Sands. *The Feynman Lectures on Physics: Volume Three.* Addison Wesley, Boston, 1963.

[9] A. R. Edmonds. *Angular Momentum in Quantum Mechanics.* Princeton University Press, Princeton, NJ, 1957.

[10] M. E. Rose. *Elementary Theory of Angular Momentum.* Wiley, New York, 1957.

[11] M. Tinkham. *Group Theory and Quantum Mechanics.* McGraw Hill, New York, 1964.

[12] E. U. Condon and G. H. Shortley. *The Theory of Atomic Spectra.* Cambridge Uiversity Press, New York, 1951.

[13] P. A. M. Dirac. *The Principles of Quantum Mechanics, Third Edition.* Oxford University Press, New York, 1974.

[14] R. R. Ernst, G. Bodenhausen, and A. Wokaun. *Principles of Nuclear Magnetic Resonance in One and Two Dimensions.* Oxford University Press, Oxford, 1987.

[15] O. W. Sørensen, G. W. Eich, M. H. Levitt, G. Bodenhausen, and R. R. Ernst. Product operator formalism for the description of NMR pulse experiments. *Progress in NMR Spectroscopy*, 16:163, 1983.

[16] H. A. Buckmaster, R. Chatterjee, and Y. H. Shing. Application of tensor operators in analysis of EPR and ENDOR spectra. *Physica Status Solidi*, 13:9, 1972.

[17] G. J. Bowden and W. D. Hutchison. Tensor operator formalism for multiple quantum NMR. I. Spin-1 nuclei. *Journal of Magnetic Resonance*, 67:404, 1986.

[18] A. Abragam and M. Goldman. *Nuclear Magnetism: Order and Disorder.* Oxford University Press, New York, 1982.

[19] G. J. Bowden, J. P. D. Martin, and F. Separovic. Tensorial sets for coupled pairs of spin-1/2 nuclei. *Molecular Physics*, 70:581, 1990.

[20] B. C. Sanctuary. Multipole NMR 10. Multispin, multiquantum, multinuclear operator bases. *Journal of Magnetic Resonance*, 61:116, 1985.

[21] B. C. Sanctuary. Multipole NMR 11. Scalar spin coupling. *Molecular Physics*, 55:1017, 1985.

[22] G. Bodenhausen, H. Kogler, and R. R. Ernst. Selection of coherence-transfer pathways in NMR pulse experiments. *Journal of Magnetic Resonance*, 58:370, 1984.

[23] A. Abragam. *Principles of Nuclear Magnetism.* Oxford University Press, Oxford, 1961.

[24] M. H. Levitt. *Spin Dynamics: Basics of Magnetic Resonance.* Wiley, New York, 2002.

[25] C. P. Slichter. *Principles of Magnetic Resonance.* Harper and Row, New York, 1963.

[26] M. Mehring. *High resolution NMR in solids.* Springer, Berlin, 1982.

[27] J. M. Ziman. *Elements of Advanced Quantum Theory.* Cambridge University Press, Cambridge, 1975.

[28] Q. H. He, W. Richter, S. Vathyam, and W. S. Warren. Intermolecular multiple-quantum coherences and cross-correlations in solution NMR. *Journal of Chemical Physics*, 98:6779, 1993.

[29] W. S. Warren, W. Richter, A.H. Andreotti, and B.T. Farmer. Generation of impossible cross peaks between bulk water and biomolecules in solution NMR. *Science*, 262:2005, 1993.

[30] B. Cowan. *Nuclear Magnetic Resonance and Relaxation*. Cambridge University Press, Cambridge, 1997.

4
Introductory magnetic resonance

Both electrons and atomic nuclei have quantised angular momentum, with an associated magnetic dipole moment. Just as a top precesses when its angular momentum experiences the reorienting torque due to gravity, so electrons and nuclei would be expected to precess when placed in a magnetic field, and to precess at a frequency characteristic of field strength and magnetic dipole moment. In the laboratory magnetic fields available in the 1930s, that frequency was expected to lie in the radiowave part of the electromagnetic spectrum, with the electron frequency being typically three orders of magnitude higher than for most nuclei. How might such precession be observed directly? The idea of Dutch physicist C. J. Gorter was to disturb precessing quantum magnets by resonant perturbation using electromagnetic radiation oscillating at the nuclear precession frequency. His attempt to see this effect for atomic nuclei via heating effects in the surrounding sample were unsuccessful [1, 2]. In 1938, after a visit from Gorter [3], Isador Rabi demonstrated the first evidence for nuclear magnetic resonance through the deflection of atoms in a molecular beam apparatus [4, 5]. The discovery of the magnetic resonance phenomenon in condensed matter, as an absorption of electromagnetic energy from a surrounding radiofrequency antenna, was achieved for electrons by Evgeny Zavoisky in 1944 [6], and for hydrogen nuclei, independently by Felix Bloch [7] and Edward Purcell [8] in 1945.

Of these two types of magnetic resonance, nuclear magnetic resonance (NMR) is the more ubiquitous in application. There are three reasons why 'electron magnetic resonance', more commonly known as electron spin resonance (ESR) or electron paramagnetic resonance (EPR) is a less powerful spectroscopic tool than NMR. The first is the need for an unpaired electron, a feature of free radicals or inorganic complexes possessing a transition metal ion. The second is the very short relaxation time of the resonantly-excited electron spin state, the lifetime before return to thermal equilibrium. For the electron in free radicals or transition metals, this is on the order of or less than a few microseconds, resulting in very broad spectral lines. By contrast, nuclear relaxation time may be as long as seconds, permitting sub-hertz spectral resolution and permitting complex manipulations of spins in the time domain of their quantum evolution. And finally, there are the comparative natures of the interactions of the electron and nuclear spins with their atomic and molecular environments. Each has interactions that provide insight regarding structure and dynamics, but of the two, the nuclear spin interaction, the so-called nuclear spin Hamiltonian, is arguably richer, and made the more powerful by the very high spectral resolution afforded by long nuclear relaxation times.

For unpaired electrons the microsecond or sub-microsecond relaxation times leave little opportunity for translational motion measurement. While some such measurements using electrons have been reported, and will be described in later chapters, the main focus of the book is the NMR phenomenon, where long 'relaxation times' allow for sophisticated measurements of a wide range of translational dynamics.

4.1 Introductory remarks

4.1.1 The NMR orchestra

The scope of magnetic resonance is daunting, partly because the interactions of atomic nuclei with their atomic and molecular surroundings are subtle and diverse, and partly because the ways in which the nuclei can be made to respond to those terms by virtue of externally applied electromagnetic fields is almost boundless. The designer of magnetic resonance pulse sequences has at his or her disposal the depth of complexity possible when a composer writes a score for an orchestra. Any book about magnetic resonance faces an obvious problem. Do we attempt to describe the principles governing our 'musical instruments', the nuclei and their interactions, or do we write of the 'orchestral compositions', the encyclopedic collection of possible measurements? And if we seek to do both we must confront our limitations. While we need to at least describe the musical instruments, we cannot write of all possible music.

This book addresses a very limited, though interesting, repertoire, the use of magnetic field gradients to measure translational motion. Rather than pretending to be comprehensive, Chapter 4 merely outlines some of those elements of NMR essential to understanding that repertoire, in particular the information able to be encoded in the phases of the spins and how transmission and reception of resonant radiowaves gives access to that information. To cover the principles of NMR with any sense of completeness requires a complete monograph, rather than a single chapter. For this purpose the reader can find a number of excellent texts [9–18]. Our criterion is simple: to lay out the principles needed to understand the task at hand, the use of NMR to track the migrations of parent molecules by the radiofrequency emissions of the nuclei contained within.

4.1.2 Coherence and the spin echo

At the heart of the remarkable power of magnetic resonance lies the idea of coherence [15]. Coherence in classical physics tell us the degree to which waves are correlated in either space or time or in sense of polarisation, waves being characterised by their 'phase', the correspondence with the angular coordinates of a cyclic phenomenon. In quantum mechanics phase is even more significant, all physical properties resulting in an inherent phase that advances with increasing time. That is the principle underlying the Schrödinger equation. When we deal with nuclear spins whose de Broglie waves are very much shorter than any internuclear spacing, the idea of particle-wave interference is unimportant. But there is another sense it which phase and coherence are manifest. The nuclear spin states have an associated quantum phase, which evolves according to the interaction of the nucleus with surrounding fields. And when we are dealing with countless numbers of nuclei in the vast ensembles of atoms and molecules in macroscopic matter, then the relative phases of the nuclei are just as important to the superposed quantum components contributing to our measurements as are the relative phases of classical waves to the superposition that results when those waves collide. Well-ordered phase relationships across an ensemble of nuclear spin states are known as coherences. Nuclear spin coherences can be long-lived, on the order of many seconds. And using radiofrequency pulses and bursts of applied magnetic field, we may manipulate coherences almost at will, even to the extent that we may move from a process of decoherence, in which nuclei apparently lose their phase registration, to recoherence, where phases are brought back in step once more. That is the principle of the spin echo.

Of all the developments that have assisted NMR measurements of translational dynamics, none is as important as the discovery of the spin echo by Erwin Hahn in 1950 [19]. Indeed, it can be argued that the power of the spin echo underpins all of modern NMR. The formation of the echo, with its inherent time-reversal properties, not only causes lost signals to re-appear, it also has the effect of removing some nuclear spin interactions while retaining others. Hahn pointed out that in the case where the magnetic field was inhomogeneous, the signal obtained by method was sensitive to translations of spin-bearing molecules, providing, for example, a means of measuring random Brownian motion. We will examine these ideas in detail in the following discussion.

In Chapter 3 we laid out the quantum mechanical basis for a description of spins, including an introduction to the way in which density matrices may be used to handle statistical ensembles of nuclei. We introduced the various interactions that nuclei spins experience due to fields, noting the predominant role of the Zeeman interaction, the response of a spin to a surrounding magnetic field. The scope of Chapter 4 is as follows. First we return to the Zeeman interaction, describing the mechanics of the resonance process whereby spin states may be manipulated using radiofrequency fields. Next we detail the signal detection process and the role of Fourier transformations in magnetic resonance spectroscopy. We then return to the details of the nuclear spin Hamiltonian and its implications for the NMR spectrum, as well as the way in which fluctuations of those interactions induce spin relaxation, the re-establishment of the thermal equilibrium ensemble state via T_1 relaxation, and the irreversible loss of phase coherence known as T_2 relaxation. Finally the chapter outlines some of the fundamental radiofrequency pulse sequence components and corresponding spin ensemble responses, which we will meet throughout the book.

4.2   Resonant excitation

So far, amongst the possible terms of the nuclear spin Hamiltonian, we have met just the static Zeeman interaction. Next we will consider the effect of an oscillating magnetic field that may be turned on and off at will, the so-called resonant radiofrequency field. This, we will find, is the device by which we can disturb ρ_0 from equilibrium and hence detect the evolution of the spin ensemble in the various terms of the remaining spin Hamiltonian.

4.2.1 The rotating frame transformation

One of the most powerful tools in the description of magnetic resonance concerns the use of a frame of reference that rotates around the static field B_0. Here we examine some consequences of rotation.

Reference frame transformations for an anticlockwise rotation by ϕ about the z-axis in the case of a ket, a density matrix, or some observable operator A correspond to the passive view and are therefore respectively written $\exp(i\phi I_z)|\Phi\rangle$, $\exp(i\phi I_z)$ $\rho \exp(-i\phi I_z)$, and $\exp(i\phi I_z)A\exp(-i\phi I_z)$[1].

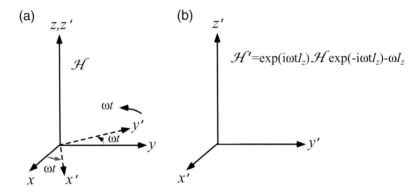

Fig. 4.1 Depiction of primed frame of reference rotating about the unprimed frame z-axis at angular speed ω from the perspective of the (a) unprimed frame and (b) primed frame.

Next, we need to consider the consequences for spin evolution of a time-dependent anticlockwise frame rotation about the z-axis, represented by the operator $\exp(i\omega t I_z)$. This frame rotation is shown in Fig. 4.1. We write the rotated frame ket as $|\Psi'\rangle = \exp(i\omega t I_z)|\Psi\rangle$ and the rotated frame Hamiltonian as $\mathcal{H}_{\text{"rotated"}} = \exp(i\omega t I_z)$ $\mathcal{H}\exp(-i\omega t I_z)$. To understand the evolution we must return to the Schrödinger equation, working in the stationary frame where we can be certain of its applicability. Writing in frequency units (i.e. dropping $\hbar$) we have

$$i\frac{\partial}{\partial t}\exp\left(-i\omega t I_z\right)|\Psi'(t)\rangle = \mathcal{H}\exp\left(-i\omega t I_z\right)|\Psi'(t)\rangle$$

$$= \exp\left(-i\omega t I_z\right)\mathcal{H}_{\text{"rotated"}}\exp\left(i\omega t I_z\right)\exp\left(-i\omega t I_z\right)|\Psi'(t)\rangle \tag{4.1}$$

whence

$$i\frac{\partial}{\partial t}|\Psi'(t)\rangle = \left(\mathcal{H}_{\text{"rotated"}} - \omega I_z\right)|\Psi'(t)\rangle \tag{4.2}$$

Notice that if we are to retain the same form for the Schrödinger equation in the primed frame, then the primed-frame Hamiltonian must be modified by the subtraction of the term ωI_z so that

---

[1]The latter two 'sandwich operations' are easy to understand. Working from right to left, as always in our sequential application or operators, and starting from our rotated frame applicable to the description, $\exp(-i\phi I_z)$ takes us back to the starting (unrotated) frame, the known operator in this starting frame is then applied, following which we return to the rotated frame via $\exp(i\phi I_z)$.

$$\mathcal{H}' = \exp\left(i\omega t I_z\right)\mathcal{H}\exp\left(-i\omega t I_z\right) - \omega I_z \tag{4.3}$$

The idea that additional terms arise in rotating frames of reference is familiar to us in Newtonian mechanics. Retaining Newton's laws in a rotating frame involves introduction of a centrifugal force and a Coriolis force. The additional energy term that arises from requiring Schrödinger's equation to apply in a rotating frame should not surprise us.

Suppose that our Hamiltonian is due to a static magnetic field B_0 along the z-axis so that $\mathcal{H}_{\text{Zeeman}} = -\gamma B_0 I_z$. Then, from eqn 4.3, in the rotating frame

$$\mathcal{H}'_{\text{Zeeman}} = -\gamma\left(B_0 + \omega/\gamma\right) I_z \tag{4.4}$$

The additional term looks exactly like a z-axis field, ω/γ, with sign sensitive to the sign of ω. If the rotation sense is clockwise, that is in the same sense as the precession of the spin phases for positive gyromagnetic ratio γ, then the effective magnetic field long the z-axis is reduced to $B_0 - |\omega|/\gamma$. The reason for the reduction in apparent longitudinal field is easily understood since the apparent precessional rotation of the spin phases about the z-axis is reduced by ω. If we are to preserve the Schrödinger equation then we must interpret the magnetic field as having been reduced accordingly.

4.2.2 The resonant radiofrequency field

Suppose we add to the Zeeman interaction arising from the static field B_0, an interaction due to a field, $2B_1$, oscillating at frequency, ω, along a laboratory-frame transverse axis, which we will take to be x. Then the new Hamiltonian is

$$\mathcal{H}_{\text{lab}} = -\gamma B_0 I_z - 2\gamma B_1 \cos\left(\omega t\right) I_x \tag{4.5}$$

It is helpful to represent the linearly polarised oscillatory field as two counter-rotating circularly polarised components each of amplitude B_1 as shown in Fig. 4.2. Hence we may rewrite eqn 4.5

$$\mathcal{H}_{\text{lab}} = -\gamma B_0 I_z - \gamma B_1 \exp\left(i\omega t I_z\right) I_x \exp\left(-i\omega t I_z\right) - \gamma B_1 \exp\left(-i\omega t I_z\right) I_x \exp\left(i\omega t I_z\right) \tag{4.6}$$

To better understand eqn 4.6 it is helpful to make a coordinate transformation to the frame of reference rotating at frequency ω about the z-axis in the same (clockwise) sense as the spin phases. We find

$$\mathcal{H}_{\text{rot}} = -\gamma\left(B_0 - \omega/\gamma\right) I_z - \gamma B_1 I_x - \gamma B_1 \exp\left(-i2\omega t I_z\right) I_x \exp\left(i2\omega t I_z\right) \tag{4.7}$$

Note that $\mathcal{H}_{\text{rot}}$ should not be confused with $\mathcal{H}_{\text{"rotated"}}$. It is the true effective Hamiltonian in the rotating frame and is the $\mathcal{H}'$ of eqn 4.3.

The final term in eqn 4.7, the so-called counter-rotating component, is fluctuating at 2ω. Suppose we make the B_1 field oscillate at a radiofrequency close to the Larmor precession rate $\omega_0 = \gamma B_0$. If the amplitude of this RF field, B_1, is much smaller than the static field B_0, then $\omega \sim \omega_0 \gg \gamma B_1$ and so the counter-rotating component fluctuates much more rapidly than its size in frequency units. This results in the spins

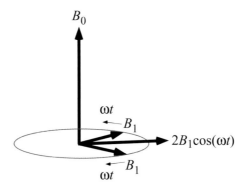

Fig. 4.2 Linearly polarised oscillating magnetic field, $2B_1 \cos(\omega t)$, represented as two counter-rotating circularly polarised fields.

experiencing only the fluctuating field's average value of zero, so that the rotating-frame Hamiltonian simplifies to retain only the co-rotating component and

$$\mathcal{H}_{\text{rot}} = -\gamma \left(B_0 - \omega/\gamma\right) I_z - \gamma B_1 I_x \tag{4.8}$$

Equation 4.8 has a very simple interpretation. At resonance, $\omega = \omega_0$ and the apparent longitudinal field vanishes, thus leaving the effective RF field along the rotating frame x-axis. Earlier, it was shown that the evolution arising from a static magnetic field, is equivalent to a rotation of spin phase about the field axis. As a consequence, the effect of a resonant RF field is to cause the to nutate about the B_1 axis as shown in Fig. 4.3. For example, a pulse of duration t_p, such that $\gamma B_1 t_p = \pi/2$, would reorient the spin magnetisation to the y-axis of the rotating frame. Such a pulse is known as a $90°$ or $\pi/2$ pulse. Equivalently a $180°$ pulse would invert the equilibrium magnetisation from I_z to $-I_z$. This disturbance from equilibrium is the NMR phenomenon.

Of course, the I_x operator, which dominates the rotating-frame Hamiltonian at resonance, is just the linear combination of the raising and lowering operators, $\frac{1}{2}\left(I_+ + I_-\right)$. This tells us that, in the case $I = 1/2$, evolution of the system with time corresponds to an inter-conversion of each spin between $\left|\frac{1}{2}\right\rangle$ and $\left|-\frac{1}{2}\right\rangle$, at a rate $\omega_1 = \gamma B_1$.

We can now interpret the resonant re-orientation process for a starting thermal equilibrium density matrix, which, in the laboratory frame, is $\rho^{\text{lab}}(0) \sim I_z$. We begin by transforming our frame of reference from the laboratory to the rotating frame. That requires a passive view rotation of $-\omega_0 t$ about z. In the frame rotating clockwise at ω_0 about the z-axis

$$\rho^{\text{rot}}(0) = \exp\left(-i\omega_0 t I_z\right) \rho^{\text{lab}}(0) \exp\left(i\omega_0 t I_z\right) \sim I_z \tag{4.9}$$

meaning that the starting density matrix is identical in the laboratory and rotating frames. Hence the rotating-frame description of the nutation process is given by an active view rotation by $-\omega_1 t$ about x, namely

$$\begin{aligned}
\rho^{\text{rot}}(t) &\sim \exp\left(i\omega_1 t I_x\right) I_z \exp\left(-i\omega_1 t I_x\right) \\
&\sim I_z \cos\left(\omega_1 t\right) + I_y \sin\left(\omega_1 t\right)
\end{aligned} \tag{4.10}$$

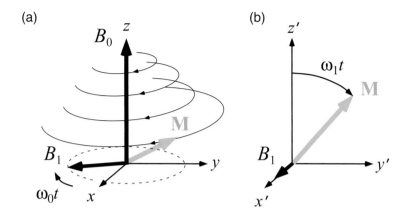

Fig. 4.3 Evolution of nuclear spins in the presence of a resonant RF magnetic field. The ensemble of spins is represented by a magnetisation vector. (a) Evolution in the laboratory frame, in the presence of a longitudinal magnetic field, $\mathbf{B}_0$, and a transverse rotating field, $\mathbf{B}_1$. On resonance the rotation rate ω is identical to $\omega_0 = \gamma B_0$ and the magnetisation vector simultaneously precesses about $\mathbf{B}_0$ at ω_0 and about $\mathbf{B}_1$ at ω_1. (b) as for (a) but in the rotating frame where $\mathbf{B}_1$ is stationary. The effective longitudinal field at resonance is zero and only the precession about $\mathbf{B}_1$ is apparent.

Rewriting the density matrix in the laboratory frame requires that we transform back via $\exp(i\omega_0 t I_z)\rho^{\mathrm{rot}}(0)\exp(-i\omega_0 t I_z)$ to yield

$$\rho^{\mathrm{lab}}(t) \sim I_z \cos(\omega_1 t) + I_y \cos(\omega_0 t)\sin(\omega_1 t) + I_x \sin(\omega_0 t)\sin(\omega_1 t) \tag{4.11}$$

leading to the laboratory-frame observables, $M_\zeta(t) = N\gamma\,\mathrm{Tr}\left(I_\zeta\,\rho^{\mathrm{lab}}(t)\right)$,

$$
\begin{aligned}
M_x &= M_0 \sin(\omega_0 t)\sin(\omega_1 t)\\
M_y &= M_0 \cos(\omega_0 t)\sin(\omega_1 t)\\
M_z &= M_0 \cos(\omega_1 t)
\end{aligned}
\tag{4.12}
$$

where $M_0 = N\gamma\,\mathrm{Tr}\left(I_z\,\rho_0\right)$. This combination of precession about the laboratory-frame z-axis and nutation about the rotating frame x-axis is shown in Fig. 4.3.

Note that these passive view transformations, back and forth between the laboratory and rotating frames, always result in implicit local coordinates. Specifically, the x- and y-directions always correspond to the local frame, whether rotating or laboratory-frame coordinates and in future we will drop the prime label for the rotating frame. We now know how to make the transformation between the laboratory and rotating frames. Clearly our description is much easier in the rotating frame and in most of what follows we will so confine our description.

Suppose that the RF field were not precisely on resonance. In that case the rotating-frame effective magnetic field would have a z-component $(B_0 - \omega/\gamma)$ and be inclined out of the transverse plane. This would lead to an oblique precession as shown in Fig. 4.4. In the extreme case where $\gamma B_1 \ll (B_0 - \omega/\gamma)$, the RF field would be entirely ineffective. This understanding leads to two important consequences. Accurate and simple reorientations of the spin density matrix between the Cartesian axes of the rotating frame requires that γB_1 not only be much larger than the off-resonant field, but also larger than any remaining terms in the spin Hamiltonian. Since the duration of an RF pulse is on the order of the time taken to produce the desired turn angle, via $(\pi/2)(\gamma B_1)^{-1}$, the bandwidth is therefore $\sim \gamma B_1$ and our requirement is equivalent to saying that bandwidth of the RF pulse is larger than all spin Hamiltonian terms other than the Larmor precession in the Zeeman field. This ensures that during application of the RF pulse, the term $\gamma B_1 I_x$ dominates the Hamiltonian and we may neglect the influence of all other terms over the pulse interval t_p. In this case the depiction of resonant re-orientation by a simple active view rotation operator is faithful. Second, when weak RF pulses are used, the on-resonant condition becomes more stringent, and so allows that we might selectively perturb just a part of the NMR spectrum. The subject of selective excitation pulses is covered in Chapter 5.

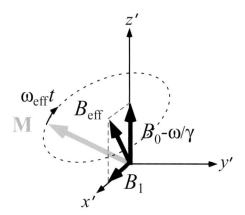

Fig. 4.4 Precession of the nuclear magnetisation vector in the rotating frame when the RF field is off-resonance. The direction of the effective field is determined by the amplitude γB_1 and by the offset frequency $\Delta\omega = \gamma B_0 - \omega$.

It is now easy to visualise the influence of a short burst of resonant RF field, B_1, known in NMR as the RF pulse. If the duration of the pulse is t, the magnetisation will rotate by an angle $\omega_1 t$ about the direction of B_1 in the rotating frame. This is the effect illustrated in Fig. 4.3. At this point in the discussion an obvious question arises. What is the meaning of the direction of B_1 in the laboratory frame where our experiment is performed? Clearly the instantaneous direction will depend on the instant that we start the RF pulse, and that is quite arbitrary. The significance of this direction becomes apparent, however, when a second pulse is applied at some later time. The first pulse establishes an arbitrary reference axis in the rotating frame and, provided that the second pulse oscillates in phase with the first, this axis system will be maintained. Labelling the rotating frame direction of the first pulse as x then the second pulse will also correspond to a field along the x-axis. This immediately suggests that the rotating-frame RF field direction can be oriented at will simply by changing the phase of the pulses. In most NMR experiments these phases are shifted in multiples of 90°. We denote an RF pulse with duration t as $\theta_{\pm x}$ or $\theta_{\pm y}$, where $\theta = \omega_1 t$ (expressed in degrees) and the subscript indicates the phase according to the chosen direction in the rotating frame.

NMR can be performed with an almost infinite array of pulse train possibilities. The capacity of the experimenter to change both the duration and phase of RF pulses, coupled with the existence of a host of environmentally sensitive terms in the nuclear spin interactions, leads to the essential richness of this branch of spectroscopy.

4.2.3 Quantum view of nutation

Having learned to deal with resonant nutation as an active view rotation of the spin system about the direction of the RF pulse in the rotating frame of reference, it is helpful to look in detail at what happens to the eigen kets, ensemble density matrix, and the expectation values for I_x, I_y, and I_z, at various stages during the evolution process.

Single spin-$\frac{1}{2}$

For simplicity, consider the case of a spin-$\frac{1}{2}$ nucleus in the up state along $\mathbf{B}_0$, before the application of an RF pulse ($-\omega_1 t I_x$ in the rotating frame), with nutation angle $-\omega_1 t_p$. Then the rotating-frame description of the pulse evolution requires an active view rotation operator, $U^{\text{nut}}(t_p) = \exp(i\omega_1 t_p I_x)$, as

$$
\begin{aligned}
U^{\text{nut}}(t_p)|\tfrac{1}{2}\rangle &= \exp\left(i\omega_1 t_p I_x\right)\begin{bmatrix}1\\0\end{bmatrix} \\
&= \begin{bmatrix} \cos\left(\frac{1}{2}\omega_1 t_p\right) & i\sin\left(\frac{1}{2}\omega_1 t_p\right) \\ i\sin\left(\frac{1}{2}\omega_1 t_p\right) & \cos\left(\frac{1}{2}\omega_1 t_p\right) \end{bmatrix}\begin{bmatrix}1\\0\end{bmatrix} \\
&= \begin{bmatrix} \cos\left(\frac{1}{2}\omega_1 t_p\right) \\ i\sin\left(\frac{1}{2}\omega_1 t_p\right) \end{bmatrix}
\end{aligned}
\tag{4.13}
$$

Using the angular momentum operators from Table 3.1 it is easy to show the following expectation values after this nutation

$$
\begin{aligned}
\langle I_z \rangle &= \frac{1}{2}\cos\left(\omega_1 t_p\right) \\
\langle I_y \rangle &= \frac{1}{2}\sin\left(\omega_1 t_p\right) \\
\langle I_x \rangle &= 0
\end{aligned}
\tag{4.14}
$$

This looks, to all intents and purposes, like the rotation of the spin vector by an angle $-\omega_1 t_p$ about the x-axis in the rotating frame.

However, it is important to realise that throughout the experiment we retain the description of the spin system in terms of its starting I_z eigenstates. What this means is that no matter how simple the description, in terms of the spin being definitely pointing along an axis inclined to z, in the original frame, the RF pulse has produced a superposition state of 'up' and 'down'. Let us take the example of a 90° RF pulse for which $\omega_1 t_p = \pi/2$. The resulting state 'spin pointing along y in the rotating frame', is in fact the superposition $\frac{1}{\sqrt{2}}|\frac{1}{2}\rangle + \frac{i}{\sqrt{2}}|-\frac{1}{2}\rangle$ in terms of the starting eigenbasis. Figure 4.5 shows how we might depict this.

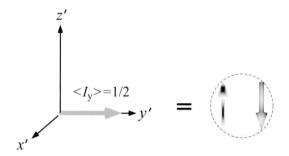

Fig. 4.5 State of spin which starts as $+\frac{1}{2}$ eigenstate of I_z following a 90°_x RF pulse. The expectation value $\langle I_y\rangle = -\frac{1}{2}$ but in terms of the original eigenbasis the final state is a coherent superposition of $+\frac{1}{2}$ and $-\frac{1}{2}$.

And what happens subsequent to the 90° nutation? In the rotating frame, where no Hamiltonian is felt once the RF pulse stops, the spin state remains constant. But in the laboratory frame the Larmor precession continues unabated so that we may write, using the active view operator for a clockwise rotation $-\omega_0 t$ about z, $U^{\text{Larmor}}(t) = \exp(i\omega_0 t I_z)$,

$$U^{\text{Larmor}}(t)U^{\text{nut}}(t_p)|\tfrac{1}{2}\rangle = \begin{bmatrix} \exp\left(i\omega_0 t/2\right) & 0 \\ 0 & \exp\left(-i\omega_0 t/2\right) \end{bmatrix} \begin{bmatrix} \frac{1}{\sqrt{2}} \\ \frac{i}{\sqrt{2}} \end{bmatrix}$$

$$= \frac{1}{\sqrt{2}} \begin{bmatrix} \exp\left(i\omega_0 t/2\right) \\ i\exp\left(-i\omega_0 t/2\right) \end{bmatrix} \tag{4.15}$$

This is a coherent superposition state, in which the magnitudes of the amplitudes remain unchanged while their relative phases rotate at a rate ω_0.

Ensemble spin-$\frac{1}{2}$

To handle the case of the ensemble of spins, with the initial state being thermal equilibrium in the Zeeman field, we turn to the density matrix description, and starting with the state

$$\rho(0_-) = \begin{bmatrix} p_\uparrow & 0 \\ 0 & p_\downarrow \end{bmatrix} \tag{4.16}$$

where $p_\uparrow = \overline{|a_{1/2}|^2}$ and $p_\downarrow = \overline{|a_{-1/2}|^2}$. That only the spin excess can contribute to any measurement is made clear by rewriting ρ as

$$\rho(0_-) = \frac{1}{2}\begin{bmatrix} 1 & 0 \\ 0 & 1 \end{bmatrix} + \frac{1}{2}\begin{bmatrix} (p_\uparrow - p_\downarrow) & 0 \\ 0 & -(p_\uparrow - p_\downarrow) \end{bmatrix} \tag{4.17}$$

Now let us apply an ideal 90° RF pulse about the rotating frame x-axis, again using active view operators. The final density matrix is $\rho(0_+)$ given by

$$\rho(0_+) = U^{\text{nut}}(t_p)\, \rho(0_-)\, U^{\text{nut}}(t_p)^{-1}$$

$$= \begin{bmatrix} 1/\sqrt{2} & i/\sqrt{2} \\ i/\sqrt{2} & 1/\sqrt{2} \end{bmatrix} \begin{bmatrix} p_\uparrow & 0 \\ 0 & p_\downarrow \end{bmatrix} \begin{bmatrix} 1/\sqrt{2} & -i/\sqrt{2} \\ -i/\sqrt{2} & 1/\sqrt{2} \end{bmatrix} \tag{4.18}$$

$$= \frac{1}{2} \begin{bmatrix} 1 & -i(p_\uparrow - p_\downarrow) \\ i(p_\uparrow - p_\downarrow) & 1 \end{bmatrix}$$

noting $p_\uparrow + p_\downarrow = 1$. The off-diagonal elements of the density matrix proportional to the spin excess represent states of 1-quantum coherence. Indeed what we have at time 0_+ is a mixed state ensemble in which a population $p_\uparrow$ is a sub-ensemble of kets $|\psi\rangle = \frac{1}{\sqrt{2}}\left|\frac{1}{2}\right\rangle + \frac{i}{\sqrt{2}}\left|-\frac{1}{2}\right\rangle$ and a population $p_\downarrow$ is a sub-ensemble of kets $\frac{i}{\sqrt{2}}\left|\frac{1}{2}\right\rangle + \frac{1}{\sqrt{2}}\left|-\frac{1}{2}\right\rangle$, corresponding to spin states pointing respectively along the rotating frame y and $-y$ axes.[2]

There is no need to calculate the effect of subsequent Larmor precession in the laboratory frame since by now it should be obvious. The spin vectors lying along y and $-y$ simply rotate at ω_0 in the transverse plane. To check this we need only apply the Larmor evolution sandwich $\rho(t) = U^{\text{Larmor}}(t)\,\rho(0_+)\,U^{\text{Larmor}}(t)^{-1}$. More interesting is the case where not every spin experiences the same magnetic field, for example when the Zeeman magnetic field is inhomogeneous. Staying in the rotating frame, we can represent this by considering the precession that arises from an offset frequency $\Delta\omega_0$, namely

$$\rho(t) = \begin{bmatrix} e^{\frac{i}{2}\Delta\omega_0 t} & 0 \\ 0 & e^{-\frac{i}{2}\Delta\omega_0 t} \end{bmatrix} \frac{1}{2} \begin{bmatrix} 1 & -i(p_\uparrow - p_\downarrow) \\ i(p_\uparrow - p_\downarrow) & 1 \end{bmatrix} \begin{bmatrix} e^{-\frac{i}{2}\Delta\omega_0 t} & 0 \\ 0 & e^{\frac{i}{2}\Delta\omega_0 t} \end{bmatrix}$$

$$= \frac{1}{2} \begin{bmatrix} 1 & -i(p_\uparrow - p_\downarrow)\exp(-i\Delta\omega_0 t) \\ i(p_\uparrow - p_\downarrow)\exp(i\Delta\omega_0 t) & 1 \end{bmatrix} \tag{4.19}$$

A first look at relaxation

Suppose that we have a distribution of offsets $f(\Delta\omega_0)$. The terms $\exp(i\Delta\omega_0 t)$ represent the integral

$$F(t) = \int_{-\infty}^{\infty} f(\Delta\omega_0)\exp(i\Delta\omega_0 t)\, d\Delta\omega_0 \tag{4.20}$$

where $F(t)$ is the Fourier transform of $f(\Delta\omega_0)$. For a symmetric distribution of $\Delta\omega_0$ about the mean Larmor frequency, that is about zero in the rotating frame, $\exp(i\Delta\omega_0 t)$ represents a decay from unity, a progressive loss of phase coherence in the spin ensemble due to the distribution of field offsets and a consequent continuous decay of transverse magnetisation. This describes what is known as the T_2^* relaxation process. With time-reversal, something that is possible using the spin echo, this inhomogeneous broadening decay may be reversed.

However, other processes contribute to a decay that is inherently stochastic in nature and hence irreversible. Look at the density matrix off-diagonal terms of the form $\overline{|a_\uparrow||a_\downarrow|e^{i(\phi_\uparrow - \phi_\downarrow)}}$, arising from quantum superposition states $|\Psi\rangle = |a_\uparrow|e^{i\phi_\uparrow}|\uparrow\rangle + |a_\downarrow|e^{i\phi_\downarrow}|\downarrow\rangle$. These contribute the single-quantum coherence that is manifest as

---

[2]To convert one to the other apply the spin=1/2 rotation matrix $R_z(\pi)$.

transverse magnetisation. Spin–spin relaxation processes involve energy-conserving 'flip-flop' terms in which the phases $\phi_\uparrow$ and $\phi_\downarrow$ fluctuate, thus causing $\overline{|a_\uparrow||a_\downarrow|e^{i(\phi_\uparrow - \phi_\downarrow)}}$ to decay. At the same time, energy-absorbing spin–lattice processes involves changes in the magnitudes $|a_\uparrow|$ and $|a_\downarrow|$ as the spins regain thermal equilibrium, a state in which, for any spin, either $|a_\uparrow|$ or $|a_\downarrow|$ is zero. Both spin–lattice and spin–spin processes cause the decay of the transverse magnetisation. Only spin–lattice processes can contribute to the restoration of thermal equilibrium and the growth of $\langle I_z \rangle$. For this reason the transverse relaxation rate T_2^{-1} always exceeds the longitudinal rate T_1^{-1}.

4.2.4 Classical descriptions of resonant reorientation

Classical precession

Precession and nutation are very familiar phenomena in classical physics, a simple example being provided by the case of a spinning wheel whose axle is suspended at a pivot point by a string, as shown in Fig. 4.6. The gravitational force acting through the centre of mass provides a torque, τ, whose direction is normal to the direction of the wheel's angular momentum, thus resulting in an azimuthal precession about the vertical axis. The magnitude of the torque is $mgd \sin\theta$, where d is the distance from the centre of mass to the pivot and θ is the polar angle to the vertical made by the axle, and hence the angular momentum vector. Of course Newton's second law requires that the rate of change of angular momentum is equal to that torque, as

$$\Delta \mathbf{L} = \tau \Delta t \qquad (4.21)$$

Figure 4.6 shows that the change in azimuthal orientation of the angular momentum in time Δt is $\Delta\phi = \Delta L / L \sin\theta$, and so the precession frequency is

$$\begin{aligned} \omega_0 &= \frac{\Delta\phi}{\Delta t} = \frac{mgd \sin\theta \Delta t}{L \sin\theta \Delta t} \\ &= \frac{mgd}{L} \end{aligned} \qquad (4.22)$$

The spinning wheel example is very similar to the case of a magnetic dipole moment μ with co-linear angular momentum, placed in a vertical magnetic field B_0, as shown in Fig. 4.7. The only difference is that the torque is now $\mu B_0 \sin\theta$. For exact analogy with the wheel geometry, we depict a dipole with negative magnetogyric ratio, i.e. with the angular momentum vector oppositely directed to the dipole moment. Following the same reasoning as for the wheel we have

$$\begin{aligned} \omega_0 &= \frac{\Delta\phi}{\Delta t} = \frac{\mu B_0 \sin\theta \Delta t}{L \sin\theta \Delta t} \\ &= \frac{\mu B_0}{L} \\ &= \gamma B_0 \end{aligned} \qquad (4.23)$$

Note that in both cases, the precession frequency, ω_0, is independent of polar angle θ.

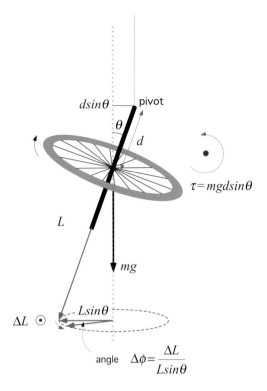

Fig. 4.6 Precession of a spinning mechanical wheel with angular momentum L, while suspended from a pivot on a string. The tension in the string exerts a torque $\tau = mg \sin \theta$ about the wheel centre. This torque, which is normal to the plane of the diagram, causes a rate of change of angular momentum, also normal to the plane, of $\Delta L / \Delta t = \tau$. This results in the angular momentum vector **L** precessing as shown in the lower circle, at a rate independent of θ.

Classical resonant nutation

We have seen that for the precession of a magnetic dipole with angular momentum, the introduction of a transverse magnetic field co-rotating at the Larmor frequency γB_0 causes resonant re-orientation. Is there an analogy for resonant re-orientation for the wheel? Indeed there is, as is shown in Fig. 4.8. Here we introduce another torque, τ_1, normal to both the angular momentum vector and the torque provided by gravity. The mechanism for achieving this is to apply a transverse force, F_1, to the axle which follows the precessional motion. To do so the force must rotate along with the precession, an effect which could result if we were to pull horizontally on a string attached to the end of the axle (at a distance $2d$ from the pivot), at the same time moving our hand in a circular path so that we stayed in step with the circular motion. Now the direction of the torque is such as to cause a change in polar inclination, θ, of the angular momentum L as $\Delta \theta = \Delta L' / L$. The magnitude of the torque is $\tau_1 = F_1 2d$, leading to a nutation frequency

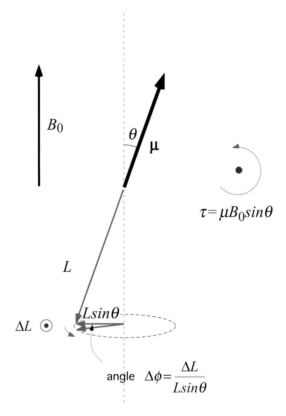

Fig. 4.7 Classical dipole with angular momentum immersed in a magnetic field that exerts a torque $\mu B_0 \sin \theta$, normal to the plane of the diagram, and hence a rate of change of angular momentum, also normal to the plane, of $\Delta L / \Delta t = \tau$. This results in the angular momentum vector $\mathbf{L}$ precessing, as shown in the lower circle, at a rate independent of θ.

$$\begin{aligned} \omega_1 &= \frac{\Delta \theta}{\Delta t} \\ &= \frac{F_1 2d}{L} \end{aligned} \tag{4.24}$$

The nutation process is quite independent of the precession around the polar axis which continues unabated at frequency ω_0. It results in a continuous polar reorientation of the angular momentum vector at a nutation frequency ω_1 which is independent of polar angle.

4.2.5   Semi-classical description

In the case of independent spin-$\frac{1}{2}$ nuclei, the motion of the ensemble of spins may always be described in terms of the precession of the spin magnetisation vector $\mathbf{M}$. In such a model the macroscopic angular momentum vector is simply $\mathbf{M}/\gamma$. By equating the torque to the rate of change of angular momentum we obtain

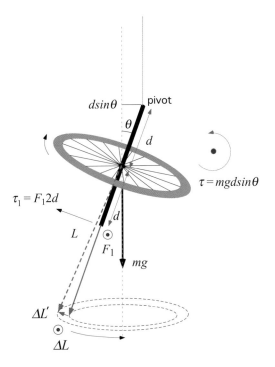

Fig. 4.8 As for Fig. 4.6, but with an additional resonant nutational torque τ_1 applied normal to the precessional torque τ and circulating in reference to the precession at exactly the Larmor frequency.

$$\frac{d\mathbf{M}}{dt} = \gamma \mathbf{M} \times \mathbf{B} \qquad (4.25)$$

The solution to eqn 4.25 when $\mathbf{B}$ is a magnetic field of amplitude B_0 corresponds to a precession of the magnetisation about the field at rate $\omega_0 = \gamma B_0$, the Larmor frequency. The resonance phenomenon results on application of a transverse magnetic field oscillating at ω_0. To obtain the expression for the spin evolution we need retain only the circularly polarised component of the oscillating transverse field that is rotating in the same sense as the spin precession, namely

$$\mathbf{B}_1(t) = B_1 \cos\left(\omega_1 t\right)\mathbf{i} - B_1 \sin\left(\omega_1 t\right)\mathbf{j} \qquad (4.26)$$

where $\mathbf{i}$, $\mathbf{j}$ and $\mathbf{k}$ are unit vectors along the x-, y- and z-axes, respectively. Then, under the same starting condition, $\mathbf{M}(0) = M_0\mathbf{k}$, eqns 4.25 and 4.26 return the result derived from quantum mechanics, eqns 4.12.

4.2.6 Relaxation and the Bloch equations

The effect of a resonant RF pulse is to disturb the spin system from its thermal equilibrium state. In due course, that equilibrium will be restored by a process known as spin–lattice relaxation. As the name implies, the process involves an exchange of energy between the spin system and the surrounding thermal reservoir, known as the 'lattice', with which it is in equilibrium. The equilibrium is characterised by a state of polarisation with magnetisation M_0 directed along the longitudinal magnetic field, B_0. The restoration of this equilibrium is therefore alternatively named longitudinal relaxation. The phenomenological description of this process is given by the equation

$$\frac{dM_z}{dt} = -\frac{M_z - M_0}{T_1} \tag{4.27}$$

with solution

$$M_z(t) = M_z(0) + M_0(1 - \exp(-t/T_1)) \tag{4.28}$$

T_1 is known as the spin–lattice or longitudinal relaxation time. At room temperature it is typically in the range 0.1 to 10 seconds for protons in dielectric materials.

At first sight it may appear that the time constant T_1 will also describe the lifetime of transverse magnetisation resulting from the application of an RF pulse. In fact transverse relaxation, which is characterised by the time constant T_2, is the process whereby nuclear spins come to thermal equilibrium among themselves. It is therefore known also as spin–spin relaxation. While indirect energy exchange via the lattice may play a role, additional direct processes are also responsible. This leads to the result $T_2 < T_1$. As indicated in the earlier quantum mechanical description, transverse magnetisation corresponds to a state of phase coherence between the nuclear spin states. This means that transverse relaxation, unlike longitudinal relaxation, is sensitive to interaction terms that cause the nuclear spins to dephase. As we shall see later in Section 4.5, this may lead to T_2 relaxation being exceedingly rapid in comparison with T_1, where the interaction between the nuclear spins fluctuates very slowly, as in the case of solids or rigid macromolecules. T_2 values are usually in the range 10 ms to 10 s.

The phenomenological description for transverse relaxation is written

$$\frac{dM_{x,y}}{dt} = -\frac{M_{x,y}}{T_2} \tag{4.29}$$

with solution

$$M_{x,y}(t) = M_{x,y}(0)\exp(-t/T_2) \tag{4.30}$$

It should be emphasised that the exponential description applies in the case where the interaction terms responsible for transverse relaxation are weak. This is the regime of Bloembergen, Purcell, and Pound (BPP) theory [20], an approach that works well for spins residing in liquid-state molecules. However, for solids and macromolecules undergoing very slow motions, the decay is more complicated than that represented by eqn 4.30. For most NMR translational motion measurements we are confined to observe slowly relaxing spins for which the phenomenological approach is entirely appropriate.

Combining eqns 4.25, 4.27, and 4.29 in the rotating frame yields a set of relationships known as the Bloch equations [21–23]

$$\frac{dM_x}{dt} = \gamma M_y \left(B_0 - \omega/\gamma\right) - \frac{M_x}{T_2}$$
$$\frac{dM_y}{dt} = \gamma B_1 - \gamma M_x \left(B_0 - \omega/\gamma\right) - \frac{M_y}{T_2} \qquad (4.31)$$
$$\frac{dM_z}{dt} = -\gamma M_y B_1 - \frac{M_z - M_0}{T_1}$$

These will provide a valuable reference in describing many phenomena important in NMR.

4.3   Signal detection

4.3.1   Free precession and Faraday detection

Resonant nutation of the spins disturbs the spins from equilibrium, but what constitutes the signal in NMR? To understand this point it is helpful to imagine a nutation in which the magnetisation is reoriented to leave a component in the transverse (x-y) plane of the rotating frame, a simple example being the case of a 90°_x pulse that causes the magnetisation to lie along the rotating frame y-axis. Following the application of this RF field the spins now experience only their natural spin Hamiltonian and are said to be in 'free-precession'. Because of the dominant Zeeman Hamiltonian, in the laboratory frame the transverse magnetisation appears to rotate at the Larmor frequency ω_0, though other weaker terms in the spin Hamiltonian may produce more subtle spectral variations.

That Larmor precession of non-equilibrium macroscopic nuclear magnetisation in the laboratory frame provides the means for signal detection. An antenna coil (for example a solenoid with its axis orthogonal to the z-axis as in Fig. 4.9) can detect such free-precession via Faraday induction. The output of that coil is an oscillating voltage. We will not be concerned here with the details of the absolute signal amplitude, which will depend on the dimensions or the RF antenna coil, and its electrical properties such as numbers of turns and resistance. However, with Faraday induction, one important factor which emerges is that the signal strength will be proportional to the Larmor precession frequency, γB_0. Combining this effect with the available magnetisation given by eqn 3.89, it is clear that the total sensitivity of the NMR signal will be proportional to $\gamma^3 B_0^2$. This places a premium on high magnetic field strength and large γ.

There are other methods of detecting the nuclear magnetisation, which are sensitive to magnetic field rather than the time rate of change of magnetic field, examples being the squid magnetometer [24, 25] or atomic magnetometer [26, 27]. For these the relevant sensitivity parameter is $\gamma^2 B_0$. These methods have special advantages for very low field ($< \mu T$) NMR. Also important for enhancing sensitivity by significantly enhancing nuclear magnetisation above the thermal equilibrium value are the so-called hyperpolarisation methods, such as dynamic nuclear polarisation [28, 29] and spin exchange [30–32]. The latter method has particular importance in the study of translational motion in gases and will be discussed further in Chapter 6.

With its gyromagnetic ratio the largest of any stable nucleus, its high isotopic abundance, its ubiquity in the hydrogen atoms of biological tissue, organic molecules, and synthetic and natural materials, the spin-$\frac{1}{2}$ proton is the predominant choice for

NMR applications using thermal polarisation and Faraday detection. Notwithstanding that emphasis, other stable nuclei such as $^2H, ^{13}C, ^{15}N, \, ^{31}P$, and ^{129}Xe present significant specific advantages as we shall see later.

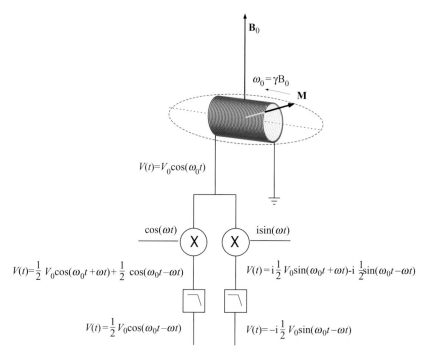

Fig. 4.9 Faraday detection of emf resulting from magnetisation vector **M** precessing at $\omega_0 = \gamma B_0$ in a solenoid coil. Also shown is the heterodyne detection scheme, whereby the complex difference signal $\frac{1}{2} V_0 \exp(-i (\omega_0 - \omega) t)$ is obtained by a combination of independent mixers and low-pass filters assigned respectively to real and imaginary channels.

Neglecting relaxation for the moment, let us write down an expression for the evolution of laboratory-frame density matrix for the free precession that follows application of an infinitesimally short 90°_x RF pulse to the thermal equilibrium ensemble $\rho_0 \sim I_z$. Using the same approach used to generate eqn 4.11, the subsequent time-dependence at time t after the pulse is switched off can be written

$$\rho^{\mathrm{lab}}(t) \sim I_y \cos{(\omega_0 t)} + I_x \sin{(\omega_0 t)} \qquad (4.32)$$

This density matrix, which represents transverse magnetisation in clockwise rotation about the z-axis, has two orthogonal components of magnetisation that oscillate in quadrature phase at the Larmor frequency. If we were to observe this precession not in the laboratory frame but in a frame rotating clockwise at ω, the result would be

$$\rho^{\mathrm{rot}}(t) \sim I_y \cos{((\omega_0 - \omega) t)} + I_x \sin{((\omega_0 - \omega) t)} \qquad (4.33)$$

Now imagine that we were able to make a measurement that could be represented by a complex signal operator $N\gamma [I_x + iI_y]$. Of course measurements are always real numbers, and so the assignment of real and imaginary parts to the signal means that we must detect in two independent channels. But allowing this, such a measurement on the rotating frame density matrix of eqn 4.33 would yield

$$\mathrm{Tr}\left(\left[I_x + iI_y\right]\rho^{\mathrm{rot}}(t)\right) \sim i\,\mathrm{Tr}\left(I_y^2\right)\cos\left(\left(\omega_0 - \omega\right)t\right) + \mathrm{Tr}\left(I_x^2\right)\sin\left(\left(\omega_0 - \omega\right)t\right) \quad (4.34)$$

Using eqns 3.89 and 3.90 we may then write

$$N\gamma\,\mathrm{Tr}\left(\left[I_x + iI_y\right]\rho^{\mathrm{rot}}(t)\right) = iM_0\exp\left(-i\left(\omega_0 - \omega\right)t\right) \quad (4.35)$$

The coefficient i tells us that at $t = 0$, when $\rho^{\mathrm{rot}} \sim I_y$, the signal is at a maximum in the quadrature channel. We will see that just such a rotating frame measurement is possible in practice using RF heterodyne detection.

Radiofrequency receivers work by mixing the signal voltage with a the output from a reference RF oscillator, a process known as heterodyning. This method of detection is inherently phase-sensitive. This means that by separately mixing the signal with two heterodyne references, each 90° out of phase, we obtain separate in-phase and quadrature-phase output signals, which are then ascribed to real and imaginary parts of a complex number. Of course, the actual Faraday detection of Larmor precession in a single coil[3] corresponds to a voltage oscillation of the form $V_0\cos\left(\omega_0 t\right)$. Heterodyning with two quadrature channels, assigned respectively to real and imaginary parts, is equivalent to the multiplication

$$V_0\cos\left(\omega_0 t\right)\exp\left(i\omega t\right) = \frac{1}{2}V_0\left[\exp\left(-i\left(\omega_0 - \omega\right)t\right) + \exp\left(i\left(\omega_0 + \omega\right)t\right)\right] \quad (4.36)$$

The receiver filters reject the sum frequency term and return only the difference $\frac{1}{2}V_0\exp(-i\left(\omega_0 - \omega\right)t)$. This is exactly proportional to the rotating-frame signal generated by the operator $N\gamma\left[I_x + iI_y\right]$. The fact that the negative frequency difference, $\left(\omega_0 - \omega\right)$, is measured is irrelevant, merely reflecting the clockwise nature of the spin precession. The important point is that we really are able to measure $[I_x + iI_y]$ in a frame of reference rotating at the heterodyne mixing frequency. Thus we place ourselves naturally in the rotating frame. We not only use a rotating-frame depiction of the RF excitation process but use it equally in the case of detection. Henceforth, the rotating frame will provide the natural perspective from which to discuss magnetic resonance phenomena.

Note that for simple precession in a magnetic field, where the mixing reference is oscillating at the Larmor frequency ω_0, we obtain dc quadrature output signals. At any other mixing frequency ω, the signal will oscillate at the offset frequency $\Delta\omega = \omega_0 - \omega$. The absolute sign attributed to that difference is an arbitrary choice to be made, but the ability to detect both real and imaginary parts of the signal lies at the heart of our ability to distinguish opposite signs of that offset frequency, to have access to both positive and negative domains of $\Delta\omega$. In practice, the absolute phase of the heterodyne receiver will not be identical to that of the precessing magnetisation. For

---

[3]Some RF antennas employ two orthogonal coils, which independently provide the cosine and sine components of the induced emf. This method can in principle be used to determine the absolute sign of the Larmor precession, unlike heterodyning a single component signal from a single coil.

that reason the signal contains an arbitrary phase factor, $\exp(i\phi)$. This factor can easily be removed in the signal processing as we shall show in the next section.

As a final consideration, let us return to the idea of the multiple quantum coherence discussed in Chapter 3, asking the question, what is the order of coherence p, which is observed in the NMR Faraday detection process? To understand this question we need to return to the definition of our observable as $\text{Tr}\left([I_x + iI_y]\,\rho^{\text{rot}}(t)\right) = \text{Tr}\left(I_+ \rho^{\text{rot}}(t)\right)$, where in the case that we wish to distinguish multiple spins, we replace I_+ by $\sum_{i=1}^{K} I_{i+}$. A non-zero signal is only possible when $\rho^{\text{rot}}(t) \sim I_-$, so clearly the detected coherence in NMR is $p = -1$.

4.3.2 Fourier transformation and the spectrum

The measurement of a complex time-domain signal, $S(t)$, lends itself to simple interpretation using the Fourier transformations

$$F\{S(t)\} = s(\omega)$$
$$= \int_{-\infty}^{\infty} S(t) \exp\left(i\omega t\right) dt$$
$$F^{-1}\{s(\omega)\} = S(t)$$
$$= \frac{1}{2\pi} \int_{-\infty}^{\infty} s(f) \exp\left(-i\omega t\right) d\omega$$

(4.37)

or

$$F\{S(t)\} = s(f)$$
$$= \int_{-\infty}^{\infty} S(t) \exp\left(i2\pi f t\right) dt$$
$$F^{-1}\{s(f)\} = S(t)$$
$$= \int_{-\infty}^{\infty} s(f) \exp\left(-i2\pi f t\right) df$$

(4.38)

The use of cyclic frequency $f = \omega/2\pi$ allows for a more symmetric form for the Fourier integral pair and we shall sometimes follow such a description in this book.

Suppose we consider the action of Fourier transformation on a signal corresponding to an oscillation at fixed offset frequency and with relaxation included, i.e.

$$S(t) = S_0 \exp\left(i\phi\right) \exp\left(i\Delta\omega t\right) \exp\left(-t/T_2\right) \qquad t \geq 0$$
$$= 0 \qquad t < 0$$

(4.39)

where S_0 is the signal immediately following the pulse, a number that is simply proportional to M_0. The primary NMR signal is measured in the time domain as an oscillating, decaying electromotive force (emf) induced by the magnetisation in free precession. It is therefore known as the free induction decay (FID). By Fourier transformation the same signal may be represented in the frequency domain. This process is shown in Fig. 4.10 for the case $\phi = 0$. The result in the real part of the domain is

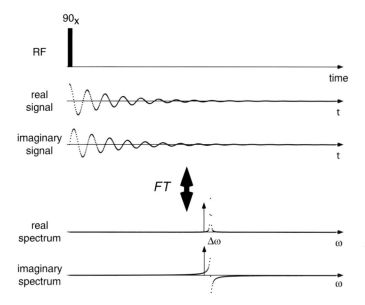

Fig. 4.10 Free induction decay (FID) following a single 90° RF pulse. The real and imaginary parts of the signal correspond to the in-phase and quadrature receiver outputs. The signal is depicted with receiver phase $\phi = 0$ and, on complex Fourier transformation, gives real absorption and imaginary dispersion spectra at the offset frequency $\Delta\omega = \omega_0 - \omega$.

a Lorentzian lineshape of cyclic frequency offset $\Delta\omega/2\pi$ and full-width-half-maximum (FWHM) $1/\pi T_2$, the so-called absorption spectrum,

$$s_{\text{abs}}(\omega) = \frac{T_2}{1 + (\omega - \Delta\omega)^2 T_2^2} \tag{4.40}$$

The imaginary lineshape is known as the dispersion spectrum and for this function the width is somewhat greater.

$$s_{\text{disp}}(\omega) = \frac{(\omega - \Delta\omega)\, T_2^2}{1 + (\omega - \Delta\omega)^2 T_2^2} \tag{4.41}$$

It is an elementary exercise in calculus to show that the real and imaginary parts of $F(S(t))$ are given by

$$
\begin{aligned}
\text{Re}\left\{F\left\{S(t)\right\}\right\} &= \cos\phi \frac{T_2}{1 + (\omega - \Delta\omega)^2 T_2^2} + \sin\phi \frac{(\omega - \Delta\omega)\, T_2^2}{1 + (\omega - \Delta\omega)^2 T_2^2} \\
\text{Im}\left\{F\left\{S(t)\right\}\right\} &= \sin\phi \frac{T_2}{1 + (\omega - \Delta\omega)^2 T_2^2} - \cos\phi \frac{(\omega - \Delta\omega)\, T_2^2}{1 + (\omega - \Delta\omega)^2 T_2^2}
\end{aligned} \tag{4.42}
$$

Note that the imaginary dispersion spectrum exists in the case $\phi = 0$, only because the signal is acquired for positive time only. It is hard to imagine what sampling at 'negative time' could possibly mean in the context of the simple experiment being discussed here, so the existence of the dispersion spectrum seems quite natural. However, where the dephasing of the magnetisation does not result from an irreversible T_2 process, but instead from some coherent interaction, it is possible to produce rephasing as well as dephasing effects. We will return to this point, and the concept of negative time acquisition, in the discussion of spin echoes in Section 4.6.2.

The area under the absorption spectrum is precisely the height of the FID at $t = 0$. The advantage of using the peak integral rather than the FID height in order to determine the NMR signal amplitude is twofold. Of course, integration in the frequency domain allows individual spectral components to be analysed. But, in addition, such integration over a defined frequency width provides a bandpass filtering of the signal, the only noise contribution arising from that lying within the defined spectral limits. Hence, even if only a single spectral component exists, it is highly advantageous to use the area under the spectral peak to determine signal amplitude. To carry out this integration it is necessary to first ensure that the real part of the spectrum is in pure absorption mode. That requires either setting the receiver phase so that $\phi = 0$ or post-processing the data, the latter being the more obviously convenient approach. Where $\phi \neq 0$ the real and imaginary parts of the transformed signal contain admixtures of absorption and dispersion parts and the spectrum may be restored to the 'correct phase' by simply multiplying by $\exp(-i\phi)$. This is equivalent to the operation

$$
\begin{aligned}
\mathrm{Re}\left\{F\left\{S(t,0)\right\}\right\} &= \cos\phi\,\mathrm{Re}\left\{F\left\{S(t,\phi)\right\}\right\} + \sin\phi\,\mathrm{Im}\left\{F\left\{S(t,\phi)\right\}\right\} \\
\mathrm{Im}\left\{F\left\{S(t,0)\right\}\right\} &= \cos\phi\,\mathrm{Im}\left\{F\left\{S(t,\phi)\right\}\right\} - \sin\phi\,\mathrm{Re}\left\{F\left\{S(t,\phi)\right\}\right\}
\end{aligned}
\tag{4.43}
$$

Phasing the spectrum is an important operation in NMR spectroscopy and can be performed either manually or automatically in the computer by inspecting the real transform and adjusting ϕ until the result is a perfect absorption spectrum in the real part (maximum spectral integral) or a perfect dispersion spectrum in the imaginary part (zero spectral integral).

4.3.3 Digital Fourier transformation

In practice, NMR signals are digitised for subsequent processing and are represented by a finite set of N time-domain points sampled at dwell-time interval T. When data are digitised in this manner then the detection bandwidth is the inverse of the sampling interval, namely $1/T$, and the separation of points in the frequency domain is $1/NT$, with values ranging between $-\frac{1}{2}N\,(1/NT)$ and $\left(\frac{1}{2}N - 1\right)(1/NT)$. The digital Fourier transformation is given by

$$
s(n/NT) = \sum_{-N/2}^{N/2-1} S(mT) \exp\left(-i2\pi mn/N\right)
\tag{4.44}
$$

the conjugate transformation being given by

$$
S(mT) = \sum_{-N/2}^{N/2-1} s(n/NT) \exp\left(i2\pi mn/N\right)
\tag{4.45}
$$

For an FID signal represented by $S(mT)$, where m ranges from 0 to $N - 1$, the application of eqn 4.44 involves a shift transformation $m' = m - N/2$ and hence a sign alternation between adjacent points in the frequency domain, an effect easily corrected in the computer. In fact, as we shall see later, it is possible to acquire signals at negative time using spin echoes. A further consequence of the finite sampling range means in effect that the data is multiplied by a hat function given by

$$
\begin{aligned}
H(t) &= 1 & 0 \leq t \leq NT \\
&= 0 & NT < t
\end{aligned}
\tag{4.46}
$$

which means that the frequency-domain spectrum is convoluted with the digital Fourier transform of $H(t)$, namely,

$$
F\{H(t)\} = \frac{NT \sin(2\pi f NT)}{2\pi f NT} + i\frac{NT(1 - \cos(2\pi f NT))}{2\pi f NT}
\tag{4.47}
$$

where $f = k(1/NT)$ and k is an integer ranging from $-\frac{1}{2}N$ to $\left(\frac{1}{2}N - 1\right)$. It is immediately clear that this apparent convolution introduces no real broadening, since $F\{H(t)\}$ is unity for $k = 0$ and zero elsewhere.

4.4   Intrinsic spin interactions

In Chapter 3 we met the spin Hamiltonian involving the Zeeman interaction, the chemical shift, the internuclear spin–spin scalar coupling, the electric quadrupole interaction, and the internuclear dipolar interaction. Here we reprise the spin Hamiltonian in the context of a dominant zeroth-order Zeeman interaction in which, for most cases, only secular parts of the remaining terms play a role as first-order perturbations. We will then outline some simple NMR experiments where the additional effect of RF pulses allows us to manipulate the evolution pathways of spin coherence such that desired terms in the spin Hamiltonian may be observed through the final detected magnetisation.

4.4.1   Resolution in NMR

The visibility of fine details

The Zeeman interaction arising from the static magnetic field, B_0, dominates the energy of the nuclear spins. It is a feature of NMR that many other much weaker interaction terms can still be observed, a consequence of the remarkable coherence times that spin ensembles exhibit. For example, because the T_2 times for nuclear spins in liquids may be of order seconds, the fundamental linewidth $1/\pi T_2$ may be eight or nine orders of magnitude smaller than the Larmor frequency. This means that very fine features in the Hamiltonian may be detected. The situation is not so favourable in solids, where the strong dipolar interactions between the spins serve to broaden the linewidth considerably, so destroying the transverse magnetisation coherence a few tens of microseconds after its formation.

The influence of interactions superposed on the applied longitudinal Zeeman field are of considerable importance in NMR and provide some extraordinary insights regarding molecular chemical structure, molecular rotational dynamics, and intramolecular dynamics, as well as molecular proximity, secondary structure, and organisation. However, in focusing on NMR measurements of translational dynamics, we will be interested in these various interactions to the extent that they enable molecular identification or enhance (or possibly limit) the timescales over which motion may be observed. For this reason we provide only a limited overview, leaving the reader to seek a more complete discussion of spin interactions specific to molecular chemistry in the many excellent texts on NMR spectroscopy. There is, however, one member of the wide class of possible spin interactions that is of particular relevance to translational motion measurement, namely the degree to which the magnetic field has spatial dependence, whether through imperfection in the polarising magnet, through magnetic inhomogeneity arising from local sample structure, or through deliberate application by the experimenter of additional inhomogeneous magnetic fields.

Magnetic field inhomogeneity

The manufacturers of NMR magnets go to great lengths to provide a homogeneous magnetic field in the region of the sample space, incorporating first-, second-, and higher-order shim coils for fine correction. However, it is inevitable that some variation in B_0 across the sample will occur, although typically this leads to a broadening of no more than a few hertz in the proton NMR spectrum of a homogeneous sample of favourable geometry, for example a sample water contained in a long cylinder coaxial with the magnetic field. However, in many real samples where there is material inhomogeneity, for example where liquid is interspersed within a porous solid matrix of different diamagnetic susceptibility, local inhomogeneities in magnetic field will exist. The fractional variation in magnetic field throughout the sample will be on the order of the fractional diamagnetic susceptibility differences, typically a few ppm. In a superconducting magnet, such a spread can represent several kilohertz of extra linewidth.

The additional broadening caused by field inhomogeneity arises from a spread in the frequency spectrum, the spins experiencing a distribution of Larmor precession frequencies that results in a spreading of spin phases as time progresses. Following an RF excitation pulse, such spreading causes a decay of the FID more rapid than that due to T_2 effects alone. The resultant time constant is often labelled T_2^*, but it is important to appreciate that coherence loss due to relaxation is inherently random and irreversible whereas that caused by field imperfection is ordered and, given an appropriate pulse sequence, can be 'undone'. This leads to the important distinction between inhomogeneous and homogeneous broadening of the NMR spectral line. In practice, coherence loss that can be refocused by an appropriate pulse sequence is said to arise from inhomogeneous broadening. This broadening is typically caused by local differences in Hamiltonian between spin packets. Homogeneous broadening, by contrast, is common to each spin packet and is essentially irreversible. It arises from random motion of the spins. Note that the localised sub-ensemble of spins in an inhomogeneous magnetic field is often termed an 'isochromat' of the spin system, in other words a sub-ensemble of spins labelled by a single Larmor frequency.

The pulse sequence that provides the facility to distinguish reversible and reversible coherence loss is the spin echo, discovered by Erwin Hahn [19]. The echo principle underlies most of the methods for translational motion measurement discussed in this book. As we shall see later, by pulsed application of suitable magnetic fields, tailored to give a spatially dependent Larmor frequency, it is possible to use the echo to measure translational motion with great precision.

4.4.2   Chemical shift

First discovered in 1950 [33, 34], the chemical shift which arises from the diamagnetic shielding arising from the electrons that surround atomic nuclei in atoms and molecules, imparts to the NMR spectrum information about chemical structure. We focus here on the secular part of the chemical shift. For spin i

$$\mathcal{H}_{\mathrm{CS}_0} = -\delta_i \omega_0 I_{iz} - \frac{1}{2}\left(3\cos^2\beta - 1\right)\left(\delta_{zz} - \delta_i\right)\omega_0 I_{iz} \qquad (4.48)$$

where δ_i is the isotropic chemical shift equal to $\mathrm{Tr}\,(\underline{\delta})$ and β is the polar angle between the polarising field direction and the principal axis system of the shift tensor. The additional term is known as the anisotropic chemical shift and transforms under rotation as a second rank tensor, hence the role of the Legendre polynomial $P_2\left(\cos\beta\right) = \frac{1}{2}\left(3\cos^2\beta - 1\right)$. It leads to a broadened spectrum as shown in Fig. 4.11. In the liquid state, where molecules tumble rapidly, this term averages to zero, leaving only the isotropic part of the chemical shift. Chemical shifts depend strongly on the atomic number and are of order a few parts per million in ^{1}H, but several hundred parts per million in ^{13}C or up to thousands in ^{129}Xe.

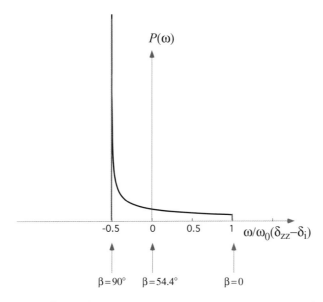

Fig. 4.11 Anisotropic chemical shift spectrum resulting from a random (isotropic) distribution of orientations ($P(\beta) = 1/4\pi$). The singularity for polar angle $\beta = 90°$ arises because of increasing solid angle near the equator.

4.4.3   Scalar coupling

The scalar spin–spin coupling, $\mathcal{H}_{\mathrm{scalar}} = 2\pi J\,\mathbf{I}_i \cdot \mathbf{I}_j$, is an indirect interaction between nuclei, which arises via the mediation of electrons in the molecular orbital. The nuclear spin causes a slight electron polarisation, which, because of electron delocalisation, is transmitted to neighbouring nuclei. $\mathcal{H}_{scalar}$ provides the finest structural detail in the liquid-state nuclear spin Hamiltonian It is typically between 1 and 50 Hz for protons. The spin–spin interaction, acting in conjunction with the chemical shift, imparts a characteristic signature to the high-resolution spectrum. Because the spin–spin interaction requires a molecular orbital, it acts only through the medium of covalent bonds. For this reason the scalar coupling is always intramolecular. The indirect nature of $\mathcal{H}_{scalar}$ leads to its rotational invariance. Note that we here simplify the notation used in Chapter 3 to write $i = 1$, $j = 2$, $J = J_{ij}$ and to express $\mathcal{H}$ in angular frequency units, the equivalent angular frequency energy associated with the scalar coupling prefactor being $2\pi J\hbar = Jh$.

When considering the role of the scalar coupling operator for a two-spin system it is helpful to write the combined chemical shift and scalar coupling Hamiltonian as

$$\begin{aligned}
\mathcal{H} &= -\Delta\omega_1 I_{1z} - \Delta\omega_2 I_{2z} + 2\pi J \mathbf{I}_1 \cdot \mathbf{I}_2 \\
&= -\Delta\omega \left(I_{1z} + I_{2z} \right) - \tfrac{1}{2}\delta \left(I_{1z} - I_{2z} \right) + 2\pi J \mathbf{I}_1 \cdot \mathbf{I}_2
\end{aligned} \tag{4.49}$$

where $\Delta\omega_1 = \delta_1\omega_0$, $\Delta\omega_2 = \delta_2\omega_0$ and we have centred the resonance frequency by setting $\tfrac{1}{2}(\Delta\omega_1 + \Delta\omega_2) = \Delta\omega$, the average offset frequency in the rotating frame, and simplified notation by $\delta = (\Delta\omega_1 - \Delta\omega_2)$. The operator $\mathbf{I}_1 \cdot \mathbf{I}_2$ commutes with $I_{1z} + I_{2z}$ but not with $I_{1z} - I_{2z}$. Hence the chemical shift, δ, plays a crucial role in the visibility of the scalar coupling interaction.

A convenient representation for two spins involves a simple product of kets $|m_1, m_2\rangle = |m_1\rangle|m_2\rangle$, where the column vectors are formed by simple outer products. However, in this representation the spin operator $\mathbf{I}_1 \cdot \mathbf{I}_2$ certainly contains off-diagonal components, $I_{1x}I_{2x}$ and $I_{1y}I_{2y}$, as well as the diagonal term $I_{1z}I_{2z}$. In practice, this off-diagonal part of the Hamiltonian only mixes product states $\left|\tfrac{1}{2}, -\tfrac{1}{2}\right\rangle$ and $\left|-\tfrac{1}{2}, \tfrac{1}{2}\right\rangle$ while $\left|\tfrac{1}{2}, \tfrac{1}{2}\right\rangle$ and $\left|-\tfrac{1}{2}, -\tfrac{1}{2}\right\rangle$ remain eigenstates. The states $|\psi_1\rangle = \left|-\tfrac{1}{2}, -\tfrac{1}{2}\right\rangle$ and $|\psi_4\rangle = \left|\tfrac{1}{2}, \tfrac{1}{2}\right\rangle$ are separated by $2\Delta\omega$, while states $|\psi_2\rangle$ and $|\psi_3\rangle$ involve superpositions of the product states, with admixture coefficients $\cos\tfrac{1}{2}\phi$ and $\sin\tfrac{1}{2}\phi$, where $\phi = \tan^{-1}(Jh/\delta)$. The set of eigenstates and their associated energy levels are are shown in Fig. 4.12. To obtain the spectra, one calculates the single-spin transition probabilities associated with the raising or lowering operators $J_{1\pm} + J_{2\pm}$. These allow for transition $E_4 - E_3$ and $E_3 - E_1$, with respective intensities proportional to $\left(\cos\tfrac{1}{2}\phi - \sin\tfrac{1}{2}\phi\right)^2$ and for $E_4 - E_2$ and $E_2 - E_1$, with respective intensities proportional to $\left(\cos\tfrac{1}{2}\phi + \sin\tfrac{1}{2}\phi\right)^2$.

In the limit $Jh \ll \delta$, where $\tfrac{1}{2}\phi \to 0$, we may write the scalar coupling interaction in its secular form as $\mathcal{H}_{scalar_0} = 2\pi J \, I_{1z}I_{2z}$. Here the Hamiltonian is diagonal in the $|m_1, m_2\rangle$ product representation and plays the role for spin 1 of an additional magnetic field proportional to the $m_2 = \pm\tfrac{1}{2}$ eigenvalue of I_{2z}, thus splitting the two chemically shifted resonance lines of the 1 and 2 spins by Jh angular frequency units. This weak coupling case is shown in the uppermost spectrum of Fig. 4.12, where the intensities of all four lines are seen to be equal.

For nuclei with identical resonant frequencies, $Jh \gg \delta$ and $\tfrac{1}{2}\phi \to 45°$. Here $\mathbf{I}_1 \cdot \mathbf{I}_2$ is invisible. In that case, known as the AA coupled spin system, $\cos\tfrac{1}{2}\phi = \sin\tfrac{1}{2}\phi = \tfrac{1}{\sqrt{2}}$ and the triplet $|\psi_1\rangle$, $|\psi_3\rangle$, $|\psi_4\rangle$ and singlet $|\psi_2\rangle$ are, respectively, eigenstates of $2\pi J \mathbf{I}_1 \cdot \mathbf{I}_2$,

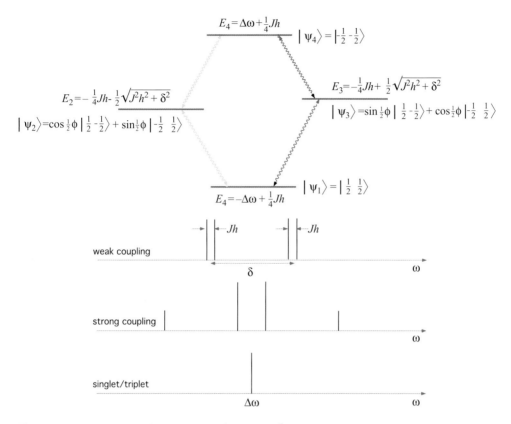

Fig. 4.12 Energy levels for a system of two spin-$\frac{1}{2}$ nuclei coupled by the scalar interaction, $2\pi J\mathbf{I}_1 \cdot \mathbf{I}_2$.

with energy eigenvalues of $\pi Jh\left[I(I+1) - \frac{3}{2}\right]$,[4] where for the triplet, $I = 1$, and for the singlet, $I = 0$. These are total angular momentum states formed by the coupling of angular momentum vectors $\mathbf{I}_1 + \mathbf{I}_2$, to form a total angular momentum vector $\mathbf{I}$. Clearly the total angular momentum states form a diagonal eigenbasis for $\mathbf{I}_1 \cdot \mathbf{I}_2$, while the product states are an eigenbasis for $\gamma B_{01}I_{1z} + \gamma B_{02}I_{2z}$. Only transitions between the Zeeman separated triplet state levels are observable[5] and these are all identically shifted so that no effect of the scalar coupling is seen.

For the intermediate strong coupling case $Jh \sim \delta$, known as the AB spin system, neither the product nor total angular momentum representations provide an eigenbasis and the general superposition of product states shown in Fig. 4.12 applies. Here the single-spin transition probabilities for $E_4 - E_3$ and $E_3 - E_1$ are unequal to those for $E_4 - E_2$ and $E_2 - E_1$, as seen in the central spectrum.

---

[4]This is simple to show from $\mathbf{I}^2 = (\mathbf{I}_1 + \mathbf{I}_2)^2 = \mathbf{I}_1^2 + \mathbf{I}_2^2 + 2\mathbf{I}_1 \cdot \mathbf{I}_2$, noting $\mathbf{I}_1^2 = I_1(I_1 + 1) = \frac{3}{4}$

[5]The reason being that the NMR transition involves an $I_{i\pm} + I_{j\pm}$ operator, which commutes with $\mathbf{I}_i \cdot \mathbf{I}_j$.

The simple weak coupling interpretation has the resonance frequency of spin i split into a doublet because of the coupling to a single nucleus elsewhere in the molecule, the strength of this splitting reducing with increasing bond distance. Similarly, a nearby group of two nuclei with identical chemical shift and scalar couplings will have possible spin states $\left|\frac{1}{2}, \frac{1}{2}\right\rangle, \left|-\frac{1}{2}, \frac{1}{2}\right\rangle, \left|\frac{1}{2}, -\frac{1}{2}\right\rangle$, and $\left|-\frac{1}{2}, -\frac{1}{2}\right\rangle$, leading to a triple peak of intensity ratio 1:2:1. In a similar binomial fashion, coupling to three identical nuclei leads to a quadruplet with intensity ratio 1:3:3:1. By this means the multiplicity of coupled nuclei of common chemical shift may be determined. By means of chemical shift and scalar couplings, it is possible to identify the local chemical character of the molecular orbital surrounding that nucleus, as well as the intramolecular connectivity to nuclei at common chemical sites. The discovery of the chemical shift and the spin–spin coupling in the 1950s has revolutionised modern chemistry.

4.4.4 Quadrupole coupling

For nuclei with $I > 1/2$, the single-spin Hamiltonian contains a term representing the interaction of the nuclear quadrupole moment with the electric field gradient associated with the surrounding molecular orbital. Quadrupole interactions are of considerable significance in solid state NMR of inorganic materials, given that the periodic table contains a large number of stable $I > 1/2$ nuclei. In the present context, where we are concerned with the fluidic state, deuterium ($I = 1$) NMR provides a means of obtaining information about molecular orientation in anisotropic soft matter or liquid crystalline states. Note that symmetry considerations dictate that the quadrupole moment of a spin-$\frac{1}{2}$ nucleus is identically zero.

Again we write down only the secular part of the quadrupole interaction in the zeroth order $\{|m\rangle\}$ basis of the dominant Zeeman interaction. Using (X, Y, Z) for the principal axis frame of the electric field gradient and retaining (x, y, z) for the laboratory frame, in which the B_0 field is along z, the rotation to take (X, Y, Z) to (x, y, z) comprises polar angle θ and azimuthal angle ϕ. Then the secular component of $\mathcal{H}_Q$ is [9]

$$\mathcal{H}_{Q_0} = \frac{3eV_{ZZ}Q}{4I(2I-1)} \frac{1}{2} \left(3\cos^2\theta - 1 + \eta\cos 2\phi \sin^2\theta\right)\left(3I_z^2 - \mathbf{I}^2\right) \tag{4.50}$$

with Q being the nuclear quadrupole moment of spin-I, V_{ZZ} the principal axis diagonal element of the electric field gradient tensor, $V_{\alpha\beta} = \partial^2 V / \partial\alpha\partial\beta$, and η the asymmetry parameter $(V_{XX} - V_{YY})/V_{ZZ}$. In many molecules, the local electronic orbital is inherently axially symmetric so that $\eta = 0$ and

$$\mathcal{H}_{Q_0}^0 = \frac{3eV_{ZZ}Q}{4I(2I-1)} \frac{1}{2} \left(3\cos^2\theta - 1\right)\left(3I_z^2 - \mathbf{I}^2\right) \tag{4.51}$$

$\frac{1}{2}\left(3\cos^2\theta - 1\right)$ being the second rank Legendre polynomial $P_2(\cos\theta)$.

The effect of the spin operator $\left(3I_z^2 - \mathbf{I}^2\right)$ is to shift the energies of the $2I + 1$ Zeeman levels so that the transition frequencies are no longer equal. This results in spectrum of $2I$ different resonances, equally spaced around the Larmor frequency and

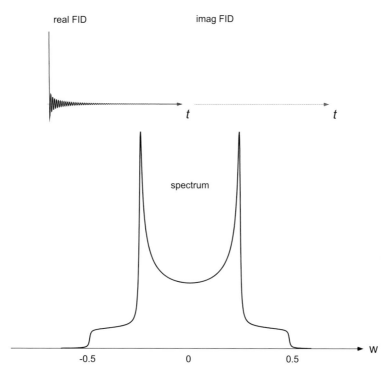

Fig. 4.13 Quadrupolar spectrum for $I = 1$ resulting from a random (isotropic) distribution of orientations $(P(\cos\theta, \phi) = 1/4\pi)$. By contrast with Fig. 4.11, the singularities for polar angle $\theta = 90°$ have been smoothed by inclusion of some relaxation in the associated FID. Note that symmetry of the spectrum implies a pure cosine FID (zero imaginary). The frequency scale is in units $3eV_{ZZ}Q/4I(2I-1)$.

separated by $2\omega_Q = 6\frac{3eV_{zz}Q}{4I(2I-1)}P_2(\cos\theta))^6$ For example, in spin-1, the single resonance line of the Larmor spectrum would be split in two for a particular value of ω_Q. In a nematic liquid crystal with director aligned at a fixed direction to the magnetic field, the NMR spectrum for the deuterium nucleus, with $I = 1$, would be a doublet with a unique quadrupole splitting. In a polydomain nematic or in the solid state, the distribution of director angles α would ensure a 'powder broadened line', as shown in Fig. 4.13.

Quadrupole interaction frequencies for deuterons in organic molecules in the solid state are on the order of 100 kHz. However, in material environments where molecules may tumble, the interaction is significantly reduced. Indeed, in simple isotropic liquids, molecular tumbling causes θ and hence $\mathcal{H}_{Q_0}$ to fluctuate. Provided that this motion is more rapid than the quadrupolar interaction strength, a condition, which is true for all but the most viscous liquids, the quadrupolar Hamiltonian is averaged to zero.

[6] The splitting is set here to $2\omega_Q$ for convenience. It allows us to use ω_Q as the precession frequency of transverse magnetisation under the influence of the quadrupole interaction, so that for $I = 1$ the spectrum consists of two lines, at $\pm\omega_Q$.

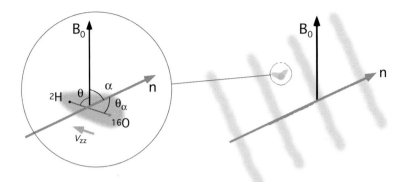

Fig. 4.14 Motional averaging of quadrupole interaction due to water molecule tumbling in a lyotropic lamellar phase with director **n**. The principal axis of the deuterium quadrupole interaction makes angle $\theta(t)$ with the magnetic field and angle $\theta_\alpha(t)$ with respect to the director. Rapid fluctuations of $\theta_\alpha(t)$ leave the effective quadrupole interaction scaled by $\overline{P_2(\cos\theta_\alpha(t))}P_2(\cos\alpha)$.

In anisotropic liquids, the motional averaging is incomplete. An example would be a liquid crystal for which there exists a local nematic director inclined at angle α to the magnetic field B_0, as shown in Fig. 4.14. If this director is fixed in space or fluctuating only slowly by comparison with the quadrupolar frequency, then the tumbling motion leads to a projection of the averaged quadrupole interaction along the director, with a second angular transformation to the laboratory Zeeman frame. The orientation (θ, ϕ) of the electric field gradient (bond) axis with respect to B_0 can be decomposed into $(\theta_\alpha, \phi_\alpha)$, describing the bond axis with respect to the director, and the polar angle α by which the director is inclined to the magnetic field. Allowing for averaging of the azimuthal angle ϕ_α[7] the spherical harmonic addition theorem may be used to factorise as [36]

$$\overline{P_2(\cos\theta)} = \overline{P_2(\cos\theta_\alpha)}P_2(\cos\alpha) \qquad (4.52)$$

$\overline{P_2(\cos\theta_\alpha)}$ thus defines a motionally averaged local order parameter that scales down the effective size of quadrupole interaction strength. The angular dependence of that scaled interaction now follows the local director axis.

Next we look at another source of line broadening in the solid state, one which is inherently more complex than the quadrupole interaction and which plays an important role for all nuclear spins.

4.4.5 Internuclear dipolar interactions

The last term in the spin Hamiltonian arises from through-space dipolar interactions between nuclei. It is of immense importance in magnetic resonance because of the ubiquity of these interactions in all matter, and is seen in all nuclei of non-zero spin angular momentum. In the case of protons, for which the nuclear dipole moment is

---

[7]This is tantamount to a requirement of axial symmetry of the material with respect to the nematic director. Where the sample is biaxial, the asymmetry effects are observable [35].

largest, these interactions may be on the order of $100\,\mathrm{kHz}$, comparable with quadrupole interaction strengths in spin-1 and higher spin nuclei. Again, dipolar interactions are but a first-order perturbation in the presence of a Zeeman field of hundreds of megahertz. When calculating the spin energy levels and hence the NMR spectrum, we may, to a good approximation, retain just the secular part of the dipolar interaction between like spins, i and j, and write

$$\mathcal{H}_{\mathrm{D}}^{0} = \frac{\mu_0 \gamma^2 \hbar}{4\pi} \sum_{i<j} \frac{1}{r_{ij}^3} \frac{1}{2} \left(1 - 3\cos^2\theta_{ij}\right) \left[3I_{iz}I_{jz} - \mathbf{I}_i \cdot \mathbf{I}_j\right] \tag{4.53}$$

where $\mathbf{r}_{ij}$ is the inter-nuclear vector that makes a polar angle θ_{ij} with the magnetic field direction. Note again the role of the Legendre polynomial $P_2(\cos\theta_{ij})$ in determining the dipolar interaction strength. As with the chemical shift anisotropy and the quadrupole interaction, the dipolar interaction transforms under rotation as a second rank tensor. The complexity of $\mathcal{H}_{\mathrm{D}}^{0}$ arises for a number of reasons. First, there is the bilinear nature of the spin operator, similar to the spin–spin scalar coupling but with both $I_{iz}I_{jz}$ and $\mathbf{I}_i \cdot \mathbf{I}_j$ components. As with the weak scalar coupling case, the $I_{iz}I_{jz}$ term leads to a splitting of the resonance, a simple doublet in the case of spin-$\frac{1}{2}$ But in nearly all materials there is a multiplicity of i to j dipolar interactions contributing in the sum and, for each i spin in the sum, numerous j term splittings of differing width. This spread arises because of the various internuclear separations and θ_{ij} orientations associated with proximate intermolecular and intramolecular spins. The net result is an approximately Gaussian spectrum associated with a Gaussian signal decay of the form

$$S(t) = S(0)\exp\left(-\frac{1}{2}M_2 t^2\right) \tag{4.54}$$

where M_2 is the second moment of the linewidth given by

$$M_2 = -\operatorname{Tr}\left(\left[\mathcal{H}_{\mathrm{D}}^{0}, I_x\right]^2\right) / \operatorname{Tr}\left(I_x^2\right) \tag{4.55}$$

In most rigid solids the decay of this Gaussian for proton NMR will be on the order of microseconds, consistent with the dipolar linewidth on the order of of $100\,\mathrm{kHz}$.[8] Note that these solid-state dipolar interactions are static and represent a form of inhomogeneous broadening. In principle, and given the right pulse sequence, the associated decay can be reversed and the linewidth narrowed. An example of a pulse sequence that does this will be discussed in Section 4.6.2.

The effect of either dipolar or quadrupolar interactions is dramatically different in the case of liquids. Here molecular tumbling leads to rapid fluctuations in θ_{ij} and provided that these are fast compared with the dipolar linewidth in frequency units, and isotropic in character, then $P_2(\cos\theta_{ij})$ and hence H_D^0 is averaged to zero, this narrowing the NMR linewidth. However, the off-diagonal (non-secular) terms in $\mathcal{H}_{\mathrm{D}}$ of $\mathcal{H}_{\mathrm{Q}}$ are thereby caused to fluctuate, inducing transitions between nuclear energy levels and thus contributing to both T_1 and T_2 relaxation. Provided field inhomogeneity is made sufficiently small then it is the decay associated with T_2 relaxation that sets the

---

[8]For other nuclei it will be scaled (down) by the respective gyromagnetic ratio factors.

spectral linewidth in the liquid state, typically on the order of 1 Hz and thus narrow enough to reveal fine details of remaining terms in the nuclear spin Hamiltonian. That decay, being exponential, leads to a Lorentzian lineshape of finite width. By contrast with the solid state, the decay of the liquid-state signal arises from stochastic fluctuations and leads to irreversible decay and hence homogeneous broadening of the spectral line.

4.5 Fluctuations, spin relaxation, and motional averaging

The multiplicity of interactions experienced by nuclear spins each result in deterministic quantum phase evolution. But where those interactions result from multiple sources, as in the case of the dipolar interaction whose sum may involve the multiple possibilities of intra- and intermolecular internuclear vectors, a distribution of phase shifts arises. That distribution represents a loss of phase coherence, with consequent decay of the FID or, more generally, loss of the non-equilibrium spin ensemble coherence. In that sense, such dephasing lies at the heart of spin–spin (T_2) relaxation. However, T_2 is a rather slippery concept. In principle, given the right pulse sequence, deterministic phase shifts may be reversed, a phenomenon observed by Hahn in his discovery of the spin echo. To the extent that phase-spreading is not recoverable, we might say that 'true' T_2 decay remains; that which was observed in the simple FID being an 'effective' T_2. These distinctions will be clearer once we embark on a description of specific pulse sequences in the next section.

The factor that mitigates against such recovery of spin phase will be stochastic fluctuations in the spin interactions. Fluctuations may arise from lattice vibrations, from molecular translational and rotational diffusion, or from internal vibrations and segmental motions of molecules. All of these cause those spin interactions sensitive to orientation and/or internuclear spacing to vary with a complex dynamical spectrum. Predominant are the internuclear dipolar interaction and, in the case of $I > 1/2$, the nuclear quadrupole interaction. Of course, if the motion is isotropic and rapid enough, then dipolar and quadrupolar interactions will be averaged to zero. Rapid enough in this context means, in the Heisenberg uncertainty sense, by comparison with the size of the dipolar and quadrupolar interactions (typically 100 kHz). In such a sense we may define an 'NMR liquid', namely one for which T_2 decay is slow because the molecular tumbling is sufficiently fast.[9] Herein lies a subtle idea. In the absence of motion, a distribution of spin interactions leads to rapid T_2 decay but one which is pulse sequence dependent. Fluctuations will render spin phase shifts irreversible and so increase the irreversible part of T_2 while at the same time reducing the effective size of the spin interactions as the degree of stochastic excursion increases. The ultimate limit of 'extreme motional averaging' is one in which there is only one T_2, albeit long and robustly pulse sequence independent. This is almost invariably the case for liquids comprising small molecules at room temperature, where isotropic tumbling is associated with rotational correlation times on the order of picoseconds.

However, fluctuations, while having the potential to diminish the spin dephasing associated with a distribution of spin interactions present in the solid state, also have

---

[9]Nothing quite so baffles the general physics community as the idea of motional averaging and its consequence of slow NMR signal decays arising from fast motion.

the capacity to induce transitions between spin states. Thus they lie at the heart of T_1 relaxation, the process of return to thermal equilibrium associated with the re-establishment of a Boltzmann distribution of spin state populations. For this reason T_1 will be generally very long in a solid, while T_2 is short. At the other extreme, simple liquids are characterised by T_2 and T_1 both long, identically equal in the 'motion-narrowed limit'. Understanding this transition from solid to liquid in the relaxation sense is important for two reasons. First, in materials where spin translational motion is permitted, for example complex fluids or soft-condensed matter, a hierarchy of dynamics may exist in which both slow and fast motions are present. For biological tissue, polymeric materials, liquid crystals, fluid-imbibed porous media, and colloids, there may be a sense in which both solid *and* liquid descriptors may apply. But, most importantly, as we shall later discover, T_1 and T_2 relaxation provide a fundamental limiting timescale for the measurement of translational motion. For these reasons it is helpful to expand a little on the fundamental processes involved. But we shall not delve deeply. Relaxation is an immensely interesting and complex topic, worthy of many chapters of discussion. Indeed, relaxation may in its own right be used as a probe of translational dynamics, although doing so requires a modelling process to interpret the way in which spin interactions translate the effects of translational motion to the relaxation time. This aspect of magnetic resonance is well treated in a number of other volumes [9, 37–40]

In spin-$\frac{1}{2}$ nuclei, the dominant interaction causing spin relaxation arises from the dipolar Hamiltonian, whereas in higher spin systems quadrupolar interactions are significant. Because of the pre-eminent role of the proton in NMR, our discussion will concentrate on the effect of $\mathcal{H}_\mathrm{D}$.

4.5.1 Spin–lattice relaxation

Spin–lattice relaxation is quite naturally described in the laboratory frame, where the difference in energy levels is dominated by the longitudinal Zeeman field and the dipolar interaction can be regarded as weak by comparison. This permits the use of time-dependent perturbation theory [9, 10, 41]. By a simple argument, Hebel and Slichter [42] showed that the spin–lattice relaxation rate could be written

$$\frac{1}{T_1} = \frac{1}{2} \frac{\sum_{nm} W_{nm} (E_n - E_m)^2}{\sum_n E_n^2} \tag{4.56}$$

where the transition rate between the two states, n and m, is given by perturbation theory as

$$W_{nm} = \frac{1}{\hbar^2} \int_0^\infty \exp\left[i\left(E_n - E_m\right)(\tau)/\hbar\right] \overline{\langle n| \hbar\mathcal{H}_\mathrm{D}(0) |m\rangle \langle m| \hbar\mathcal{H}_\mathrm{D}(\tau) |n\rangle} \, d\tau + c.c. \tag{4.57}$$

Evaluating eqn 4.57 requires decomposition of $\mathcal{H}_\mathrm{D}(t)$ into spatial and spin operators. The matrix elements of spin operators imply specific selection rules for n and m, and so determine which energy differences are relevant. Unsurprisingly, given the bilinear nature of the dipolar (or quadrupolar) Hamiltonian, the relevant separations of $n = m \pm q$, where q is 1 or 2. The respective spatial operator parts of $\mathcal{H}_\mathrm{D}(0)$ and $\mathcal{H}_\mathrm{D}(\tau)$

appear in correlation functions $G^q(\tau)$, which are Fourier transformed by the integral in eqn 4.57 to produce spectral density functions $J^{(q)}(\omega)$. Clearly the relevant frequencies are ω_0 and $2\omega_0$. Following some algebraic manipulation, one finds [9]

$$\frac{1}{T_1} = \left(\frac{\mu_0}{4\pi}\right)^2 \gamma^4 \hbar^2 \frac{3}{2} I(I+1) \left[J^{(1)}(\omega_0) + J^{(2)}(2\omega_0)\right] \tag{4.58}$$

4.5.2 Spin-spin relaxation

The case of transverse relaxation is not amenable to such an approach, since here we are dealing with a process that is naturally described in the rotating-frame, where the transverse magnetisation is stationary. In this frame of reference there is no large zeroth order Hamiltonian that will dominate the dipolar interaction. Instead the behaviour of the magnetisation is best handled using the density operator formalism. Following eqn 3.53, the density matrix in the rotating frame, $\rho^*(t)$, obeys

$$i\frac{d\rho^*(t)}{dt} = [\mathcal{H}_D^*(t), \rho^*(t)] \tag{4.59}$$

where $\mathcal{H}_D^*$ is the transformed dipolar Hamiltonian, $\exp(i\omega_0 t I_z)H_D(t)\exp(-i\omega_0 t I_z)$. Once the evolution of $\rho^*(t)$ is calculated, all relevant spin properties, such as the decay of the transverse magnetisation, can be determined. The problem with an equation such as 4.59 is that it cannot easily be integrated. To first order, however [9],

$$\frac{d\rho^*(t)}{dt} = -i[H_D^*(t), \rho^*(0)] - \int_0^t [H_D^*(t), [H_D^*(t'), \rho^*(0)]]dt' \tag{4.60}$$

Taking the ensemble average, noting that $\overline{H_D^*(t)} = 0$, replacing $t' = t-\tau$, and extending the integration limit to ∞ , one obtains [9] the vital result from which the decay of M_y (ie $Tr[\rho^* I_y]$) can be calculated, namely,

$$\frac{d\rho^*(t)}{dt} = -\int_0^\infty \overline{[H_D^*(t), H_D^*(t-\tau), \rho^*(0)]]}d\tau \tag{4.61}$$

This linear differential equation leads to exponential relaxation, a key property of Bloembergen, Purcell and Pound (BPP) theory [20]. Equation 4.61 contains oscillatory phase factors due to the transformation of H_D to the rotating frame and so, like eqn 4.57, can be shown to to comprise the Fourier spectra of dipolar correlation functions, $G^q(\tau)$. The various matrix element evaluations in the calculation of dM_y/dt lead to

$$\frac{1}{T_2} = \left(\frac{\mu_0}{4\pi}\right)^2 \gamma^4 \hbar^2 \frac{3}{2} I(I+1)[\frac{1}{4}J^{(0)}(0) + \frac{5}{2}J^{(1)}(\omega_0) + \frac{1}{4}J^{(2)}(2\omega_0)] \tag{4.62}$$

The assumptions involved in the various steps leading to eqn 4.62 are delicate, but the key assumption underpinning BPP theory is $\overline{\langle H_D^2 \rangle}^2 \tau_c^2 \ll 1$ (or $M_2 \tau_c^2 \ll 1$. Thus the relevant timescale for fast motion in the BPP sense is the 'precession period in the dipolar field'. This is simply the inverse of the dipolar linewidth in the absence of motion, $M_2^{1/2}$. The regime of motion in which $M_2\tau_c^2 \ll 1$ is termed 'motionally narrowed'.

One other relaxation time can be calculated using the spectral density approach. This is the rotating-frame relaxation time, $T_{1\rho}$, which describes the rate at which transverse magnetisation decays in the presence of an RF field, B_1. Provided the rotating frame Zeeman energy of the spins in the RF field exceeds the residual dipolar interaction (i.e. the NMR linewidth) the magnetisation is said to be 'spin-locked' and the perturbative treatment of relaxation, which applies in the case of the T_1 process along the B_0 Zeeman field in the laboratory frame, also applies for the $T_{1\rho}$ process along the B_1 Zeeman field in the rotating frame [41]. The result is derived from an expression akin to eqn 4.57, but where m and n refer to eigenstates along I_y and $H_D(t)$ is now $H_D * (t)$, the dipolar interaction in the rotating frame. The relevant matrix elements in eqn 4.57 cause selection rules $n - m = 0, \pm1, \pm2$, and their evaluation yields

$$\frac{1}{T_{1\rho}} = \left(\frac{\mu_0}{4\pi}\right)^2 \gamma^4\hbar^2\frac{3}{2}I(I+1)[\frac{1}{4}J^{(0)}(2\omega_1) + \frac{5}{2}J^{(1)}(\omega_0) + \frac{1}{4}J^{(2)}(2\omega_0)] \qquad (4.63)$$

As expected, the expression for $T_{1\rho}$ equals that for T_2 in the limit as $\omega_1 \to 0$ although in practice the spin locking condition breaks down as the RF field amplitude is decreased.

In order to appreciate the physical significance of eqns 4.58 and 4.62, it is helpful to evaluate them for a simple isotropic rotational diffusion model, an excellent representation of the fluctuations in dipolar interactions that occur in most liquids. Here the $J^{(q)}$ are given by [9]

$$J^{(0)}(w) = \frac{24}{15r_{ij}^6}\frac{\tau_c}{1+\omega^2\tau_c^2}$$

$$J^{(1)}(w) = \frac{4}{15r_{ij}^6}\frac{\tau_c}{1+\omega^2\tau_c^2}$$

$$J^{(2)}(w) = \frac{16}{15r_{ij}^6}\frac{\tau_c}{1+\omega^2\tau_c^2} \qquad (4.64)$$

where τ_c is the rotational correlation time (for reorientation of the rank 2 spatial tensor). The result of substituting these spectral density functions in the expression for T_1 and T_2 is shown in Fig. 4.15. The most obvious feature is the existence of two distinct motional regimes separated by a minimum in T_1 when the correlation time is of order the Larmor period. The regime corresponding to $\tau_c^{-1} \gg \omega_0$ is characterised by identity of T_1 and T_2 and, in accordance with the reduction in the homogeneous linewidth, $1/\pi T_2$, as T_2 increases, is termed 'extreme narrowed'. Such a regime applies typically for small molecules in the liquid state (such as water at room temperature), where correlation times are of order 10^{-12} s to 10^{-14} s.

The divergence of T_1 and T_2 below the T_1 minimum is characteristic of highly viscous liquids, concentrated flexible polymers, and semi-rigid polymers, where rotational tumbling of the inter-nuclear vector r_{ij} is slowed or restricted. Molecular motion in which $M_2^{1/2} \ll \tau_c^{-1} \ll \omega_0$ is identified by $T_2 \ll T_1$.

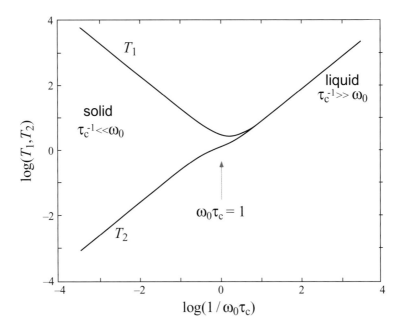

Fig. 4.15 T_1 and T_2 relaxation as a function of correlation time τ_c at fixed Larmor frequency, ω_0.

All three expressions, eqns 4.64, correspond to a dipolar interaction between a pair of like spins. Where a spin experiences a dipolar interaction with more than one other spin, these expressions can be scaled up proportionately. Similarly, if the interaction arises between spins in different molecules in relative translational motion some effective value for $1/r_{ij}^6$ is required. Furthermore, the expressions can be simply modified for dipolar interactions between unlike spins (such as ^{13}C and ^1H) or to allow for different types of perturbative interactions such as the quadrupole interaction. Such formulae are covered in detail elsewhere [9, 10]. In all such analyses the motional features apparent in Fig. 4.15 remain the same.

4.5.3   Motional averaging

When the reorientation of the dipolar interaction slows sufficiently that $M_2\tau_c^2 \gtrsim 1$, the line-narrowing assumption inherent in the BPP theory breaks down. However, the perturbation theory approach that yields the spin–lattice relaxation behaviour is still quite good, since $\gamma B_0 \gg M_2^{-1/2}$. Consequently the T_1 formula, eqn 4.56, is applicable in the slow motion limit. In the case of solids, it is clear that T_1 will be long (usually many seconds) and in insulators will arise from the spectrum of lattice vibrations.

For T_2 in solids and semi-solids, a different approach is required. A nice treatment, due to Anderson and Weiss [43], represents the fluctuating dipolar fields as a a time-dependent Larmor frequency, $\Delta\omega(t)$, which varies randomly such that the time-averaged and ensemble-averaged mean square values are the same, namely, $\overline{\Delta\omega^2}$. Note that $\Delta\omega(t)$ is measured in the rotating frame. It represents the offset from the average longitudinal field precession frequency, ω_0. The normalised free induction decay, $G(t) = S(t)/S(0)$, is therefore,

$$G(t) = \overline{\exp(i \int_0^t \Delta\omega(t')dt')} \tag{4.65}$$

There are a great many problems in physics where such an average of phase factors are required to be calculated, but in magnetic resonance eqn 4.65 is ubiquitous. We entered this discussion from a consideration of fluctuating dipolar interactions. But any randomly varying local field, any fluctuating term in the spin Hamiltonian, will require us to address the matter of phase factors with exponents $i \int_0^t \Delta\omega(t')dt'$. The Anderson–Weiss method is therefore of major significance in dealing with a wide class of magnetic resonance problems for which the effect of time-varying interactions is to be calculated.

Two assumptions make the problem tractable. The first is that the distribution of $\Delta\omega(t)$ is Gaussian, in which case the distribution of $X(t) = \int_0^t \Delta\omega(t')dt'$ is also Gaussian[10] with probability distribution

$$P(X) = [2\pi\overline{X^2}]^{-1/2} \exp(-\frac{1}{2}X^2/\overline{X^2}). \tag{4.66}$$

This leads to the result

$$\overline{\exp(i \int_0^t \Delta\omega(t')dt')} = \int P(X)\exp(iX)dX$$

$$= \exp(-\frac{1}{2}\overline{X^2}) \tag{4.67}$$

The second assumption is that the local field fluctuation is described in the usual manner by a correlation function[11]

$$\overline{\Delta\omega(t)\Delta\omega(t-\tau)} = \overline{\Delta\omega^2}g(\tau) \tag{4.68}$$

This allows one to calculate $\overline{X^2}$ as $2\overline{\Delta\omega^2} \int_0^t (t-\tau)g(\tau)d\tau$ so that

$$G(t) = \exp(-\overline{\Delta\omega^2} \int_0^t (t-\tau)g(\tau)d\tau) \tag{4.69}$$

This Anderson–Weiss expression is an extremely useful result, applicable in a wide range of dynamical situations. For example we shall use this result in Chapter 6 to deal with spin relaxation in microscopically inhomogeneous media. But one of the most important insights provided is an understanding of slow and fast motion limits.

---

[10]This follows from the Central Limit Theorem which states that the averaged sum of a sufficiently large number of identically distributed independent random variables each with finite mean and variance will be approximately normally distributed.

[11]Note that for dipolar interactions, $g(\tau)$ is a sort of 'normalised, q-averaged' $G^q(t)$.

In the slow motion limit, $\overline{\Delta\omega^2}\tau_c^2 \gg 1$, $g(\tau)$ in eqn 4.69 is approximately unity and so as $t \to 0$

$$G(t) = \exp(-\frac{1}{2}\overline{\Delta\omega^2}t^2) \qquad (4.70)$$

Strictly speaking such a Gaussian decay is due to inhomogeneous broadening and cannot strictly be termed 'relaxation', since the signal may be recovered by using appropriate pulse sequences as discussed in Section 4.6.2. Nonetheless it tells us that that signal decay is rapid when the motion is slow. In the fast motion limit $\overline{\Delta\omega^2}\tau_c^2 \ll 1$, $g(\tau)$ decays rapidly in eqn 4.69 and the integral approximates $t \int_0^\infty g(\tau)d\tau$, leading to a FID given by $\exp(-\overline{\Delta\omega^2}\tau_c t)$. Such an exponential decay corresponds to a transverse relaxation rate $1/T_2 = \overline{\Delta\omega^2}\tau_c$. Note that the shorter τ_c, the longer the relaxation time T_2. Fast motion leads to slow relaxation decay. This contrary aspect of motional averaging is often counterintuitive to those who do not live in the world of magnetic resonance.

Before leaving the Anderson–Weiss model we note that it reproduces two familiar results in the slow and fast motion limits of the dipolar interaction. We start with slow motion. Noting $\overline{\Delta\omega^2} = M_2$, eqn 4.69 is precisely the solid-state Gaussian decay behaviour of eqn 4.54. For the fast motion limit, an exponential decay is found and, using the definition of the second moment in eqn 4.55,

$$\frac{1}{T_2} = M_2\tau_c$$

$$= \left(\frac{\mu_0}{4\pi}\right)^2 \gamma^4\hbar^2\frac{3}{2}I(I+1)\frac{1}{r_{ij}^6}\tau_c. \qquad (4.71)$$

Hence the T_2 relaxation rate is equivalent to the dominant zero-frequency term of the BPP theory (eqn 4.62), namely

$$\frac{1}{T_2} \approx \left(\frac{\mu_0}{4\pi}\right)^2 \gamma^4\hbar^2\frac{3}{2}I(I+1)\left[\frac{1}{4}J^{(0)}(0)\right] \qquad (4.72)$$

What happened to the remaining terms? In short, they are absent because, like those that determine the T_1 relaxation rate, they arise from resonant transitions that are unrepresented in the Anderson–Weiss model. But for fluctuations slower than the Larmor frequency ($\omega_0\tau_c \gg 1$), but still fast enough to result in motional narrowing ($M_2^{1/2}\tau_c \ll 1$), the Anderson–Weiss model works extremely well.

4.6 Pulse sequences

4.6.1 Basic spin manipulation

Among the enormous number of pulse sequences that are used in modern NMR are four that have their origins in the early beginnings of the subject but the usefulness of which is enduring. These are the inversion recovery sequence, the simple Hahn echo [19], the Carr–Purcell echo train [44, 45], and the stimulated echo [46]. Each concerns the manipulation of the spin system under the influence of the Zeeman

Hamiltonian and the T_1 and T_2 relaxation processes. Each can be understood using a simple vector description of the nuclear magnetisation. For more sophisticated pulses sequences, and especially when Hamiltonian terms bilinear in the spin operator play a role, the vector picture lets us down. We simply cannot understand such phenomena without recourse to quantum mechanics. That doesn't make for difficulty. On the contrary, a simple and effective quantum description of spin evolution and resulting NMR signals can be found by using the density matrix expressed in terms of spin operators.

In the sense that NMR fundamentally concerns quantum coherence, the idea of 'coherence transfer' is central to any understanding of evolution under the influence of the spin Hamiltonian. Here the precession diagrams introduced in Chapter 3 are particularly helpful. We start with the basics of precession under static magnetic fields and resonant RF fields as shown in Fig. 4.16. The cases illustrated correspond to the rotating frame of reference, with transverse magnetisation precessing under a Zeeman offset and, for the interchange of I_z with I_x and I_y, under a resonant RF pulse.

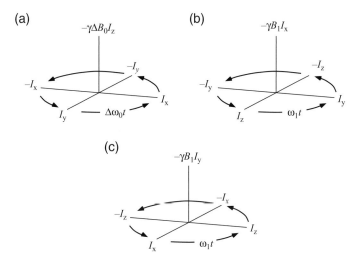

Fig. 4.16 Rotating-frame precession diagrams for states of the density matrix in the presence of a Zeeman interaction: (a) evolution of I_y under the off-resonant Zeeman interaction, $-\Delta\omega_0 I_z$; (b) I_z, under the influence of the RF field $-\omega_1 I_x$; and (c) I_x under $-\omega_1 I_y$.

Later in this section we visit the more sophisticated sequences involving coherence transfer amongst the 'non-vectorial' components of Liouville space. But even for the simplest sequence involving Zeeman terms alone, the density matrix approach will often give us the quickest route to understanding. In consequence, we will present such a description in parallel with our simple vector diagrams.

Usually, each application of a particular pulse sequence to the NMR spin system is not performed in single shot but in rapid succession, the resulting signal from each experiment being successively added. It is this aspect of NMR methodology that we consider first.

Signal averaging

Because of the low sensitivity of NMR, it is customary to co-add signals from N successive experiments in order to enhance the signal-to-noise ratio. Successive addition has the effect that the signals add coherently while the noise adds in random phase. Because the noise power is additive, the root mean-squared (rms) noise amplitude is proportional to $N^{1/2}$. As a result the signal-to-noise ratio improves as $N/N^{1/2}$ or $N^{1/2}$. There is, however, a limitation to the rate at which successive additions can be performed. To retain full signal strength the nuclear spin system must be allowed to recover its z-axis magnetisation between experiments, and this recovery requires a time delay of several T_1-relaxation times. This recovery imposes a limit to the number of co-additions that are possible in a given time duration.

Of course it is possible to repeat excitation pulses more rapidly than the time taken for full recovery, in which case the spins establish an equilibrium longitudinal magnetisation before each excitation pulse that is less than the thermal equilibrium value. This effect is known as partial saturation [47]. The time over which the longitudinal magnetisation recovers before the next excitation pulse is the simply the pulse repetition time. Given a repetition time T_R and longitudinal relaxation time T_1, the equilibrium magnetisation immediately before the RF pulse for a repetitive single θ_x pulse experiment is

$$M_z = M_0 \frac{\exp(-T_R/T_1)}{1 - \cos\theta \exp(-T_R/T_1)} \qquad (4.73)$$

The method by which this result is obtained is instructive since it is generally applicable to more complicated pulse sequences. The dynamic equilibrium is established by equating the z-magnetisation just before the RF pulse, $M_z(0_-)$ to the z-magnetisation, $M_z(T_R)$, that remains from the preceding sequence, following relaxation. In the derivation of eqn 4.73, a simplifying assumption is made, namely that the role of transverse magnetisation can be neglected because of irreversible decay after each excitation and signal acquisition. Using the symbol $M_z(0_+)$ to represent the z-magnetisation immediately following the RF pulse, we can write [14]

$$M_z(0_+) = M_z(0_-) \cos\theta \qquad (4.74)$$

and

$$M_z(T_R) = M_z(0_+) \exp(-T_R/T_1) + M_0(1 - \exp(-T_R/T_1)) \qquad (4.75)$$

Equating $M_z(T_R)$ and $M_z(0_-)$, eqn 4.73 follows directly.

The initial amplitude of the transverse magnetisation just after the θ pulse is

$$M_y = M_0 \frac{\exp(-T_R/T_1}{1 - \cos\theta \exp(-T_R/T_1)} \sin\theta \qquad (4.76)$$

The maximum partial saturation signal amplitude is therefore not obtained for $\theta = 90°$, except where T_R/T_1 is large. An optimum condition can be found by adjusting the nuclear turn angle θ to a value known as the Ernst angle, θ_E, for which

$$\cos\theta_E = \exp(-T_R/T_1) \qquad (4.77)$$

For T_R/T_1 of around three, the relaxation between pulses is fairly complete and the optimum turn angle close to $90°$. The signal-to-noise advantage in a repetitive pulse experiment, in which T_R is chosen to be much shorter than this and in which θ is set to θ_E, is around $\sqrt{2}$.

It should, however, be noted that use of the Ernst angle requires care. Incomplete transverse relaxation between successive experiments can lead to unexpected signals appearing in multiple pulse sequences [14, 48]. Pulsed magnetic field gradient experiments, however, are not necessarily subject to this effect because the gradient pulses used during acquisition of the signal cause a destruction of magnetisation coherence. This feature facilitates the use of rapidly repeated low turn-angle pulse experiments.

Whatever the repetition delay and nuclear turn-angle employed, the T_1 relaxation time provides a fundamental limit to the available signal-to-noise ratio.

Phase cycling

Because the phase of the signal depends on the RF pulse phase, it becomes possible to distinguish the NMR FID signal from any background interference by RF phase alternation. For example, incrementing the RF phase by 180° will lead to signal inversion. Thus a successive phase alternation in 180° steps linked to successive addition and subtraction of the incoming signal from the data sum will lead to coherent superposition of the FID while the background interference is nulled. The process of addition and subtraction can be thought of as an alternation of the receiver phase so that the phase cycle can be written (0°,0°)-(180°,180°), where the RF transmitter and receiver phases are given in brackets. This particular cycle is called coherent noise cancellation.

The (0°,0°)-(180°,180°) sequence represents the simplest possible form of phase cycling. Other more sophisticated schemes enable the spectroscopist to correct for phase and amplitude anomalies in quadrature detection [49], transverse magnetisation interference due to rapid pulse repetition [50], and echo artifacts due to pulse amplitude errors. Among these phase and amplitude errors in quadrature detection are the primary cause of artifacts in NMR imaging and they can be nicely corrected in the 4-pulse CYCLOPS sequence of Hoult and Richards [49], a sequence which is so useful that it is generally incorporated as a subcycle of all other phase cycles.

When the amplitudes of the quadrature signal channels are unmatched or the relative phases are not precisely 90° apart, the complex Fourier transformation results in foldback artifacts. By successively swapping the channels used to acquire the real and imaginary signals, equivalent to a 90° transmitter and receiver phase-shift, the error is compensated to first order as shown in Fig. 4.17. The CYCLOPS sequence also incorporates the usual 0°/180° alternation of coherent noise cancellation, giving a four-pulse cycle (0°,0°)-(90°,90°)-(180°,180°)-(270°,270°).

Generally, phase cycling procedures are performed via the data acquisition software. Most modern NMR spectrometers select the phase of the RF transmitter by digital means so that the RF pulse phase can be simply manipulated through the software, usually in steps of 90°, although smaller steps are also available for some experiments involving multiple quantum filters. Similarly it is possible to 'adjust the receiver phase' in 90° steps. In practice, this is usually performed not in the phase sensitive detector, but after digitisation, by means of a software 'trick'. Whereas a shift of 180° corresponds to negation of both the incoming real and imaginary signals, a shift of 90° corresponds to negation of the incoming real followed by an interchange

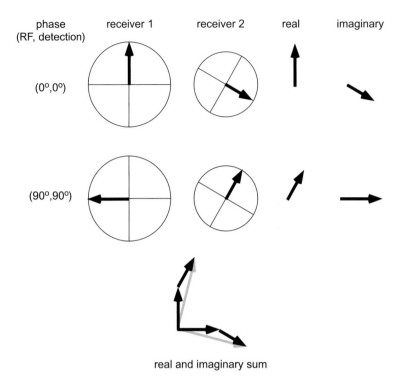

phase (RF, detection)   receiver 1   receiver 2   real   imaginary

(0°,0°)

(90°,90°)

real and imaginary sum

Fig. 4.17 The CYCLOPS method of phase cycling. The two receivers are shown as having different gains and phase setting not precisely in quadrature. By alternating both the transmitter and receiver phase between $0°$ and $90°$, the real and imaginary signals are routed through both receivers and the summed resultants have matched amplitudes and quadrature phases to first order.

of the real and imaginary signals. It is of course this interchange that is so essential to the success of the CYCLOPS cycle.

Inversion recovery

Inversion recovery [51] is used for the measurement of T_1-relaxation times, as well as for the selective suppression of unwanted spin signals. The RF pulse sequence and resulting magnetisation trajectories are shown in Fig. 4.18. The first $180°$ RF pulse inverts the magnetisation vector, so subjecting the system to the most severe disturbance from equilibrium. Spin–lattice relaxation proceeds for a time t, following which a $90°_x$ pulse is used to inspect the remaining longitudinal magnetisation. The signal amplitude is described by

$$M_y(t) = M_0(1 - 2\exp(-t/T_1)) \tag{4.78}$$

A noteworthy feature of this method is the crossover from a negative signal (proportional to $-M_0$) at $t = 0$ to a positive signal (proportional to M_0) as $t \to \infty$. This crossover through zero magnetisation occurs at $t = 0.6931T_1$ and can be exploited to good effect. First it is clear that a measurement of time for the crossover null can yield T_1 in a single measurement. Second, the method may be used to remove the

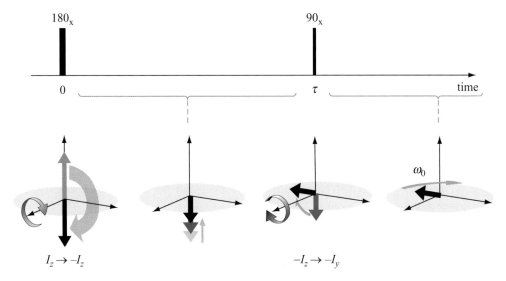

Fig. 4.18 Inversion recovery pulse sequence and magnetisation trajectories. At each RF pulse the grey arrow represents the magnetisation prior to the pulse, while the black vector represents the magnetisation after the pulse. Also shown is the evolution of the density matrix under the action of the pulse sequence.

contribution from a spin with specific T_1-value by applying a pulse at a time interval $0.6931T_1$ before the experimental pulse sequence. Such signal suppression is particularly important in biological experiments where water suppression is desired. While the inversion recovery method of signal suppression is simple, it can lead to distortions in the NMR spectrum. Other more sophisticated and effective methods for selective signal suppression include binomial pulse sequences [52] and SUBMERGE [53].

4.6.2 Echoes

Spin echo

Magnetic field inhomogeneity causes nuclear spins to precess at differing Larmor frequencies according to their location in the sample, an effect which forms the basis of the NMR imaging method. Even in conventional NMR experiments, where no magnetic field gradient is deliberately applied, the inhomogeneity of the polarising magnet will result in a field spread across the sample of ΔB_0. This spread causes a dephasing of transverse magnetisation following a 90_x° RF pulse. Transverse magnetisation phase coherence therefore lasts only for a time of order $(\gamma \Delta B_0)^{-1}$, a transience which apparently constrains the length of time over which this magnetisation can be manipulated or detected. Many years ago Erwin Hahn [19] discovered that this loss of phase coherence was inherently reversible. Application of a second 180° RF pulse after a time delay τ will cause refocusing at 2τ, as shown in Fig. 4.19, and the resulting phase coincidence is known as a spin echo, with the phase of the 180° pulse determining the sign of the echo signal.

At this point it is useful to introduce the idea of precession using the density matrix picture. Equation 3.53 describes the evolution of the density matrix in terms of operators involving the Hamiltonian. We will not solve these equations explicitly in the case of the spin echo but we will describe the successive evolution stages. To help visualise this process we can use rotating frame 'precession diagrams', like those shown in Fig. 4.16. Here the Hamiltonian term is in each case a Zeeman interaction, in (a) caused by an offset, ΔB_0, in magnetic field, applied along the z-axis, and giving a Hamiltonian $\Delta\omega_0 I_z$; in (b) caused by an RF pulse θ_x; and in (c) caused by an RF pulse θ_y.

In thermal equilibrium the spin system density matrix has a state of longitudinal polarisation so that the density matrix is proportional to I_z. The first RF pulse, applied along the x-axis in the rotating frame, corresponds to a Hamiltonian $\gamma B_1 I_x$, which causes a cyclic precession of I_z to I_y and back to I_z at a rate γB_1, a process which is illustrated in Fig. 4.16. 90° rotation is achieved by terminating the pulses at a time when ρ has evolved into I_y. Following the RF pulse, the new density matrix is now subject to the local Zeeman field offset.[12] This next stage in the evolution, the precession due to the Hamiltonian, $\Delta B_0 I_z$, is shown in Fig. 4.16. The evolution diagram for the second RF pulse can be easily worked out by cyclically permuting the vectors of Fig. 4.16. A 180°_y pulse inverts the I_x terms in the density matrix, but leaves the I_y terms unchanged. If the time of initial evolution in the field offset ΔB_0 was τ, the final evolution for an equal time τ in the same field offset ΔB_0 will therefore cause a perfect restoration of the I_y density matrix state that existed immediately after the first RF pulse, a process which is also shown in Fig. 4.19. It is important to note that the phase of the 180° inversion pulse is relevant. A 180°_x pulse, for example, results in an echo of negative sign.

We can follow the evolution of the density matrix polarisation state by using a succession of arrows to represent the time progression, where the Hamiltonian term under which ρ evolves is written above the arrow. For example, the evolution progression for the Hahn echo would have the following appearance:

$$I_z$$
$$\xrightarrow{-(\pi/2)I_x} \quad I_y$$
$$\xrightarrow{-\Delta\omega_0\tau I_z} \quad I_y\cos\phi + I_x\sin\phi$$
$$\xrightarrow{-(\pi)I_y} \quad I_y\cos\phi - I_x\sin\phi$$
$$\xrightarrow{-\Delta\omega_0\tau' I_z} \quad I_y\cos\phi\cos\phi' + I_x\cos\phi\sin\phi' - I_x\sin\phi\cos\phi' + I_y\sin\phi\sin\phi'$$

$$(4.79)$$

where ϕ and ϕ' are the precessional phase shifts, $\Delta\omega_0\tau$ and $\Delta\omega_0\tau'$, τ and τ' being evolution times in the dephasing and rephasing intervals of the echo sequence. Note that magnetic field inhomogeneity leads to a distribution of offsets, $\Delta\omega_0$, and hence a dephasing of the transverse magnetisation. However, the effect on the density matrix

---

[12]Of course, this field was also present during the RF pulse but we are able to neglect its effect by ensuring that the RF field strength is very much greater than ΔB_0.

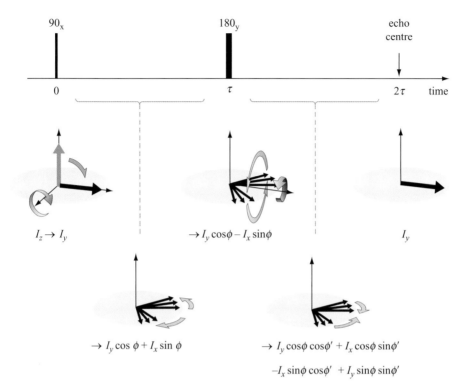

Fig. 4.19 Spin-echo pulse sequence showing the evolution of magnetisation and the density matrix. Note that the 180°_x pulse inverts the phase of each spin isochromat (i.e. $\phi \to -\phi$) so that perfect refocusing occurs at time $t = 2\tau$ where $\phi = \phi'$. The diagram shows a sample of three isochromats.

of the 180° RF pulse $(-(\pi)I_y)$ is to make it appear as though the time interval τ (and hence the phase evolution ϕ) were reversed in sign. This 'time-reversal' property lies at the heart of all echo methods.

Consider the case where the intervals τ and τ' are equal so that $\phi = \phi'$. Then the final state of the density matrix is

$$I_y \cos^2 \phi + I_x \cos \phi \sin \phi - I_x \sin \phi \cos \phi + I_y \sin^2 \phi = I_y \tag{4.80}$$

The 180°_y RF pulse (denoted $-(\pi)I_y$) causes the phases to be inverted (ie angle $\Delta\omega_0\tau$ becomes $-\Delta\omega_0\tau$) so that the effect of the next evolution period τ' is to restore the the polarisation to I_y at the time $\tau' = \tau$. Of course, I_y is an observable under the transverse magnetisation detection operator $I_x + iI_y$. In other words

$$Tr([I_x + iI_y][I_y]) = iTr(I_y^2)$$
$$= i\frac{1}{3}(2I + 1)I(I + 1) \tag{4.81}$$

We may use eqn 3.88 to relate the expectation value, $\langle I_x + iI_y \rangle$, of our detection operator to the initial expectation value of longitudinal angular momentum, $\langle I_z \rangle_0 = Tr(I_z\rho_0)$, namely

$$\langle I_x + iI_y \rangle = i \langle I_z \rangle_0 \tag{4.82}$$

Note again that all observables are in fact real numbers. The use of the complex numbers here is simply a device that we choose to utilise in order to independently manage the in-phase and quadrature channels. What eqn 4.82 states is that our signal derives from the full $\langle I_z \rangle_0$ angular momentum of the initial thermal equilibrium state and that it is detected in the quadrature channel, in other words, along I_y in the rotating frame.

Suppose the second RF pulse were 90°_x, identical to the first.[13] The density matrix treatment makes the analysis simple.

$$
\begin{aligned}
I_z & \\
\xrightarrow{-(\pi/2)I_x} \quad & I_y \\
\xrightarrow{-\Delta\omega_0\tau I_z} \quad & I_y \cos\phi + I_x \sin\phi \\
\xrightarrow{-(\pi/2)I_x} \quad & -I_z \cos\phi + I_x \sin\phi \\
\xrightarrow{-\Delta\omega_0\tau' I_z} \quad & -I_z \cos\phi + I_x \sin\phi \cos\phi' - I_y \sin\phi \sin\phi'
\end{aligned}
\tag{4.83}
$$

The echo appears where the intervals τ and τ' are equal, so that $\phi = \phi'$ and the final state of the density matrix is

$$-I_z \cos\phi + I_x \sin\phi \cos\phi - I_y \sin^2\phi \tag{4.84}$$

Noting $Tr(I_z I_x) = Tr(I_y I_z) = Tr(I_y I_x) = 0$, the detection operator gives

$$
\begin{aligned}
Tr([I_x + iI_y]\rho) &= Tr([I_x + iI_y]\overline{[-I_z \cos\phi + I_x \sin\phi \cos\phi - I_y \sin^2\phi]}) \\
&= Tr(I_x^2)\overline{\sin\phi\cos\phi} - iTr(I_y^2)\overline{\sin^2\phi} \\
&= -\frac{1}{2}iTr(I_y^2)
\end{aligned}
\tag{4.85}
$$

The overbar represents the averaging needed to account for the distribution of offsets resulting from the magnetic field inhomogeneity. For a uniform distribution of precessional phase shifts, $\phi = \Delta\omega_0\tau$, this average gives $\overline{\sin\phi\cos\phi} = 0$ and $\overline{\sin^2\phi} = 1/2$ and hence leads to an echo signal $-\frac{1}{2}iTr(I_y^2)$, exactly -1/2 of that found for the echo obtained using the 180°_y refocusing pulse.

In spin systems experiencing only Zeeman Hamiltonians, the spin-echo sequence refocuses all dephasing due to inhomogeneous broadening, chemical shift, and heteronuclear scalar spin–spin interactions. Because the 180° RF pulse inverts all spins and so leaves the interaction term $\pi J 2\mathbf{I}_1 \cdot \mathbf{I}_2$ invariant, homonuclear scalar spin–spin interactions are not refocused and therefore remain to modulate the echo. Over and above this modulation, residual attenuation of the echo is due to spin–spin relaxation

---

[13]Historically, this was the first echo experiment performed by Hahn.

alone and, in principle, a plot of echo amplitudes $M_y(2\tau)$ obtained from differing τ values can be used to yield T_2 according to

$$M_y(2\tau) = M_0 \exp(-2\tau/T_2) \tag{4.86}$$

Earlier we saw that sampling of the decaying NMR signal at positive times only leads to the existence of a dispersion signal in the imaginary domain following Fourier transformation. Figure 4.20 suggests a mechanism by which 'negative' time may be sampled. If we take the centre of the echo as our time origin and begin sampling the signal at this point, the resulting spectrum is the same as that shown earlier for the single-pulse experiment, with the exception that T_2 relaxation will have attenuated the signal at $t = 0$ and hence reduced the area under the absorption peak by the same factor. [14] Furthermore the dispersion spectrum will still exist.

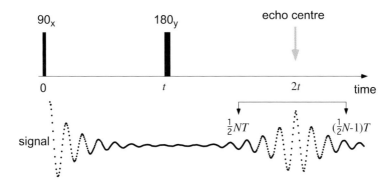

Fig. 4.20 Echo signal in a simple spin echo (real part only shown). Note that the echo centre represents a local time origin. Because of echo symmetry, sampling of the full echo leads to a spectrum without dispersion.

Suppose, however, that we were to sample the entire echo, dividing our sampling time period equally before and after the echo, with $t = 0$ at the echo centre. In this case the digitised time values run between $-\frac{1}{2}NT$ and $(\frac{1}{2}N - 1)T$, and, as before, the Fourier-domain points run between $-\frac{1}{2}N(1/NT)$ and $(\frac{1}{2}N - 1)(1/NT)$. Neglecting T_2 effects and presuming a signal phase $\phi = 0$, it is clear that the in-phase and quadrature signals are, respectively, symmetric and antisymmetric, leading to an entirely real spectrum. If $\phi \neq 0$ then the spectrum is multiplied by $\exp(i\phi)$ and will contain both real and imaginary parts, although the real spectrum can be recovered by the usual 'phasing' process. There is, however, a crucial difference between this situation the case of $t > 0$ sampling. When full echo sampling is used the resulting modulus spectrum for $\phi \neq 0$ exactly reproduces the real spectrum for $\phi = 0$, a restoration process which is impossible in the case of data sampled only at positive times because of the existence of the dispersion component in the spectrum.

---

[14]For the purpose of the present argument we will neglect the role of bilinear spin interactions such as homonuclear couplings, which are non-refocusable via the Hahn echo and remain to modulate the echo envelope.

CPMG

The phase coherence recovered in the nuclear spin echo is subsequently lost for $t > 2\tau$. Successive recoveries are, however, possible if a train of additional 180° RF pulses is used, as suggested by Carr and Purcell in 1954 [44]. The choice of phase with which these pulses are applied is important if the cumulative effects of small turn-angle errors is to be avoided. In the Meiboom–Gill modification [45] to the Carr–Purcell train, the use of quadrature 180°_y pulses provides the appropriate compensation. The envelope of the echoes in a Carr–Purcell–Meiboom–Gill (CPMG) sequence is determined by T_2 decay alone, and so it is possible to determine T_2 in a single experiment. The method is illustrated in Fig. 4.21.

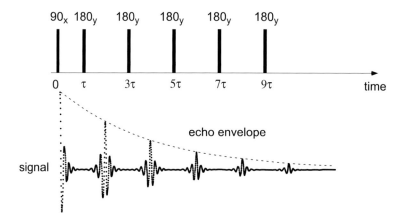

Fig. 4.21 CPMG pulse sequence exhibiting multiple spin echoes at times $2n\tau$, modulated by a T_2 relaxation envelope.

The production of multiple echoes in the CPMG pulse sequence suggests an obvious application in signal averaging. Co-addition of echoes within a train leads to signal-to-noise enhancement [54] in addition to that obtained by addition of the independent experiments separated by the T_1 recovery period, T_R.

Stimulated echo

In many materials, especially those with molecules undergoing motion that is slow compared with the period of the nuclear Larmor precession, the transverse relaxation time T_2 is considerably shorter than T_1. The fact that the 'lifetime of spin polarisation' exceeds the 'lifetime of first-order spin coherence' can be turned to advantage where one wishes to 'store' coherence over a long time interval. Suppose that the transverse magnetisation existing at some point of time, τ, is required to be stored for later recall. An example might be where one wishes to use this magnetisation at a later time to see how far the molecules containing the NMR nuclei have moved. The method used is shown in Fig. 4.22, where a single 90°_x pulse applied after a time t has the effect of rotating the y-component of magnetisation into longitudinal polarisation along the z-axis, a state in which only T_1 relaxation will occur. Of course any x-magnetisation will be unaffected, so that only half the transverse magnetisation can be stored in this way. Recall is made at some later stage using another 90°_x pulse. As shown, this leads to an echo at time τ after the last RF pulse. This stimulated echo [19] is of particular importance where the translational motions of molecules are being measured using pulsed field gradient spin-echo methods.

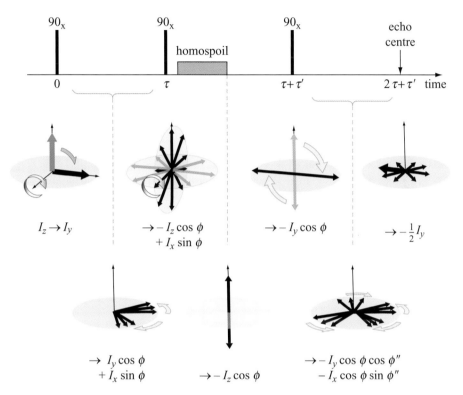

Fig. 4.22 stimulated-echo pulse sequence incorporating a 'homospoil' gradient pulse during the 'z-storage' period τ'. The final density matrix is obtained from an ensemble average over the isochromats $-I_y \overline{\cos\phi\cos\phi''} - I_x \overline{\cos\phi\sin\phi''}$. At $\tau'' = \tau$, $\phi'' = \phi$ and $\overline{\cos^2\phi} = \frac{1}{2}(1 + \overline{\cos(2\phi)}) = 1/2$, while $\overline{\cos\phi\sin\phi} = \frac{1}{2}\overline{\sin(2\phi)} = 0$. Note that the final echo state represents a vector $-\frac{1}{2}I_y$ along with an ensemble spread of $-\frac{1}{2}(I_y \cos(2\phi) + I_x \sin(2\phi))$, a set of isochromats distributed in the transverse plane whose net contribution is zero.

Let us analyse the stimulated echo using the same density matrix method employed earlier. The pulse sequence chosen consists of three 90°_x RF pulses separated by time intervals τ, τ', and τ''. Then

$$I_z$$
$$\xrightarrow{-(\pi/2)I_x} I_y$$
$$\xrightarrow{-\Delta\omega_0\tau I_z} I_y \cos\phi + I_x \sin\phi$$
$$\xrightarrow{-(\pi/2)I_x} -I_z \cos\phi + I_x \sin\phi$$

$$\xrightarrow{-\Delta\omega_0\tau'I_z} \quad -I_z\cos\phi + I_x\sin\phi\cos\phi' - I_y\sin\phi\sin\phi'$$

$$\xrightarrow{-(\pi/2)I_x} \quad -I_y\cos\phi + I_x\sin\phi\cos\phi' + I_z\sin\phi\sin\phi'$$

$$\xrightarrow{-\Delta\omega_0\tau''I_z} \quad -I_y\cos\phi\cos\phi'' - I_x\cos\phi\sin\phi'' + I_x\sin\phi\cos\phi'\cos\phi''$$

$$-I_y\sin\phi\cos\phi'\sin\phi'' + I_z\sin\phi\sin\phi' \tag{4.87}$$

The detection operator $I_x + iI_y$ gives

$$Tr(I_x + iI_y\rho) = Tr(I_x^2)[-\overline{\cos\phi\sin\phi''} + \overline{\sin\phi\cos\phi'\cos\phi''}]$$
$$+iTr(I_y^2)[\overline{\cos\phi\cos\phi''} - \overline{\sin\phi\cos\phi'\sin\phi''}]$$

$$\tag{4.88}$$

and yields echoes $-\frac{1}{2}iTr(I_y^2)$ at $\tau'' = \tau$, $-\frac{1}{2}iTr(I_y^2)$ at $\tau'' = \tau - \tau'$, and $\frac{1}{2}iTr(I_y^2)$ at $\tau'' = \tau' - \tau$. The first of these is the desired stimulated echo. But the stimulated-echo RF pulse sequence also generates two additional spin echoes. These are, respectively, the echo of the initial pulse FID caused by the second pulse and the echo of that echo caused by the third pulse. Special care is needed to avoid interference between the stimulated echo and the two spin echoes. One effective approach is the use of a homogeneity-spoiling (homospoil) magnetic field gradient pulse applied during the 'z-storage' period between the second and third RF pulses, as shown in Fig. 4.22. This has the effect of destroying the unwanted transverse magnetisation without influencing the magnetisation that has been stored along the z-axis.

Density matrix precessions for coupled spins under $I_{1z}I_{2z}$

Let us revisit the spin–spin (scalar coupling) interaction of eqn 4.49. In the weak coupling case, $Jh \ll \delta$ ($\frac{1}{2}\phi \to 0$), we may write the scalar coupling interaction in its secular form as $H_{scalar_0} = 2\pi JI_{1z}I_{2z}$. Just as the Hamiltonian operator may be written, for coupled spins, in the product basis so may the density matrix, meaning that an operator such as I_{1z} is implicitly an outer product of I_{1z} and the identity for the second spin. Also, since we are dealing with a two-spin problem, we must be careful to always express our density matrix in terms of both spins, so that the starting thermal equilibrium state is $\sim (I_{1z} + I_{2z})$, the θ RF pulse in the rotating frame is the rotation operator $\exp(i\theta(I_{1x} + I_{2x}))$, and the density matrix after a 90_x° pulse is $\sim (I_{1y} + I_{2y})$. Subsequent evolution under the influence of the Zeeman terms, the RF pulses, and the secular part of the scalar coupling, $I_{1z}I_{2z}$, are shown in Fig. 4.23.

As a simple exercise, let us follow the evolution of the density matrix under the influence of the scalar coupling and under the RF pulses of a spin echo. Note that we will need to allow for the effect of both the chemical shift terms present in eqn 4.49. In other words, we allow each spin to have different Larmor frequencies $\delta_1\omega_0$ and $\delta_2\omega_0$. Note that because the Zeeman and scalar coupling terms commute in this weak coupling case, we are entitled to apply each evolution sequentially. The evolution progression for the Hahn echo, up to and including the effect of the 180_y° pulse, would have the following appearance:

$$I_{1z} + I_{2z}$$

$$\xrightarrow{-(\pi/2)(I_{1x}+I_{2x})}$$

$$\xrightarrow{-\delta_1\omega_0\tau I_{1z}+\delta_2\omega_0\tau I_{2z}}$$

$$\xrightarrow{-2\pi JI_{1z}I_{2z}}$$

$$I_{1y} + I_{2y}$$

$$I_{1y}\cos\phi_1 + I_{2y}\cos\phi_2 + I_{1x}\sin\phi_1 + I_{2x}\sin\phi_2$$

$$I_{1y}\cos\phi_1 cos(\pi J\tau) - 2I_{1x}I_{2z}\cos\phi_1\sin(\pi J\tau)$$
$$+I_{1x}\sin\phi_1 cos(\pi J\tau) + 2I_{1y}I_{2z}\sin\phi_1\sin(\pi J\tau)$$
$$+I_{2y}\cos\phi_2 cos(\pi J\tau) - 2I_{1z}I_{2x}\cos\phi_2\sin(\pi J\tau)$$
$$+I_{2x}\sin\phi_2 cos(\pi J\tau) + 2I_{1z}I_{2y}\sin\phi_2\sin(\pi J\tau)$$

$$\xrightarrow{-(\pi)(I_{1y}+I_{2y})}$$

$$I_{1y}\cos\phi_1 cos(\pi J\tau) - 2I_{1x}I_{2z}\cos\phi_1\sin(\pi J\tau)$$
$$-I_{1x}\sin\phi_1 cos(\pi J\tau) - 2I_{1y}I_{2z}\sin\phi_1\sin(\pi J\tau)$$
$$+I_{2y}\cos\phi_2 cos(\pi J\tau) - 2I_{1z}I_{2x}\cos\phi_2\sin(\pi J\tau)$$
$$-I_{2x}\sin\phi_2 cos(\pi J\tau) - 2I_{1z}I_{2y}\sin\phi_2\sin(\pi J\tau)$$

$$(4.89)$$

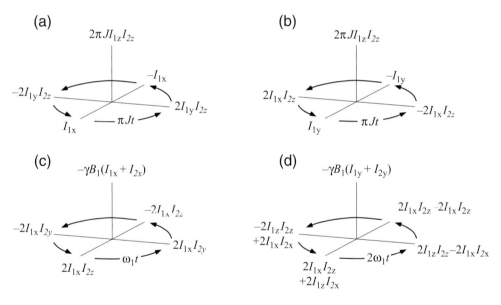

Fig. 4.23 Precession diagrams for states of a scalar coupled spin-$\frac{1}{2}$ pair. The bilinear interaction $2\pi JI_{1z}I_{2z}$ causes the magnetisation states to evolve into two-spin coherences. Note that the RF pulse consists of a sum of spin 1 and spin 2 operators if it is applied non-selectively (i.e. with broad bandwidth). (a) and (b) show evolution of I_x and I_y under the bilinear interaction into states comprising single-quantum coherence of spin 1 in antiphase with spin 2. (c) and (d) show evolution of these antiphase states under the influence of a broadband RF pulse. Note that the precession rate in (d) is $2\omega_1$, meaning that a 45° RF pulse causes a precession of the total spin density matrix state, $(2I_{1x}I_{2z} + 2I_{1z}I_{2x})$, into $(2I_{1z}I_{2z} - 2I_{1x}I_{2x})$. $2I_{1x}I_{2x}$ is an equal admixture of double and single-quantum coherence.

At this point we can see that the evolutions of the 1 and 2 spins are the same, apart from their different Larmor frequencies. Hence we may focus our attention on the 1 spin. The chemical shift evolution in the equal period τ after the 180°_y pulse leads to additional $\cos\phi_1$ and $\sin\phi_1$ modulation. The echo arises from the sum $\cos^2\phi_1 + \sin^2\phi_1 = 1$. Allowing for the scalar coupling evolution as well, these $\cos^2\phi_1$ and $\sin^2\phi_1$ contributions come from each of the four 1-spin terms, respectively, as $I_{1y}\cos^2\phi_1 cos^2(\pi J\tau)$, $-I_{1y}\cos^2\phi_1 sin^2(\pi J\tau)$, $I_{1y}\sin^2\phi_1 cos^2(\pi J\tau)$, and $-I_{1y}\sin^2\phi_1 sin^2(\pi J\tau)$. Thus the full density matrix at the point of the echo is $(I_{1y} + I_{2y})\cos(2\pi J\tau)$.

The modulation of the echo occurs because the 180°_y pulse has the effect of flipping both the 1 and 2 spins, leaving the scalar coupling Hamiltonian term, $2\pi J I_{1z} I_{2z}$, unchanged. The effect of the modulation by $\cos(2\pi J\tau)$ will be apparent in the successive echo amplitudes of a CPMG train. Fourier transformation will lead to a spectrum split into a doublet separated by J in cyclic frequency. Indeed, the very first observation of J-coupling was made using just such an echo train to remove field inhomogeneity effects, while leaving the spin–spin coupling revealed [55, 56].

The solid echo

Echoes arise when it is possible to devise a pulse sequence in which, at some point, undesired prior evolutions of the density matrix are made to undergo a time reversal so that subsequent evolution results in a perfect cancellation or 'refocusing'. In the case of the spin echo and stimulated echo, the evolutions that we seek to refocus are those due to a spread of Zeeman interactions, arising, for example, from inhomogeneous magnetic fields. In this section we demonstrate another echo phenomenon where the undesired evolution arises from a spread of internuclear dipolar interactions. These will typically occur in a solid where each nucleus experiences, with its many neighbours, static dipolar interactions characterised by a distribution of internuclear vectors of varying orientation and length. The dipolar interaction (eqn 4.53) involves a bilinear spin operator, $\omega_{ij}[3I_{iz}I_{jz} - \mathbf{I_1} \cdot \mathbf{I_2}]$, where ω_{ij} is a 'dipolar precession frequency', which we use to characterise the strength of the interaction for the particular spin pair, i and j. For the purpose of this exercise we may neglect the term, $\mathbf{I_i} \cdot \mathbf{I_j}$, which commutes with the initial density matrix term $I_{iy} + I_{jy}$, and with the term $I_{ix}I_{jz} + I_{iz}I_{jx}$, which evolves from $I_{iy} + I_{jy}$ under the influence of $I_{iz}I_{jz}$.

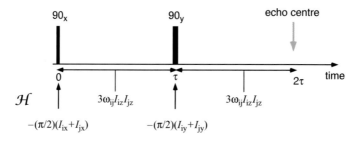

Fig. 4.24 Solid echo pulse sequence showing the active Hamiltonian terms at different times during the sequence.

Of course, for each nucleus i there will exist a distribution of ω_{ij}, arising from the various internuclear separations and θ_{ij} orientations associated with the various j spins, resulting in a coherence dephasing. That in turn leads to decaying signal, sometimes called the 'Bloch decay', as represented by eqn 4.54. It is that dephasing that the solid echo seeks to reverse.

Let us consider the echo formation from the standpoint of the product operator formalism, using the precession diagrams shown in Fig. 4.23. The dipolar interaction causes precession of the initial I_{iy} state of the density matrix into $I_{ix}I_{jz}$, and hence a dephasing of I_{iy} into an admixture $-I_{iy}\cos(\omega_{ij}\tau) + 2I_{ix}I_{iz}\sin(\omega_{ij}\tau)$. Note that ω_{ij} will be of order $M_2^{1/2}$. The time-reversal process will therefore require an RF pulse that will convert $I_{ix}I_{iz}$ to $-I_{ix}I_{iz}$, in effect inverting the phase. This is achieved by a 90°_y RF pulse in the following sequence in which, following our previous practice, we use ϕ to represent the phase acquired. In the case of the dipolar interaction, the phase angle after time τ is $\phi = \frac{3}{2}\omega_{ij}\tau$. Hence

$$
\begin{aligned}
I_{iz} + I_{jz} \\
\xrightarrow{-(\pi/2)(I_{ix}+I_{jx})} \quad & I_{iy} + I_{jy} \\
\xrightarrow{3\omega_{ij}I_{iz}I_{jz}} \quad & (I_{iy} + I_{jy})\cos\phi - (2I_{ix}I_{jz} + 2I_{iz}I_{jx})\sin\phi \\
\xrightarrow{-(\pi/2)(I_{iy}+I_{jy})} \quad & (I_{iy} + I_{jy})\cos(\phi) + (2I_{ix}I_{jz} + 2I_{iz}I_{jx})\sin(\phi) \\
\xrightarrow{3\omega_{ij}I_{iz}I_{jz}} \quad & (I_{iy} + I_{jy})\cos\phi\cos\phi' \\
& -(2I_{ix}I_{jz} + 2I_{iz}I_{jx})\cos\phi\sin\phi' \\
& +(2I_{ix}I_{jz} + 2I_{iz}I_{jx})\sin\phi\cos\phi' \\
& +(I_{iy} + I_{jy})\sin\phi\sin\phi'
\end{aligned}
\tag{4.90}
$$

The echo appears where the intervals τ and τ' are equal, so that $\phi = \phi'$ and the final state of the density matrix is $I_{iy} + I_{jy}$, exactly the magnetisation following the initial 90°_x pulse. Note that the detection operator in this multispin problem is now $[(I_{ix} + I_{jx} + i(I_{iy} + I_{jy})]$.

It is a simple, though tedious, extension to combine dephasing due to inhomogeneous magnetic fields and the distribution of (secular) dipolar interactions. Provided that the local magnetic field experienced by two spins, i and j, coupled by the dipolar interaction is the same, then the Zeeman operator involves $I_{iz} + I_{jz}$. This operator commutes with the full dipolar Hamiltonian so that the respective dephasings may be treated independently. The requirement of similar Zeeman interactions for i and j is satisfied for inhomogeneous magnetic fields because spins coupled by the dipolar interaction are within a few nanometers of each other and hence experience practically identical local fields.[15]

Because the 90°_y echo pulse is only capable of refocusing half the Zeeman dephasing, the echo turns out to be only half as big. However, by subtracting the signal of a $90^\circ_x - \tau - 90^\circ_x - \tau$ from that of the $90^\circ_x - \tau - 90^\circ_y - \tau$ sequence, both Zeeman and secular dipolar dephasings may be fully refocused.

---

[15]Of course a very different situation applies for intramolecular scalar couplings, where the chemical shift can make proximate spins experience different local fields.

The solid echo sequence equally works for high spin nuclei experiencing a distribution of quadrupolar interactions ($\sim I_z^2$), although the algebra is slightly more complicated. For example, for deuterium in the solid state the respective quadrupolar interactions may vary from nucleus to nucleus as a result of variations in the orientation of the electric field gradient, as would be the case in a powder or polydomain liquid crystal. Note that when dealing with the spin-1 deuteron in the solid state we can generally neglect dipolar interactions because the deuteron gyromagnetic ratio is so much smaller than that for the spin-$\frac{1}{2}$ proton.

4.6.3 Multiple pulse line-narrowing

In the liquid state, the rapid fluctuation of the dipolar interaction caused by molecular tumbling results in an average Hamiltonian in which the dipolar interaction is effectively zero. In order to produce this averaging it is necessary that the fluctuation rate, τ_c, be greater than the dipolar linewidth, $M_2^{1/2}$. In 1966 Ostroff and Waugh [57] and Mansfield and Ware [58] discovered that by application of a suitable train of RF pulses, the effective dipolar spin Hamiltonian could be made to fluctuate in a controlled way, thus leading to dipolar line-narrowing, sometimes referred to as 'decoupling'. The ability to slow the decay of the transverse magnetisation and hence to 'narrow the line' arises from the fact that, in the solid state, the dephasing of the FID is caused by static interactions that constitute an inhomogeneous broadening. Given the right pulse sequence these phase shifts can always be reversed, the solid echo being a simple example of such an approach.

A thorough description of multiple pulse line-narrowing is given in the book by Mehring [12], and the reader wishing to understand this subject in depth will find this text particularly helpful. Here we briefly review the essentials. The key element in the method is to introduce a periodic time-dependence in the dipolar Hamiltonian, and then to strobe the signal acquisition synchronously with the period of this Hamiltonian. Time-dependence is caused by a sequence of RF pulses, since these rotate the spin orientations and hence the vectors $\mathbf{I}_i$ and $\mathbf{I}_j$ in eqn 4.53. Of course, these pulses not only affect the Hamiltonian but also cause a sudden transformation in the density matrix as well. We can think of this by saying that in the rotating frame the total evolution operator is a product, $U_{RF}(t)U_D(t)$, where $U_{RF}(t)$ is the evolution operator representing the cumulative effect of the RF pulses and $U_D(t)$ is the evolution operator for the 'toggled' dipolar interaction. In order to separate the evolutions in this way, the appropriate time-dependent form for the dipolar Hamiltonian turns out to be

$$\mathcal{H}_D^*(t) = U_{RF}^{-1}(t)\mathcal{H}_D U_{RF}(t) \tag{4.91}$$

A key point to note is that, in comparison with density matrix evolution, the order of the U operators is reversed.

Before finding $U_D(t)$, we consider the role of $U_{RF}(t)$ in the product $U_{RF}(t)U_D(t)$. The longer-term influence of U_{RF} can be removed by making the RF pulse sequence cyclic, such that its associated evolution operator, $U_{RF}(t)$, is unity where t is a multiple of the cycle time, t_c. Provided the signal is sampled stroboscopically at multiples of this period, the net effect on the density matrix is due only to the evolution caused by $U_D(t)$. The manner in which this fluctuation is introduced must have the symmetry necessary to cancel the dipole–dipole interaction and so make $U_D(t)$ equal to the identity operator.

Average Hamiltonian

Each time an RF pulse is applied the Hamiltonian suddenly changes with the time dependence given by eqn 4.91. If we imagine that the RF pulses are infinitesimally narrow then $\mathcal{H}_D^*(t)$ becomes 'piecewise constant' such that

$$\mathcal{H}_D^*(t) = \mathcal{H}_{D_k}^* \tag{4.92}$$

for $(\tau_1 + \tau_2 + ... + \tau_{k-1}) < t < (\tau_1 + \tau_2 + ... + \tau_k)$. To understand how the evolution of the density matrix occurs, we return to the Liouville equation 3.53. The relevant evolution operator is

$$U_D(t) = \exp(-i\mathcal{H}_{D_n}^* \tau_n)....\exp(-i\mathcal{H}_{D_1}^* \tau_1) \tag{4.93}$$

where

$$t = \sum_{k=1}^{n} \tau_n \tag{4.94}$$

Of course the order of these successive evolution operators is crucial. Because a product of unitary transformations is itself unitary, the product in eqn 4.93 can be written as a single operator

$$U(t) = \exp(-i\overline{\mathcal{H}}_D(t)t) \tag{4.95}$$

$\overline{\mathcal{H}}_D(t)$ is the average Hamiltonian [59]. In the case of the dipolar interaction, we attempt to make this zero. $\overline{\mathcal{H}}_D(t)$ is not simply the time average of the various $\mathcal{H}_{D_k}^*$, although such an average is one of the leading terms. It can be shown that $\overline{\mathcal{H}}_D(t)$ comprises a sum of terms, $\overline{\mathcal{H}}_D^{(0)} + \overline{\mathcal{H}}_D^{(1)} + \overline{\mathcal{H}}_D^{(2)}) + ...$, where, for example,

$$\overline{\mathcal{H}}_D^{(0)} - \frac{1}{t}\{\mathcal{H}_{D_1}^* \tau_1 + \mathcal{H}_{D_2}^* \prime_2 + ... + \mathcal{H}_{D_n}^* \tau_n\}$$

$$\overline{\mathcal{H}}_D^{(1)} = -\frac{i}{t}\{[\mathcal{H}_{D_2}^* \tau_2, \mathcal{H}_{D_1}^* \tau_1] + [\mathcal{H}_{D_3}^* \tau_3, \mathcal{H}_{D_1}^* \tau_1] + [\mathcal{H}_{D_3}^* \tau_3, \mathcal{H}_{D_2}^* \tau_2] + ...\}$$

$$\tag{4.96}$$

The ideal RF pulse sequence is one that renders all terms $\overline{\mathcal{H}}_D^{(k)}$ zero. Removal of the zeroth-order term is relatively straightforward, and can be achieved with a cycle of four pulses. Provided this cycle is symmetric in the sense $\mathcal{H}_D^*(t) = \mathcal{H}_D^*(t_c - t)$, then all odd-order terms also vanish. Removal of higher, even-order terms can, in principle, be achieved by including more pulses in the cycle, but often the cumulative effect of RF pulse imperfections leads to additional decay of the transverse magnetisation.

WHH-4 and MREV-8 sequences

One of the first line-narrowing sequences proposed was the WHH cycle [60] shown in Fig. 4.25. The dipolar Hamiltonian[16] starts in the form given by eqn 4.53. The trick

---

[16]We mean here the secular part, denoted $\mathcal{H}_D^0$ in eqn 4.53.

(a) WHH-4

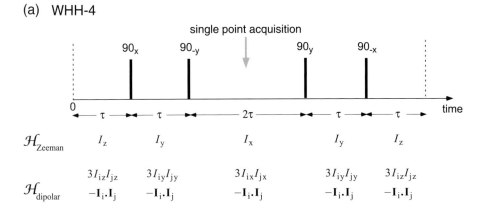

(b) MREV-8

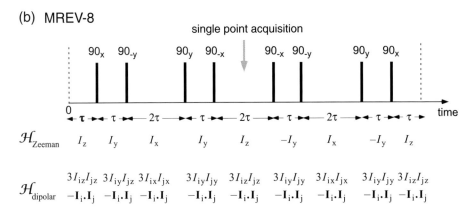

Fig. 4.25 Pulse sequences for (a) WHH-4 and (b) MREV-8 dipolar line-narrowing. The states of the Zeeman and dipolar Hamiltonians are shown below each respective precession frequency. The averaged $\mathcal{H}_{dipolar}$ is zero in both sequences, while the averaged Zeeman Hamiltonian is $\frac{1}{3}\Delta\omega_0(I_x + I_y + I_z) = \frac{1}{\sqrt{3}}\Delta\omega_0 I'_z$ for WHH-4 and $\frac{1}{3}\Delta\omega_0(I_x + I_y) = \frac{\sqrt{2}}{3}\Delta\omega_0 I''_z$ for MREV-8.

involved in this cycle is to cause the $I_{iz}I_{jz}$ terms in $\mathcal{H}_D$ to progressively change to $I_{iy}I_{jy}$ then $I_{ix}I_{jx}$ then $I_{iy}I_{jy}$, returning to $I_{iz}I_{jz}$ at the end of the cycle. Of course the term $\mathbf{I}_i \cdot \mathbf{I}_j$ is invariant under the rotations caused by the pulses.

The zeroth-order average Hamiltonian term, $\overline{\mathcal{H}}_D^{(0)}$, will vanish according to eqn 4.96 if the respective times spent in these states is τ, τ, 2τ, τ, and τ. Note that because of the evolution matrix order reversal represented by eqn 4.91, the effect of the RF pulses on the angular momentum operators I_{iz}, etc, must also be calculated in reverse order. For the WHH sequence the RF pulses are antisymmetric such that pulse 4 is the inverse of pulse 1 and pulse 3 is the inverse of pulse 3. The piecewise constant $\mathcal{H}^*_{D_k}$ terms are therefore

$$\mathcal{H}^*_{D_1} = \mathcal{H}_D$$
$$\mathcal{H}^*_{D_2} = U^{-1}_{RF}(1)\mathcal{H}_D U_{RF}(1)$$
$$\mathcal{H}^*_{D_3} = U^{-1}_{RF}(1)U^{-1}_{RF}(2)\mathcal{H}_D U_{RF}(2)U_{RF}(1)$$
$$\mathcal{H}^*_{D_4} = U^{-1}_{RF}(1)U^{-1}_{RF}(2)U^{-1}_{RF}(3)\mathcal{H}_D U_{RF}(3)U_{RF}(2)U_{RF}(1)$$
$$= U^{-1}_{RF}(1)\mathcal{H}_D U_{RF}(1)$$
$$\mathcal{H}^*_{D_5} = U^{-1}_{RF}(1)U^{-1}_{RF}(2)U^{-1}_{RF}(3)U^{-1}_{RF}(4)\mathcal{H}_D U_{RF}(4)U_{RF}(3)U_{RF}(2)U_{RF}(1)$$
$$= \mathcal{H}_D \tag{4.97}$$

So far we have considered the Hamiltonian in the rotating frame to be dominated by the secular part of the dipolar interaction. If this is largely removed by multiple pulse line-narrowing then the weaker chemical shift terms can be revealed. Of course these terms will also be affected by the RF pulse train, but because the symmetry of the Zeeman and dipolar interactions differ, the average Hamiltonian is not necessarily zero. Writing the total (isotropic and anisotropic) chemical shift Hamiltonian, $\mathcal{H}_{CS}$, of eqn 4.48 as $\delta_{CS}I_z$ and including the effect of any resonant offset term, $\Delta\omega_0 I_z$, the total rotating-frame Zeeman Hamiltonian is

$$\mathcal{H}_Z = (\delta_{CS}I_z + \Delta\omega_0 I_z) \tag{4.98}$$

It is a simple exercise to show that $\overline{\mathcal{H}}^{(0)}_Z$ is $\frac{1}{3}(\delta_{CS} + \Delta\omega_0)(I_x + I_y + I_z)$, which results in a precession about a tilted axis I'_z in the rotating frame, with effective frequency $\frac{1}{\sqrt{3}}(\delta_{CS} + \Delta\omega_0)$. The chemical shift and resonance offset do not vanish under the WHH-4 sequence but are reduced by a factor of $\frac{1}{\sqrt{3}}$. Normally, in solid state samples, the chemical shift is 'buried' in the much larger dipolar linewidth. Where information about $\mathcal{H}_{CS}$ is sought in these systems, retention of the isotropic and anisotropic chemical shift, albeit somewhat attenuated, is a valuable feature of multiple pulse line-narrowing experiments.

The WHH-4 cycle does not remove $\overline{\mathcal{H}}^{(2)}_D$. Furthermore, it suffers from the effects of RF pulse imperfections. An noticeable improvement results from using an eight-pulse cycle such as the MREV-8 method proposed independently by Mansfield [61, 62] and by Rhim, Elleman, and Vaughan [63]. Although MREV-8 also does not remove $\overline{\mathcal{H}}^{(2)}_D$, it has the added advantage of being less adversely affected by the effects of finite RF pulse width, RF inhomogeneity, and pulse-phase deviations. This particular sequence is shown in Fig. 4.25. In the MREV-8 cycle, the chemical shift and resonance offsets are scaled by $\frac{\sqrt{2}}{3}$ in comparison with the $\frac{1}{\sqrt{3}}$ factor in WHH-4.

Note that a simple 'pulse and collect' experiment would involve the line-narrowing sequence being preceded by a preparation (excitation) RF pulse, but the WHH-4 or MREV-* train can be inserted in a variety of more complex pulse sequences, where removal of dipolar interactions is required.

4.6.4 Multiquantum pathways

In Chapter 3, the concept of multiple-quantum coherence was introduced. The importance of such coherences in the measurement of spin translational motion arises for several reasons. First, the rate of Larmor precession in an applied magnetic field is determined by the coherence order p, so that by utilising higher-order coherences, a greater degree of phase migration can be acquired under magnetic field gradients, an effect we will discuss in greater detail in Chapter 11. Second, and of particular interest in the measurement of diffusion over long times, singlet states that combine zero-quantum coherence and longitudinal two-spin order can be generated, and these experience very long relaxation times, hence extending the timescale over which translational dynamics can be measured [64–66]. Finally, the selection of specific multiple-quantum coherence pathways in the evolution of the spin system, so-called 'multiple-quantum-filtering', can provide molecular selectivity based on spin interactions.

In the following sections we revisit multiple-quantum coherence in the context of the simple scalar-coupled two-spin system discussed earlier, outlining a simple way of visualising these coherences in terms of the energy level diagrams. We then turn to a brief discussion of the use of multiple quantum filtering in the selection of specific coherence pathways.

Multiple-quantum coherences in coupled spin systems

Let us analyse the scalar coupled two-spin problem discussed in Section 4.4.3 in the weak coupling limit, $Jh << \delta$. The energy levels of Fig. 4.12 can be simply represented in product state form as shown in Fig. 4.26, where we use the notation $\uparrow$ and $\downarrow$ to represent spin states $\frac{1}{2}$ and $-\frac{1}{2}$, respectively. For primary understanding, we write down the density matrix for the weakly coupled case as in eqn 4.99. The diagonal terms such as $\overline{a_{\uparrow\uparrow}a_{\uparrow\uparrow}*}$ simply represent the populations of the states and, following Levitt [18], we illustrate these as shown in part (a) of Fig. 4.26.

$$
\rho = \begin{bmatrix} a_{\uparrow\uparrow} \\ a_{\uparrow\downarrow} \\ a_{\downarrow\uparrow} \\ a_{\downarrow\downarrow} \end{bmatrix} \begin{bmatrix} a_{\uparrow\uparrow}* & a_{\uparrow\downarrow}* & a_{\downarrow\uparrow}* & a_{\downarrow\downarrow}* \end{bmatrix}
$$

$$
= \begin{bmatrix} \overline{a_{\uparrow\uparrow}a_{\uparrow\uparrow}*} & \overline{a_{\uparrow\uparrow}a_{\uparrow\downarrow}*} & \overline{a_{\uparrow\uparrow}a_{\downarrow\uparrow}*} & \overline{a_{\uparrow\uparrow}a_{\downarrow\downarrow}} \\ \overline{a_{\uparrow\downarrow}a_{\uparrow\uparrow}*} & \overline{a_{\uparrow\downarrow}a_{\uparrow\downarrow}*} & \overline{a_{\uparrow\downarrow}a_{\downarrow\uparrow}*} & \overline{a_{\uparrow\downarrow}a_{\downarrow\downarrow}} \\ \overline{a_{\downarrow\uparrow}a_{\uparrow\uparrow}*} & \overline{a_{\downarrow\uparrow}a_{\uparrow\downarrow}*} & \overline{a_{\downarrow\uparrow}a_{\downarrow\uparrow}*} & \overline{a_{\downarrow\uparrow}a_{\downarrow\downarrow}} \\ \overline{a_{\downarrow\downarrow}a_{\uparrow\uparrow}*} & \overline{a_{\downarrow\downarrow}a_{\uparrow\downarrow}*} & \overline{a_{\downarrow\downarrow}a_{\downarrow\uparrow}*} & \overline{a_{\downarrow\downarrow}a_{\downarrow\downarrow}} \end{bmatrix}
\tag{4.99}
$$

These populations are responsible for states of z-magnetisation of the spin system, but are not directly observable in the receiver coil of the NMR apparatus. The NMR observable is the single quantum coherence, $p = -1$, represented by the terms $\overline{a_{\uparrow\uparrow}a_{\uparrow\uparrow}*}$, $\overline{a_{\downarrow\downarrow}a_{\downarrow\uparrow}*}$, $\overline{a_{\downarrow\downarrow}a_{\uparrow\uparrow}*}$, and $\overline{a_{\uparrow\uparrow}a_{\uparrow\uparrow}*}$, in which there exists a coherent superposition between up and down states of a single spin. Two examples of $p = \pm 1$ quantum coherences are shown in Fig. 4.26 (b), where the grey arrows indicate the coherence sign. The –1 quantum coherence involves a flip down of one spin. Notice that only one spin at a time is active, the direct detection probability for dual-spin processes being vanishingly small. In Table 3.4 these single-quantum coherences are identified by the spherical tensor operators, $T^1_{\pm1}(10)$ and $T^1_{\pm1}(01)$, with associated product operator forms $\mp\frac{1}{\sqrt{2}}I_{i\pm} \otimes \underline{\underline{1}}$ and $\mp\frac{1}{\sqrt{2}}\underline{\underline{1}} \otimes I_{j\pm}$.

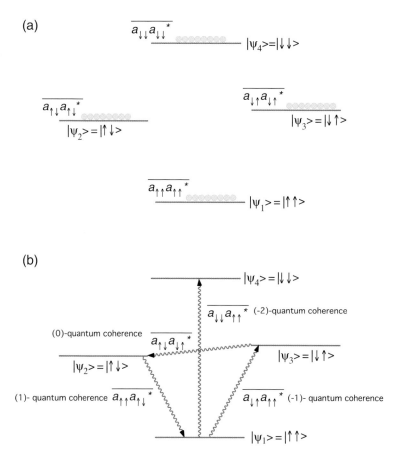

Fig. 4.26 Energy levels for scalar coupled spins in the weak coupling limit. (a) shows density matrix elements corresponding to populations (the four diagonal elements) while (b) shows a sample of multiple-quantum coherences.

Also shown in Fig. 4.26(b) is an example of a double-quantum coherence and a zero-quantum coherence. In practice, excitation of these coherent superposition states often involves the presence of two terms, for example a double-quantum coherence as $\overline{a_{\uparrow\uparrow}a_{\downarrow\downarrow}*} + \overline{a_{\downarrow\downarrow}a_{\uparrow\uparrow}*}$, or $[I_{i+}I_{j+} + I_{i-}I_{j-}]$, or a zero-quantum coherence as $\overline{a_{\uparrow\downarrow}a_{\downarrow\uparrow}*} - \overline{a_{\downarrow\uparrow}a_{\uparrow\downarrow}*}$, or $\frac{1}{\sqrt{2}}[I_{i-}I_{j+} - I_{i+}I_{j-}]$. Note that zero-quantum coherence is but one example of a $p = 0$ state, longitudinal one-spin magnetisation, I_{iz}, or longitudinal two-spin order, $I_{iz}I_{jz}$, being others.

States of the weakly coupled spin system in the product operator basis
In Chapter 2, the Liouville basis involving product operators of (I_{ix}, I_{iy}, I_{iz}) and (I_{jx}, I_{jy}, I_{jz}) was outlined, with details of the operator basis set given in Tables 3.4 and 3.5. Using this basis set, any states of the density matrix given in eqn 4.99 can be rewritten as a linear superposition of the operators from Tables 3.4 and 3.5. By

this means, the product operator basis provides a convenient language to describe the evolution of the coupled spin system.

In other words, rather than describe the spin system in terms of elements such as $\overline{a_{\downarrow\downarrow}a_{\downarrow\uparrow}{}^{*}}$, this element could be re-expressed in terms of amplitudes of I_{jx}, I_{jy}, $I_{iz}I_{jx}$, and $I_{iz}I_{jy}$, as given in eqn 3.63. This product basis set can be conveniently represented pictorially, as shown by Sørensen [67] in the manner of Figs 4.26 and 4.27. The diagrams follow directly from Tables 3.4 and 3.5. States diagonal in the product operator basis are presented purely by population differences. Off-diagonal states are associated with coherences, and are represented pictorially by wavy lines, which are are solid or dashed depending on the phase of the coherence.

Multiple quantum filters

All NMR experiments starting from thermal equilibrium and ending with Faraday detection represent a transition from $p = 0$ to $p = -1$. The pathway by which this transition occurs may involve migrations through various orders of quantum coherence, depending on the actions of the spin Hamiltonian and the RF pulses [68]. In the case of uncoupled spins, quadrupolar interactions may permit the formation of coherences with $|p| > 1$, while spins coupled by the dipolar or scalar interaction may form not only $|p| > 1$ but zero-quantum coherences as well. By appropriate selection procedures it is possible to filter the evolution pathway so that only elements of the density matrix that have passed through a particular coherence state can contribute to the detected NMR signal. Such a process is known as multiple quantum filtering. The advantage of such a filter is that it permits the experimenter to screen out contributions to the signal that have not been directly associated with a particular spin Hamiltonian term. For example, a double quantum filter, in the case of a quadrupolar nucleus, will allow contributions only from spins that are experiencing non-zero quadrupole interactions, a signature for molecular ordering. Alternatively, a zero quantum filter or double quantum filter can be used to discriminate spins coupled by a bilinear interaction, such as the scalar coupling. More importantly, because the highest order of quantum coherence possible for K coupled spin-$\frac{1}{2}$ nuclei is K, a K-quantum filter acts as a spin-counting tool, allowing signals only from such a pool of interacting spins.

The standard trick used for coherence selection is the use of a phase cycle on the RF pulses [68]. Recall from eqn 3.65 that a rotation, the action of an RF pulse, may change the order but not the rank of a spherical tensor. Suppose we represent a sequence of RF pulses by evolution operators, U_i, so that each operator causes a transition of the density matrix of order p between different orders of coherence as

$$U_i \rho_p(t_{i-}) U_i^{-1} = \sum_{p'} \rho_{p'}(t_{i+}) \tag{4.100}$$

where t_{i-} and t_{i+} refer to times just before and after the RF pulse evolution, respectively. For each evolution, the change in coherence order is Δp_i. The idea is illustrated schematically in Fig. 4.28. A phase shift ϕ_i in the ith RF pulse is represented by

$$U_i(\phi_i) = \exp(-i\phi_i \sum I_z) U_i(0) \exp(i\phi_i \sum I_z) \tag{4.101}$$

where $\sum I_z$ is the total z-component of angular momentum for the K coupled spins. According to eqn 3.54 we find

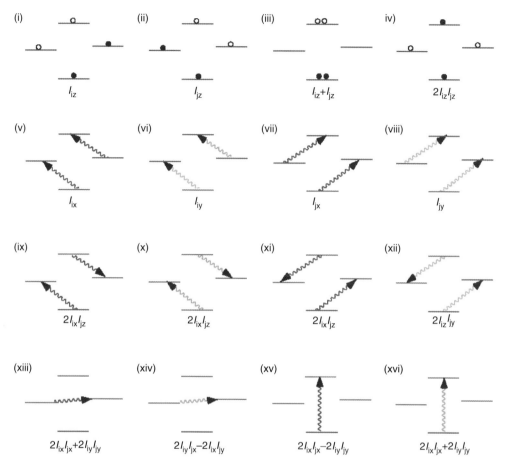

Fig. 4.27 (i) to (xii) Pictorial representations of product operators representing single quantum magnetisation and longitudinal magnetisation in a system of two coupled nuclei with $I = l/2$, after Sørensen *et al.* [67]. The x and y magnetisation components are represented by wavy lines in the energy-level diagram (dashed lines for y-components). Populations are represented by open symbols for states that are depleted, filled symbols for states that are more populated. (xiii) to (xvi) Linear combinations of product operators that represent pure zero and double-quantum coherence. The wavy lines are solid or dashed depending on the phase of the coherence. (Adapted from Sorensen *et al.* [67].)

$$U_i(\phi_i)\rho_p(t_{i-})U_i(\phi_i)^{-1} = \sum_{p'} \rho_{p'}(t_{i+}) \exp(-i\Delta p_i \phi_i) \qquad (4.102)$$

These successive phase shifts are transmitted to the final signal.

Suppose we focus our attention on a particular RF pulse i and perform N_i experiments such that the phase for that pulse is successively incremented as

$$\phi_{k_i} = k_i 2\pi/N_i \qquad (4.103)$$

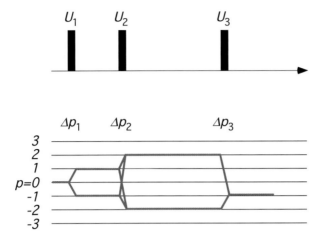

Fig. 4.28 Schematic representation of changes in coherence order during a multiple RF pulse experiment. The starting and finishing orders are $p = 0$ and $p = -1$, corresponding to thermal equilibrium magnetisation and detectable single-quantum coherence, respectively. At each RF pulse the change of coherence order is $\Delta p_i = p' - p$.

where k_i ranges from 0 to $N_i - 1$. Then if we sum the signal contributions, S_{k_i}, arising from each phase shift, each carries a phase term $\exp(-i\Delta p_i \phi_{k_i})$. Extracting the desired signal is done by a Fourier analysis as

$$S = \frac{1}{N_i} \sum_{k_i=0}^{N_i-1} S_{k_i} \exp(i\Delta p_i \phi_{k_i}) \qquad (4.104)$$

Inclusion of the factor $\exp(i\Delta p_i \phi_i)$ on detection simply involves the multiplication of the signal by the appropriate complex number, in other words setting the acquisition phase. This cycle of RF pulse phase and acquisition phase selects coherence transfer pathways involving Δp_i for the ith RF pulse. To ensure a unique Δp_i is selected, N_i must at least be equal to the maximum number of changes in phase. Where the phases of each RF pulse are cycled, the successive phase change is represented by a vector, $\mathbf{\Delta p}$, where each component corresponds to an RF pulse. The basic principles of the selectivity process remain the same.

An alternative approach to multiple quantum filtering is to utilise the fact that the rate of Larmor precession of coherence order p is $p\Delta\omega$, where $\Delta\omega$ is the frequency of single-quantum coherence in the rotating frame. If a magnetic field gradient is applied, $\Delta\omega$ is spatially dependent so that, across the sample, a wide uniform distribution of phase shifts can be engendered, resulting in the sum of signal contributions being zero. This effect is known as 'homospoiling'. Since the rate of homospoiling depends on p, careful choice of gradient amplitude during the various time evolution periods of any pulse sequence can be used to select for multiple-quantum coherence order.

4.6.5   Multidimensional NMR

The complex NMR signal, $S(t)$, corresponding to coherence order $p = -1$, is collected at successive points of a time domain t. However, any pulse sequence involving more than one RF pulse contains other time intervals during which the spin density matrix may evolve. In 1971 Jean Jeener [69] suggested the possibility of using this evolution time as a second dimension for spectroscopy. The consequences of this suggestion have transformed NMR and magnetic resonance imaging (MRI) [14].

Suppose that some evolution time associated with a particular pulse sequence interval is varied in successive independent experiments so that each signal acquired now becomes a 2-D function of the evolution time t_1 and the acquisition time t_2. The idea can be extended to higher dimensions by inserting additional evolution intervals. By convention we always ascribe the highest index to the acquisition dimension, so that in 3-D spectroscopy, the two evolution dimensions are t_1 and t_2, while the acquisition dimension is t_3. Then, just as the 1-D FID, $S(t)$, can be Fourier transformed to yield a 1-D spectrum $s(\omega)$, so a 2-D array of time-domain data, $S(t_1, t_2)$ can be Fourier transformed to provide a 2-D spectrum $s(\omega_1, \omega_2)$, as

$$s(\omega_1, \omega_2) = \int_{-\infty}^{\infty} \int_{-\infty}^{\infty} S(t_1, t_2) \exp(i\omega_1 t_1) \exp(i\omega_2 t_2) dt_1 dt_2 \qquad (4.105)$$

In the next chapter, where magnetic field gradients are discussed, we will see that phase evolution of the spin system can equally occur under the influence of field gradient pulses, where the equivalent to 'advancing the evolution time' will be 'advancing the gradient pulse area' [54]. In this case the Fourier spectrum directly relates to the distribution of physical displacements, in other words, the magnetic resonance image.

The schematic for a 2-D NMR spectroscopy experiment is shown in Fig. 4.29, in which separate preparation, evolution, mixing, and detection intervals are identified. The experiment is performed by collecting a succession of FIDs, sampled over time variable t_2, for a sequence of different evolution times t_1, so as to produce a 2-D matrix of time-domain data with rows labelled by t_2 and columns by t_1. This matrix may be Fourier transformed to give a 2-D spectrum. with rows labelled by f_2 and columns by f_1. Note that the preparation period is used to prepare the state of coherence whose evolution is desired to be investigated during the period t_1. While the mixing segment of the pulse sequence must be used to ultimately transform the coherence back to the detectable $p = -1$, it may also allow for possible storage of magnetisation before ultimate recall for detection.

The process of obtaining a phased 2-D spectrum is a little more subtle than in the case of 1-D NMR. In particular, if the evolution and acquisition contributions to the signal each generate a dispersive component, then absorption and dispersion become 'twisted together' in the 2-D spectrum, the real and imaginary parts of the 2-D spectrum both containing absorptive and dispersive parts, thus making the the process of 'phasing the spectrum' in 2-D intractable [18]. There are various means of avoiding this problem, for example acquiring a complete echo signal in the acquisition domain, comprising both negative and positive time intervals, avoids dispersion in acquisition. Alternatively, by a clever sampling in the evolution domain in which 1-D spectra from

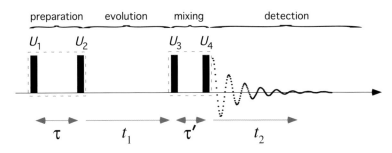

Fig. 4.29 Schematic representation of a 2-D NMR experiment. The preparation period in the first dashed box is represented, for simplicity, by two RF pulses with evolution operations U_1 and U_2 separated by time period τ. The first of these would normally be a 90° RF pulse. The period marked 'evolution' is the particular evolution singled out to be the variable time period, separate experiments being performed at different values of t_1. During the mixing period, the spin system goes through a further evolution, perhaps a transfer of coherence, or simply kept 'on hold' by storage as z-magnetisation, before the final recall as $p = -1$ coherence and FID detection over the time variable t_2.

separate cosine and sine amplitude modulations are subsequently combined to form a complex signal, dispersion twist can be avoided at the second Fourier transform. These matters are discussed in a number of texts dealing with NMR spectroscopy [14, 18] and will not be commented on further here.

The power of multidimensional NMR spectroscopy is that two or more spectroscopic domains allow for the correlation of different spectral properties or for the measurement of how certain spectral properties change under time. These two possibilities are known as correlation and exchange spectroscopy, respectively. And most importantly for this book, the analysis of the time-dependencies of the two domains need not involve Fourier transformation alone. In particular, where the signal behaviour under time advance is an exponential decay, as in relaxation, as opposed to an oscillation arising from spin Hamiltonian terms, then the more appropriate means of analysis may be inverse Laplace transformation. Multidimensional inverse Laplace spectroscopy is discussed in Chapter 10.

Exchange spectroscopy

In a 2-D exchange experiment, two identical evolution processes on the same identically prepared spin system are compared after allowing for an interval between the evolution in which spin magnetisation is kept 'in storage'. This storage interval is known as the mixing time τ_m, and the purpose of the exchange experiment is to determine if some spectral property changes with time. It is therefore a probe of geometric or structural dynamics of molecules or molecular assemblies [14].

Figure 4.30 shows an example of a simple exchange spectroscopy pulse sequence in the case of a spectrum comprising two lines, each of which corresponds to sites which chemically exchange. Below the pulse sequence are representations of the 2-D matrices for the time and frequency domain, in the first case where the mixing time

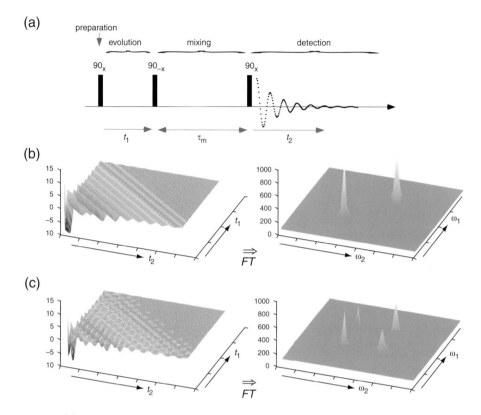

Fig. 4.30 (a) Pulse sequence for a simple 2-D exchange spectroscopy experiment in which the transverse magnetisation produced by a 90°_x RF pulse is allowed to evolve in the spin Hamiltonian during the evolution interval t_1, before being stored along the z-axis for later recall after a mixing time τ_m. (b) The time-domain data and frequency-domain 2-D spectrum for a system with two frequencies associated with sites A and B, for which τ_m is sufficiently short that no exchange between A and B occurs. (c) For longer τ_m with exchange evident as off-diagonal peaks in the 2-D spectrum.

is short compared with the exchange time, and in the second where the times are comparable. The appearance of off-diagonal peaks in the 2-D spectrum is the indication that site exchange has occurred. Such exchange experiments can be generalised to any spectral property indicative of chemical site, translational position, or orientation, where changes in site occupancy, position, or orientation lead to a spectral change. In each case the evolution and detection components of the pulse sequence must be tailored to the desired spectral feature.

Correlation spectroscopy

The purpose of a multidimensional correlation experiment is to reveal proximity or connectivity between specific nuclei in the NMR spectrum [14]. These connectivities might be, for example, carbon–proton direct bonding (heteronuclear single-quantum coherence spectroscopy—HSQC), protons experiencing inter-nuclear through-space dipolar interactions (nuclear Overhauser effect spectroscopy—NOESY) or protons experiencing inter-nuclear J-couplings (correlation spectroscopy—COSY). The last of these was first proposed by Jeener and first performed by Aue, Bartholdi, and Ernst in 1976. It is instructive to look at this experiment in a little more detail.

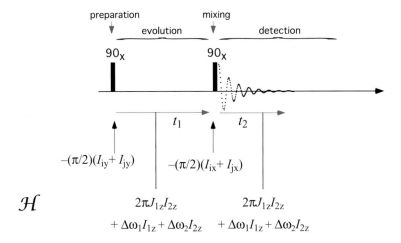

Fig. 4.31 Pulse sequence for a simple 2-D COSY spectroscopy experiment in which the transverse magnetisation produced by a 90°_x RF pulse is allowed to evolve in the spin Hamiltonian during the evolution interval t_1 before a second 90°_x RF pulse mixes and transforms coherences before a final detection period.

The COSY experiment is easy to understand in the weak coupling case represented by eqn 4.49. Here the scalar interaction is represented by $2\pi J I_{1z} I_{2z}$, and we naturally use the standpoint of the product operator formalism with the precession diagrams shown in Fig. 4.23. The scalar interaction causes precession of the initial I_{1y} state of the density matrix into $I_{1x} I_{2z}$ and hence a dephasing of I_{2y} into an admixture $-I_{2y}\cos(\pi J t_1) + 2I_{1x} I_{2z}\sin(\pi J t_1)$. Hence

$I_{1z} + I_{2z}$

$\xrightarrow{-(\pi/2)(I_{1x}+I_{2x})}$ $I_{1y} + I_{2y}$

$\xrightarrow{2\pi J I_{1z} I_{2z}}$ $(I_{1y} + I_{2y})\cos(\pi J t_1) - (2I_{1x}I_{2z} + 2I_{1z}I_{2x})\sin(\pi J t_1)$

$\xrightarrow{\Delta\omega_1 I_{1z}+\Delta\omega_2 I_{2z}}$ $[I_{1y}\cos(\Delta\omega_1 t_1) + I_{1x}\sin(\Delta\omega_1 t_1)$
$+I_{2y}\cos(\Delta\omega_2 t_1) + I_{2x}\sin(\Delta\omega_2 t_1)]\cos(\pi J t_1)$
$-[2(I_{1x}\cos(\Delta\omega_1 t_1) - I_{1y}\sin(\Delta\omega_1 t_1))I_{2z}$
$+2I_{1z}(I_{2x}\cos(\Delta\omega_2 t_1) - I_{2y}\sin(\Delta\omega_2 t_1))]\sin(\pi J t_1)$

$\xrightarrow{-(\pi/2)(I_{1x}+I_{2x})}$ $[-I_{1z}\cos(\Delta\omega_1 t_1) + I_{1x}\sin(\Delta\omega_1 t_1)$
$-I_{2z}\cos(\Delta\omega_2 t_1) + I_{2x}\sin(\Delta\omega_2 t_1)]\cos(\pi J t_1)$

$$-[2(I_{1x}\cos(\Delta\omega_1 t_1) + I_{1z}\sin(\Delta\omega_1 t_1))I_{2y}$$
$$+2I_{1y}(I_{2x}\cos(\Delta\omega_2 t_1) + I_{2z}\sin(\Delta\omega_2 t_1))]\sin(\pi J t_1)$$

$$\xrightarrow{\text{retaining detectable}} [I_{1x}\sin(\Delta\omega_1 t_1) + I_{2x}\sin(\Delta\omega_2 t_1)]\cos(\pi J t_1)$$
$$-[2I_{1z}I_{2y}\sin(\Delta\omega_1 t_1) + 2I_{1y}I_{2z}\sin(\Delta\omega_2 t_1)]\sin(\pi J t_1)$$

$$\xrightarrow{2\pi J I_{1z}I_{2z}} [I_{1x}\sin(\Delta\omega_1 t_1) + I_{2x}\sin(\Delta\omega_2 t_1)]\cos(\pi J t_1)\cos(\pi J t_2)$$
$$+[2I_{1y}I_{2z}\sin(\Delta\omega_1 t_1) + 2I_{1z}I_{2y}\sin(\Delta\omega_2 t_1)]\cos(\pi J t_1)\sin(\pi J t_2)$$
$$-[2I_{1z}I_{2y}\sin(\Delta\omega_1 t_1) + 2I_{1y}I_{2z}\sin(\Delta\omega_2 t_1)]\sin(\pi J t_1)\cos(\pi J t_2)$$
$$+[I_{2x}\sin(\Delta\omega_1 t_1) + I_{1x}\sin(\Delta\omega_2 t_1)]\sin(\pi J t_1)\sin(\pi J t_2)$$

$$\xrightarrow{\text{retaining detectable}} [I_{1x}\sin(\Delta\omega_1 t_1) + I_{2x}\sin(\Delta\omega_2 t_1)]\cos(\pi J t_1)\cos(\pi J t_2)$$
$$+[I_{2x}\sin(\Delta\omega_1 t_1) + I_{1x}\sin(\Delta\omega_2 t_1)]\sin(\pi J t_1)\sin(\pi J t_2)$$

$$\xrightarrow{\Delta\omega_1 I_{1z} + \Delta\omega_2 I_{2z}} [I_{1x}\sin(\Delta\omega_1 t_1)\cos(\Delta\omega_1 t_2) - I_{1y}\sin(\Delta\omega_1 t_1)\sin(\Delta\omega_1 t_2)$$
$$+I_{2x}\sin(\Delta\omega_2 t_1)cos(\Delta\omega_2 t_2) - I_{2y}\sin(\Delta\omega_2 t_1)\sin(\Delta\omega_2 t_2)]$$
$$\times\cos(\pi J t_1)\cos(\pi J t_2)$$
$$+[I_{2x}\sin(\Delta\omega_1 t_1)\cos(\Delta\omega_2 t_2) + I_{1x}\sin(\Delta\omega_2 t_1)\cos(\Delta\omega_1 t_2)$$
$$-I_{2y}\sin(\Delta\omega_1 t_1)\sin(\Delta\omega_2 t_2) - I_{1y}\sin(\Delta\omega_2 t_1)\sin(\Delta\omega_1 t_2)]$$
$$\times\sin(\pi J t_1)\sin(\pi J t_2)$$

$$(4.106)$$

Note that the 'retaining detectable' steps involve removing those density matrix terms that are unable to return to observable single-quantum coherence ($p = -1$) under the spin Hamiltonian. The final terms that remain neatly divide into two parts. The first three lines involve diagonal components (ω_1, ω_1) and (ω_2, ω_2) of the spectrum, since spin 1 experiences $\Delta\omega_1$ during both the evolution (t_1) and detection (t_2) periods, while spin 2 experiences only $\Delta\omega_2$. The second three terms have spin 1 experiencing $\Delta\omega_2$ during the evolution period and $\Delta\omega_1$ during detection, with similar crossing over for spin 2. These terms lead to off-diagonal peaks at (ω_1, ω_2) and (ω_2, ω_1) and are a direct consequence of the scalar coupling between spins 1 and 2. The point in the pulse sequence at which the magnetisation transfer occurs between the two sets of spins is the mixing step associated with the second RF pulse.

References

[1] C. J. Gorter. Negative result of an attempt to detect nuclear magnetic spins. *Physica*, 3:995, 1936.

[2] C. J. Gorter and L. J. F. Broer. Negative result of an attempt to observe nuclear magnetic resonance in solids. *Physica*, 9:591, 1942.

[3] J. S. Rigden. *Rabi, Scientist and Citizen*. Sloan Foundation Series: Basic Books, New York, 1987.

[4] I. I. Rabi, J.R. Zacharias, S. Millman, and P. Kusch. A new method of measuring nuclear magnetic moment. *Phys. Rev.*, 53:318, 1938.

[5] N. F. Ramsey. Early history of magnetic resonance. *Bulletin of Magnetic Resonance*, 7:95, 1985.

[6] E. K. Zavoisky. Paramagnetic absorption for solutions in parallel fields. *Zhurn. Eksperiment. Teoret. Fiziki*, 15:253, 1945.

[7] F. Bloch, W. Hansen, and M. E. Packard. Nuclear induction. *Phys. Rev.*, 69:127, 1946.

[8] E. M. Purcell, H. G. Torrey, and R. V. Pound. Resonant absorption by nuclear magnetic moments in a solid. *Phys. Rev.*, 69:37, 1946.

[9] A. Abragam. *Principles of Nuclear Magnetism*. Clarendon Pres, Oxford, 1961.

[10] C. P. Slichter. *Principles of Magnetic Resonance (3rd edition)*. Springer, New York, 1996.

[11] T. C. Farrar and E. D. Becker. *Pulse and Fourier Transform NMR*. Elsevier, Amsterdam, 1971.

[12] M. Mehring. *High Resolution NMR in Solids*. Springer, Berlin, 1982.

[13] A. Bax. *Two-Dimensional Nuclear Magnetic Resonance in Liquids*. Delft University Press, Dordrect, 1984.

[14] R. R. Ernst, G. Bodenhausen, and A. Wokaun. *Principles of Nuclear Magnetic Resonance in One and Two Dimensions*. Clarendon Press, Oxford, 1987.

[15] M. Munowitz. *Coherence and NMR*. Wiley-Interscience,New York, 1988.

[16] M. Goldman. *Quantum Description of High-Resolution NMR in Liquids*. Oxford University Press, New York, 1999.

[17] E. D. Becker. *High Resolution NMR: Theory and Chemical Applications (3rd edition)*. Academic Press, New York, 1999.

[18] M. H. Levitt. *Spin Dynamics (2nd edition)*. Wiley, New York, 2008.

[19] E. L. Hahn. Spin echoes. *Phys. Rev.*, 77:746, 1950.

[20] N. Bloembergen, E. M. Purcell, and R. V. Pound. Relaxation effects in nuclear magnetic resonance absorption. *Phys. Rev.*, 73:679, 1948.

[21] F. Bloch. Nuclear induction. *Phys. Rev.*, 70:460, 1946.

[22] F. Bloch and R. K. Wangsness. The differential equations of nuclear induction. *Phys. Rev.*, 78:82, 1950.

[23] H. C. Torrey. Bloch equations with diffusion terms. *Phys. Rev.*, 104:563, 1956.

[24] N. Q. Fan, M. B. Heney, J. Clarke, D. Newitt, L. L. Waard, E. L. Hahn, A. Bielecki, and A. Pines. Nuclear magnetic resonance with DC squid preamplifiers. *IEEE Transactions on Magnetics*, 25:1193, 1989.

[25] R. McDermott, A. H. Trabesinger, M. Muck, E. L. Hahn, A. Pines, and J. Clarke. Liquid-state NMR and scalar couplings in microtesla magnetic fields. *Science*, 295:2247, 2002.

[26] V. V. Yashchuk, J. Granweh, D. F. Kimball, S. M. Rochester, A. H. Trabesinger, J. T. Urban, D. Budker, and A. Pines. Hyperpolarized Xenon nuclear spins detected by optical atomic magnetometry. *Phys. Rev. Lett.*, 93:160801, 2004.

[27] S. J. Xu, V. V. Yashchuk, M. H. Donaldson, S. M. Rocheste, D. Budker, and A. Pines. Magnetic resonance imaging with an optical atomic magnetometer. *PNAS*, 103:12668, 2006.

[28] A. Abragam, M. A. H. McCausland, and F. N. H. Robinson. Dynamic nuclear polarization. *Phys. Rev. Lett.*, 2:449, 1959.

[29] T. Maly, G. T. Debelouchina, V. S. Bajaj, S. Vikram, K. N. Hu, C. G. Joo, M. L. Mak-Jurkauskas, J. R. Sirigiri, R. Jagadishwar, P. C. A. van der Wel, J. Herzfeld,

R. J. Temkinand, and R. G. Griffin. Dynamic nuclear polarization. *J. Chem. Phys.*, 128:052211, 2008.

[30] W. Happer. Optical pumping. *Rev. Mod. Phys.*, 44:169, 1972.

[31] B. C. Grover. Noble gas NMR detection through noble gas rubidium hyperfine contact interaction. *Phys. Rev. Lett.*, 40:391, 1978.

[32] D. Raftery, H. Long, T. Meersmann, P. J. Grandinetti, L. Reven, and A. Pines. High field NMR of absorbed xenon polarized by laser pumping. *Phys. Rev. Lett.*, 66:584, 1991.

[33] W. C. Proctor and F. C. Yu. The dependence of a nuclear magnetic resonance frequency on a chemical compound. *J. Chem. Phys.*, 484:3831, 1968.

[34] W. C. Dickinson. Dependence of the ^{19}F nuclear resonance position on chemical compound. *Phys. Rev.*, 77:736, 1968.

[35] L. A. Madsen, T. J. Dingemans, M. Nakata, and E. T. Samulski. Thermotropic biaxial nematic liquid crystals. *Phys. Rev. Lett.*, 92:145505, 2004.

[36] Z. Luz and S. Meiboom. Nuclear magnetic resonance studnmies of smectic liquid crystals. *J.Chem. Phys.*, 59:275, 1973.

[37] D. Wolf. *Spin-Temperature and Nuclear-Spin Relaxation in Matter*. Clarendon Press, Oxford, 1979.

[38] R. Lenk. *Fluctuations, Diffusion, and Spin Relaxation*. Elsevier, Amsterdam, 1986.

[39] B. Cowan. *Nuclear Magnetic Resonance and Relaxation*. Cambridge University Press, Cambridge, 2005.

[40] J. Kowalewski and L. Maler. *Nuclear Spin Relaxation in Liquids*. CRC Press, New York, 2006.

[41] D. C. Look and I. J. Lowe. Nuclear magnetic dipole-dipole relaxation along static and rotating magretic field-application to gypsum. *J. Chem. Phys.*, 44:2995, 1966.

[42] L. C. Hebel and C. P. Slichter. Nuclear spin relaxation in normal and superconducting aluminium. *Phys. Rev.*, 113:1504, 1959.

[43] P. W. Anderson and P. R. Weiss. Exchange narrowing in paramagnetic resonance. *Rev. Mod. Phys.*, 25:269, 1953.

[44] H. Y. Carr and E. M. Purcell. Effects of diffusion on free prcession in nuclear magnetic resonance experiments. *Phys. Rev..*, 94:630, 1954.

[45] S. Meiboom and D. Gill. Modified spin echo method for measuring nuclear relaxation times. *Rev. Sci. Instr.*, 29:688, 1958.

[46] L. K. Wanlass and J. Wakabaya. Stimulated spin echo measurement of cross relaxation times in neutron-irradiated calcite. *Phys. Rev. Lett.*, 6:271, 1961.

[47] R. Freeman and H. D. W. Hill. Fourier transform study of NMR spin-lattice relaxation by progressive saturation. *J. Chem. Phys.*, 54:3367, 1971.

[48] H. Y. Carr. Steady state free precession in nuclear magnetic resonance. *Phys. Rev.*, 112:1693, 1958.

[49] D. I. Hoult and R. E. Richards. Critical factors in design of sensitive high-resolution nuclear magnetic resonance spectrometers. *Proc. Roy. Soc. Series A*, 344:311, 1975.

[50] C. J. Turner and S. L. Patt. Artifacts in 2D NMR. *J. Magn. Reson.*, 85:492, 1989.

[51] R. L. Vold, J. S. Waugh, M. P. Klein, and D. E. Phelps. Measurement of spin relaxation in complex systems. *J. Chem. Phys.*, 484:3831, 1968.

[52] P. J. Hore. Solvent suppression in Fourier transform nuclear magnetic resonance. *J. Magn. Reson.*, 55:283, 1983.

[53] D. M. Doddrell, G. J. Galloway, W. M. Brooks, J. Field, J. M. Bulsing, M. G. Irving, and H. Baddeley. Water signal elimination in vivo using suppression by mistimed echo and repetitive gradient episodes. *J. Magn. Reson.*, 70:176, 1986.

[54] P. T. Callaghan. *Principles of Nuclear Magnetic Resonance Microscopy.* Oxford, New York, 1991.

[55] E. L. Hahn and D. E. Maxwell. Spin echo measurements of nuclear spin-spin coupling. *Phys. Rev.*, 85:7621, 1952.

[56] H. S. Gutowsky, D. W. McCall, and C. P. Slichter. Nuclear magnetic resonance multiplets in liquids. *J. Chem. Phys.*, 21:279, 1953.

[57] E. D. Ostroff and J. S. Waugh. Multiple spin echoes and spin-locking in solids. *Phys. Rev. Lett.*, 16:1097, 1966.

[58] P. Mansfield and D. Ware. Nuclear resonance line narrowing in solids by repeated short pulse RF irradiation. *Phys. Lett.*, 22:133, 1966.

[59] U. Haeberlen and J. S. Waugh. Coherent averaging effects in magnetic resonance. *Phys. Rev.*, 175:453, 1968.

[60] J. S. Waugh, L. M. Huber, and U. Haeberlen. Approach to high-resolution NMR in solids. *Phys. Rev. Lett.*, 20:180, 1968.

[61] P. Mansfield. Symmetrized pulse sequences in high resolution NMR in solids. *J. Phys. C*, 4:1444, 1971.

[62] P. Mansfield, M. J. Orchard, D. C. Stalker, and K. H. B. Richards. Symmetrized multipulse nuclear-magnetic-resonance experiments in solids—measurement of chemical shift shielding tensor in some compounds. *Phys. Rev. B*, 7:90, 1973.

[63] W. K. Rhim, D. D. Elleman, and R. W. Vaughan. Analysis of multiple pulse NMR in solids. *J. Chem. Phys.*, 59:3740, 1973.

[64] M. Carravetta, O. G. Johannessen, and M. H. Levitt. Beyond the t_1 limit: singlet nuclear spin states in low magnetic fields. *Phys. Rev. Lett.*, 92:153003, 2004.

[65] R. Sarkar, P. R. Vasos, and G. Bodenhausen. Singlet-state exchange NMR spectroscopy for the study of very slow dynamic processes. *J. Am. Chem. Soc.*, 129:328, 2007.

[66] R. Sarkar, P. Ahuja, P. R. Vasos, and G. Bodenhausen. Measurement of slow diffusion coefficients of molecules with arbitrary scalar couplings via long-lived spin states . *Chem. Phys. Chem.*, 9:2414, 2008.

[67] O. W. Sørensen, G. W. Eich, M. H. Levitt, G. Bodenhausen, and R. R. Ernst. Product operator formalism for the description of NMR pulse experiments. *Progress in NMR Spectroscopy*, 16:163, 1983.

[68] G. Bodenhausen, H. Kogler, and R. R. Ernst. Selection of coherence-transfer pathways in NMR pulse experiments. *J. Magn. Reson.*, 85:370, 1984.

[69] J. Jeener. *Ampere International Summer School.* Basko Polje, Yugoslavia, 1971.

5

Magnetic field gradients and spin translation

When the very first NMR experiments were performed on liquids, the role of the laboratory magnet in determining the linewidth of the spectrum was evident. Inhomogeneity of the magnetic field led to a range of Larmor frequencies and hence the natural relaxation-limited linewidth was buried under the much larger inhomogeneous broadening. Only when the spin echo was used to refocus the effects of phase-spreading under the inhomogeneous field was it possible to see the slowly decaying exponential of the echo envelope. This was the decay arising from relaxation, a time-domain attenuation representing the Fourier conjugate of the homogeneous spectral linewidth.

And so magnetic field inhomogeneities were seen from the beginning as a mixed blessing. From a frequency-domain perspective they were a nuisance, a resolution-limiting imperfection, to be minimised by the use of correction coils and strategically placed magnetic metal foils known as shims. Shimming, which is, in its modern version, the independent adjustment of currents in multiple coils representing various orders of magnetic field harmonics, is a familiar ritual in setting up high resolution NMR spectrometers. But from a time-domain perspective the field inhomogeneity was a help rather than a hindrance. Its very existence made possible the echo phenomenon, and the shape of the echo was determined by the field distribution. But, more importantly, it was realised by Erwin Hahn [1] that translational motion of spin-bearing molecules over the duration of the echo formation led to imperfect refocusing and hence to an echo attenuation. Thus the echo became a window on molecular motion itself, and the sensitivity to that motion was governed by the local gradient in the magnetic field. In other words, inhomogeneous fields were useful.

Of course, a spatially varying magnetic field is the fundamental tool of magnetic resonance imaging. It may seem surprising that so much time elapsed since the discovery of NMR in 1945 to the suggestion of MRI in 1973 [2, 3], before the practitioners of magnetic resonance realised that using magnetic field gradients to measure molecular motion naturally had a counterpart in using magnetic field gradients to determine molecular position. Such are the strange twists and turns of scientific discovery. In the opinion of this author, the key precursor to MRI was the integration of the computer with the NMR spectrometer. As so often is the case, new technology generates new science, the two being irrevocably intertwined.

5.1 Gradient fields and Maxwell's equations

The physical principle underlying the usefulness of the inhomogeneous magnetic field is the Larmor equation, $\omega_0 = \gamma B_0$. In other words, a field $B_0(\mathbf{r})$ leads to a spatially varying frequency $\omega_0(\mathbf{r})$, and frequency, along with its associated phase shifts, is what we measure in magnetic resonance.

Generally, for an optimally shimmed magnetic resonance system, we find it convenient to separate the main field, $B_0\hat{\mathbf{k}}$,[1] which is taken to be uniform and oriented along z, from any spatially varying part $\mathbf{B}(\mathbf{r})$. This latter is usually deliberately applied as an additional field by a special set of coils whose currents are independently controlled. Consider what Maxwell's equations tell us about this spatially dependent magnetic field, in the absence of any local currents, the usual case for the magnetic resonance sample where all conductors are external to the sample space. For local current density $\mathbf{J} = 0$ and for static or slowly varying fields where the displacement current $\partial\mathbf{D}/\partial t$ may be neglected,

$$\nabla \cdot \mathbf{B} = 0$$
$$\nabla \times \mathbf{B} = 0. \tag{5.1}$$

Expressing the divergenceless nature of the field in Cartesian coordinates we have

$$\frac{\partial B_x}{\partial x} + \frac{\partial B_y}{\partial y} + \frac{\partial B_z}{\partial z} = 0 \tag{5.2}$$

The curl equation leads to equalities of the form $\partial B_x/\partial y = \partial B_y/\partial x$. A little algebra leads then to $\nabla^2\mathbf{B} = 0$, showing that the magnetic field obeys the Laplace equation. This useful result facilitates the use of the various Laplace eigenmode solutions in designing magnetic field gradients in the presence of known boundary conditions.

Equation 5.2 tells us that a gradient in the field, such as a term $\partial B_z/\partial z$, cannot exist in isolation but requires the existence of terms $\partial B_y/\partial y$, and $\partial B_x/\partial x$. These latter are known as concomitant fields, and the effect is illustrated in Fig. 5.1. We will show that, in the presence of a much larger magnetic field, B_0, oriented along z, only the parallel field term $\partial B_z/\partial z$ makes a significant contribution. However, in low fields these concomitant field effects may be very important.

Suppose we attempt to create a spatial variation in the Larmor frequency by adding to $B_0\hat{\mathbf{k}}$ a gradient field $\mathbf{B}(\mathbf{r})$, which, in the vicinity of the sample centre, is zero, with local gradient $\nabla\mathbf{B}(\mathbf{r})$. Off-centre, the total field is

$$\mathbf{B}_{tot}(x, y, z) = (\mathbf{r} \cdot \nabla B_x)\hat{\mathbf{i}} + (\mathbf{r} \cdot \nabla B_y)\hat{\mathbf{j}} + (B_0 + \mathbf{r} \cdot \nabla B_z)\hat{\mathbf{k}}. \tag{5.3}$$

Of course the spins care not a jot about the orientation of $\mathbf{B}_{tot}(x, y, z)$, taking its local direction to be their quantisation axis and precessing about that axis accordingly. In other words, the spins are sensitive to $|\mathbf{B}_{tot}|$ so that the proper form of the Larmor equation should be

$$\omega_0 = \gamma|\mathbf{B}_{tot}|. \tag{5.4}$$

---

[1] We use unit vectors $\hat{\mathbf{i}}$, $\hat{\mathbf{j}}$, and $\hat{\mathbf{k}}$ for the respective x-, y-, and z-axes.

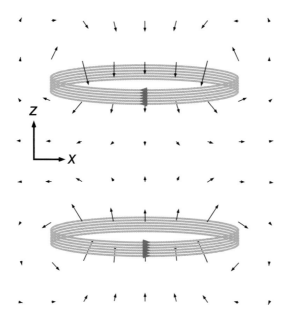

Fig. 5.1 Magnetic field vectors in the plane $y = 0$ and in the vicinity of Maxwell pair gradient coil set, designed to produce a linear magnetic field gradient, $B_z/\partial z$. The effect of the concomitant gradient, $\partial B_x/\partial x$ is apparent.

Let us evaluate $|\mathbf{B}_{tot}|$ when B_0 is much larger than the gradient fields, $x\partial B_x/\partial x$, and so on. Then

$$
\begin{aligned}
|\mathbf{B}_{tot}| &= [(\mathbf{r} \cdot \nabla B_x)^2 + (\mathbf{r} \cdot \nabla B_y)^2 + (B_0 + \mathbf{r} \cdot \nabla B_z)^2]^{1/2} \\
&\approx B_0 + \mathbf{r} \cdot \nabla B_z + \frac{1}{2}B_0^{-1}[(\mathbf{r} \cdot \nabla B_x)^2 + (\mathbf{r} \cdot \nabla B_y)^2 + (\mathbf{r} \cdot \nabla B_z)^2] \\
&\approx B_0 + \mathbf{r} \cdot \nabla B_z
\end{aligned}
\tag{5.5}
$$

The reason the concomitant fields have disappeared is that they are orthogonal in direction to the main field and so slightly tilt its direction, but influence its magnitude only to second order.

So, the case of a dominant uniform field is easy to deal with. We may ignore concomitant fields and apply gradients in B_z at will, for example using coils that produce, in effect, pure $\partial B_z/\partial x$, $\partial B_z/\partial y$ and $\partial B_z/\partial z$. Indeed, we may write

$$
\mathbf{G} = \nabla B_z
\tag{5.6}
$$

and

$$
\omega_0(\mathbf{r}) = \gamma B_0 + \mathbf{r} \cdot \mathbf{G}.
\tag{5.7}
$$

What then for low B_0 fields, where $B_0 \sim B(\mathbf{r})$?[2] Here we must return to the magnitude form of the Larmor precession, eqn 5.4. The gradients which matter then are those in the magnitude of the total field and so

---

[2] $B(\mathbf{r})$ means $|\mathbf{B}(\mathbf{r})|$.

$$\mathbf{G} = \nabla |\mathbf{B}_{tot}|$$
$$= \frac{(\mathbf{B}_{tot} \cdot \nabla)\mathbf{B}_{tot}}{|\mathbf{B}_{tot}|} \tag{5.8}$$

and

$$\omega_0(\mathbf{r}) = \gamma |\mathbf{B}_{tot}| + \mathbf{r} \cdot \mathbf{G}. \tag{5.9}$$

Note that the vector $\mathbf{G}$ points in the direction of the variation of the magnetic field along its line, $(\mathbf{B}_{tot} \cdot \nabla)\mathbf{B}_{tot}$ [4], and this direction plays the role of the gradient of the magnetic field in a single fixed direction evident in the strong B_0 field case of eqns 5.5 and 5.7. One might reasonably ask whether migrating spins could follow the changing direction of a local magnetic field, remaining in states quantised with respect to that direction. Such an ability to track the field will depend on the adiabatic condition being fulfilled. In other words, provided the rate of precession of the spins about the local field is much greater than the rate of change of orientation of that field due to molecular motion, the local quantisation persists. Such a requirement is almost invariably satisfied.

We will revisit the condition $B_0 \sim B(\mathbf{r})$ in later chapters. But for the rest of Chapter 5, let us assume that $B_0 \gg B(\mathbf{r})$ so that eqns 5.6 and 5.7 apply.

5.2 Phase evolution of spin isochromats

5.2.1 Magnetisation phase

In Chapter 4 it was shown that the emf arising from the Faraday detection process in magnetic resonance could be quadrature mixed with an RF field oscillating at frequency, ω, close to the Larmor precession frequency ω_0, resulting in in-phase and quadrature signal components oscillating at difference frequency $\omega_0 - \omega$. In quantum mechanical terms, this detection is associated with an observable $N\gamma(I_x + iI_y)$ operating on the rotating frame density matrix ρ^{rot} as $N\gamma Tr(\rho^{rot}(I_x + iI_y))$. For any state of the rotating frame density matrix written in an orthogonal tensor basis of spin operators,[3] only I_x and I_y components can contribute to the signal. We saw in Chapter 4 how the Larmor precession following a $90°_x$ RF pulse led to a signal

$$N\gamma Tr([I_x + iI_y]\rho^{rot}(t)) = iM_0 \exp(-i(\omega_0 - \omega)t) \tag{5.10}$$

where the minus sign is associated with clockwise precession, but the frequency scale may be arbitrarily chosen, for convenience, to be positive. The phase of the signal is determined in part by the initial RF pulse, which assigned the spins to I_y at $t = 0$ and hence the signal at $t = 0$ to the quadrature channel, and in part by the subsequent time-evolution due to the off-resonant nature of the mixing frequency, ω. Indeed, any pulse sequence will, at the final detection stage, lead to some particular absolute phase at the detection origin and some time-dependent phase depending on the detection frequency offset at that position in the NMR spectrum.

Suppose we have settled on some particular RF pulse sequence and signal detection time origin which yield, at $t = 0$, a complex signal $S(0)$. Now imagine that we carry

---

[3]That is, orthogonal under the trace operation.

out some variation on our NMR experiment that results in a new signal $\exp(i\phi)S(0)$. In that case the phase change $\exp(i\phi)$ may be obtained by taking a simple ratio of the two signals. Thus the concept of an absolute phase change is easily utilised by reference to the unperturbed signal. This idea applies equally to the signal contributions of sub-ensembles of spins where, as we shall see in subsequent sections, phase changes may differ.

5.2.2 Spin isochromats in an inhomogeneous field

Avogadro's number is so extraordinarily large that it allows for astonishing numerical descriptions. For example, there are more molecules of water in a glass of water than there are glasses of water in all the oceans. Or, perhaps even more remarkably, the numbers of carbon atoms removed in one revolution of a car tyre is approximately the same as the age of the universe in seconds. In short, the size of Avogadro's number means that we may consider extraordinarily small volumes of material and still have sufficient numbers of nuclear spins for a density matrix description to apply very well. Consider, for example, a very small volume of $(1\,\text{nm})^3$. For typical materials this volume will contain on the order of 10^2 nuclear spins, sufficient for such a volume to be represented by a sub-ensemble obeying well-defined statistical averages.

Suppose that our spin system is subject to an inhomogeneous magnetic field. How big would the magnetic field gradient have to be to resolve the spin precession difference across $1\,\text{nm}$? Of course, that will be in part determined by our available frequency resolution, which in turn is limited by the transverse relaxation time. For $T_2 \sim 1\,\text{s}$ we could allow for a resolution $\sim 1\,\text{Hz}$. For protons, a $1\,\text{Hz}$ frequency separation across $1\,\text{nm}$ corresponds to a magnetic field gradient of around $25\,\text{T m}^{-1}$, while for other nuclei this must be scaled upwards as the magnetogyric ratio, γ, reduces. A $25\,\text{T m}^{-1}$ gradient is very large by the standards of MRI, but achievable in the highest pulsed magnetic field gradient systems used for laboratory diffusion measurement. In practice, the best resolution achieved in diffusion measurement, where the signal arises from the entire sample, is around $10\,\text{nm}$. For MRI however, the signal arises separately from each resolvable volume element (voxel). Hence spatial resolution is limited by the signal-to-noise available as the size of the voxel shrinks, leading to a resolution limit of around 10 microns.

Let us return to the concept of an isochromat of an inhomogeneous magnetic field, as described in Chapter 4. In MRI we might allow that the length scale for which the isochromat concept applies might be on the order of microns, whilst for very high magnetic field gradient NMR, a nanometre scale might be more appropriate. So long as this length scale is smaller than the feature to be resolved, we may treat the isochromats as part of an integrable continuum.

5.2.3 Phase evolution in the presence of a field gradient

In this section we introduce the idea of 'phase-encoding' of the spin magnetisation. In general, the ensemble of nuclear spins will be presented by a magnetisation vector (M_x, M_y, M_z). Only the $M_+ = M_x + iM_y$ components can contribute to the NMR signal, S. When a field gradient in B_0 exists, the magnetisation component M_z is unaffected, while M_+ experiences a local phase evolution, dependent on position $\mathbf{r}$. Each isochromat evolves according to its local phase precession rate.

Behaviour of an isochromat

Let us assume the high field limit $B_0 \gg B(\mathbf{r})$, which commonly applies in MRI and in laboratory high field magnets. Then $\mathbf{G} = \nabla B_z$ and the local Larmor frequency for an isochromat of the spin system at position $\mathbf{r}$ may be written as in eqn 5.7.

In the rotating frame, with the detection frequency set on-resonance to ω_0, the isochromat contributes a transverse magnetisation $M_+(\mathbf{r},t)d\mathbf{r}$ given by

$$M_+(\mathbf{r},t) = M_+(\mathbf{r},0)\exp(-i\gamma\mathbf{r}\cdot\mathbf{G}t) \tag{5.11}$$

where $M_+(\mathbf{r},t)$ is the local magnetisation density at $\mathbf{r}$. The minus sign in eqn 5.11 arises because under a Zeeman field $\mathbf{r}\cdot\mathbf{G}$ and hence Zeeman interaction $-\gamma\mathbf{r}\cdot\mathbf{G}I_z$, the density matrix operator, I_x, responsible for the measurement of M_x at $= 0$, precesses to $I_x\cos(\gamma\mathbf{r}\cdot\mathbf{G}t) - I_y\sin(\gamma\mathbf{r}\cdot\mathbf{G}t)$ at time t (see Fig. 4.16) . In other words, suppose we start with the transverse magnetisation along the rotating-frame x axis so that $M_+(0) = M_0$. Later in time it will have evolved to $M_0\cos(\gamma\mathbf{r}\cdot\mathbf{G}t) - iM_0\sin((\gamma\mathbf{r}\cdot\mathbf{G}t))$, meaning $M_x = M_0\cos(\gamma\mathbf{r}\cdot\mathbf{G}t)$ and $M_y = -M_0\sin(\gamma\mathbf{r}\cdot\mathbf{G}t))$.

Of course, the sign of the phase term is really a matter for arbitrary choice, since in a magnetic field gradient the offset frequency may be above or below resonance, depending on which side of the gradient field centre our isochromat is located. Consequently, we will mostly drop the minus sign, returning to it only when going back to fundamentals, as in the discussion of the Bloch–Torrey equation in Section 5.4.4.

Note that we have not allowed for T_2 relaxation over the detection period. This is easily handled by including an additional $-t/T_2$ term in the exponent, but for simplicity we will ignore the effect of relaxation here, a situation which would apply if the phase-spreading rate due to the magnetic field gradient, $\gamma\mathbf{r}\cdot\mathbf{G}$, greatly exceeds the relaxation rate T_2^{-1}.

The magnetisation helix and k-space

Let us picture our isochromats as spread along a particular axis, which for convenience, we will label z. Figure 5.2 shows a simple pulse sequence in which a 90°_x pulse is used to cause all spin isochromats to have their magnetisation vectors lying in the transverse plane along the y-axis. We will take the origin of time to be immediately following the 90°_x pulse, such that each isochromat at $t = 0$ has identical phase.

Now imagine that a magnetic field gradient of magnitude G is applied along the z-axis. In other words $\mathbf{G} = G\hat{\mathbf{k}}$ and for each isochromat the subsequent time evolution is given by

$$M_+(z,t) = M_+(z,0)\exp(i\gamma zGt) \tag{5.12}$$

where, as discussed in the previous section, we choose for notational simplicity to use a positive argument for the gradient field precession. Figure 5.2 depicts, for each isochromat, the Argand plane in which their phase evolves.[4] Because at any time t the accumulated phase is proportional to the displacement of the isochromat along z, the

---

[4]The magnetisation helix concept was first introduced by Saarinen and Johnson [5].

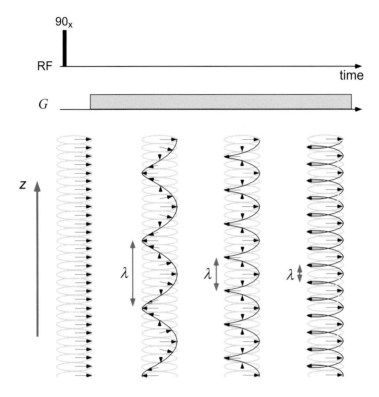

Fig. 5.2 Evolution of isochromat phase under the influence of a magnetic field gradient, G, along the z-axis following a 90°_x RF pulse. Immediately after the RF pulse the isochromats at different positions along the gradient axis start in phase. With time, a helix of magnetisation phase is gradually wound.

overall phase evolution pattern takes the form of a helix whose wavelength λ decreases with time as

$$\lambda = \frac{2\pi}{\gamma G t} \tag{5.13}$$

This magnetisation helix is extremely helpful in providing insight regarding the phase behaviour of spins under the influence of a magnetic field gradient. In analogy with the physics of diffraction, Mansfield [2] introduced the concept of a reciprocal space vector, $\mathbf{k}$, given by

$$\mathbf{k} = \gamma \delta \mathbf{g} \tag{5.14}$$

in units of angular spatial frequency, radians m^{-1} or

$$\check{\mathbf{k}} = (2\pi)^{-1} \gamma \mathbf{G} t \tag{5.15}$$

in units of cyclic spatial frequency, m^{-1}. Note that the use of the 'check' (˘) symbol is not at all standard and is introduced in this book to distinguish the two forms in common use, for which universally, and perhaps confusingly, the symbol **k** is used. The cyclic frequency form has some advantages in providing transform symmetry as we shall see. But whether one chooses angular or cyclic frequency forms is ultimately a matter of taste.

Under such a gradient of arbitrary direction, the phase evolution of any isochromat at position **r** can be written

$$M_+(\mathbf{r}, t) = M_+(\mathbf{r}, 0) \exp(i\mathbf{k} \cdot \mathbf{r}) \tag{5.16}$$

It is clear that **k**-space may be traversed by moving either in time or in gradient magnitude. However, the direction of this traverse is determined by the direction of the gradient **G**.

5.3 Magnetic resonance imaging

Both the NMR measurement of translational motion and the imaging of nuclear spin positions, otherwise known as magnetic resonance imaging or MRI, are based on ideas that are closely linked. Both utilise magnetic field gradients. Both draw on spatial Fourier relationships. And the two methods can be intertwined, for example in the use of sample slice selection or in the spatial localisation of translational dynamics. Consequently it is sensible that some of the essential physics of MRI is reviewed here. However, given that a more detailed description of imaging methods was the subject of an earlier book by this author [6], there seems little reason to repeat those details here and readers are referred to this and other texts dealing with MRI methods [7–10].

5.3.1 Spatial Fourier relations

Suppose that we reveal the gradient-related phase of the isochromat, by normalising with the total sample signal at $t = 0$, $M_+(0)$. Let the local normalised spin density be $\rho(\mathbf{r})$. Then $M_+(\mathbf{r}, 0)/M_+(0) = \rho(\mathbf{r})$ and the total normalised signal S_N is $\int M_+(\mathbf{r}, t)/M_+(\mathbf{r}, 0)d\mathbf{r}$, where the symbol $d\mathbf{r}$ is used to represent volume integration. In the formalism of **k**-space, and using the concept of the Fourier transform and its inverse,

$$S_N(\mathbf{k}) = \int \rho(\mathbf{r}) \exp(i\mathbf{k} \cdot \mathbf{r})d\mathbf{r} \tag{5.17}$$

$$\rho(\mathbf{r}) = (2\pi)^{-1} \int S_N(\mathbf{k}) \exp(-i\mathbf{k} \cdot \mathbf{r})d\mathbf{k} \tag{5.18}$$

or

$$S_N(\check{\mathbf{k}}) = \int \rho(\mathbf{r}) \exp(i2\pi\check{\mathbf{k}} \cdot \mathbf{r})d\mathbf{r} \tag{5.19}$$

$$\rho(\mathbf{r}) = \int S_N(\check{\mathbf{k}}) \exp(-i2\pi\check{\mathbf{k}} \cdot \mathbf{r})d\check{\mathbf{k}} \tag{5.20}$$

Equations 5.17, 5.18, 5.19, and 5.20, state that the signal, $S_N(\mathbf{k})$, and the spin density, $\rho(\mathbf{r})$, are mutually conjugate. These are the fundamental relationships of NMR imaging.[5]

---

[5]Note that the use of cyclic rather than angular frequency units for **k** necessitates the inclusion of the 2π factor in the exponent. The advantage of this choice is that it symmetrises the forward and inverse Fourier relations and avoids the need for further 2π coefficients in front of these integrals.

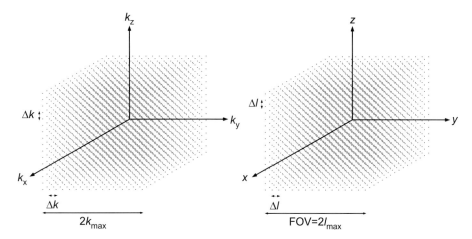

Fig. 5.3 3-D matrices for **k**-space (left) and real space (right).

The process of MRI thus requires that one samples the signal $S_N(\mathbf{k})$ in a discrete **k**-space, for example on a Cartesian grid as represented by a 3-D matrix of N^3 data points, where each dimension $\check{k}_x$, $\check{k}_y$, and $\check{k}_z$ ranges from $-\frac{1}{2}N\Delta\check{k}$ to $(\frac{1}{2}N - 1)\Delta\check{k}$, with the stepping interval being $\Delta\check{k} = 2\check{k}_{max}/N$. On digital Fourier transformation this leads to a conjugate real space matrix representing the image, $\rho(\mathbf{r})$, represented by a 3-D matrix of N^3 data points, where in each dimension x, y, and z ranges from $-\frac{1}{2}N\Delta l$ to $(\frac{1}{2}N - 1)\Delta l$, with the stepping intervals being $\Delta l = 2l_{max}/N$. These conjugate matrices are shown in Fig. 5.3. The Fourier relation dictates that the resolution for each component in real space is determined by the sampling range of each corresponding component in $\check{\mathbf{k}}$-space, so that

$$\Delta l = \frac{1}{2\check{k}_{max}} \tag{5.21}$$

while the field of view (FOV) for each component in real space, for example $2l_{max}$, is given by

$$FOV = \frac{1}{\Delta\check{k}}$$
$$= \frac{N}{2\check{k}_{max}} \tag{5.22}$$

5.3.2 Trajectories in k-space

The manner in which **k**-space is sampled depends on the particular pulse sequence used, an almost infinitely varied range of possibilities being possible. Whether sampled on a Cartesian grid, a polar grid, or via any other coordinate system, the principles of sampling and image reconstruction are encapsulated in the Fourier relations of eqns 5.17 and 5.18. Of course, a traverse through **k**-space implies a stepping of either gradient or time variables. The former is known as phase encoding and the latter, frequency encoding. Figures 5.4 and 5.5 show examples of two different sampling schemes, the first a purely phase-encoding approach and the second involving a combination of phase and frequency encoding [11].

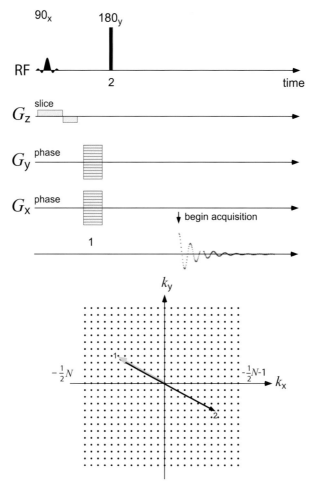

Fig. 5.4 Two-dimensional imaging pulse sequence along with **k**-space trajectory, in the case of pure phase encoding. The sequence starts with a soft RF pulse, which excites only a plane of spins normal to the z-axis. Subsequently, gradient pulses in both the x and y directions cause a phase traverse (1) to a point in the (k_x, k_y) plane determined by the particular amplitudes chosen for G_x and G_y, as well as their fixed duration. Next a 180° RF pulse inverts all the phases of the spins (2), allowing a spin echo to be formed. Acquisition of the signal begins at the centre of the echo and an FID is obtained in the absence of any magnetic field gradient, thus allowing full spectral resolution. The sequence must be repeated N^2 times to sweep out all points in **k**-space.

The frequency encoding method of using time as the variable involves sampling the signal at successive intervals under a steady 'read-out' gradient. Thus the frequency

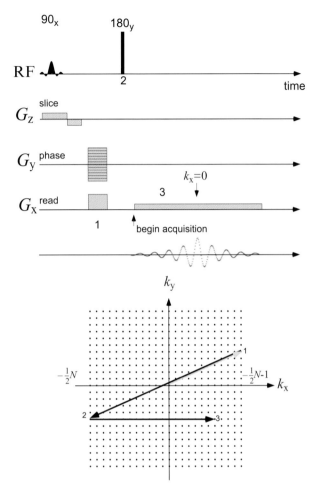

Fig. 5.5 Two-dimensional imaging pulse sequence along with **k**-space trajectory, in the case of combined phase and read encoding. The sequence starts with a soft RF pulse, which excites only a plane of spins normal to the z-axis. Subsequently, gradient pulses in both the x and y directions cause a phase traverse (1) to a point at the edge ($k_x = k_{max}$) of the (k_x, k_y) plane determined by the particular amplitudes chosen for G_x and G_y. Next a 180° RF pulse inverts all the phases of the spins (2), allowing a spin echo to be formed under an applied readout gradient (3). Acquisition of the signal takes place along a k_x line and at a k_y value set by the initial phase-encoding pulse. This signal, which is obtained right through the positive and negative k_x, is obtained in the presence of the readout magnetic field gradient, thus masking any spectral resolution. The sequence must be repeated only N times to sweep out all points in **k**-space.

encoding gradient is sometimes called the read gradient. Given that the suggested digital sampling implies a traverse through both negative and positive **k**-space, the reader may wonder how this is possible when time is the variable. The practical solution is provided by the spin echo, in which the natural origin of time is the echo centre so that negative time corresponds to the time displacement before that origin. For frequency encoding, $S_N(\mathbf{k})$ is measured in the time domain while the Fourier transform, which yields $\rho(\mathbf{r})$, is therefore in the frequency domain. In this simple sense we can say that there is a correspondence between real space and frequency, and between reciprocal space and time.

For phase-encoding, a fixed time interval is used, during which the phase gradient is applied and the spin phases are allowed to evolve. Actually, as is clear from eqn 5.15, both $\mathbf{G}$ and t are involved in determining $\mathbf{k}$-space. For each phase gradient value shown in Fig. 5.4, one signal acquisition is possible.[6] In the case of the 2-D imaging scheme of Fig. 5.5, where a readout gradient is used, N independent phase-encoding steps are required to traverse $\mathbf{k}$-space, while in Fig. 5.4 where pure phase encoding is used, N^2 steps are required. While pure phase encoding is clearly less efficient, it does have particular virtues. First the lack of a read gradient means that the FID is available for spectral resolution, making it possible to spatially localise any feature of the NMR spectrum. Second, it is possible to use single-point ($t = 0$) time sampling, meaning that there is none of the perturbation apparent in the readout direction associated with relaxation effects, inhomogeneous broadening, or spectral interference. Thus these image-distorting contributions can be avoided when one is desirous of sampling spectral shifts due to the phase gradients alone. For that reason pure phase encoding has particular application either in chemical imaging for liquids where spectral effects are of interest, or in the imaging of solids where relaxation and spectral broadening effects are severe.

5.3.3 Selective excitation

Soft and hard pulses

Selective excitation involves applying an RF pulse that affects only a specific region of the NMR frequency spectrum. By this means, only nuclei of a certain chemical shift may be disturbed or, when the spectral properties of the spins are dominated by the spread of Larmor frequencies in the presence of a magnetic field gradient, the selective RF pulse may be used to excite only those spins within some specified layer of the sample. We shall refer to these respective uses as chemical selection and slice selection.

Efficient and precise selective excitation is a vital component of most NMR imaging techniques. The principle underlying the excitation of spins in a specified region of the spectrum is as follows. The bandwidth of frequencies contained in an excitation pulse is inversely proportional to the pulse duration. For example, if the 90° pulse has a duration, T, of order 1 ms, then only those spins with a resonant frequency within a 1 kHz bandwidth of the radiofrequency will be stimulated in an appreciable manner. In normal NMR spectroscopy the pulse duration is made sufficiently short that the associated bandwidth covers the chemical shifts of all spins of a given nuclear species. Clearly the bandwidth of the pulse is simply related to the RF amplitude. A non-selective (broadband) 90° pulse will have a very much larger magnitude, B_1, than a selective (narrowband) 90° RF pulse, since the turn angle is determined by the product $\gamma B_1 T$.

---

[6]In multi-echo schemes, such as echo planar imaging [12] or RARE [13], multiple phase encoding steps may be achieved in a single echo train.

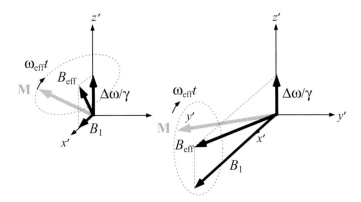

Fig. 5.6 Soft pulse effect showing different magnetisation nutation in rotating frame for spins that are off-resonant. On the left is the soft pulse, where the RF field amplitude is comparable with the field offset, and the spins are only weakly disturbed from equilibrium. On the right, the hard pulse with a much larger RF field amplitude results in a nutation that is closer to ideal for off-resonant spins.

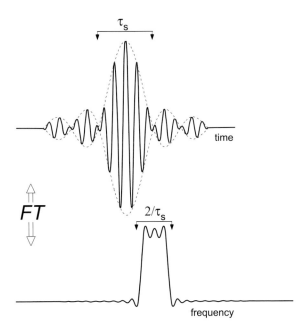

Fig. 5.7 Amplitude-modulated oscillatory waveform and its Fourier transform for the case of a double lobe sinc modulation. The spectrum shows some oscillatory deviation from the ideal rectangular response, due to the sinc lobe truncation.

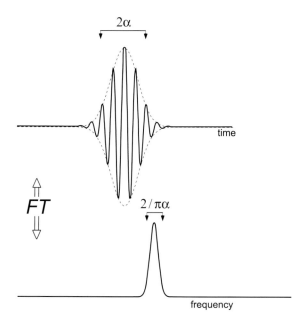

Fig. 5.8 Amplitude-modulated oscillatory waveform and its Fourier transform for the case of a Gaussian modulation. The spectrum is also Gaussian.

The crucial factor in determining the effect on other spins is the relative size of the RF field precession frequency, $\omega_1 = \gamma B_1$, and the resonance offset, $\Delta\omega_0$. Suppose that an RF pulse is applied with a reduced bandwidth, but with a magnitude and duration so as to cause 'on-resonant' spins to precess through $90°$ into the transverse plane. Spins with Larmor frequencies far from the RF excitation frequency $(\Delta\omega_0 \gg \omega_1)$ will experience an effective field in the rotating frame considerably tilted towards the longitudinal axis as shown in Fig. 5.6. Unlike the on-resonant spins they will therefore be tipped through an angle less than $90°$. Intense broadband excitation pulses are termed 'hard' pulses, while the weak, narrowband pulse is termed 'soft'.

The simplest form of soft pulse is obtained by simply reducing the amplitude and extending the duration of the usual rectangular time-domain profile. The corresponding frequency spectrum of this pulse is given by the Fourier transform, namely the sinc function shown in Fig. 5.7. Clearly the weak rectangular pulse suffers from having side lobes, so that, while the majority of the excitation is close to the central frequency, extensive excitation due to the lobes occurs over a wide bandwidth. One solution is to 'soften' the edges of the RF pulse, for example by the use of Gaussian shaping in the time domain, as shown in Fig. 5.8. Gaussian pulses are very effective in chemical shift selective excitations.

Evolution during selective excitation

Rectangular excitation requires $90°$ or $180°$ precessions for spins within a well-defined range normal to the slice plane and zero precession for spins outside. This implies that the frequency response of the RF pulse should have a rectangular profile and so, in the time domain, should be modulated in *sinc* ($\sin(at)/(at)$) form, the Fourier transform of the hat function. However, such a model assumes that the nuclear spins behave as a linear system whereas a linear response [14] can only be true for small excitation angles. Consider a weak RF field $\omega_1(t) = \gamma B_1(t)$ applied along the rotating frame x-axis over the time period $-T < t < T$, during which a gradient G_z is applied along the z-axis. The magnetisation in this linear response approach may be written

$$M_+(z) = iM_0 \exp(-i\gamma G_z zT) \int_{-T}^{T} \omega_1(t) \exp(i\gamma G_z zdt)dt \qquad (5.23)$$

The integral is simply the Fourier transform, or spectrum, of the RF pulse of finite duration. The frequency scale for this spectrum is $\gamma G_z z$, so that the integral represents the amplitude of the RF corresponding to the Larmor frequency of the spin plane labelled by z. Equation 5.23 states that the magnetisation contribution to the FID, $M_+(z)$, from isochromats at z, is proportional to the amplitude of the RF spectrum at z. If we want to excite a rectangular slice then we will need a rectangular spectrum. Second it tells us that the magnetisation for a spin plane normal to z has a net phase shift, $\gamma G_z zT$, which is a nuisance since the phase shift will vary across a slice of finite thickness. It is, however, easy to remove, since all we have to do is apply an opposite sign z gradient of magnitude $-G_z$ for a time T.[7]

Despite the bold approximations inherent in its use, the linear assumption is very helpful and moderately accurate when applied to finite turn angles, as discussed in reference [6]. There, numerical solutions to the Bloch equations suggest that for tip angles of up to 90°, the results are not too dissimilar to the linear prediction, hence justifying this rather simple but illuminating approach.

While the description of selective 90° pulses using a linear response theory is similar to the exact solution, for 180° selective pulses the effect of non-linearity distortions is far more severe and artefacts result [15, 16]. However, when a selective 180° pulse is used for phase inversion, for example as part of a spin-echo scheme, these phase errors are less severe and most out-of-slice magnetisation generated by the pulse tends to lie along the rotating frame z-axis. Hence, slice-selective 180° inversion pulses based on sinc modulation are used routinely in MRI to good effect.

5.4   Translational motion encoding

Powerful insight is provided by the evolving magnetisation helix associated with transverse magnetisation precession of spin isochromats distributed along the axis of a uniform magnetic field gradient. The helix graphically illustrates the relationship between spin phase and molecular position. It suggests an inherent wavelength λ and spatial frequency $k = 2\pi/\lambda$ associated with the distribution of spin precessions. And it allows us to visualise, in a very natural manner, the effect of the migration of spin-bearing molecules over the duration of precession.

Suppose we reverse the sign of the magnetic field gradient at some time τ following its imposition at $t = 0$, as shown in Fig. 5.9. Provided the spins remain in their original

---

[7]Or alternatively we can invert all the spin phases with a 180° RF pulse and then apply the identical sign gradient for this same time T.

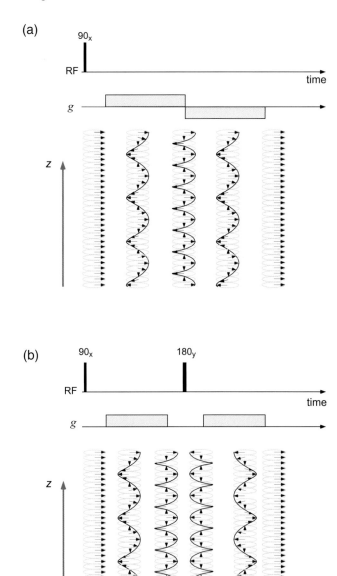

Fig. 5.9 (a) Gradient echo and (b) spin echo under the influence of gradient pulses of amplitude g. In (a), sign reversal of the gradient causes the magnetisation helix to unwind. In (b) the 180° RF pulse inverts the phase of the helix, so that a subsequent gradient pulse of the same sign has the effect of rewinding. The echo occurs when all the spin isochromats come into phase registration.

positions, the effect of the gradient reversal will be to unwind all the phases so that, once the reversed gradient has acted for an equal period τ, the isochromats will have been restored to their original states of identical phase and an echo will have been formed. That echo could take the form of the gradient echo as in Fig. 5.9(a) or the spin echo of Fig. 5.9(b). In the latter case, rather than reversing the gradient direction, the spin phases are reversed by a 180° RF pulse, so that unwinding occurs under a same-sign gradient.[8]

Figure 5.9 makes it clear that perfect echo formation is only possible if the spin-bearing molecules remain fixed in position. Any translational motion will cause perturbations to the phase evolution so that the history of the motion becomes embedded in the final phase distribution. In general we observe that the use of time-varying gradients allows insight regarding molecular translation. The key determinant of the phases embedded in the spin system following a period of motion in the presence of these field gradients depends crucially on the combined effects of the time dependence of the gradients and the time dependence of molecular translation.

Historically, the two simple variants of translational motion of interest in considering the effect of magnetic field gradients were unrestricted self-diffusion and coherent streamline flow. For these particular cases, quite simple theory is available to predict the NMR signal under the most general gradient waveforms. However, the power of NMR is that it permits in principle the elucidation of more complex motion, of restricted or timescale-dependent diffusion, of incoherent or periodic flow, of spatially inhomogeneous dynamics, or of fluid dispersion with specific spatio-temporal structure. For such applications the time dependence of the gradient waveform may need to be specially tailored to circumstance, and an appropriate theoretical treatment expressed. In the remainder of this chapter we will lay out the basics of this approach.

In embarking on this exercise, it needs to be admitted up front that there does not exist an exact theory that gives analytic closed form expressions for any motion under any gradient. However, we can come sufficiently close to such expressions that simple numerical implementations can bridge the final step. Before addressing that bridge, we first review the case where precise mathematical description is indeed possible.

5.4.1 Time-varying gradients and phase factors

The first step in writing down the phase evolution for a moving spin under the influence of time varying gradient, $\mathbf{g}(t)$, is pretty obvious. We will ignore transverse relaxation at this point, noting that it can be easily incorporated when needed. To distinguish gradients used for motion measurement from those used for imaging, we use a lower case $\mathbf{g}(t)$ for the former. The distinction is arbitrary of course, since any gradient, whether intended for imaging or motion measurement, can engender sensitivity to both. However, when $\mathbf{g}(t)$ is used we will normally be dealing with an experiment where an echo is formed at time t, which, as we shall see, will require $\int_0^t \mathbf{g}(t')dt' = 0$.

Consider a spin isochromat labelled by starting position $\mathbf{r}_0$, such that the position of the spin at later time t is $\mathbf{r}(t)$. Then the signal contribution from that isochromat is

[8]We will see later how to treat these approaches interchangeably using the idea of an 'effective gradient'.

$$M_+(\mathbf{r}_0, t) = M_+(\mathbf{r}_0, 0) \exp(i\gamma \int_0^t \mathbf{g}(t') \cdot \mathbf{r}(t')dt') \tag{5.24}$$

To estimate the total signal we would need to sum the contributions from all starting positions, $\mathbf{r}_0$. However, in the case of self-diffusion or other complex motions such as dispersive flow, there exists a distribution of possible molecular trajectories for every starting position. This requires us, even before summing over starting positions, to deal with the ensemble average implied by that. In other words, labelling each spin in the sub-ensemble by j, eqn 5.24 would need to be modified to read

$$M_+(t) = \int d\mathbf{r}_0 M_{+j}(\mathbf{r}_0, 0) \langle \exp(i\gamma \int_0^t \mathbf{g}(t') \cdot \mathbf{r}_j(t')dt') \rangle_{j-ensemble} \tag{5.25}$$

The evaluation of ensemble averages of phase factors is a central theme in physics. We saw this earlier in Chapter 4 in dealing with the Anderson–Weiss treatment of relaxation caused by time-varying fields. In that case the use of a Gaussian phase approximation (GPA) allowed for easy evaluation and such an approach has direct application in helping us deal with a number of problems associated with molecular motion under magnetic field gradients. For the moment however, it is enough to make the point that exact evaluation of eqn 5.25 is not trivial.

Effective gradients, the spin echo and gradient echo
The phase shift experienced at time t by a nuclear spin j following the path $\mathbf{r}_j(t')$ in a gradient $\mathbf{g}(t')$ is

$$\phi_j(t) = \gamma \int_0^t \mathbf{g}(t') \cdot \mathbf{r}_j(t')dt' \tag{5.26}$$

This simple relation takes no account of the influence of RF pulses on the spin phase. For example, a 180_y° pulse inverts all prior phase shifts, an effect which is equivalent to that which would result if we had used a negative value of gradient up to the time of the RF pulse. This suggests that we can take account of RF pulses by defining an 'effective gradient', $\mathbf{g}^*(t)$, which has the actual sign at the current time t, but is historically inverted by all prior 180_y° pulses. In the case of 90_y° RF pulses, the value of $\mathbf{g}^*$ is transformed to zero since such RF pulses leave magnetisation along the z-axis are immune to precession. The effect is illustrated in Fig. 5.10. More generally, $\mathbf{g}^*$ is defined by the equivalence of its Hamiltonian,

$$\mathcal{H}(t) = -\gamma \mathbf{g}^*(t) \cdot \mathbf{r}_j(t) I_{jz}$$
$$= -\gamma \mathbf{g}(t) \cdot \mathbf{r}_j(t) U_{RF}^\dagger I_{jz} U_{RF} \tag{5.27}$$

where U_{RF} is the ordered product of evolution operators associated with the history of prior RF pulses.

Let us try to simplify eqn 5.25 so that we can see how it applies in some special cases. First, let's allow that the spin does not move, whence $\mathbf{r}_j(t) = \mathbf{r}_0$. Then

$$M_+(t) = \int d\mathbf{r}_0 M_+(\mathbf{r}_0, 0) \exp(i\gamma \mathbf{r}_0 \cdot \int_0^t \mathbf{g}^*(t')dt') \tag{5.28}$$

At any time t such that $\int_0^t \mathbf{g}^*(t')dt' = 0$, $M_+(t) = M_+(0)$ and an echo is formed. Thus the condition for the echo is that the time integral of the effective gradient is

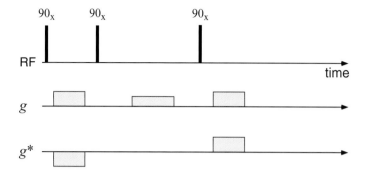

Fig. 5.10 A stimulated-echo sequence in which magnetic gradient pulses appear in the dephasing period before the second (90_x°) RF pulse, in the rephasing period after the third (90_x°) RF pulse, and as a homospoiling gradient between the second and third RF pulses. Note the effective gradient, g^*, which is only sensitive to gradient applied when the spins have transverse magnetisation.

zero. In the case of the gradient echo, no RF pulses are used and $\mathbf{g}^*(t) = \mathbf{g}(t)$, while for the spin echo, the role of RF pulses must be accounted for in determining $\mathbf{g}^*(t)$.

The normalised echo signal

Up to this point we have been accounting for the total signal by summing over isochromats as $\int d\mathbf{r}_0 M_+(\mathbf{r}_0, 0)$. Such an integral can be used to allow for non-uniform spin density across $\mathbf{r}_0$. We could also allow for relaxation effects by incorporating appropriate decays $\exp(-t/T_2)$ and $\exp(-t/T_1)$ during periods when the spin magnetisation resides in the transverse plane or along the longitudinal axis. But in many cases we are only concerned to understand the role of the applied gradient $\mathbf{g}^*(t')$, in its ability to impart phase shifts to the total magnetisation as a result of spin translational motion. The easiest way to demonstrate that is to use a signal normalised to the case $\mathbf{g}^*(t') = 0$, namely

$$E(t) = \frac{M_+(t)_{\mathbf{g}^*(t') \neq 0}}{M_+(t)_{\mathbf{g}^*(t')=0}} \tag{5.29}$$

This signal, when used to measure motion instead of position, will be ideally sampled when all phase shifts due to starting position alone will be zero, in other words the echo condition $\int_0^t \mathbf{g}^*(t')dt' = 0$. Hence we may refer to $E(t)$ as the *normalised echo amplitude*.

Note that this normalisation of relaxation, such that all translational motion is associated with gradient-induced phase shifts, is only true if relaxation effects are not correlated with position. In other words we require that relaxation is a uniform process intrinsically associated with each nuclear spin and not related to where the spin-bearing molecule resides within the sample. An example of a positional correlation where relaxation is coupled with geometry would be where additional relaxation mechanisms occur at surfaces bounding the fluid. The effects of such surface relaxation will be considered in the next chapter. For the rest of Chapter 5, however, the uniform relaxation assumption will apply so that relaxation normalisation is straightforward.

5.4.2   Coherent spin motion: velocity, acceleration, and jerk

In the special case that the molecules all have a common motional behaviour (which specifically precludes Brownian motion), eqn 5.26 yields a simple result.[9] For example, suppose that the motion comprises some constant velocity, $\mathbf{v}$, then $\mathbf{r}_j(t')$ is $\mathbf{r}_j(t') + \mathbf{v}t'$. Where an acceleration term is present, then $\mathbf{r}_j(t')$ is $\mathbf{r}_j(t') + \mathbf{v}t' + \frac{1}{2}\mathbf{a}t'^2$.

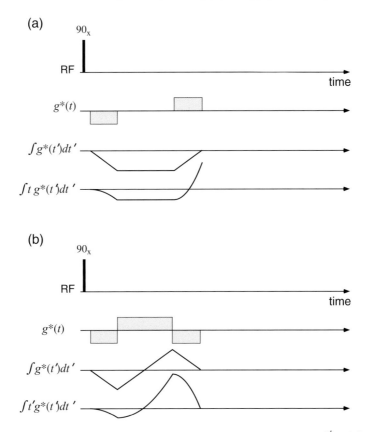

Fig. 5.11 Effective gradient sequences resulting in the echo condition $\int_0^t g^*(t')dt' = 0$, but with first moment $\int_0^t t'g^*(t')dt'$ non-zero in (a) and zero in (b).

It is clear that the resulting phase shifts involve successively higher moments of $\mathbf{g}^*(t)$. The zeroth moment, $\int_0^t \mathbf{g}^*(t')dt'$, is required to be zero if the final phase shift is to depend only on the motion and not on the starting positions of the spins. This is, of course, the condition for a spin echo to be formed! Furthermore, by choosing a specific time dependence for $\mathbf{g}^*(t)$, we can make the echo sensitive to either velocity, acceleration, or the next higher time derivative of displacement (sometimes known as 'jerk'). In particular, the final phase modulation of the spin echo will consist, in the case of the moment expansion, of

---

[9] Here we use the effective gradient $\mathbf{g}^*(t)$ to apply to any gradient waveform and RF pulse sequence.

$$E(t) = \exp(i\gamma \mathbf{v} \cdot \int_0^t t' \mathbf{g}^*(t')dt' + i\gamma \mathbf{a} \cdot \int_0^t t'^2 \mathbf{g}^*(t')dt' + ...) \tag{5.30}$$

Examples of pulse sequences in which the effective gradient obeys the echo condition are shown in Fig. 5.11. In Fig. 5.11(a) the echo will be sensitive to velocity while in (b) it is sensitive to acceleration but produces no phase shift for steady flow.

5.4.3 The Carr–Purcell analysis of diffusion effects

We now turn to the problem of Brownian motion. One of the earliest attempts to deal with the effect of molecular self-diffusion in the presence of time-varying gradients was provided by Carr and Purcell [17]. In their treatment, unrestricted self-diffusion is pictured as a succession of discrete hops with motion resolved in one dimension, the direction of the field gradient. The independence of motion in other directions means that we treat dimensions separately, obtaining total displacements where necessary by adding quadratically according to the Pythagoras theorem. The method is ideally suited to piecewise-constant gradient waveforms. In the following section we meet a treatment more suited to continuously varying gradients. Nonetheless, the Carr–Purcell approach is sufficiently illuminating to warrant our interest.

The particle hopping model
The principle behind the Carr–Purcell approach is the definition of a fundamental diffusion jump with mean time τ_s and rms displacement in one dimension, ξ. A molecule has equal probability of jumping to the left or right so that the distance travelled after n jumps at time $t = n\tau_s$ is given by

$$Z(n\tau_s) = \sum_i^n \xi a_i \tag{5.31}$$

where a_i is a random number equal to ± 1. Z represents the z-axis displacement of molecules from their respective origins. Thus

$$\overline{Z^2(n\tau_s)} = \sum_i^n \sum_j^n \xi^2 \overline{a_i a_j} \tag{5.32}$$

where the over bar represents an ensemble average. Because a_i randomly varies between ± 1, $\overline{a_i a_j} = 0$ unless $i = j$, so that all cross terms cancel. Thus

$$\overline{Z^2(n\tau_s)} = \sum_i^n \xi^2 \overline{a_i^2} = \xi^2 \sum_i^n 1 = n\xi^2 \tag{5.33}$$

Defining the self-diffusion coefficient as

$$D = \xi^2 / 2\tau_s \tag{5.34}$$

we obtain

$$\overline{Z^2(t)} = 2Dt \tag{5.35}$$

and for three dimensions,

$$\overline{R^2(t)} = 6Dt \tag{5.36}$$

Signal attenuation under a steady gradient

Next we calculate the influence of this motion on the coherence of transverse magnetisation in the case where a steady magnetic field gradient, $g = \partial B/\partial z$, is present, following the method originally used by Carr and Purcell in their classic paper on spin echoes [17]. The only motion that will concern us is that along the field gradient axis, z. For convenience we consider the influence of diffusion on the transverse magnetisation of spins originating at z. Then the local Larmor frequency after m jumps is

$$\omega(m\tau_s) = \gamma B_0 + \gamma g z + \gamma g \sum_{i=1}^{m} \xi a_i \tag{5.37}$$

so that the cumulative phase angle after time $t = n\tau_s$ is

$$\phi(t) = \gamma B_0 t + \gamma g z t + \gamma \sum_{m=1}^{n} g \tau_s \sum_{i=1}^{m} \xi a_i \tag{5.38}$$

The first term in eqn 5.38 represents the constant average Larmor precession and is of no particular interest here. The second is the phase spread due to starting positions of spins. We shall return to that effect later. But for the moment we shall only be concerned with the third, phase deviation term, which varies randomly across the ensemble and so causes irreversible dephasing. Note that it contains two parts. The second (outer) sum tells us the local Larmor frequency after m hops. The first sum adds the cumulative precessions resulting from that local frequency with local dwell time τ_s. The dephasing part of eqn 5.38 may be rewritten

$$\Delta\phi(t) = \gamma g \tau_s \xi \sum_{i=1}^{n} (n+1-i) a_i \tag{5.39}$$

The proof of this relationship is best demonstrated pictorially, as in Fig. 5.12, where each row in the triangle indicates the successive, random, value of a_i up to time $m\tau_s$, and the horizontal sum of these elements gives the phase shift that occurs in the interval τ_s, while the spin is at the current location. The total phase shift accumulating at time t is given by the area of the triangle multiplied by $\gamma g \tau_s \xi$.

What we wish to calculate is $\overline{\exp(i\Delta\phi)}$, the coefficient by which the ensemble-averaged transverse magnetisation will be phase modulated as a result of the diffusional motion in the presence of the gradient. Of course $\Delta\phi(t)$ varies randomly over the ensemble. We will assume that its distribution, $P(\Delta\phi)$, is Gaussian, a good assumption when dealing with the sum of very many randomly varying quantities, a result known as the 'central limit theorem' [18]. Thus

$$\overline{\exp(i\Delta\phi)} = \int_{-\infty}^{\infty} P(\Delta\phi) \exp(i\Delta\phi) d(\Delta\phi) \tag{5.40}$$

Since $P(\Delta\phi)$ is given by the normalised Gaussian function, $(1/2\pi\overline{(\Delta\phi)^2})^{1/2} \exp(-\Delta\phi^2/2\overline{(\Delta\phi)^2})$, eqn 5.40 yields

$$\overline{\exp(i\Delta\phi)} = exp(-\overline{(\Delta\phi)^2}/2) \tag{5.41}$$

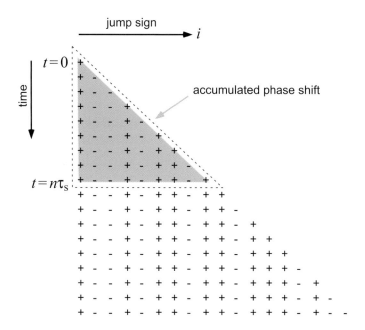

Fig. 5.12 Cumulative phase diagram in which the sign at the end of each row indicates whether the jump in the current τ_s interval is positive or negative with respect to the gradient direction. The precessional phase shift that results during that interval is determined by the net position along the gradient axis and is therefore calculated by summing jumps along the row. The cumulative phase shift at time t is given by the sum within the shaded triangle.

This coefficient represents a simple attenuation of the magnetisation and hence the signal at time t. In order to evaluate $\overline{\Delta\phi^2}$, we square eqn 5.39 and take an ensemble average. Once again, cross terms disappear. Thus

$$\overline{(\Delta\phi)^2} = \gamma^2 g^2 \tau_s^2 \xi^2 \sum_{i=1}^{n}(n+1-i)^2$$

$$= \gamma^2 g^2 \tau_s^2 \xi^2 \sum_{j=1}^{n} j^2$$

$$= \tfrac{1}{3}\gamma^2 g^2 \tau_s^2 \xi^2 n^3 \tag{5.42}$$

where the last sum is evaluated assuming that n is large. (Note that this is the standard mean-squared phase shift for a triangular section of steps, as shown in Fig. 5.12, where n represents the numbers of steps in the orthogonal sides.) Substituting eqn 5.34, we find that the signal attenuation due to diffusion is expressed by the coefficient

$$\overline{\exp(i\Delta\phi)} = \exp(-\tfrac{1}{3}\gamma^2 g^2 D t^3) \tag{5.43}$$

This t^3 dependence is characteristic of self-diffusion in the presence of a steady gradient.

Of course eqn 5.43 only tells us about phase spreading due to migrations of molecules, based on a common displacement from $z = 0$, and ignoring the second term in eqn 5.38. In practice, our sample is of finite size and so starting positions for molecules are distributed as $\rho(z)$ along the gradient axis.[10] Hence the total signal phase will need to account for this initial distribution as

$$\overline{\exp(i\phi)} = \left(\int_{-\infty}^{\infty} \rho(z) \exp(i\gamma gzt)dz \right) \overline{\exp(i\Delta\phi)}$$

$$= \left(\int_{-\infty}^{\infty} \rho(z)exp(i\gamma gzt)dz \right) \exp(-\tfrac{1}{3}\gamma^2 g^2 Dt^3) \tag{5.44}$$

The term in brackets is the decay known as T_2^*, and arises from the field inhomogeneity (in this case imposed by gradient g) across the sample. It merely serves to mask the underlying attenuation due to diffusion alone. The means of removing this dephasing associated with the starting position of spins is of course, the echo.

Before moving on, it is worth noting that the decay of the signal caused by the positional phase spread, $\int_{-\infty}^{\infty} \rho(z) \exp(i\gamma gzt)dz$, can be used to good effect in suppressing unwanted signals, a process known as 'homospoiling'.

Signal attenuation under an echo with steady gradient

Suppose we now consider the dephasing effect that remains at the time of formation of an echo at time $t = 2\tau$ when a steady gradient is present. Because of the echo formation we may ignore all phase shifts associated with starting position z and deal only with phase shifts $\Delta\phi$.

The phase step diagram for this is shown in Fig. 5.13. Here, the 180_y° pulse has reversed all phase shifts that existed before time $t = \tau$, where τ is the separation of the first and second RF pulses. Clearly the second evolution period contains a section that completely cancels the net phase shift that occurred before the 180° pulse. The residual phase shift at the time of the echo is given by the sum of the shaded triangular region. This region now involves two uncorrelated triangular segments, each with orthogonal sides of n steps. Using our previous observation concerning $\overline{(\Delta\phi)^2}$ in such triangular sections, we can see that the value of $\overline{(\Delta\phi)^2}$ that applies in the case of the echo is exactly double that which applied in Fig. 5.13 (and twice that which occurs during the period t between the first two RF pulses.) Consequently, for the echo at $t = 2\tau$,

$$\overline{\exp(i\Delta\phi)} = \exp(-\tfrac{2}{3}\gamma^2 g^2 D\tau^3)$$

$$= \exp(-\tfrac{1}{12}\gamma^2 g^2 D(2\tau)^3) \tag{5.45}$$

where the second form expresses this attenuation in terms of the total time 2τ from the first RF pulse to the echo centre. This constant gradient spin-echo method has been used to obtain self-diffusion coefficients in simple liquids with an accuracy of order 0.1 [19].

---

[10]In Chapter 1, the notation $p(\mathbf{r}, 0)$ was used to signify the initial probability density of finding a molecule at position $\mathbf{r}$. Here we use $\rho(\mathbf{r})$ to describe the density of molecules in the sample. Normally these will be identically the same quantities.

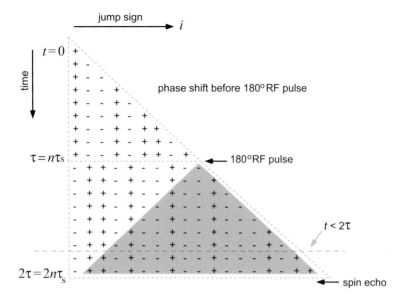

Fig. 5.13 Cumulative phase diagram for spin echo formation in a steady magnetic field gradient. Note that the 180° RF pulse inverts all prior phase shifts. The net phase shift is the sum of that occurring before and after the 180° RF pulse and corresponds to the shaded area.

Signal attenuation due to diffusion leading up to the echo

Note that eqn 5.45 is valid only at the echo centre. It is interesting to recalculate the diffusive attenuation factor during the time $\tau < t < 2\tau$ leading up to the echo. We can easily do this by reference to Fig. 5.13. Instead of the shaded double triangular region being of height τ, it has the lesser height $t - \tau$. Furthermore, the phase cancellation between the upper and lower regions on the left is incomplete, leaving a small uncompensated triangle of phase of height $2\tau - t$ on the bottom left. Note that these two residual triangular regions are uncorrelated and contribute to the signal attenuation independently. Using the arguments of the previous section concerning $\overline{(\Delta\phi)^2}$, we can see that

$$\overline{\exp(i\Delta\phi)} = \exp\left(-\tfrac{1}{3}\gamma^2 g^2 D[(2\tau - t)^3 + 2(t - \tau)^3]\right)$$
$$= \exp\left(-\gamma^2 g^2 D(2\tau^3 - 2\tau^2 t + \tfrac{1}{3}t^3)\right) \qquad (5.46)$$

Equation 5.46, as required, reproduces the correct result at the echo centre, at $t = 2\tau$. However, this attenuation function has the remarkable property that its maximum is at $t = \sqrt{2}\tau$ and not $t = 2\tau$. What can this mean?

Of course the signal envelope that is conventionally used to define the echo is based on the refocusing of static dephasing effects. In this perspective, the echo is defined by the dephasing and rephasing associated with the starting positions z in the magnetic field gradient. For the case of the steady gradient, the effect of the dephasing due to starting positions was given by the factor $\left(\int_{-\infty}^{\infty} \rho(z) \exp(i\gamma g z t) dz\right)$. For an echo, we may rewrite that term in eqn 5.44 using the idea of a phase reversal at $t = \tau$, so that the time variable t in the interval $\tau < t < 2\tau$ may be replaced by $(t - 2\tau)$, and

$$\overline{\exp(i\phi)} = \left(\int_{-\infty}^{\infty} \rho(z) \exp(i\gamma g z[t - 2\tau]) dz \right) \overline{\exp(i\Delta\phi)}$$

$$= \left(\int_{-\infty}^{\infty} \rho(z) exp(i\gamma g z[t - 2\tau]) dz \right) \exp\left(-\gamma^2 g^2 D(2\tau^3 - 2\tau^2 t + \tfrac{1}{3}t^3) \right)$$

$$(5.47)$$

If the first term dominates the signal attenuation away from $t = 2\tau$, then of course the echo maximum will occur at time $t = 2\tau$. But if, for whatever reason, the static dephasing term produces only a small phase spread, for example if the molecules are confined in a small region of space of size d such that $\gamma g d[t - 2\tau] \ll \pi$, while D is sufficiently large that the diffusive attenuation term dominates, then indeed an echo will be seen at $t = \sqrt{2}\tau$ rather than at $t = 2\tau$. Such an effect has been noted by Zanker *et al.* [20] for diffusion of gases contained in small sample tubes.

Signal attenuation under a steady gradient CPMG sequence

Equation 5.45 is easily generalised to a multiple, Carr–Purcell–Meiboom–Gill, echo sequence. For m echoes there are m such triangular phase step regions in the diagram equivalent to Fig. 5.13. Consequently the attenuation coefficient is

$$\overline{\exp(i\Delta\phi)} = \exp(-\frac{2m}{3}\gamma^2 g^2 D t^3)$$

$$= \exp(-\tfrac{2}{3}\gamma^2 g^2 D t^2 (mt))$$

$$(5.48)$$

Now the dependence of the attenuation on the total time mt is linear rather than cubic as in the case of the simple spin echo. This means that the attenuation effect at some total time, mt, may be arbitrarily reduced by decreasing the interval t between echoes and increasing m, the number of echoes. This was the method suggested by Carr and Purcell to eliminate additional relaxation of the transverse magnetisation due to diffusive motion occurring in the presence of field inhomogeneities.

Note that in all these expressions for $\overline{\exp(i\Delta\phi)}$, no account is taken of relaxation effects. Provided, however, we divide our magnetisation by that which applies when no magnetic field gradients are applied, then we may identify $\overline{\exp(i\Delta\phi)}$ with the normalised signal, $E(t)$.

5.4.4   The Bloch–Torrey equation for diffusion and flow

A later and more general approach to providing a description of spin phase evolution under any time-dependent gradient was made by Torrey [21]. The effect of molecular self-diffusion and velocity advection was accounted for by introducing an additional term in the Bloch equations (4.31), so that self-diffusion and flow is represented as a 'transport of magnetisation'. The magnetisation is effectively treated as fluid, and represented in an Eulerian sense by a vector $\mathbf{M}(\mathbf{r}, t)$. Using the substantive derivative to relate the rate of change of transverse magnetisation to precession, transverse relaxation, and diffusive replenishment in a magnetisation gradient, the equation for the x-component contains additional terms as,

$$\frac{DM_x}{Dt} = \gamma M_y(B_0 - \omega/\gamma) - \frac{M_x}{T_2} + \nabla \cdot \underline{\underline{D}} \cdot \nabla M_x \qquad (5.49)$$

or

$$\frac{\partial M_x}{\partial t} = \gamma M_y(B_0 - \omega/\gamma) - M_x/T_2 + \nabla \cdot \mathbf{D} \cdot \nabla M_x - (\mathbf{v} \cdot \nabla)M_x \qquad (5.50)$$

Note that the diffusion tensor is used in this description. We will have occasion to return to anisotropic diffusion later but for the moment, we will, for simplicity, assume a common diffusion scalar coefficient, D, and evaluate eqn 5.50 in the rotating frame, where the Zeeman Hamiltonian, on resonance, arises from the magnetic field gradient terms alone and

$$\frac{\partial M_+}{dt} = -i\gamma \mathbf{r} \cdot \mathbf{g}^*(t)M_+ - \frac{M_+}{T_2} + D\nabla^2 M_+ - (\mathbf{v} \cdot \nabla)M_+ \qquad (5.51)$$

where $M_+ = M_x + iM_y$.

It is clear that M_+ is a function of both $\mathbf{r}$ and t, and eqn 5.51 is solved by making the substitution [21, 22]

$$M_+(\mathbf{r}, t) = A(t)\exp\left(-i\gamma \mathbf{r} \cdot \int_0^t \mathbf{g}^*(t')dt'\right)\exp(-t/T_2) \qquad (5.52)$$

We will see that $A(t)$ is a (generally complex) modulation factor with a modulus less than unity, determined by the spreading motion of the spin-bearing molecules, and an argument that derives from net translational displacement. The phase factor term $\exp\left(-i\gamma \mathbf{r} \cdot \int_0^t \mathbf{g}^*(t')dt'\right)$ arises from the helical phase distribution 'burned into' the spin system by the magnetic field gradient, and under the condition for a gradient echo at time t, $\int_0^t \mathbf{g}^*(t')dt' = 0$, so that, at the echo centre, this term becomes unity. Finally, the term $\exp(-t/T_2)$ accounts for the relentless spin–spin relaxation experienced by the spins while the magnetisation is resident on the transverse plane. Consequently, at the echo centre $M_+(\mathbf{r}, t)$ reduces to $A(t)\exp(-t/T_2)$, independent of the local coordinates.

Substitution of eqn 5.52 in eqn 5.51 leads to

$$\frac{\partial A(t)}{\partial t} = -\left[D\gamma^2\left(\int_0^t \mathbf{g}^*(t')dt'\right)^2 - i\gamma\mathbf{v} \cdot \int_0^t \mathbf{g}^*(t')dt'\right]A(t) \qquad (5.53)$$

with solution [21, 22]

$$A(t) = \exp\left[-D\gamma^2 \int_0^t \left(\int_0^{t'} \mathbf{g}^*(t'')dt''\right)^2 dt'\right]\exp\left[i\gamma\mathbf{v} \cdot \int_0^t \int_0^{t'} \mathbf{g}^*(t'')dt''dt'\right] \qquad (5.54)$$

Under the echo condition $\int_0^t \mathbf{g}^*(t')dt' = 0$, $A(t)$ is simply the normalised echo attenuation $E(t)$. We will use the Bloch–Torrey result to evaluate a number of echo-attenuation expressions for cases where the motion comprises simple diffusion and flow.

At any time other than the echo condition, the Bloch–Torrey factorisation of eqn 5.52 contains a time-dependent term in addition to $A(t)$, namely the phase factor $\exp\left(-i\gamma\mathbf{r}\cdot\int_0^t \mathbf{g}^*(t')dt'\right)$. This term is akin to a structure factor, a Fourier phase factor rising from the effect of gradient dephasing due to the spin positions in the magnetic field gradient, and one which must be evaluated by averaging over all $\mathbf{r}$. In that sense we might regard it as being similar to the structure factor seen in the bracketed terms of eqns 5.44 and 5.47, but note the subtle difference. The spin positions in the exponent of eqns 5.44 and 5.47 are *starting* positions. Crucially, $\mathbf{r}$ in the Bloch–Torrey factor, $\exp\left(-i\gamma\mathbf{r}\cdot\int_0^t \mathbf{g}^*(t')dt'\right)$, represents, by contrast, the local position vector in the Eulerian fluid, for which any density function, $\rho(\mathbf{r})$, over which this phase factor is integrated to represent the whole sample, will depend on the time of evaluation. Only at the echo centre will the expressions coincide. When $\int_0^t \mathbf{g}^*(t')dt' = 0$ this troublesome, time-dependent, exponential term is unity, and the distribution of the values of $\mathbf{r}$ irrelevant. But at any other time, the ensemble average of this exponential term contributes an additional time-dependent attenuation. For example, if the Bloch–Torrey expression for $A(t)$ is used to estimate the diffusion attenuation component in advance of the echo, eqn 5.47 is not reproduced. So long as we use Bloch-Torrey to evaluate our signal only at the echo centre, we are safe.

5.5 Pulsed gradient spin-echo NMR: diffusion and flow

In each of the cases analysed using the Carr–Purcell particle hopping model, the gradient was considered to be time-independent. Clearly the attenuation of the spin echo or CPMG train that arises from diffusive dephasing under the influence of a steady gradient may be conveniently used to measure molecular self-diffusion. This method has been effective in providing very precise self-diffusion coefficients in a variety of simple liquids. However, the method is inconvenient where the motion is slow. This is because the steady gradient spreads the Larmor spectrum at all times, and in particular during the period of RF pulse transmission and during the period of signal detection. This means that the maximum gradient that can be applied is limited by the transmitter and receiver bandwidths. McCall, Douglass, and Anderson [23] suggested in 1963 that the gradient might be applied in the form of rectangular pulses inserted respectively in the dephasing and rephasing parts of the echo sequence, but gated off during RF pulse transmission and signal detection. This pulsed gradient spin-echo (PGSE) sequence was first demonstrated by Stejskal and Tanner in 1965 [24].[11]

5.5.1 The Stejskal–Tanner experiment

The basic pulse sequence for the Stejskal and Tanner experiment is shown in Fig. 5.14, first as a spin echo in (a) and then as a stimulated echo in (b). From the perspective of

---

[11]The pulsed gradient method for measuring spin translational motion is sometimes referred to as pulsed field gradient NMR (PFG NMR). Given that pulsed magnetic field gradients are variously used in magnetic resonance applications for imaging and for coherence selection purposes, we prefer the name pulsed gradient spin echo, with its acronym PGSE NMR, where the role of the echo, however formed, and so central to the measurement of dynamics, is explicit. The name pulsed gradient spin-echo NMR was originally suggested by Stilbs and Moseley [25, 26].

motion-induced phase shifts, the two sequences are equivalent, but the way that relaxation acts to attenuate the echo signal is clearly different, the spin echo experiencing T_2 throughout the evolution of magnetisation in the transverse plane, and the stimulated echo allowing for a slower T_1 relaxation during the period in which magnetisation is stored along the longitudinal axis. There is no universally agreed nomenclature for the acronyms describing these methods. To this author, PGSE seems to present a convenient label for both types of echo.

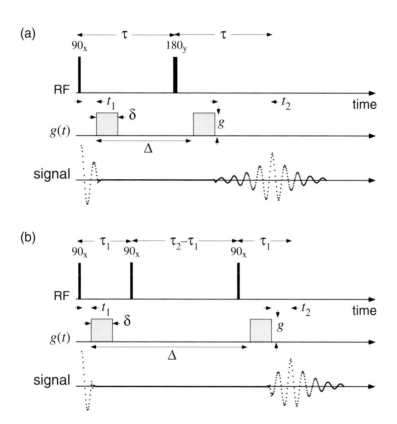

Fig. 5.14 (a) Pulsed gradient spin-echo (PGSE) sequence, with gradient amplitude g, pulse duration δ, and gradient pulse spacing Δ. τ is the time between the 90° and 180° RF pulses and corresponds to half the spin-echo formation time, T_E. (b) stimulated-echo version of PGSE sequence.

One of Stejskal's important contributions was the use of a propagator description of generalised motion [27]. As we shall see later, this allows a very natural and insightful formalism in the special circumstance that the gradient pulse duration is short. For the moment though, we will outline the effect of simple self-diffusion and flow under the condition of finite duration gradient pulses.

Carr–Purcell derivation

The cumulative phase step diagram for the PGSE experiment in the spin echo version is shown in Fig. 5.15, where $\delta = m\tau_s$ and $\Delta = p\tau_s$.

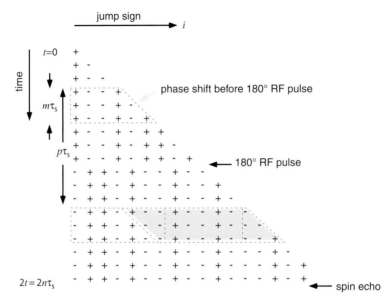

Fig. 5.15 Cumulative phase diagram for spin-echo formation in the PGSE experiment. The net phase shift is the sum of that occurring before and after the 180° RF pulse and corresponds to the shaded area. Note that $\delta = m\tau_s$ and $\Delta = p\tau_s$.

The effect of the 180° RF pulse is to cause the inversion of prior phase shifts, leading to cancellation of phase shifts associated with the unshaded areas. The net phase shift is therefore obtained by summing two uncorrelated triangular regions, each with mean square phase shift $\frac{1}{3}\gamma^2 g^2 \tau_s^2 \xi^2 m^3$, along with one uncorrelated rectangular region with mean square phase shift $\gamma^2 g^2 \tau_s^2 \xi^2 m^2 (p-m)$. The net mean square phase shift is therefore

$$\overline{\Delta\phi^2} = \gamma^2 g^2 \tau_s^2 \xi^2 m^2 (p - m + \tfrac{2}{3}m)$$
$$= 2\gamma^2 g^2 \delta^2 D(\Delta - \delta/3). \tag{5.55}$$

Using eqn 5.41, we obtain the well-known Stejskal–Tanner relation [24] for the attenuation of the echo amplitude,

$$S(g)/S(0) = \exp(-\gamma^2 g^2 \delta^2 D(\Delta - \delta/3)) \tag{5.56}$$

where, as before, we label the ratio of the echo amplitude at gradient g to that at zero gradient, as the normalised echo attenuation, $E(g)$. Note that the chosen variable here is the gradient amplitude g. In fact, a very convenient experimental procedure is to keep all the experimental times fixed, and to vary the magnetic field gradient amplitude. In this way, $E(g)$ has relaxation effects removed, except in as much as overall signal attenuation due to relaxation reduces the available signal to noise ratio. Of course, the timescale over which diffusion occurs is set by the period, Δ, separating the start of the gradient pulses, a parameter to be varied if the time dependence of diffusion is to be probed. Even where $E(g, \delta)$ is measured, care in arranging the pulse sequence can allow for normalisation of relaxation effects.

Equation 5.56 provides a precise description of the influence of self-diffusion in the PGSE experiment and is the basis of a considerable literature pertaining to this technique [24, 28–32]. An example of self-diffusion measurement is shown in Fig. 5.16, in which spectrally separated molecular species in the same liquid mixture have their respective echo attenuations plotted as a function of squared gradient amplitude. The spectral separation is possible because the echo signal is sampled in the absence of the magnetic field gradient. The PGSE method may be used to precisely measure diffusion down to 10^{-14} m^2 s^{-1}, some five orders of magnitude slower than that exhibited in Fig. 5.16.

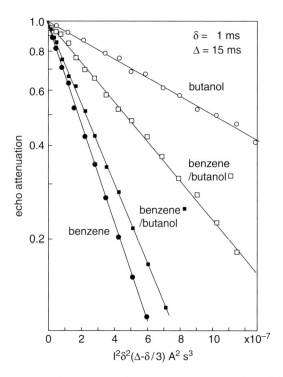

Fig. 5.16 Echo-attenuation data for pure benzene, pure butanol, and for the benzene and butanol components in an equimolar mixture. The self-diffusion coefficients are, respectively, $(2.23\pm0.02)\times10^{-9}$, $(0.43\pm0.01)\times10^{-9}$, $(1.83\pm0.02)\times10^{-9}$, and $(0.90\pm0.01)\times10^{-9}$ m^2 s^{-1}. (From reference [33].)

Note that the exponent in the Stejskal–Tanner equation is proportional to the mean-squared displacement of the molecules over an effective timescale $(\Delta - \delta/3)$. The attenuation of the echo arises because of the incoherent nature of the phase shifts across the nuclear ensemble. Suppose by contrast that there is an additional coherent motion due to each molecule being additionally displaced by an identical amount. This of course corresponds to the superposition of a uniform velocity $\mathbf{v}$, so that the extra displacement moved per incremental step is $\mathbf{v}\tau_s$. The effect is easy to calculate. Consider the PGSE phase step diagram of Fig. 5.15. The shape of the diagram which allows for a constant velocity component will be exactly the same. However, all steps will now have an additional, identical magnitude contribution, with negative sign before the 180° RF pulse and positive sign after. The mean phase shift corresponds to the sum in the shaded area. This is $\gamma\tau_s\mathbf{g}\cdot(\mathbf{v}\tau_s)pm$ or $\gamma\delta\mathbf{g}\cdot\mathbf{v}\Delta$. Consequently a phase shift of $\exp(i\gamma\delta\mathbf{g}\cdot\mathbf{v}\Delta)$ is common to all spins in the ensemble, and may be factorised out of the signal, leaving the diffusive contribution exactly as before. The combined effect of diffusion and flow is therefore a phase shift due to flow and an attenuation due to diffusion given by

$$E(\mathbf{g}) = \exp(i\gamma\delta\mathbf{g}\cdot\mathbf{v}\Delta - \gamma^2 g^2\delta^2 D(\Delta - \delta/3)) \qquad (5.57)$$

Equation 5.57 is an idealisation. It makes no allowance for the interfering effects of steady background gradients, nor for the effects of finite rise times in the gradient pulses, instead assuming perfect rectangular pulse shapes. Later we will address these perturbing effects. But for the moment, let us see how the Bloch–Torrey result yields the idealised Stejskal–Tanner relation.

Bloch–Torrey derivation

We return to the simple spin-echo version of the PGSE experiment shown in Fig. 5.14 (a), whose gradient echo equivalent, based on effective gradient g^*, is shown in Fig. 5.17. Note that if there are small residual gradients due to polarising magnet inhomogeneity then the centre of this echo will be at a time τ after the 180_y° pulse, the rephasing time being equal to the dephasing period separating the 90_y° and 180_y° pulses.

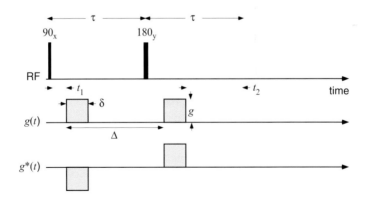

Fig. 5.17 As in Fig. 5.14(a), but showing the effective gradient $g^*(t)$.

To evaluate the spin phase distributions we need to evaluate a number of Bloch–Torrey integrals using the timings indicated. Note that we call the time delay before the first gradient pulses t_1. We will also assume a vector gradient, though at this stage we will be dealing with isotropic diffusion and only the squared magnitude of $\mathbf{g}^*$ will enter the diffusive contribution to the phase spread. At this point any weak background gradients will be ignored. The relevant integrals are

$$\int_0^{2\tau} \mathbf{g}^*(t')dt' = 0 \tag{5.58}$$

and

$$
\begin{aligned}
\int_0^t \int_0^{t'} \mathbf{g}^*(t'')dt''dt' &= -\int_{t_1}^{t_1+\delta} \mathbf{g}(t' - t_1)dt' - \int_{t_1+\delta}^{t_1+\Delta} \mathbf{g}\delta dt' \\
&\quad + \int_{t_1+\Delta}^{t_1+\Delta+\delta} [-\mathbf{g}\delta + \mathbf{g}(t' - t_1 - \Delta)]dt' \\
&\quad + \int_{t_1+\Delta+\delta}^{2\tau} [-\mathbf{g}\delta + \mathbf{g}\delta]dt' \\
&= -\mathbf{g}\delta\Delta
\end{aligned}
\tag{5.59}
$$

and

$$
\begin{aligned}
\int_0^t \left(\int_0^{t'} \mathbf{g}^*(t'')dt'' \right)^2 dt' &= -\int_{t_1}^{t_1+\delta} g^2(t' - t_1)^2 dt' - \int_{t_1+\delta}^{t_1+\Delta} g^2\delta^2 dt' \\
&\quad + \int_{t_1+\Delta}^{t_1+\Delta+\delta} [g^2\delta^2 - 2g^2(t' - t_1 - \Delta) + g^2(t' - t_1 - \Delta)^2]dt' \\
&= g^2\delta^2(\Delta - \delta/3)
\end{aligned}
\tag{5.60}
$$

Equation 5.58 is simply the condition for the formation of an echo. Equations 5.59 and 5.60 taken together with eqn 5.58 reproduce the earlier expression for the velocity-induced echo phase shift and diffusion-induced echo attenuation given in eqn 5.56. Because we chose to make the effective gradient negative for the first pulse, the sign of the velocity phase shift term is negative. By contrast, in the Carr–Purcell derivation we chose to negate the phases after the 180° pulse, so that the velocity term was positive. So long as we understand the consequences of these phase conventions, we should be able to understand the absolute sign of the phase shift.

Phase evolution during the pulse sequence
Before leaving the Bloch–Torrey analysis it is instructive to plot the evolution of the phase shift integrals that generate sensitivity to position (eqn 5.58) and velocity (eqn 5.59), as well as the evolution of the phase-spreading integrals associated with diffusion (eqn 5.60). These are shown in Fig. 5.18.

The effect of finite gradient pulse rise and fall times
Perfectly rectangular gradient pulses are an idealisation hard to achieve in practice. The changing magnetic field associated with the rising gradient pulse can act to induce

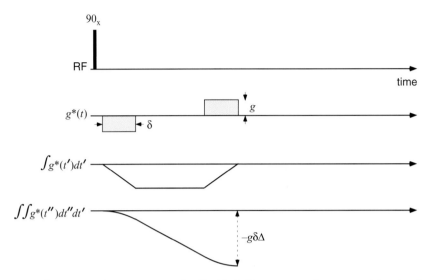

Fig. 5.18 Effective gradient sequence for Stejskal–Tanner experiment resulting in the echo condition $\int_0^t g^*(t')dt' = 0$, along with the velocity phase term $\int_0^t \int_0^{t'} g^*(t'')dt''dt'$.

eddy currents in the conductors surrounding the NMR sample space, for example the room temperature bore or cooled surfaces in the vicinity of a superconducting magnet. These eddy currents themselves produce magnetic fields and gradients that act back on the sample. By using a self-screened design for the gradient coil, such effects may be reduced, but in practice a significant reduction in eddy currents can be achieved just by switching the gradient pulse on and off at a controlled rate, for example by using a linear ramp.

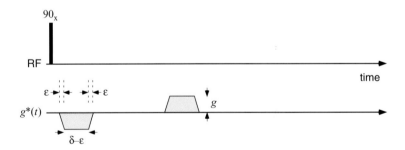

Fig. 5.19 Effective gradient sequence for Stejskal–Tanner experiment with finite rise and fall times.

Again the Bloch–Torrey approach may be easily used to generate the normalised echo signal when such a ramp is used [34, 35]. A useful timing nomenclature is to allow for a gradient pulse whose total area is $g\delta$, with rise and fall times ϵ, and hence constant gradient duration $\delta - \epsilon$, as shown in Fig. 5.19. The phase shift due to flow is unaffected by this variation in pulse shape, depending only on pulse area, while the attenuation due to diffusion is slightly modified as

$$E(\mathbf{g}) = \exp(i\gamma\delta\mathbf{g} \cdot \mathbf{v} - \gamma^2 Dg^2\delta^2(\Delta - \delta/3)[1 + \frac{\epsilon}{\Delta - \delta/3}(\frac{\epsilon^2}{30\delta^2} - \frac{\epsilon}{6\delta})]) \qquad (5.61)$$

The correction term in the square brackets is so weak as to be practically negligible, except where high-precision diffusion measurements are required. For example, even for a slow ramp comprising 20% of the pulse duration along with a relatively short $\Delta \sim 2\delta$, $\epsilon/\delta \sim 0.2$ and $\epsilon/\Delta \sim 0.1$, yields a correction term of around 0.3 %.

5.5.2 The role of background gradients

The influence of a steady background gradient
The Bloch–Torrey method is ideal for calculating the subtle perturbations to the Stejskal–Tanner relation that arise from gradient artifacts. Here we consider the effect of a weak steady background gradient, $\mathbf{G}_0$. The derivation is tedious but straightfor-

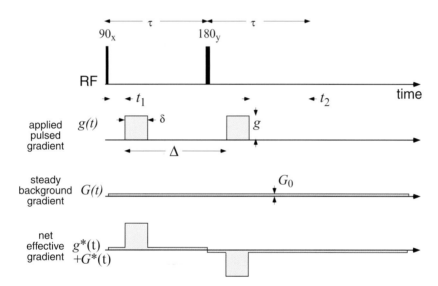

Fig. 5.20 As in Fig. 5.14(a), but showing the effective gradient $\mathbf{g}^*(t)$. Also shown is a weak background gradient G_0. Because $\mathbf{g}^*(t)$ and $\mathbf{G}_0$ and may act in different directions, the two gradients are shaded different colours.

ward and the exact form of the signal is given by [24, 27]

$$\begin{aligned} E(\mathbf{g}) = \ & \exp(i\gamma\delta\mathbf{g} \cdot \mathbf{v}\Delta + i\gamma\tau^2\mathbf{G}_0 \cdot \mathbf{v} \\ & -\gamma^2 D[g^2\delta^2(\Delta - \delta/3) + \tfrac{2}{3}G_0^2\tau^3 \\ & -\mathbf{g} \cdot \mathbf{G}_0\delta\{t_1^2 + t_2^2 + \delta(t_1 + t_2) + \tfrac{2}{3}\delta^2 - 2\tau^2\}]) \end{aligned} \qquad (5.62)$$

with t_1 again being the delay between the 90°_x RF pulse and the start of the first gradient pulse, and t_2 being the delay between the end of the second gradient pulse and the centre of the echo. This result is useful in dealing with an apparatus where the residual gradients are poorly shimmed or where a steady background gradient is deliberately used to stabilise the echo.[12]

Stimulated echo

The measurement of self-diffusion by PGSE NMR is limited by the loss of phase coherence due to transverse relaxation. One nice method of avoiding this problem is to store the spatially-encoded magnetisation along the longitudinal axis by means of a second 90° pulse, as shown in Fig. 5.14(b). The magnetisation is recalled at a later time and rephased in a stimulated echo. Over the storage period the spins are subject to T_1 relaxation, which is generally slower than T_2. The expression equivalent to eqn 5.62 for the case of the stimulated echo is [36]

$$
\begin{aligned}
E(\mathbf{g}) \;=\; & \exp(i\gamma\delta\mathbf{g}\cdot\mathbf{v}\Delta + i\gamma\tau_1\mathbf{G}_0\cdot\mathbf{v}\Delta \\
& -\gamma^2 D[g^2\delta^2(\Delta - \delta/3) + G_0^2\tau_1^2(\tau_2 - \tfrac{1}{3}\tau_1) \\
& -\mathbf{g}\cdot\mathbf{G}_0\delta\{t_1^2 + t_2^2 + \delta(t_1 + t_2) + \tfrac{2}{3}\delta^2 - 2\tau_1\tau_2\}])
\end{aligned}
\tag{5.63}
$$

For small background gradients this expression is entirely equivalent to that of the spin-echo expression, the only penalty paid being the loss of a factor of 2 in the signal intensity due to projection of only half the transverse magnetisation along the z-axis during the storage period. Comparison of eqns 5.62 and 5.63 illustrates the reduced influence of $\mathbf{G}_0$ in the latter expression because of the replacement of τ^3 and τ^2 by $\tau_1^2(\tau_2 - \tfrac{1}{3}\tau_1)$ and $\tau_1\tau_2$, respectively. The stimulated-echo method thus has the additional advantage that the magnetisation is not only protected from T_2 relaxation during the storage period but is also experiences reduced precessional dephasing caused by residual gradients.

5.5.3 Echo schemes to reduce the effect of background gradients

When measuring molecular diffusion in porous media, the diamagnetic susceptibility difference between the matrix and the imbibed fluid can lead to quite large inhomogeneous magnetic fields in the pore space. This has the effect of significantly perturbing the measurement of diffusion using pulsed gradients, both because of the cross terms apparent in eqns 5.62 and 5.63, and because diffusion in these inhomogeneous fields may lead to additional spin relaxation and loss of signal intensity.[13] Note that in an experiment where the diffusive attenuation is measured by incrementing the pulsed gradient amplitude g, only the cross terms bother us, since the quadratic (G_0^2) term in the echo-attenuation exponent arising from the background gradient alone contributes a fixed attenuation.

CPMG and spin-locking

One approach to the reduction of residual gradient influence is to employ a CPMG echo train but with a single pair of gradient pulses inserted in appropriately spaced locations between the 180° RF pulses [37]. While the CPMG train method does not

---

[12]See Chapter 12.

[13]This effect is discussed in detail in Section 6.4.

suffer from the factor-of-two signal-to-noise ratio loss of the stimulated echo, it does suffer T_2 relaxation. An alternative method of retaining the full signal amplitude while avoiding transverse relaxation is to employ spin-locking pulses over the waiting period between the two gradient pulses in a normal spin echo [38]. Under a sustained spin-lock pulse the magnetisation decays as $T_{1\rho}$, which, for sufficiently powerful RF fields, can be made much longer than T_2.

13-interval sequence

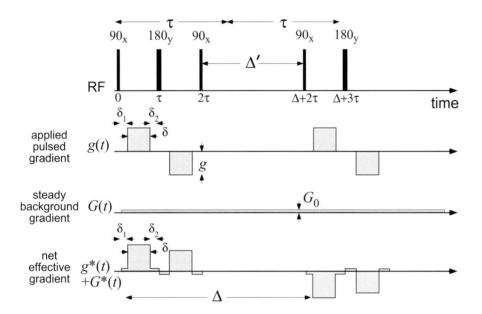

Fig. 5.21 13-interval gradient pulse scheme of Cotts *et al.*, designed to minimise the effect of background gradients. Note that Δ' as shown in eqn 5.64 differs from the conventional definition of Δ.

The role of spin echoes in reducing the effect of background gradients was analysed in detail by Karlicek and Low and led to a suggestion that the pulsed magnetic field gradients be applied with alternating sign in respective dephasing and rephasing spin-echo segments [39]. This idea was followed by the best known of the schemes to reduce the effect of background gradients, the so-called '13-interval' sequence of Cotts *et al.* [40]. The pulse sequence is based on a stimulated echo, as shown in Fig. 5.21, but with the diffusion encoding period divided into a spin-echo dephasing and rephasing segment, with bipolar gradients. The latter have the effect of refocusing background gradient dephasing at the point of both the 90° RF storage pulse and the stimulated echo, but cumulatively adding the desired dephasing of the pulsed gradients, thanks to the use of bipolar pulses. The effective gradients also shown in Fig. 5.21 make clear how the sequence works. Ignoring the influence of flow, diffusive effect of the background gradients is given by

$$E(\mathbf{g}) = \exp\left(-\gamma^2 D \left[g^2\delta^2(4\Delta' + 6\tau - 2\delta/3) + \frac{4\tau^2}{3}G_0^2 - \mathbf{g}\cdot\mathbf{G}_0 2\delta\tau(\delta_1 - \delta_2)\right]\right)$$

implying that the cross term vanishes under the condition $\delta_1 = \delta_2$.

Compensating a changing background field

Of course this cancellation depends on the background gradients being the same in the two encoding intervals separated by Δ'. Given that Δ' may be quite long and, as a consequence, in a heterogeneous medium, spin-bearing molecules may diffuse to regions of differing local background field, the cancellation may be imperfect. This problem is discussed by Sun *et al.* [41] and Seland *et al.* [42], who propose the use of asymmetric bipolar pulses designed to correct for this effect.

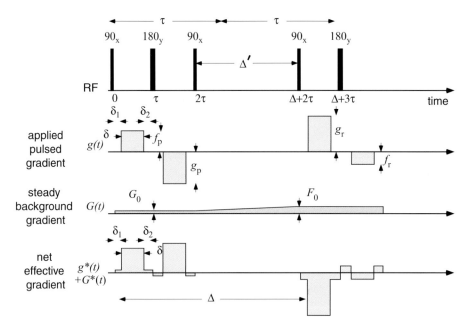

Fig. 5.22 Generalised 13-interval gradient pulse scheme of Cotts *et al.*, with four unequal pulsed field laboratory gradients, with the intensities $f_{p,r}$ and $g_{p,r}$, respectively, along with a background gradient, which changes value from G_0 to F_0 during the storage period Δ'. $f_{p,r}$ and $g_{p,r}$ can be adjusted to cancel the effect of the background field. (Adapted from Galvosas *et al.* [43].)

An exact solution has been provided by Galvosas *et al.* based on the parametrisation of Fig. 5.22 [43]. First, the pulsed gradients must obey the echo condition, namely

$$f_p + g_p = f_r + g_r \tag{5.64}$$

where the subscripts refer to the 'prepare' and 'read' segments of times 2τ when the magnetisation is in the transverse plane. In their analysis, the normalised echo attenuation at the echo centre ($t = t_e$) may be written

$$E = \exp(-\gamma^2 D[I_p + I_c + I_b]) \tag{5.65}$$

where I_p, I_c, and I_b are Bloch–Torrey-type integrals associated with the pulsed gradients, the cross term, and the background effective gradients as

$$I_p = \int_0^{t_e} dt' \left(\int_0^{t'} \mathbf{g}^*(t'')dt'' \right)^2$$

$$I_c = \int_0^{t_e} dt' \int_0^{t'} \mathbf{g}^*(t'')dt'' \int_0^{t'} \mathbf{G}^*(t'')dt''$$

$$I_1 = \int_0^{t_e} dt' \left(\int_0^{t'} \mathbf{G}^*(t'')dt'' \right)^2 \tag{5.66}$$

The result of the integrations is

$$I_p = \delta^2 \{ (\Delta' + \tau)(g_r + f_r)^2 + \tau \left[2(f_r)^2 - (f_p + f_r)(g_p - g_r) \right]$$
$$- \frac{1}{3}\delta \left[(g_p)^2 + f_p f_r + g_r(f_p - f_r) \right] \} \tag{5.67}$$

$$I_c = \delta \{ G_0[a(g_p - f_p) + 2\tau^2 f_p + \tau(\delta_2 - \delta_1)g_p]$$
$$- F_0[a(g_r - f_r) + 2\tau^2 f_r - \tau(\delta_2 - \delta_1)f_r] \} \tag{5.68}$$

$$I_b = \frac{2}{3}\tau^3(F_0^2 + G_0^2) \tag{5.69}$$

where $a = \delta_1^2 + \delta_1\delta + \delta^2/3$. Galvosas *et al.* show that the cross term can always be eliminated ($I_c = 0$) if the ratios of the pulsed gradients are set to obey [43]

$$\frac{g_p}{f_p} = \frac{a - 2\tau^2}{a + \tau(\delta_2 - \delta_1)} \tag{5.70}$$

and

$$\frac{g_r}{f_r} = 1 - \frac{2\tau^2 - \tau(\delta_2 - \delta_1)}{a} \tag{5.71}$$

Further details regarding the practical setting of ratios in an easy manner are given in reference [43].

5.5.4 More efficient encoding schemes

DIFFTRAIN

A useful variant of the stimulated-echo sequence, which allows for repeated sampling at different diffusion times Δ, has been proposed by Buckley *et al.* [44]. The key idea behind their 'DIFFTRAIN' method, shown in Fig. 5.23, is the replacement of the final $90°$ RF pulse by a small tip-angle (α) pulse, so that only a fraction, $\sin \alpha$, of the stored magnetisation is recalled for echo sampling, leaving a fraction $\cos \alpha$ remaining along the z-axis for subsequent recall. This allows a range of Δ values to be explored in a single excitation process,[14] thus significantly improving the efficiency of the experiment. Of course there is a price to be paid. So long as we are only interested in the echo amplitude as sampled by a single point in the time domain, the method works well. But should we wish to sample the FID following the echo, an essential requirement if spectral resolution is desired, then we must trade away the acquisition time (and hence the spectral resolution) against the fineness of the Δ stepping interval.

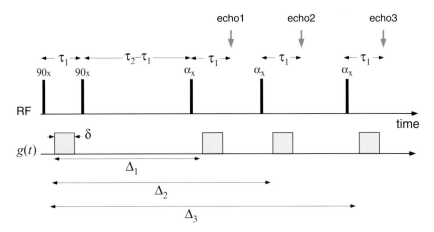

Fig. 5.23 DIFFTRAIN stimulated-echo pulse sequence in which a small tip angle-pulse α_x is used to successively recall magnetisation already encoded by the first gradient pulse. By this means, multiple echoes corresponding to different values of Δ may be obtained in a single shot.

Multiple-modulation multiple-echo
The downside of DIFFTRAIN is the sacrifice of signal-to-noise ratio inherent in the use of small tip angles. In many signal-rich applications this is a small price to pay. But in applications where signal-to-noise ratio is inherently poor, for example in bore-hole NMR well-logging operations, maximisation of available signal is paramount. A comparison familiar in MRI is to be found in the relative performance of fast low angle shot (FLASH) imaging [45], where multiple recall of magnetisation via low tip-angle pulses is used, and the echo planar imaging (EPI) [12] or rapid acquisition relaxation enhanced imaging (RARE) techniques, [13] where full signal recall is made using gradient of spin-echo trains. These comparisons are further discussed in Chapter 10.

By analogy, Y-Q Song and co-workers [46, 47] have suggested the use of echo train recycled magnetisation with large tip angle RF pulses to provide a multiple gradient

---

[14]Note that the fraction of magnetisation that remains after n recall events is $\cos^n \alpha$, so that for small tip angles, the available stored magnetisation decays quite slowly with increasing n.

step encoding in diffusion and flow measurements, an experiment termed multiple-modulation multiple-echo (MMME). However, unlike RARE or EPI, MMME uses unequal time spacings that generate a maximal number of spin echoes. For example, as discussed in Section 4.6.2, the stimulated-echo pulse sequence generates two additional spin echoes, the echo of the initial pulse FID caused by the second pulse, and the echo of that echo caused by the third pulse. Using a formalism developed by Hürlimann [48], it is possible to account for the effect of relaxation, diffusion (or flow) on multiple echo amplitudes in more complex RF and gradient pulse trains. First, however, it is helpful to design an RF sequence that generates a large number of echoes, all nicely separated. An example is shown in Fig. 5.24, the so-called '1-3-9' sequence, in which the four RF pulses of turn angles α_1 to α_4 are spaced by τ, 3τ, and 9τ. The generation of such equally spaced echoes was first pointed out by Hennig [49]. The series can be adjusted by factors of 3, as in MMME3 '1-3', MMME5 '1-3-9-27' and so on.

Under the influence of a constant gradient, the RF pulses are slice-selective, and the spatial modulation of the frequency can lead to complex echo shapes. Song and co-workers have shown that these effects can be minimised, and that echoes of nearly equal magnitude can be achieved in the MMM4 sequence by using $\alpha_1 = 54.74°$, $\alpha_2 = \alpha_3 = 70.53°$, and $\alpha_4 = 109.47°$. The use of this large turn angle compromise ensures that the echo amplitudes, while reduced somewhat (about a quarter of the full Hahn echo), are larger than achievable by low turn angle recall.

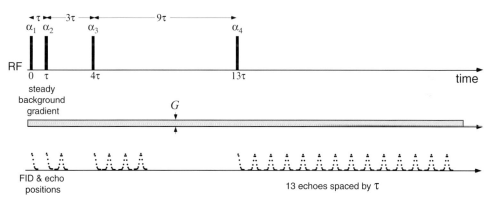

Fig. 5.24 MME4 pulse sequence in which the pulse spacings are set so as to generate 13 equally spaced echoes after the fourth RF pulse. The figure shows the positions of FIDs and echoes, though their relative amplitudes in practice depend on the RF pulse turn angles, α_1 to α_4. Each echo has its own relative sensitivity to relaxation and diffusion, factors which need to be calculated from an analysis of each coherence pathway. (Adapted from Song *et al.* [47].)

The diffusion sensitivity for each echo can be defined by a factor $b_\mathbf{Q}$, where $\mathbf{Q}$ is a vector that describes the apparent precession states of the magnetisation component contributing to that echo at each interval during the echo sequence, $Q = \pm 1$ referring to transverse magnetisation components of opposite sign precession and 0 indicating magnetisation in z-storage. Note that the natural state of precession in the transverse

Table 5.1 Diffusion sensitivity of MME4 sequence. The Q value indicates the sense of precession for magnetisation components in the transverse plane, where 0 indicates magnetisation in z-storage. The starting condition prior to α_1 is of course 0. (From Song *et al.* [47].)

Echo number	Q					$b_{\mathbf{Q}}$
1	0	1	0	0	−1	38/3
2	0	−1	1	0	−1	42
3	0	0	1	0	−1	99
4	0	1	1	0	−1	560/3
5	0	−1	−1	1	−1	126
6	0	0	−1	1	−1	162
7	0	1	−1	1	−1	704/3
8	0	−1	0	1	−1	345
9	0	0	0	1	−1	486
10	0	1	0	1	−1	2009/3
11	0	−1	1	1	−1	888
12	0	0	1	1	−1	1152
13	0	1	1	1	−1	4394/3

plane is a clockwise $Q = -1$. This is apparent in our complex number representation of transverse plane precession by the factor $\exp(-i\phi) = \exp\left(-i\gamma\mathbf{r} \cdot \int_0^t \mathbf{g}^*(t')dt'\right)$. However, subsequent RF pulses can invert phases of transverse magnetisation so as to make them appear later as $Q = 1$, the idea of 180° RF pulses causing the prior effective gradient to be reversed, being an example of that. The crucial point to understand is that, just as with the effective gradient idea, the successive states of the $\mathbf{Q}$-vector are viewed from the perspective following the final α_4 RF pulse.

The easiest way to appreciate the notation is to start with echo 13. Here α_1 generates a transverse plane component of magnetisation, with a clockwise precession $Q = -1$. The successive components after α_2 and α_3 are those remaining in the transverse plane and continuing as $Q = -1$. Then α_4 performs an inversion, all previous Q values are inverted to $+1$, and it is left to the natural -1 precession after α_4 to reverse the accumulated phases of 13 τ intervals preceding. When a component is stored along z, as denoted by $Q = 0$, the phase is stored as a $\cos\phi$ or $\sin\phi$ component, in other words a superposition of $\exp(-i\phi)$ and $\exp(i\phi)$. Thus z-storage generates the phase reversal necessary to make possible an echo after α_4. Using these simple ideas, the echo number analysis is easily performed by accumulating phase based on local Q-value weighted by the relevant duration between RF pulses.

Because the intrinsic echo amplitudes in the absence of diffusion differ, the easiest way to use MMME is to perform two experiments with two different τ values, so that the intrinsic amplitudes (provided relaxation is negligible) will be identical. Then the echo attenuation may be written in terms of the ratios of corresponding echoes as

$$\frac{E(\tau)}{E(\tau')} = \exp\left(-b_{\mathbf{Q}}\gamma^2 DG^2 \left[\tau^3 - \tau'^3\right]\right). \tag{5.72}$$

The MMME method may be extended to multiple dimensions by inclusion of pulsed gradients in other directions, retaining the steady gradient component to properly space the echoes. Examples of this approach to measure the diffusion tensor, or to combine diffusion measurements with imaging, are given in reference [47].

DANTE 'one shot' methods

Another approach to rapid diffusion measurement is to use multi-frequency magnetisation gratings, generated by pulse sequences that excite a discrete spectrum of frequencies in the presence of a magnetic field gradient. Principal among these is the DANTE sequence proposed by Morris and Freeman [50]. The sequence name (Delays alternating with nutations for tailored excitation) alludes to the repetitive circular journeys by Dante and Virgil in Dante Alighieri's *Purgatorio*, akin to the trajectories undergone by off-resonant spins. B_1 value.

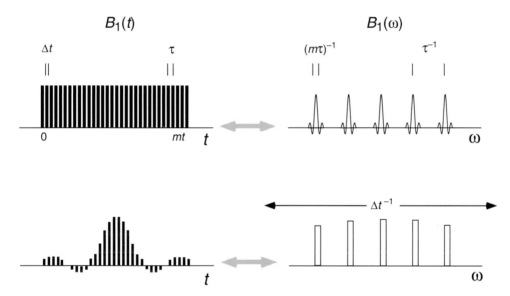

Fig. 5.25 DANTE train of RF pulses with corresponding frequency spectra. In the upper part of the diagram the RF modulation takes the form of a series of m equally spaced rectangular pulses. Because the train is finite, this is equivalent to multiplication by a Heaviside window, causing each frequency-domain spike to be convoluted with a sinc profile. In the lower part, the use of an overall sinc window implies that the local spike profile will be rectangular. Note that the spectrum has a finite extent, Δt^{-1}, determined by the width of the RF pulses.

A simple explanation is obtained by using linear response theory. Figure 5.25 shows a series of m RF pulses, of duration Δt and spacing τ, for which a spin isochromat on-resonance experiences a total flip angle $m\gamma B_1 \Delta t$. The frequency response of this train is shown alongside and consists of a comb of sidebands spaced by τ^{-1}, with each spike in the comb having the form of a sinc function of width $(mt)^{-1}$, the total comb width being of order $(\Delta t) - 1$. For a specific total turn angle, such as $\pi/2$, B_1 can be made arbitrarily large by making Δt sufficiently short, within the constraints set by the transmitter bandwidth. By choosing τ sufficiently short that the off-resonant sidebands are outside the spectral range of the spins, the selective excitation will be confined to the central on-resonant comb element, with an excitation bandwidth given by $(m\tau)^{-1}$. Because of the sharp discrete nature of the spin precession steps, it is possible to use a value of B_1 that is larger by a factor of $\tau/\Delta t$ than that used in continuous RF excitation, thus avoiding the non-linear transmission problem that is so severe at low levels of B_1. The additional transmitted power is, of course, dissipated in the outer sidebands and, for a specific total turn angle θ, represents no more in total than would be applied to the coil in a hard θ pulse at the same B_1 value. It is not necessary to restrict the DANTE sequence to equal amplitude (or phase) RF pulses. Indeed, where an ideal rectangular frequency spread is desired in the selective excitation, the train can be sinc modulated as shown in the lower part of Fig. 5.25.

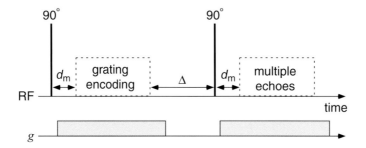

Fig. 5.26 Schematic grating-based diffusion pulse sequence in which the grating block consists of RF pulses with small tip angles. The magnetisation grating created by the excitation decays during the period Δ, and the remaining pattern is detected as a train of echoes. (Adapted from Szutkowski and Furó [51].)

Szutkowski and Furó [51] have developed an effective method for the measurement of diffusion using one-shot magnetisation grating methods, based on an earlier sequence by Zha and Lowe [52]. The principle is shown schematically in Fig. 5.26. The DANTE RF pulse sequence is applied during the 'grating encoding' period and results in a multiple echo train during the time domain of the detection period, the amplitude of each echo being proportional to the corresponding component of the grating. The delay period Δ allows molecules to diffuse, thus blurring the grating and resulting in echo attenuation. The echo train readout is sensitive to Stejskal–Tanner-like attenuation as $g^2(n\tau)^2$, where τ is the DANTE (and echo) pulse spacing. In a single-shot measurement of echo attenuation vs $g^2(n\tau)^2$, for a sample comprising both protonated water and polyethylene glycol (M_w=6000), Szutkowski and Furó were able to clearly identify bi-exponential diffusion decay [51].

5.5.5 Using pulsed gradients to measure the diffusion tensor

We now rewrite the Bloch–Torrey equation when the diffusive behaviour of the spin ensemble is represented by the tensor $\underline{\underline{D}}$,

$$\frac{\partial M_+}{\partial t} = -i\gamma \mathbf{r} \cdot \mathbf{g}^*(t) M_+ - \frac{M_+}{T_2} + \nabla \cdot \underline{\underline{D}} \nabla M_+ - (\mathbf{v} \cdot \nabla) M_+ \qquad (5.73)$$

with solution

$$A(t) = \exp\left(-\gamma^2 \int_0^t \left[\int_0^{t'} \mathbf{g}^*(t'')dt''\right]^T \underline{\underline{D}} \left[\int_0^{t'} \mathbf{g}^*(t'')dt''\right] dt'\right)$$

$$\times \exp\left(i\gamma \mathbf{v} \cdot \int_0^t \int_0^{t'} \mathbf{g}^*(t'')dt''dt'\right) \qquad (5.74)$$

where $\mathbf{g}^*(t) = \left(g_x^*(t), g_y^*(t), g_z^*(t)\right)^T$. The normalised echo attenuation for the idealised Stejskal–Tanner pulse sequence is therefore

$$E(\mathbf{g}) = \exp(i\gamma\delta\mathbf{g} \cdot \mathbf{v}\Delta - \gamma^2\delta^2\mathbf{g}^T\underline{\underline{D}}\mathbf{g}(\Delta - \delta/3)), \qquad (5.75)$$

$\mathbf{g}$ being the vector description of the magnetic field gradient pulse amplitude. Expressions for echo attenuation in the case of isotropic diffusion continue to apply by simply replacing $g^2 D$ by $\mathbf{g}^T\underline{\underline{D}}\mathbf{g}$. Note that a commonly used shorthand notation is $\underline{\underline{g}} : \underline{\underline{D}}$ where $\underline{\underline{g}}$ is formed by the outer product $\mathbf{g}^T\mathbf{g}$. In an experiment in which the gradients have components along all three axes, the diffusive attenuation is

$$E(\mathbf{g}) = \exp(-\gamma^2\delta^2[g_{xx}D_{xx} + g_{yy}D_{yy} + g_{zz}D_{zz}$$
$$+ (g_{xy} + g_{yx})D_{xy} + (g_{yz} + g_{zy})D_{yz} + (g_{zx} + g_{xz})D_{zx}](\Delta - \delta/3)).$$
$$(5.76)$$

For self-diffusion $\underline{\underline{D}} = \underline{\underline{D}}^T$ so that there exist six independent elements of the diffusion tensor to be determined. In a typical experiment, seven independent gradient directions are used to determine the six independent $D_{\alpha\beta}$ elements along with the echo amplitude at zero gradient, $E(0)$. A suitable combination is

$$\mathbf{g} = g_0\{(1,0,0),(0,1,0),(0,0,1),(1,0,1),(1,1,0),(0,1,1),(1,1,1)\}. \qquad (5.77)$$

The $\underline{\underline{D}}$ matrix is then diagonalised by a similarity transformation to find the eigenvalues corresponding to the diagonal elements in the principal axis frame. This similarity transformation of course represents the rotation required to take the lab-frame coordinates to the principal-axis frame and, as a consequence, the process of measuring the diffusion tensor not only reveals the underlying diffusion anisotropy, but also the direction of the axis corresponding to the most rapid diffusion coefficient. As a consequence the method has proven particularly useful a contrast in medical imaging [53], where the fast diffusion axis direction enables one to track the orientation of nerve fibres or other aligned tissue structure.

5.6   Pulsed gradient spin-echo NMR: general motion

Of course, diffusion and flow represent an important but nonetheless specific type of molecular motion. For more general motions it is not possible to obtain a closed form expression for the echo amplitude under all possible effective gradients, $\mathbf{g}^*(t)$ using the Bloch–Torrey approach. However, there does exist a variant of the PGSE experiment for which general analytic expressions are available. E.O. Stejskal pointed out [27] that for the particular case of narrow gradient pulses, in the limit as $\delta \to 0$, it is possible to use the propagator formalism for molecular translational dynamics, introduced in Chapter 1.

5.6.1 Narrow gradient pulse approximation and q-space

The principle underlying this description is that gradient pulses must be sufficiently narrow that we may neglect motion over their duration. In the case of unrestricted diffusion, the narrow pulse condition might naively translate to $\delta \ll \Delta$, although, as we have seen, we know how to handle unrestricted diffusion when the pulse width is finite. The meaning of narrow in this context really hinges on the existence of any characteristic times for the motion. For unrestricted steady flow there is no characteristic length or time scale, and likewise for self-diffusion, at time intervals beyond the molecular collision time. But where boundaries to motion exist, as in a porous medium or lyotropic liquid crystal, or when translational dynamics differ across a hierarchy of timescale regimes, as in the case of segmental self-diffusion for entangled polymers, then characteristic lengths and timescales play a significant role in determining the characteristics of the propagator for translational motion, and the allowed conditions for narrow pulse approximation will be so-determined.

Using the narrow pulse approach, we can see that the effect of the first gradient pulse is to impart a phase shift $\gamma \delta \mathbf{g} \cdot \mathbf{r}$ to a spin located at position $\mathbf{r}$ at the instant of the pulse, as shown in Fig. 5.27. This phase shift is subsequently inverted by the 180°_y RF pulse. Suppose that the molecule containing the spin has moved to $\mathbf{r}'$ at the time of the second gradient pulse. The net phase shift following this pulse will be $\gamma \delta \mathbf{g} \cdot (\mathbf{r}' - \mathbf{r})$. If the spins are stationary then, of course, a perfectly refocused echo will occur. Any motion of the spins will cause phase shifts in their contribution to the echo. The size of this shift is a product of two vectors, the dynamic displacement $(\mathbf{r}' - \mathbf{r})$ and a vector $\gamma \delta \mathbf{g}$.

The conditional probability and q-space

Again we define our 'echo signal' , $E(\mathbf{g}, \Delta)$, as the amplitude of the echo at its centre and, as important as these may be in determining the actual signal-to-noise ratio of our experiment, we neglect relaxation effects, which are removed by the normalisation process. We handle the ensemble average of individual spin contributions by taking each phase term $\exp(i\gamma \delta \mathbf{g} \cdot [\mathbf{r} - \mathbf{r}'])$ weighted by the probability for a spin to begin at $\mathbf{r}$ and move to $\mathbf{r}'$ at later time Δ. Using the Markovian propagator definition of Section 1.3.3, this probability is identically $p(\mathbf{r}, 0)P(\mathbf{r}|\mathbf{r}, t)$. Of course, in NMR the probabilities we are interested in concern molecules existing at a particular location along with their spin magnetisation having survived. In experiments where the magnetisation is freshly prepared, for example transverse magnetisation following a 90° RF pulse being applied to an ensemble of spins in thermal equilibrium, there is no distinction to be made, so that $M(\mathbf{r}, 0)$ and $p(\mathbf{r}, 0)$ are synonymous, and both will be simply equal to the molecular density function, $\rho(\mathbf{r})$ that characterises the sample, the

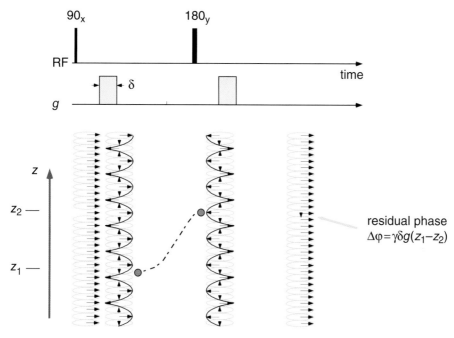

Fig. 5.27 The effect of displacement along the gradient direction in the narrow pulse depiction of the Stejskal–Tanner PGSE experiment. A spin-bearing molecule migrating by a displacement $Z = z_2 - z_1$ over the time between the gradient pulses results in a residual phase shift $\gamma \delta g Z$.

same density we measure in NMR imaging. For the rest of this chapter, we shall write the starting density condition as $\rho(\mathbf{r})$ and treat $\rho(\mathbf{r})P(\mathbf{r}|\mathbf{r}, t)$ as the probability for a spin to begin at $\mathbf{r}$ and move to $\mathbf{r}'$ at later time Δ. Thus

$$E(\mathbf{g}, \Delta) = \int \rho(\mathbf{r}) \int P(\mathbf{r}|\mathbf{r}, \Delta) \exp(i\gamma\delta\mathbf{g} \cdot [\mathbf{r} - \mathbf{r}']) d\mathbf{r}' d\mathbf{r} \qquad (5.78)$$

In fact eqn 5.78 bears a close resemblance to the expressions (such as eqn 5.17) that were used to describe the signal in **k**-space imaging. We will pursue this analogy by defining a reciprocal space q, where[15]

$$\mathbf{q} = \gamma\delta\mathbf{g} \qquad (5.79)$$

in units of angular spatial frequency, radians m^{-1}, or

$$\check{\mathbf{q}} = \frac{1}{2\pi}\gamma\delta\mathbf{g} \qquad (5.80)$$

in units of cyclic spatial frequency, m^{-1}.

---

[15]This **q**-space notation, which follows in an obvious manner from a neutron scattering analogy, was first introduced in reference [54], while the connection with neutron scattering was originally outlined in reference [28].

The reader will find in the literature both versions, both written simply as **q**. Therein lies a dilemma for the remainder of this book. Use of the angular frequency leads to simpler mathematical expressions, avoiding factors of 2π in the exponents of phase factors. Use of the cyclic frequency leads to a better comparison of structural features, $1/\check{q} = 2\pi/q$ being the relevant wavelength. In practice, one needs to be fluent with both forms, and for the rest of the book, we simplify the mathematics by using the angular frequency version. But we retain the symbol $\check{q}$ to explicitly identify cyclic frequency when desired.

Now eqn 5.78 can be rewritten

$$E(\mathbf{q}, \Delta) = \int \rho(\mathbf{r}) \int P(\mathbf{r}|\mathbf{r}', t) \exp(i\mathbf{q} \cdot [\mathbf{r}' - \mathbf{r}]) d\mathbf{r}' d\mathbf{r} \tag{5.81}$$

The neutron scattering analogy

The Pulsed Gradient Spin Echo NMR signal attenuation expression is akin to a scattering function which applies in inelastic neutron scattering where **q** is the scattering wavevector, as illustrated in Fig. 5.28. Eqn 5.81 is akin to the scattering function that applies in neutron scattering, where **q** is the scattering wavevector. PGSE and neutron scattering are closely analogous, as illustrated in Fig. 5.28, the main differences being the scale of temporal and spatial regimes sampled, and the detection of $E(\mathbf{g}, \Delta)$ in the time domain of Δ in the case of PGSE and in the frequency domain in the case of neutron scattering.

Neutron scattering results in a q-dependent signal which comprises two components. The first, known as the *coherent* part, is sensitive to relative displacements between nuclear scattering centres, and yields a diffraction pattern related to the average sample structure. The second, known as the *incoherent* part, is sensitive to time-dependent displacements of individual nuclear scattering centres, and so provides information regarding dynamics, revealed as a change in the outgoing neutron energy. Incoherent scattering is associated with spin-flips, which provide the ability in principle to localise scattering and so 'collapse the wavefunction' so that interference effects disappear.

Equation 5.81 is equivalent to the neutron scattering function for the incoherent fraction,

$$S_{incoherent} = \overline{N^{-1} \sum_{i=1} \exp(i\mathbf{q} \cdot [\mathbf{r}_i(t) - \mathbf{r}_i(0)])} \tag{5.82}$$

where the sum is taken over all N scattering centres. By contrast, the coherent neutron scattering is described by

$$S_{coherent} = \overline{N^{-2} \sum_{i=1} \sum_{j=1} \exp(i\mathbf{q} \cdot [\mathbf{r}_j(t) - \mathbf{r}_i(0)])} \tag{5.83}$$

This sensitivity to relative motion, $\mathbf{r}_j(t) - \mathbf{r}_i(0)$, makes the interpretation of coherent neutron scattering and quasi-elastic light scattering more difficult. The direct measurement of self-motion is a major advantage in PGSE NMR.

While PGSE NMR and incoherent inelastic neutron scattering are formally analogous in their mathematical expression, there exist a number of significant differences.

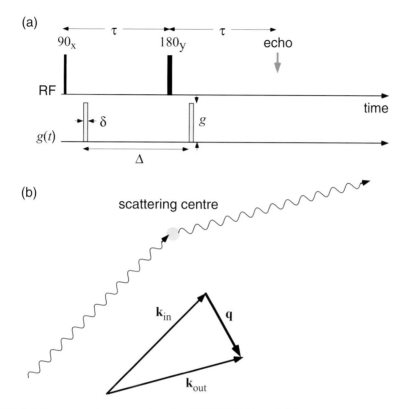

Fig. 5.28 (a) Narrow pulse approximation PGSE NMR and (b) scattering analogy in which the scattering wavevector, **q**, has magnitude $\gamma\delta g$ (in angular spatial frequency units) and orientation given by the gradient direction.

Most obvious are the very scales of temporal and spatial regimes sampled, and the detection of $E(\mathbf{g}, \Delta)$ in the time domain of Δ in the case of PGSE and in the frequency domain in the case of neutron scattering. But in a more fundamental sense, the reader should not be tempted to identify Fig. 5.28 b) with the process of applying two magnetic field gradient pulses in PGSE NMR. In no sense are the incoming and outgoing **k** vectors equivalent to the actions of the first and second gradient pulses,[16] the underlying physics of the two methods being fundamentally different. Nonetheless, it has become customary to refer to PGSE NMR as 'scattering', and this practice is harmless unless actually believed.

The averaged propagator and q-space
Clearly the phase shifts appearing in the integrand of eqn 5.81 depend only on the dynamic displacement, $\mathbf{R} = \mathbf{r}' - \mathbf{r}$. Using the definition of the averaged propagator [55] given in eqn 1.81,

---

[16]The reader will see this immediately by noting that such an assumption would imply that PGSE NMR corresponds to $\mathbf{k}_{\text{out}} = -\mathbf{k}_{\text{in}}$, in which case $\mathbf{q} = 2\mathbf{k}$, in violation of eqn 5.79.

$$\overline{P}(\mathbf{R}, t) = \int \rho(\mathbf{r}) P(\mathbf{r}|\mathbf{r} + \mathbf{R}, t) d\mathbf{r}, \tag{5.84}$$

it is easy to see that eqn 5.81 may be rewritten[17]

$$E(\mathbf{q}, \Delta) = \int \overline{P}(\mathbf{R}, \Delta) \exp(i\mathbf{q} \cdot \mathbf{R}) d\mathbf{R}, \tag{5.85}$$

or, taking Z as the component of $\mathbf{R}$ along $\mathbf{q}$,

$$E(q, \Delta) = \int_{\infty}^{\infty} \overline{P}(Z, \Delta) \exp(iqZ) dZ \tag{5.86}$$

Equation 5.85 expresses a simple Fourier relationship between $E(\mathbf{q}, \Delta)$ and $\overline{P}(\mathbf{R}, \Delta)$. The meaning of the narrow pulse PGSE experiment is now transparent. Acquisition of the signal in q-space permits us to image $\overline{P}(\mathbf{R}, \Delta)$, just as acquisition in $\mathbf{k}$-space permitted us to image $\rho(\mathbf{r})$. PGSE is an imaging experiment in its own right, probing the dynamic displacement space, $\mathbf{R}$, rather than the static displacements, $\mathbf{r}$. Note that while the echo attenuation, E, is implicitly a function of the PGSE pulse separation Δ, we shall often omit the argument Δ for simplicity.

For completeness it is helpful to write the normalised echo signals for diffusion and flow in this simple narrow pulse, q-space perspective, that is

$$E(q, \Delta) = \exp(-q^2 D \Delta) \tag{5.87}$$

and

$$E(\mathbf{q}, \Delta) = \exp(i\mathbf{q} \cdot \mathbf{v} \Delta) \tag{5.88}$$

5.6.2   Low q limit

An especially useful simplification occurs in the low q limit, that is,

$$E(\mathbf{q}, \Delta) \approx 1 + iq \int \overline{P}(Z, \Delta) Z dZ - \tfrac{1}{2} q^2 \int \overline{P}(Z, \Delta) Z^2 dZ + \dots \tag{5.89}$$

where Z is the projection of $\mathbf{R}$ along $\mathbf{q}$ and q is $|\mathbf{q}|$. The leading term linear in q represents a phase shift in the signal that informs on the mean displacement. For Brownian motion, or any other motion with no net flow, the linear term disappears and

$$E(\mathbf{q}, \Delta) \approx 1 - \tfrac{1}{2} q^2 \langle Z^2(\Delta) \rangle \tag{5.90}$$

When the low q echo-attenuation data is plotted against q^2, the initial linear decay allows $\langle Z^2(\Delta) \rangle$ to be measured directly. This represents the simplest of all possible signal analysis in the case of the narrow gradient pulse PGSE experiment. In the study of hindered and restricted diffusion, such an analysis provides a useful guide to interdependence of length and time scales.

---

[17]The use of the average propagator formalism was introduced to description of the PGSE NMR experiment by Kärger and Heink [55].

5.6.3   The meaning of k-space and q-space for time-varying gradients

In the case of a magnetic field gradient, $\mathbf{G}(t)$, that varies with time, the naive definition of **k**-space given in eqn 5.15 needs revisiting. Of course, **k** provides a means of
determining what phase has been acquired by a spin isochromat remaining stationary
at position **r**. If the gradient is time-dependent then that phase will be given by the
cumulation of incremental precessions, as described by the integral

$$\phi(t) = \gamma \int_0^t \mathbf{g}(t') \cdot \mathbf{r} dt'$$

$$= \gamma \left(\int_0^t \mathbf{g}(t') dt' \right) \cdot \mathbf{r} \qquad (5.91)$$

which tells us that

$$\mathbf{k}(t) = \gamma \left(\int_0^t \mathbf{G}(t') dt' \right) \qquad (5.92)$$

where we use angular frequency units.

For q-space, where we used eqn 5.80 to define **q** pertaining to a pair of narrow
gradient pulses, the redefinition is more subtle when a general time-varying effective
gradient, $\mathbf{g}^*(t)$, is used over an experimental interval ranging from $t' = 0$ to the echo
centre at $t' = t$. Of course, at the echo centre where the signal is to be acquired,
$\int_0^t \mathbf{g}^*(t') dt' = 0$. And in the experiment where we encode for motion, we might expect
that a definition of **q** will be provided by the requirement that the acquired phase is
given by $\gamma \mathbf{q} \cdot \mathbf{R}$, where $\mathbf{R} = \mathbf{r}(t) - \mathbf{r}(0)$. But that phase is given by

$$\phi(t) = \gamma \int_0^t \mathbf{g}^*(t') \cdot \mathbf{r}(t') dt' \qquad (5.93)$$

and it is clear that any definition of **q** will depend on a detailed knowledge of the
trajectory $\mathbf{r}(t')$! Let us take a simple example where we know that trajectory, namely
when the spin-bearing molecules are translating at a constant velocity **v** such that
$\mathbf{R}(t') = \mathbf{v}t'$. Then

$$\phi(t) = \mathbf{v} \cdot \gamma \int_0^t t' \mathbf{g}^*(t') dt' \qquad (5.94)$$

and, we may define for this special case,

$$\mathbf{q} = t^{-1} \gamma \int_0^t t' \mathbf{g}^*(t') dt' \qquad (5.95)$$

For example, if we apply eqn 5.95 to the case of gradient pulses of width δ and amplitude g, then it is easy to show by this definition that q is simply given by the area
under the gradient pulse, $q = \gamma g \delta$.

But in general, where the motion is unknown in advance, no simple definition of **q**
is possible, except in the limit when $\mathbf{g}^*(t')$ takes the form of delta functions of opposite
sign, applied at times $t' = 0$ and $t' = t$. In other words, only in the narrow gradient
pulse approximation is the scattering analogy truly meaningful.

5.7 Finite gradient pulses and generalised motion

We now turn our attention to the problem of how to describe the spin echo attenuation under quite general conditions of motion. Clearly some generally modulated effective gradient, obeying the echo condition $\int_0^t \mathbf{g}^*(t')dt' = 0$ at some time t, will not be amenable to an exact description provided by the Bloch–Torrey relations unless the motion is unbounded diffusion or flow. For example, when the gradient pulse widths of a PGSE experiment are finite in comparison with some characteristic dynamical timescale, eqn 5.78 breaks down. Such a timescale might be the time, τ_{pore}, for liquid or gas molecules in a porous medium to diffuse across a pore space. τ_{pore} provides a measure for crossover from free Brownian motion at short times to restricted motion, with strong boundary influences, at long times. Clearly the narrow gradient pulse assumptions inherent in the derivation of eqn 5.78 no longer apply if $\delta \sim \tau_{pore}$.

In the next two sections we outline two different approaches to the problem. The first, which retains the language of conditional probabilities, provides a time-domain approach to a closed form analytic expression for generalised motion. The second utilises the frequency domain and provides expressions based on the analysis of the spectrum of the velocity autocorrelation function.

5.7.1 Multiple propagator approach

The multiple propagator approach is a method by which the language of the propagator may be generalised to handle any gradient waveform, and not just the two-impulse experiment. One motivation for this approach is to decouple the physics of the restricted diffusion from that of the NMR pulse sequence employed. Ideally we would like a general expression for the echo attenuation that satisfies the following two conditions. First, the propagator description should be retained and the nature of the restricted diffusion should be embedded entirely within the mathematical form of that propagator. Second, the gradient waveform should appear within a closed form expression in a natural and obvious manner, so that the time sequence of the waveform evolution is explicit.

The method is based on a suggestion by Caprihan, Wang, and Fukushima [56], who postulated that the case of restricted diffusion under general gradient waveforms could be handled by breaking the gradient pulse into successive intervals, writing a propagator for each stage of the evolution. Their method involved quite a complex sum over a large number of independent terms. Here we adopt a similar philosophy, but using a different approach which leads to much simpler, and quite general, closed-form expressions. The chosen approach [57] explicitly demonstrates the succession of spin phase evolutions at each step of the gradient waveform and the evolutions are expressed in terms of a product of matrix operators, between which are sandwiched time-evolution terms associated with the diffusive motion. Note that the mathematical description is developed in terms of motion that is fundamentally diffusive, albeit restricted, although in principle it could be generalised to any motion where a conditional probability description was possible.

Discretising the effective gradient waveform

In order to retain the language of propagators when dealing with generalised gradient waveforms, the translational motion must be sub-divided into a sequence of discrete time intervals, with all spin phase evolution taking place at well-defined times at the boundaries of those intervals. In other words we will require the narrow pulse approximation of eqn 5.78 to apply over each evolutionary step, if not over the entire waveform duration. This means, in effect, that we must approximate the gradient waveform by a succession of impulses, such that the time integral function, $\mathbf{F}(t) = \int_0^t \mathbf{g}^*(t')dt'$, is represented by a stepwise progression. The basic scheme is shown in Fig. 5.29.

Suppose that we break the waveform into N time intervals τ, each bounded by impulses $\mathbf{q}_n$, $\mathbf{q}_{n+1}$ etc. We may discretise the waveform amplitude $\mathbf{g}^*(n\tau)$ into units of dimension $\mathbf{g}_{step}$. At time $n\tau$ the impulse will be $\mathbf{q}_n = m_n\mathbf{q}$, where $\mathbf{q} = \gamma\delta\mathbf{g}_{step}$ and m_n is some positive or negative integer, depending on the local magnitude and sign of $\mathbf{g}^*(n\tau)$, and given by

$$m_n = integ(\mathbf{g}^*(n\tau))/\mathbf{g}_{step} \tag{5.96}$$

We may write the echo amplitude at the end of the sequence as

$$E(t) = \int d\mathbf{r}_1 \int d\mathbf{r}_2 \ldots \int d\mathbf{r}_{N+1}$$
$$\rho(\mathbf{r}_1)\exp(i\mathbf{q}_1 \cdot \mathbf{r}_1)P(\mathbf{r}_1|\mathbf{r}_2,\tau)\exp(i\mathbf{q}_2 \cdot \mathbf{r}_2)P(\mathbf{r}_2|\mathbf{r}_3,\tau)$$
$$\ldots \exp(i\mathbf{q}_N \cdot \mathbf{r}_N)P(\mathbf{r}_N|\mathbf{r}_{N+1},\tau) \tag{5.97}$$

Multiple propagators expressed in the eigenfunction expansion

Using the eigenmode expansion $P(\mathbf{r}|\mathbf{r}',\tau) = \sum_{k=0}^{\infty} u_k(\mathbf{r})u_k^*(\mathbf{r}')\exp(-\lambda_k\tau)$,

$$F(t) = \int d\mathbf{r}_1 \int d\mathbf{r}_2 \quad \int d\mathbf{r}_{N|1}$$
$$\rho(\mathbf{r}_1)\exp(i\mathbf{q}_1 \cdot \mathbf{r}_1)\sum_{k_1} u_{k_1}(\mathbf{r}_1)u_{k_1}^*(\mathbf{r}_2)\exp(-\lambda_{k_1}\tau)$$
$$\exp(i\mathbf{q}_2 \cdot \mathbf{r}_2)\sum_{k_2} u_{k_2}(\mathbf{r}_2)u_{k_2}^*(\mathbf{r}_3)\exp(-\lambda_{k_2}\tau)$$
$$\ldots \exp(i\mathbf{q}_N \cdot \mathbf{r}_N)\sum_{k_N} u_{k_N}(\mathbf{r}_N)u_{k_N}^*(\mathbf{r}_{N+1})\exp(-\lambda_{k_N}\tau) \tag{5.98}$$

Regrouping the summations, a succession of matrix products emerges as

$$E(t) = \int d\mathbf{r}_1 \int d\mathbf{r}_2 \ldots \int d\mathbf{r}_{N+1}$$
$$\rho(\mathbf{r}_1)\exp(i\mathbf{q}_1 \cdot \mathbf{r}_1)\sum_{k_1,k_2,k_3\ldots k_N} u_{k_1}(\mathbf{r}_1)u_{k_1}^*(\mathbf{r}_2)\exp(-\lambda_{k_1}\tau)$$
$$\exp(i\mathbf{q}_2 \cdot \mathbf{r}_2)u_{k_2}(\mathbf{r}_2)u_{k_2}^*(\mathbf{r}_3)\exp(-\lambda_{k_2}\tau)$$
$$\ldots \exp(i\mathbf{q}_N \cdot \mathbf{r}_N)u_{k_N}(\mathbf{r}_N)u_{k_N}^*(\mathbf{r}_{N+1})\exp(-\lambda_{k_N}\tau)$$

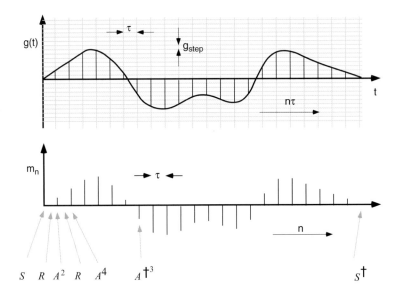

$$SRA^2RA^4RA^6RA^6RA^4RA^2RA^{\dagger 3}RA^{\dagger 6}RA^{\dagger 7}RA^{\dagger 6}RA^{\dagger 4}RA^{\dagger 3}RA^{\dagger 4}RA^{\dagger 5}RA^{\dagger 4}RA^3RA^6RA^6RA^5RA^4RA^3RA^2RA \qquad \dagger$$

Fig. 5.29 Schematic decomposition of a generalised gradient waveform into a sequence of equally time-separated impulses, the magnitudes of each of which are expressed as integer multiples, m_n, of the basic digitisation unit, g_{step}. From this train the succession of matrix operators may be written directly, starting and finishing with the S row and column vectors. The effect of the diffusive evolution is contained within the diagonal R matrices, while that of the phase evolutions is contained in the A matrices, each of which is raised to the power m_n, the integer corresponding to the impulse strength at the nth step. Negative gradients are represented by using the Hermitian transpose of A. The entire sequence shown is indicated in the matrix product below the diagram.

$$= \sum_{k_1,k_2,k_3\dots k_N} S_{k_1}(\mathbf{q}_1) R_{k_1 k_1} A_{k_1 k_2}(\mathbf{q}_2)) R_{k_2 k_2} A_{k_2 k_3}(\mathbf{q}_3)$$

$$\dots R_{k_{N-1}k_{N-1}} A_{k_{N-1}k_N}(\mathbf{q}_N)) R_{k_N k_N} S^*_{k_N}(\mathbf{q}_{N+1}) \tag{5.99}$$

where

$$S_k(\mathbf{q}) = V^{-1/2} \int d\mathbf{r} u_k(\mathbf{r}) \exp(i\mathbf{q_k} \cdot \mathbf{r}), \tag{5.100}$$

$$R_{kk} = \exp(-\lambda_k \tau), \tag{5.101}$$

and

$$A_{kk'} = \int d\mathbf{r} u^*_k(\mathbf{r}) u_{k'}(\mathbf{r}) \exp(i\mathbf{q_{k'}} \cdot \mathbf{r}) \tag{5.102}$$

and the initial density, $\rho(\mathbf{r}_1)$, may be set to the inverse of the pore volume (V), provided that the eigenfunctions are appropriately normalised.

Equation 5.99 describes a matrix multiplication process. For the moment, let us note a helpful result which follows from the orthogonality condition $\delta(\mathbf{r}' - \mathbf{r}) = \sum_k u_k(\mathbf{r})u_k^*(\mathbf{r}')$, namely

$$A_{kk''}(\mathbf{q}_A + \mathbf{q}_B) = \sum_{k'} A_{kk'}(\mathbf{q}_A)A_{k'k''}(\mathbf{q}_B) \tag{5.103}$$

From this we obtain the result that $A(n\mathbf{q}) = A(\mathbf{q})^n$.

The matrix product

With these results we may write our echo-attenuation scheme using the simple matrix product

$$E = S(\mathbf{q}_1)RA(\mathbf{q}_2)RA(\mathbf{q}_3).....RA(\mathbf{q}_N)RS^\dagger(-\mathbf{q}_{N+1}) \tag{5.104}$$

It is the recognition of the matrix algebra inherent in eqns 5.99 to 5.104 that represents the key simplifying step in the formulation of the problem. Because of the emergence of mathematical tools such as Matlab and Mathematica, the matrix product is very easy and very rapid to evaluate, and so the expression presented here provides a practical tool for rapid computation.

To illustrate how the method works, consider the case of an effective gradient waveform $\mathbf{g}^*(t)$, which begins and ends with zero amplitude and which has zero time integral at the sampling time in order to satisfy the echo condition. As shown in Fig. 5.29, the waveform may be represented by a series of equally spaced impulses. While this may seem a rather crude approximation in the domain of $\mathbf{g}^*(t)$, it provides a stepwise approximation to the more important time-integral waveform, $\mathbf{F}(t)$. Setting, for convenience, the initial and final impulses to the minimum value $\mathbf{q}$ (an unimportant requirement if N is sufficiently large),

$$E = S(\mathbf{q})R[A(\mathbf{q})]^{m_2}.....R[A(\mathbf{q})]^{m_n}.....R[A(\mathbf{q})]^{m_N}RS^\dagger(-\mathbf{q}) \tag{5.105}$$

Thus we find that, in general, any waveform may be handled, provided that we calculate just three matrices, $R(\tau)$, $S(\mathbf{q})$, and $A(\mathbf{q})$, where $\mathbf{q}$ is the smallest impulse used to digitise the waveform. Equation 5.105 is so straightforward that it can be immediately adapted to any sequence by organising the matrices as required.

Dimensions of the matrices

The matrix elements of S, R, and A are calculated, in the case of restricted diffusion within a pore, using the fundamental eigenfunction expansion for the particular geometry under study. The R matrix, which is diagonal, comprises the temporal solutions to the Fick's law differential equation, typically of the form $\exp(-\alpha_k^2 D_0 \tau/a^2)$, where α_k is an eigenvalue dependent on the boundary conditions and a gives a measure of the pore dimension. These elements rapidly decay with increasing k and hence limit the required dimension of our matrices. The size $k_{max} \times k_{max}$ of our desired matrices will thus be determined by the exponents of the elements of the R-matrix, and our wish that matrix element sums converge when we multiplying out the products. Suppose we take the example of diffusion between parallel planes.[18] For $k \times k$ matrices the R-matrix elements will decay as $\exp(-k^2\pi^2 D_0\tau/a^2)$. Provided $k^2 D_0\tau/a^2 \sim 1$,

---

[18]This case, along with those of cylinders and spherical pores are treated in Chapter 7.

this convergence condition should be reasonably satisfied and 15×15 matrices are sufficiently accurate for calculating any gradient waveform response for plane parallel pore-restricted diffusion. The relatively small size of these matrices makes multiplication of large numbers of matrices extremely rapid, particularly using software such as Matlab™.

Application to standard gradient pulse schemes

Equation 5.105 can be easily evaluated for some cases of special interest shown in Fig. 5.30. The sequences are shown using a small number of impulses simply for clarity. In practice one may calculate the echo attenuation quite rapidly using on the order of 100 impulses.

(i) Steady gradient spin echo.

The effective gradient is as shown in Fig. 5.30(a). The waveform is subdivided into $2N + 1$ intervals and $2N + 2$ impulses, so that the total effective scattering wave vector amplitude is $q_{net} = (N + 1)q$, with the pulse spacing and pulse duration identical and given by $\Delta = \delta = (N + \frac{1}{2})\tau$ and

$$E = S(q)[RA(q)]^N [RA(q)]^N RS^\dagger(q) \tag{5.106}$$

Note that the final and initial impulses have opposite sign q amplitudes, so that for this case the initial and final vectors are, respectively, $S(q_1) = S(q)$ and $S^\dagger(-q_{2N+2}) = S^\dagger(q)$.

(ii) Finite gradient pulse spin echo.

The effective gradient is as shown in Fig. 5.30(b). In this case we break the entire waveform into $2N + 1$ intervals, so that the total effective scattering wave vector amplitude is $q_{net} = (M + 1)q$, with $\Delta = (N + \frac{1}{2})\tau$ and $\delta = (M + \frac{1}{2})\tau$. Hence

$$E = S(q)[RA(q)]^M R^{N-M} [RA^\dagger(q)]^M RS^\dagger(q) \tag{5.107}$$

Note that in the narrow gradient pulse approximation, $M = 0$ and

$$E = S(q)R^{N+1}S^\dagger(q) \tag{5.108}$$

precisely consistent with eqn 5.81.

(iii) CPMG spin-echo train.

Consider the effective gradient train shown in Fig. 5.30(c). It comprises $2N$ impulses and N periods of the gradient time integral waveform. This series of impulses may be represented by the relation

$$E = S(q)[RA^\dagger(q)A^\dagger(q)RA(q)A(q)]^M RA^\dagger(q)A^\dagger RS^\dagger(-q) \tag{5.109}$$

where $M = N - 1$. This CPMG train has a square wave time-integral waveform, with period $T = 2\tau$, where τ is the time-evolution step interval. Here the final and initial impulses have equal sign so that in this case the initial and final operators are, respectively, $S(q_1) = S(q)$ and $S^\dagger(-q_{2N+1}) = S^\dagger(-q)$.

The matrix method provides a completely general way to deal with the problem of restricted diffusion, including the effects of wall relaxation, under any gradient

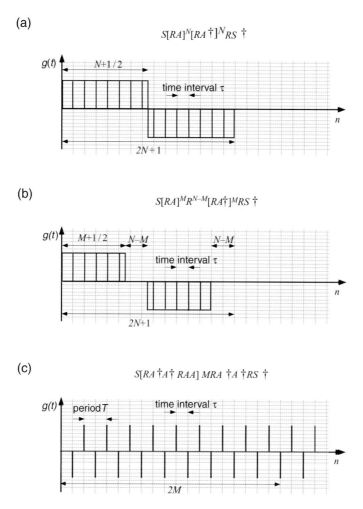

Fig. 5.30 Schematic illustration of impulse decomposition for different pulse sequences. In each case the evolution matrix product corresponding to the sequence is shown above. (a) The steady gradient spin echo (or pulsed gradient spin echo with $\delta = \Delta$) represented by a train of constant magnitude impulses, q. Note that $\Delta = (N + 1/2)\tau$ while the wave vector value corresponding to the area under the complete gradient pulse is $q_{net} = (N + 1)q$. (b) Finite pulse-width PGSE experiment in which $q_{net} = (M+1)q$, $\Delta = (N+1/2)\tau$, and $\delta = (M+1/2)\tau$. (c) CPMG PGSE sequence.

waveform, provided one can obtain an eigenfunction expansion for the propagator. As regards the finite gradient pulse-width problem in PGSE NMR, the matrix expression allows for the ready calculation of echo-attenuation expressions in a few lines of computer code. In Chapter 6, where we investigate restricted diffusion in some detail, we shall return to these matrix expressions to evaluate the PGSE NMR experiment in specific restriction geometries.

5.7.2 Generalised gradients and the frequency domain

The narrow pulse requirement of the two pulse PGSE NMR method severely limits the ability of the method to access information about motion for which the characteristic times are very short. As δ is decreased, so the scattering wavevector q decreases, making the method less and less sensitive to shorter translational displacements, an inconvenient contradiction to say the least. Even if the multiple propagator matrix method is used to allow for a theoretical description under the condition of finite duration gradient pulses, the shortest practical length of time scale Δ accessed by two-pulse PGSE is on the order of a few milliseconds. Of course if gradient amplitudes were able to increase in compensation, q values could be maintained, but higher amplitude gradient pulses bring with them greater eddy current effects and the need for finite pulse rise time.[19]

Notice that the depiction of dynamics in the PGSE description is time-domain in nature. A quite different approach to the use of gradients to measure motion was proposed in the 1980s by Janez Stepišnik [58, 59], in which the frequency domain provides the natural description. In consequence the Stepišnik method is formulated in terms of gradient waveforms of long duration but well-defined frequency characteristics. In this approach, one treats the effect of the gradient via the time dependence of the accumulated spin phase. The mathematical treatment lends itself to analogy with the measurement of rotational dipolar auto-correlation spectral density [22, 60] via the 'probe' of the spin relaxation times, where T_1 samples at positions ω_0 and $2\omega_0$, while $T_{1\rho}$ samples at ω_1 and $2\omega_1$.

The frequency-domain method, which is termed here frequency-domain modulated gradient NMR (FDMG-NMR), was first demonstrated experimentally in 1995 [61]. It has been used to great effect in the elucidation of quite complex dynamics in problems concerning restricted diffusion and fluid dispersion. But it does suffer from two 'Achilles heels'. First, it is based on the use of multiple echoes. In particular the fundamental period defining the measurement frequency results in a net stationary spin phase migration of zero. In the CPMG sequence, such a cyclic process minimises sensitivity to diffusion, and this is also the case for FDMG-NMR. Of course the use of a long echo train duration helps restore sensitivity and better define the frequency, but nonetheless the constraint remains. In consequence the most effective application of the method is for examining restricted diffusion of small molecules (for example, water) or the fluctuation dynamics of flow-driven fluid dispersion. That constitutes a fairly large area of application!

The second weakness in the method concerns the problem of how to deal with an ensemble average of phase factors, and the need to use a Gaussian phase approximation. While stochastic processes often lend themselves to this step, in the case of restricted diffusion, where highly non-Gaussian propagators result, such a simplification is controversial. As always in science, the usefulness of the theory will be determined by its

[19]The use of large steady gradients in a stimulated-echo sequence does allow a means of controlling the effective pulse duration by the bracketing of the 90° RF pulses. This method is discussed in Chapter 12. However, a disadvantage of this approach is the loss of signal amplitude due to the RF pulses being able to stimulate only a narrow slice of the sample under the frequency-spreading effect of the gradient and the loss of spectral information during detection.

ability to represent the data. As we will see in this and later chapters, judged by such a criterion, the method is very useful indeed.

Time interval of the effective gradient

The key idea that underpins the use of magnetic field gradients to measure the translational motion of spin-bearing molecules is that the individual spins acquire a phase factor, $\exp(i\phi)$, due to precession in the local magnetic field $\mathbf{B}_0(\mathbf{r}, t)$. In the presence of a uniform effective magnetic field gradient $\mathbf{g}^*(t)$, this phase is position-dependent,

$$\phi(t) = \gamma \int_0^t \mathbf{g}^*(t') \cdot \mathbf{r}(t') dt' \tag{5.110}$$

In the case of a signal acquired in a spin echo, the time integral of the effective gradient $\mathbf{g}^*(t')$ is zero at the point of echo formation, so that integrating by parts, we may re-write the integral in eqn 5.110 in terms of the local spin velocity $\mathbf{v} = \mathbf{r}/t$ as $\mathbf{g}^*(t)$,

$$\phi(t) = -\int_0^t \mathbf{F}(t') \cdot \mathbf{v}(t') dt' \tag{5.111}$$

where

$$\mathbf{F}(t) = \gamma \int_0^t \mathbf{g}^*(t') dt' \tag{5.112}$$

To evaluate the normalised amplitude E, of the spin-echo signal arising from the entire spin ensemble, one must calculate the ensemble average

$$E(t) = \langle \exp(-i \int_0^t \mathbf{F}^*(t') \cdot \mathbf{v}(t') dt') \rangle \tag{5.113}$$

Note that the concept of q-space no longer has any meaning, since the Fourier dependence of E on $\mathbf{R}$ is no longer apparent.

The Gaussian approximation

The problem now is to evaluate the ensemble average. This can be performed by expanding the exponential function in a power series and taking ensemble averages of each term as

$$\langle \exp(iA) \rangle = \sum_{r=0}^{\infty} \frac{(i)^r}{r!} \langle A^r \rangle \tag{5.114}$$

where

$$\langle A^r \rangle = \int P(A) A^r dA \tag{5.115}$$

$P(A)$ being the ensemble probability distribution over A.

The power series, eqn 5.114, can be converted back to exponential form by a method known as the cumulant expansion. To obtain this result it is helpful to introduce the characteristic function,

$$\langle \exp(ikA) \rangle = \int \exp(ikA) P(A) dA \tag{5.116}$$

It can be shown that this function may be written

$$\langle \exp(ikA) \rangle = \exp\left(\sum_{r=1}^{\infty} \frac{(ik)^r}{r!} \Xi_r\right) \tag{5.117}$$

where

$$
\begin{aligned}
\Xi_1 &= \langle A \rangle \\
\Xi_2 &= \langle (A - \langle A \rangle)^2 \rangle = \langle A^2 \rangle - \langle A \rangle^2 \\
\Xi_3 &= \langle (A - \langle A \rangle)^3 \rangle = \langle A^3 \rangle - 3\langle A^2 \rangle \langle A \rangle + 2\langle A^3 \rangle \\
\Xi_4 &= \langle (A - \langle A \rangle)^4 \rangle - 3\langle (A - \langle A \rangle)^2 \rangle^2 = \langle A^4 \rangle - 4\langle A^3 \rangle \langle A \rangle - 3\langle A^2 \rangle^2 \\
&\quad + 12\langle A \rangle^2 \langle A^2 \rangle - 6\langle A^4 \rangle
\end{aligned}
\tag{5.118}
$$

Equation 5.117 is easily verified term by term by taking successive derivatives $\partial^n / \partial k^n$, with $n = 0, 1, 2 \dots$, then setting $k = 0$. With $k = 1$, eqn 5.117 yields the cumulant expansion

$$\langle \exp(iA) \rangle = \exp\left(\sum_{r=1}^{\infty} \frac{(i)^r}{r!} \Xi_r\right) \tag{5.119}$$

Condensing the expression is easily done, provided the assumption is made that the probability distribution $P(A)$ is Gaussian, namely

$$P(A) = (2\pi\Xi_2)^{-1/2} \exp\left(-\frac{(A - \langle A \rangle)^2}{2\Xi_2}\right) \tag{5.120}$$

In this case the Fourier transform of the Gaussian $P(A)$ represented by eqn 5.116 reduces to

$$\langle \exp(ikA) \rangle = \exp\left(i\langle A \rangle k - \tfrac{1}{2}\Xi_2 k^2\right) \tag{5.121}$$

and setting $k = 1$ we have

$$\langle \exp(iA) \rangle = \exp\left(i\langle A \rangle - \tfrac{1}{2}(\langle A^2 \rangle - \langle A \rangle^2)\right) \tag{5.122}$$

That the Ξ_r terms for $r \geq 3$ disappear in the cumulant expansion under the normalised Gaussian distribution is not immediately obvious. For the $r = 3$ term

$$
\begin{aligned}
\Xi_3 &= \int (A - \langle A \rangle)^3 P(A) dA \\
&= -\Xi_2 \int (A - \langle A \rangle)^2 \frac{\partial P(A)}{\partial A} dA.
\end{aligned}
\tag{5.123}
$$

and integrating by parts we find

$$\Xi_3 = 2\Xi_2 \int (A - \langle A \rangle) P(A) dA. \tag{5.124}$$

$(A - \langle A \rangle)$ is antisymmetric about $A = \langle A \rangle$, while $P(A)$ is symmetric and so the integral is zero. For $n = 4$, and again integrating by parts,

$$\Xi_4 = \int (A - \langle A \rangle)^4 P(A) dA - 3 \left[\int (A - \langle A \rangle)^2 P(A) dA \right]^2$$

$$= -\Xi_2 \int (A - \langle A \rangle)^3 \frac{\partial P(A)}{\partial A} dA - 3\Xi_2^2$$

$$= 0, \tag{5.125}$$

and so on.

The delicate question concerns the validity of the Gaussian phase approximation. Of course, our relevant moments Ξ_1 and Ξ_2 will be derived from $\langle \int_0^t \mathbf{F}(t') \cdot \mathbf{v}(t') dt' \rangle$ and $\langle [\int_0^t \mathbf{F}(t') \cdot \mathbf{v}(t') dt']^2 \rangle$. It is immediately obvious that eqn 5.122 cannot possibly represent any PGSE NMR experiment in which the dependence of the echo decay on the gradient amplitude g is other than Gaussian. In the next chapter we will see that such non-Gaussian behaviour is very apparent in the case of restricted diffusion where the q-vector in the two gradient pulse Stejskal–Tanner experiment is sufficiently large. However, if we look at the FDMG NMR experiment from its intended standpoint as a frequency-domain analysis of motion, we will quickly come to the conclusion that the two gradient pulse Stejskal–Tanner experiment is not its ideal vehicle. On the contrary, to have a well-defined frequency we will want to use an $\mathbf{F}(t)$ waveform that is oscillatory and which comprises many periods. In effect then, each period of the cycle will contribute only small phase excursions, these gradually accumulating to generate the final signal attenuation. It is precisely under these cumulatively averaging conditions that the central limit theorem suggests the likelihood of a Gaussian distribution of phase.

FDMG signal phase shift and attenuation
Equations 5.113 and 5.122 lead directly to

$$E(t) = \exp(i\alpha(t) - \beta(t)) \tag{5.126}$$

where

$$\alpha(t) = -i \int_0^t \mathbf{F}(t') \cdot \langle \mathbf{v}(t') \rangle dt' \tag{5.127}$$

and

$$\beta = \frac{1}{2} \langle [\int_0^t \mathbf{F}(t') \cdot \mathbf{v}(t') dt']^2 \rangle - [\int_0^t \mathbf{F}(t') \cdot \langle \mathbf{v}(t') \rangle dt']^2 \tag{5.128}$$

These relations are simplified if we decompose $\mathbf{v}(t)$ into its ensemble mean $\langle \mathbf{v}(t) \rangle = \langle \mathbf{v} \rangle$ and variable component $\mathbf{u}(t)$ as

$$\mathbf{v}(t) = \mathbf{u}(t) + \langle \mathbf{v} \rangle. \tag{5.129}$$

whence

$$\alpha(t) = -i \int_0^t \mathbf{F}(t') \cdot \langle \mathbf{v} \rangle dt' \tag{5.130}$$

and

$$\beta(t) = \frac{1}{2} \int_0^t \int_0^t \mathbf{F}(t') \cdot \langle \mathbf{u}(t') \mathbf{u}(t'') \rangle \cdot \mathbf{F}(t'') dt' dt'' \tag{5.131}$$

Note that the simplification that reduces eqn 5.128 to 5.131 arises from the uncorrelated nature of $\mathbf{u}(t)$ and the mean flow $\langle \mathbf{v} \rangle$ such that $\langle \mathbf{u}(t) \cdot \langle \mathbf{v} \rangle \rangle = 0$. $\alpha(t)$ represents the phase shift due to the mean flow of the spin-bearing molecules, while $\beta(t)$ gives the signal attenuation due to the fluctuations in displacement about this mean. It involves a tensor of velocity correlation functions.

In Chapter 2 we defined the diffusion spectrum by

$$\underline{\underline{D}}(\omega) = \tfrac{1}{2} sym \int_{-\infty}^{\infty} \langle \mathbf{u}(\tau)\mathbf{u}(0) \rangle \exp(i\omega\tau) d\tau \qquad (5.132)$$

with inverse transform

$$sym \langle \mathbf{u}(\tau)\mathbf{u}(0) \rangle = \frac{1}{\pi} \int_{-\infty}^{\infty} \underline{\underline{D}}(\omega) \exp(-i\omega\tau) d\omega \qquad (5.133)$$

or, equivalently

$$sym \langle \mathbf{u}(t')\mathbf{u}(t'') \rangle = \frac{1}{\pi} \int_{-\infty}^{\infty} \underline{\underline{D}}(\omega) \exp(i\omega(t'' - t')) d\omega, \qquad (5.134)$$

Taking the (half) Fourier transform of the time-integral effective gradient

$$\mathbf{F}(\omega, t) = \int_{0}^{t} \mathbf{F}(t') \exp(i\omega t') dt', \qquad (5.135)$$

and noting the factor $\tfrac{1}{2}$ in eqn 5.131, one obtains the spin-echo-attenuation exponent [58]

$$\beta(t) = \frac{1}{2\pi} \int_{\infty}^{\infty} \mathbf{F}(\omega, t) \cdot \underline{\underline{D}}(\omega) \cdot \mathbf{F}(-\omega, t) d\omega$$

$$= \frac{1}{\pi} \int_{0}^{\infty} \mathbf{F}(\omega, t) \cdot \underline{\underline{D}}(\omega) \cdot \mathbf{F}(-\omega, t) d\omega \qquad (5.136)$$

where the symmetry of $\underline{\underline{D}}(\omega)$ is assumed.

$\mathbf{F}(\omega, t)$ extracts only the diagonal components of the diffusion tensor. Suppose we apply the effective gradient along some axis z. Then in this case we will measure the element $D_{zz}(\omega)$ as

$$\beta(t) = \frac{1}{\pi} \int_{0}^{\infty} |F(\omega, t)|^2 D_{zz}(\omega) d\omega \qquad (5.137)$$

The meaning of eqn 5.137 is clear. The product of $|F(\omega, t)|^2$ and $D_{zz}(\omega)$ in the integral means that the diffusion spectrum is sampled by the squared spectrum of the time-integral effective gradient. Figure 5.31 illustrates this schematically. As in the multiple propagator approach, the FDMG NMR method manages to decouple the description of the motion ($D_{zz}(\omega)$) from the description of the gradient waveform ($|F(\omega, t)|^2$), thus greatly simplifying the analysis.

For a precise frequency sampling, $|F(\omega, t)|^2$ should be as close to a delta function as possible and in the next section we show how this can be achieved. For the moment,

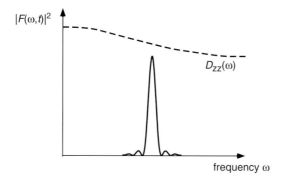

Fig. 5.31 Schematic effective gradient spectrum, $|F(\omega, t)|^2$ used to sample the diffusion spectrum $D(\omega)$.

however, it useful to touch base with an earlier result. Suppose that the fluctuation correlation time for molecular velocities is very short compared with the characteristic time ω^{-1} able to be sampled by $|F(\omega, t)|^2$. This would be the case for unrestricted Brownian motion of small molecules, where $\tau_c \sim 10^{-12} s$. For an exponential correlation function, the diffusion spectrum would feature a constant Lorentzian plateau out to 10^{12} Hz, so that in eqn 5.137, $D_{zz}(\omega)$ reduces to a constant $D_{zz}(0)$, and

$$\beta(t) = \frac{1}{\pi} D_{zz}(0) \int_0^\infty |F(\omega, t)|^2 d\omega \tag{5.138}$$

From the Parceval identity

$$\beta(t) = D_{zz}(0) \int_0^t |F(t')|^2 dt' \tag{5.139}$$

in precise agreement with the Bloch–Torrey relation 5.54.

FDMG signal for pulse sequences of interest
Let us start with the two-gradient pulse Stejskal–Tanner PGSE NMR experiment, comprising finite gradient pulses of amplitude g, width δ, separated by Δ, and with echo formation at time t . The phase spectrum follows from eqns 5.112 and 5.135 as

$$\mathbf{F}(\omega, t) = \gamma \mathbf{g} \frac{(1 - \exp(i\omega\Delta))(1 - \exp(i\omega\delta))}{\omega^2} \tag{5.140}$$

with

$$|\mathbf{F}(\omega, t)|^2 = \gamma^2 \delta^2 \Delta^2 |\mathbf{g}|^2 \left[\frac{\sin(\omega\delta/2) \sin(\omega\Delta/2)}{(\omega\Delta/2)(\omega\delta/2)} \right]^2 \tag{5.141}$$

Figure 5.32(a) shows the spectrum. Again for free diffusion of small molecules, $D_{zz}(\omega)$ is essentially constant (with value $D = D_{zz}(0)$ over the frequency range of interest. Using the standard integral $\int_0^\infty \sin^2(ax) \sin^2(bx)/x^4 dx = a^2\pi(3b - a)/6$, one obtains the Stejskal–Tanner result

$$\alpha(t) = \gamma^2 \delta^2 |\mathbf{g}|^2 D(\Delta - \delta/3) \tag{5.142}$$

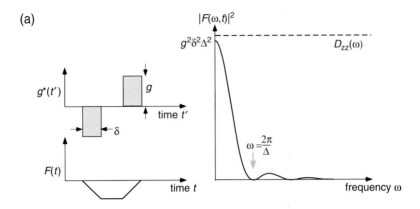

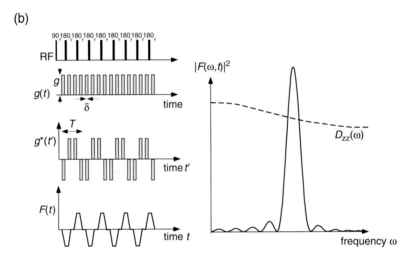

Fig. 5.32 Frequency-domain modulated gradient NMR RF and gradient pulse sequences, showing the (actual) gradient modulation waveform $g(t')$, the time integral of the effective gradient waveform $F(t)$, and the spectrum of $F(t)$. $|F(,\omega,t)|^{2}$, directly samples the diffusion spectrum. The wave forms and spectra are for (a) Stejskal–Tanner PGSE sequence, with a constant diffusion spectrum as discussed in eqn 5.138 and (b) single lobe/ac rectangular modulation. Note that pulse sequence (b) samples the diffusion spectrum at a single frequency.

The spectrum of the two pulse PGSE gradient is dominated by the zero frequency lobe, with frequency width of order Δ^{-1}. It is therefore unsuitable for extracting high-frequency information concerning $D_{zz}(\omega)$. It would be very useful to have available a gradient modulation sequence whose frequency spectrum contained a high-frequency peak that could be adjusted in position, in order to trace out the frequency dependence of $D_{zz}(\omega)$. Such a measurement could in principle locate 'edges' in the spectrum at the inverse motional correlation time, τ_c^{-1}. Our ideal gradient modulation sequence would possess a frequency spectrum that contained a single peak whose frequency could be adjusted in position in order to trace out the frequency dependence of $D(\omega)$.

Figure 5.32(b) shows a suitable gradient modulation wave form, based on a CPMG sequence with repeated gradient pulses. Also shown is the associated spectrum. Equally effective is the CPMG train using a constant gradient. The dominant sampling lobe of these idealised sequences is at $\omega = 2\pi/T$, where T is the time-integral gradient repeat period. This lobe has width of order $2\pi/NT$, where N is the total number of periods employed. With $N \geq 4$, a reasonably narrow peak can be achieved and with such a train, given gradient pulse durations $\delta \geq 10\mu s$, it is possible in principle to probe spectral densities in the frequency range 10–100 kHz. This becomes possible because, rather than using two gradient pulses, for which the attenuation effect disappears as the gradient pulse duration δ is shortened, the repetitive pulse train employs an increasing number of gradient pulses in any time interval t as the frequency is increased and T is reduced. Thus the frequency-domain analysis extends the effective timescale of the PGSE experiments downward to the sub-millisecond range.

The exact behaviour of the CPMG train of 180° RF pulses, interspaced by $T/2$, in the presence of a constant magnetic field gradient, is easily described. If the first 180° pulse follows the excitation at time $T/4$, the $F(t)$ time dependence is a sawtooth-shaped function oscillating about zero. From eqn 5.135 it can be shown that its spectrum is

$$|\mathbf{F}(\omega, t)|^2 = 4\gamma^2|\mathbf{g}|^2 \frac{8 \sin^2(\omega T/8) \sin^2(N\omega T/2)}{\omega^4 \cos^2(\omega T/4)} \tag{5.143}$$

Note that the number of 180° RF pulses must be a multiple of two. This spectrum has only one frequency peak at $\omega = 2\pi/T$, with a width depending on N. Figure 5.32(b) shows that even $N = 4$ gives a reasonably narrow peak, which can be approximated by

$$|\mathbf{F}(\omega, t)|^2 = NT\gamma^2|\mathbf{g}|^2 T^2 \delta(\omega - 2\pi/T) \tag{5.144}$$

The expected echo-attenuation factors for the pulsed gradient CPMG wave form shown in Fig. 5.32(b), along with its constant gradient counterpart, are, respectively,

$$\beta(t) = \tfrac{1}{2} NT\gamma^2|\mathbf{g}|^2\delta^2 D(2\pi/T) \tag{5.145}$$

and

$$\beta(t) = \tfrac{1}{2} NT\gamma^2|\mathbf{g}|^2 T^2 D(2\pi/T) \tag{5.146}$$

where $t = NT$.

By varying T it is possible to probe the frequency spectrum of the molecular translational dynamics, adjusting N as necessary to retain sufficient attenuation as T is reduced. Of course, the dependence of β on NT^3 shows the price paid in using many (N) periods in which the applied gradient pulse duration, which would be potentially $\sim NT$ in a two-pulse experiment, is subdivided into into durations on the order T. The dephasing factor for the two-pulse experiment is $\sim (NT)^3$ whereas for the CPMG train it is smaller by N^2. This, of course, was the very reason for which the CPMG trains was developed: to decrease sensitivity to diffusion! Hence we see that FDMG NMR comes at the price of a restriction to relatively fast moving molecules.

Of course the really interesting question concerns the way in which complex molecular translational dynamics are manifest in the spectral density $D(\omega)$. This topic we cover in the next chapter.

5.8   Phase effects of RF pulses and homospoiling

Dealing with the combined effects of gradient pulses and RF pulses is relatively straightforward in the case of the spin echo. The idea that a 180° RF pulse simply inverts spin phase, and hence negates prior effective gradients, is a helpful idea. But when dealing with multiple combinations of RF pulses, as found in the double PGSE NMR sequences discussed in Chapter 8, alternative approaches can prove useful. Here we introduce a means of tracking the effect of RF pulses on spin coherence encoded by magnetic field gradient pulses. The idea is based on the Argand plane depiction of transverse magnetisation, introduced in Chapter 4. Suppose, for convenience, we regard the first 90° RF pulse of our sequence as having a phase x. Its effect is to induce a magnetisation along the y-axis of the rotating frame. We would detect this magnetisation with arbitrary phase, depending on our receiver setting and our subsequent 'phasing of the signal' by $\exp(i\theta)$ multiplication. So we can simply choose to define y as the real axis of our Argand plane.

5.8.1   Rules for RF pulses

Now consider the effect of our first gradient pulse on a single-spin isochromat. It will induce a phase change factor $\exp(i\phi)$ depending on the position of those spins. What happens with the subsequent RF and gradient pulses follows very simply once we work out the basic rules. For example, what do various RF pulses do to a pre-existing magnetisation of $\exp(i\phi)$?

Let us take a very simple example: that of $180^\circ_{x,y}$ pulses.

-180°_x
The 180°_x pulse inverts the real part of $\exp(i\phi)$, while leaving the imaginary part unchanged. Thus $\exp(i\phi) \rightarrow -\exp(-i\phi)$

-180°_y
The 180°_y pulse inverts the imaginary part of $\exp(i\phi)$, while leaving the real part unchanged. Thus $\exp(i\phi) \rightarrow \exp(-i\phi)$. Note that exactly the same results are achieved if the phases of these (ideal) 180° pulses are reversed: $x \rightarrow -x, y \rightarrow -y$.

Next we need to consider the effect of $90^\circ_{x,y}$ pulses on gradient-encoded magnetisation. These move transverse magnetisation out of the plane, to lie along the longitudinal axis for later recall. That component which remains in the transverse plane may be considered to be 'homospoiled'; that is, dephased to a degree that it will no longer contribute to the subsequent NMR signal. This will be achieved either by the residual background field inhomogeneity, or where this is insufficient, by deliberately applying 'homospoil' gradient pulses, pulses that impart no phase shifts to the stored z component, and hence have no impact on the final signal encoding. The use of z-storage requires that we clearly identify the 'coherence pathway' that leads to our ultimate signal. Hence, in what follows we look at combinations of 'storage'–'recall' pulses.

$-90°_x - \tau - 90°_x$

The first $90°_x$ pulse stores the real part of $\exp(i\phi)$ along the negative z-axis, while the second $90°_x$ pulse recalls this component to the negative real axis. Thus $\exp(i\phi) \rightarrow -\cos(\phi) = -0.5[\exp(i\phi) + \exp(-i\phi)]$

$-90°_y - \tau - 90°_y$

. The first $90°_y$ pulse stores the imaginary part of $\exp(i\phi)$ along the negative z-axis, while the second $90°_x$ pulse recalls this component to the negative imaginary axis. Thus $\exp(i\phi) \rightarrow -i\sin(\phi) = -0.5[\exp(i\phi) - \exp(-i\phi)]$.

$-90°_x - \tau - 90°_{-x}$

The first $90°_{-x}$ pulse stores the real part of $\exp(i\phi)$ along the positive z-axis, while the second $90°_x$ pulse recalls this component to the positive real axis. Thus $\exp(i\phi) \rightarrow \cos(\phi) = 0.5[\exp(i\phi) + \exp(-i\phi)]$

$-90°_{-y} - \tau - 90°_y$

. The first $90°_y$ pulse stores the imaginary part of $\exp(i\phi)$ along the positive z-axis, while the second $90°_{-y}$ pulse recalls this component to the positive imaginary axis. Thus $\exp(i\phi) \rightarrow i\sin(\phi) = 0.5[\exp(i\phi) - \exp(-i\phi)]$.

Unlike the case of 180° RF pulses, reversal of phase with a single 90° pulse produces an outcome of two transverse plane components with opposite sign phase. This difference turns out to be crucial in PGSE NMR, where we must track the evolution of phase shifts induced by gradient pulses, ensuring the appropriate summation combinations.

The stimulated-echo formation in the absence of homospoil pulses

Care is needed whenever a stimulated echo is formed without the use of homospoil pulses. In each case above, recall to the transverse plane by the final 90° RF pulse in the stimulated-echo train leaves a superposition of $\exp(i\phi)$ and $\exp(-i\phi)$, each with amplitude $1/2$. One of these has the phase-cancelling effect needed to produce the echo, again with amplitude $1/2$. But without the action of homospoiling pulses, a very different outcome results. Homospoiling destroys remaining transverse magnetisation following z-storage. Without homospoil pulses, and under high background field homogeneity with a sufficiently short z-storage period, this transverse magnetisation will persist. In that case, the outcomes of the above four $90° - 90°$ combinations will act in the manner of a concatenated combination, their effect akin to 180° pulses or 0° null pulses. For the four combinations respectively, the effect is $180°_x$, $180°_y$, null and null. The former two result in spin echoes of full amplitude, while the latter two combinations produce no echo.

As the storage time is gradually increased, the action of background field inhomogeneity increasingly destroys the transverse magnetisation, so that echo amplitudes, for the four combinations, fall from from unity to $1/2$ in the case of the first two, and rise from zero to $1/2$ in the case of the final two. This provides a nice rule for determining whether a homospoil pulse is sufficient for purpose. The stimulated echo, however formed, should be half the amplitude of the spin echo.

5.8.2   The spin echo and stimulated echo

The rules are now in place for handling combinations of RF and gradient pulses for the spin and stimulated echoes, as shown in Fig. 5.33. Note the scheme for labelling the RF pulse phases. The initial 90° RF pulse has phase φ_0, by our convention taken as x, and we will assume that the acquisition phase is set to give a real signal for y-magnetisation. In other words we can regard its phase as coincident with that of the first RF pulse and label it as x.

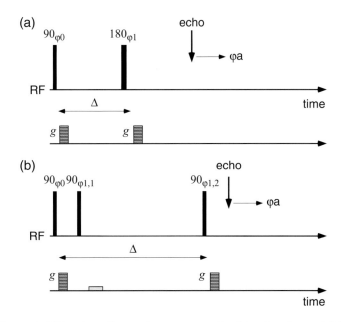

Fig. 5.33 Schematic RF and gradient pulse sequences for (a) spin echo and (b) stimulated-echo PGSE NMR experiments in which the gradient pulse area $(g\delta)$ is stepped. A homospoil gradient is included in the case of the stimulated echo.

Suppose we now track what happens in the spin echo where φ_1 is taken as y. After the first RF pulse, the phase factor is 1. After the first gradient pulse it is $\exp(i\phi)$. After the 180°_y RF pulse it is $\exp(-i\phi)$, and after then second gradient pulse it is $\exp(i\Delta\phi) = \exp(i\phi' - i\phi)$.

For the stimulated echo, let us take three examples.

$\varphi_{1,1} = -x$ **and** $\varphi_{1,2} = x$
After the first gradient pulse the phase factor is $\exp(i\phi)$. After the 90°_{-x} RF pulse subsequent homospoil and recall by the 90°_x RF pulse, it is $\cos(\phi)$. After the second gradient pulse it is $0.5[\exp(i\phi' - i\phi) + \exp(i\phi'i + i\phi)]$. The ensemble-averaging process destroys the second exponential term because it represents the cumulative phase shift of both gradient pulses averaging over all macroscopic positions of the spins, rather

than the phase difference, which is sensitive to the microscopic distance the spins have moved. The ensemble average is therefore $0.5\langle\exp(i\phi' - i\phi)]\rangle$.

$\varphi_{1,1} = x$ **and** $\varphi_{1,2} = x$

After the first gradient pulse the phase factor is $\exp(i\phi)$. After the 90°_x RF pulse, subsequent homospoil and recall by the 90°_x RF pulse, it is $-\cos(\phi)$. After the second gradient pulse it is $-0.5[\exp(i\phi' - i\phi) + \exp(i\phi' + i\phi)]$. The ensemble average is $-0.5\langle\exp(i\phi' - i\phi)]\rangle$.

$\varphi_{1,1} = y$ **and** $\varphi_{1,2} = y$

After the first gradient pulse the phase factor is $\exp(i\phi)$. After the 90°_y RF pulse, subsequent homospoil and recall by the 90°_y RF pulse, it is $-i\sin(\phi)$. After the second gradient pulse it is $-0.5[\exp(i\phi' + \phi) - \exp(i\phi' - i\phi)]$. The ensemble average is $0.5\langle\exp(i\phi' - i\phi)\rangle$.

These simple rules will prove particularly useful in considering the cumulative effects of RF and gradient pulses in longer trains.

5.9 Diffusion in the radiofrequency field

Throughout this chapter, we have considered only the case where the spatial dependence of precession arises from gradients in the polarising field, B_0. Of course there is another precession process of central importance to magnetic resonance and that is the nutation around the RF field, B_1. Because these nutations also encode phase information into the spin magnetisation, such gradients in B_1 encode for position and hence may also be used to image spin density, or to measure translational motion. The first suggested use of RF gradients for imaging was by Hoult [62], while the extensive application to diffusion and flow measurements has been made by Canet and co-workers [63, 64].

In this section we show how radiofrequency field gradient precessions relate to phase-encoding ideas covered throughout this chapter, and how such RF gradient experiments may be employed for tracking molecular motion. Finally we ask, what are advantages and disadvantages of this alternative approach?

5.9.1 How RF field gradients work

The mode of action of an RF field gradient, $\mathbf{g}_1 = \nabla B_1$, is no different in principle to that of a B_0 gradient, $\mathbf{g}_0 = \nabla B_0$, as apparent in the rotating-frame description shown in Fig. 5.34. In Fig. 5.34(a) a transverse magnetisation generated by a prior 90°_x RF pulse is subject to precessional dephasing due to a distribution of spin positions, $\mathbf{r}$, in the presence of $\mathbf{g}_0$. In Fig. 5.34(b) a longitudinal magnetisation is subject to nutational dephasing due to a distribution of spin positions, $\mathbf{r}$, in the presence of $\mathbf{g}_1$. In both situations spin phase has been encoded according to a positional coordinate and, in the same way, $\mathbf{k}$-space vectors may be defined. Furthermore, both processes may be reversed by applying the same duration gradient pulse with opposite polarity. Indeed, there is a formal equivalence between the actions of $\mathbf{g}_0$ and $\mathbf{g}_1$ [65] if the RF gradient pulse is bracketed by a pair of perfectly homogeneous 90° RF pulses as shown in Fig. 5.34(c).

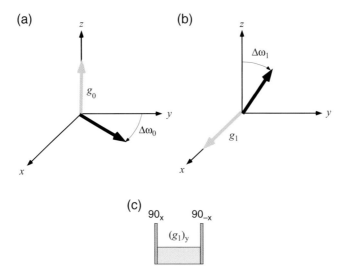

Fig. 5.34 (a) Precession at a rate $\Delta\omega_0 = \gamma\mathbf{g}_0 \cdot \mathbf{r}$ due to a gradient $\mathbf{g}_0$ in the longitudinal static magnetic field (B_0) and (b) nutation under an RF field at a rate $\Delta\omega_1 = \gamma\mathbf{g}_1 \cdot \mathbf{r}$, determined by the position of the spin in a radiofrequency (B_1) gradient, $\mathbf{g}_1$. In both cases the depiction is in the rotating frame. (c) A B_1 gradient pulse directed along the y-axis in the rotating frame, and bracketed by two homogeneous $90°$ RF pulses. The result is equivalent to a single B_0 gradient pulse of the same magnitude. ((a) to (c) are adapted from Canet [63]).

The bracketed RF gradient pulses could be used in conjunction with a regular Stejskal–Tanner pulse sequence to produce all the motion-encoding effects we have so far seen. However, there is a much simpler way to carry out the RF gradient experiment [63, 66] based on the two RF gradient pulse experiment shown in Fig. 5.35. Each pulse has duration δ, and the pulses are, as in the $\mathbf{g}_0$ case, separated by interval Δ. At the end of the second pulse a homogeneous $90°$ RF pulse is used to tip any longitudinal magnetisation into the transverse plane. There is, however, an important distinction to be made. In considering $\mathbf{g}_0$ gradients, the phase is embedded in M_+. For $\mathbf{g}_1$ gradients, we need to take account of the behaviour of M_z as well. In a step-by-step analysis of the pulse sequence of Fig. 5.35, it is possible to follow each magnetisation component in turn [63].

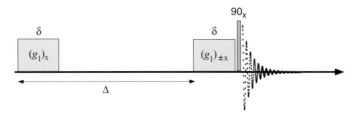

Fig. 5.35 Simple pulse sequence for measuring translational motion of spin-bearing molecules using B_1 gradients.

Let us start at time $t = 0$ and with an isochromat at position $\mathbf{r}$, with uniform magnetisation density, $M_0\hat{\mathbf{k}}$, oriented along $B_0\hat{\mathbf{k}}$, in the direction of the laboratory-frame z-axis. Assuming short RF gradient pulses, we allow that the isochromat of spins starts at $\mathbf{r}$, at the time of the first pulse, but moves to $\mathbf{r}'$ at the time of the second. Unlike the $\mathbf{g}_0$ case, where all precessions are about z, in the RF gradient case both precessions about z and nutations about x play a role. Consequently, care is needed to allow for the condition that spins are 'off-resonance', in other words, that there exists some precession angle ϕ around z in the rotating frame, over the duration between the gradient pulses [63]. The magnetisation density following the first pulse is

$$M_x(\mathbf{r}, \delta) = 0$$
$$M_y(\mathbf{r}, \delta) = M_0 \sin\theta$$
$$M_z(\mathbf{r}, \delta) = M_0 \cos\theta \tag{5.147}$$

where $\theta = \gamma \delta \mathbf{r} \cdot \mathbf{g}_1$. After the duration Δ, and allowing for relaxation of M_x and M_y by $\exp(-\Delta/T_2)$ and M_z to $M_z \exp(-\Delta/T_1) + M_0(1 - \exp(-\Delta/T_1))$

$$M_x(\mathbf{r}, \Delta) = M_0 \exp(-\Delta/T_2) \sin\theta \sin\phi$$
$$M_y(\mathbf{r}, \Delta) = M_0 \exp(-\Delta/T_2) \sin\theta \cos\phi$$
$$M_z(\mathbf{r}, \Delta) = M_0 \cos\theta \exp(-\Delta/T_1) + M_0(1 - \exp(-\Delta/T_1))$$

Choosing the second $\mathbf{g}_{1\pm}$ pulse to be positive or negative, we have, after that pulse,

$$
\begin{aligned}
M_x(\mathbf{r}, \Delta + \delta) &= M_0 \exp(-\Delta/T_2) \sin\theta \sin\phi \\
M_y(\mathbf{r}, \Delta + \delta) &= M_0 \exp(-\Delta/T_2) \sin\theta \cos\phi \cos\theta' \\
&\quad \pm M_0(1 - \exp(-\Delta/T_1)) \sin\theta' \pm M_0 \exp(-\Delta/T_1) \cos\theta \sin\theta' \\
M_z(\mathbf{r}, \Delta + \delta) &= \mp M_0 \exp(-\Delta/T_2) \sin\theta \cos\phi \sin\theta' \\
&\quad + M_0(1 - \exp(-\Delta/T_1)) \cos(\theta') + M_0 \exp(-\Delta/T_1) \cos\theta \cos\theta'
\end{aligned}
$$

where $\theta' = \gamma \delta \mathbf{r}' \cdot \mathbf{g}_1$. A phase cycle in which the contributions of positive and negative $\mathbf{g}_1$ second RF gradient pulses are co-added greatly simplifies the result to

$$M_x(\mathbf{r}, \Delta + \delta) = M_0 \exp(-\Delta/T_2) \sin\theta \sin\phi$$
$$M_y(\mathbf{r}, \Delta + \delta) = M_0 \exp(-\Delta/T_2) \sin\theta \cos\phi \cos\theta'$$
$$M_z(\mathbf{r}, \Delta + \delta) = M_0(1 - \exp(-\Delta/T_1)) \cos(\theta') + M_0 \exp(-\Delta/T_1) \cos\theta \cos\theta'$$

5.9.2 Measurement of diffusion and flow

The next step in the analysis is to calculate ensemble averages over all isochromats. Noting the homospoiling effect $\langle \sin\theta \rangle = \langle \cos\theta \rangle = \langle \sin\theta' \rangle = \langle \cos\theta' \rangle = 0$,

$$\langle M_x \rangle = 0$$
$$\langle M_y \rangle = M_0 \exp(-\Delta/T_2) \cos\phi \langle \sin\theta \cos\theta' \rangle$$
$$\langle M_z \rangle = M_0 \exp(-\Delta/T_1) \langle \cos\theta \cos\theta' \rangle$$

The difference $\theta' - \theta$ tells us about the displacement of spin-bearing molecules over the time Δ. By contrast, the sum $\theta' + \theta$ is a total (homospoiling) positional dephasing for which $\langle\sin(\theta + \theta')\rangle = \langle\cos(\theta + \theta')\rangle = 0$. Following the case of the narrow gradient pulse depiction of the Stejskal–Tanner experiment, and denoting $\mathbf{R} = \mathbf{r}' - \mathbf{r}$, then $\theta' - \theta = \gamma\delta\mathbf{R}\cdot\mathbf{g}_1$, and

$$\langle M_x \rangle = 0$$
$$\langle M_y \rangle = \tfrac{1}{2}M_0\exp(-\Delta/T_2)\cos\phi\langle\sin(\gamma\delta\mathbf{R}\cdot\mathbf{g}_1)\rangle$$
$$\langle M_z \rangle = \tfrac{1}{2}M_0\exp(-\Delta/T_1)\langle\cos(\gamma\delta\mathbf{R}\cdot\mathbf{g}_1)\rangle$$

$$(5.148)$$

These are the magnetisations before the final recall pulse. If a 90°_x recall pulse is used, the M_y magnetisation is flipped to the z-axis, while M_z is flipped to the transverse plane to contribute our signal. $\langle\cos(\gamma\delta\mathbf{R}\cdot\mathbf{g}_1)\rangle$ is the cosine transform $\int_{-\infty}^{\infty}\overline{P}(Z, \Delta)$ $\cos(qZ)dZ$, where, as before, $q = \gamma\delta g_1$.

In the case of a diffusive propagator, the result is identical to the exponential Fourier transform. Normalised to the result for $g_1 = 0$, the relaxation term disappears and, as before

$$E(q, \Delta) = \exp(-\gamma^2\delta^2 g_1^2 D\Delta)\qquad(5.149)$$

again with Δ replaced by $\Delta - \delta/3$ in the case of finite-duration gradient pulses. For net flow, one no longer obtains a phase-shifted echo signal in the RF gradient experiment, but the cosine oscillation

$$E(q, \Delta) = \cos(\gamma\delta g_1 v\Delta)\qquad(5.150)$$

5.9.3   Advantages and disadvantages of RF gradients

The principal motivation for the use of B_1 gradients concerns the ability to switch gradient pulses more easily. B_0 gradient pulses generate eddy-current-related fields during their rise and fall times, typically $\sim 100\,\mu s$. These can significantly perturb the quality of the B_0 field, resulting in resolution loss, or interference with the field-frequency lock system. Unlike B_0 gradient coils, B_1 coils are resonant circuits, allowing rise and fall times on the order of microseconds. B_1 gradients generate no eddy currents. However, they are generally much smaller in magnitude than is possible to achieve with B_0 gradients, and the need to incorporate both homogeneous and gradient B_1 coils in the same RF system presents technical challenges and places limitations on design.

References

[1] E. L. Hahn. Spin echoes. *Phys. Rev.*, 77:746, 1950.
[2] P. Mansfield and P. K. Grannell. NMR diffraction in solids. *J. Phys. C-Solid State Physics*, 6:L422, 1973.
[3] P. C. Lauterbur. Image formtion by local interactions-examples employing nuclear magnetic resonance. *Nature*, 242:190, 1973.
[4] J. Stepisnik. Violation of the gradient approximation in NMR self-diffusion measurements. *Z. Phys. Chem.*, 190:51, 1995.
[5] T. R. Saarinen and C. S. Johnson. Imaging of transient magnetization gratings in NMR-analogies with laser-induced gratings and applications to diffusion and flow. *J. Magn. Reson.*, 78:257, 1988.

[6] P. T. Callaghan. *Principles of Nuclear Magnetic Resonance Microscopy.* Oxford, New York, 1991.

[7] P. Mansfield and P. G. Morris. *NMR Imaging in Biomedicine.* Academic Press, New York, 1982.

[8] P. G. Morris. *NMR Imaging in Biology and Medicine.* Oxford University Press, Oxford, 1986.

[9] E. M. Haacke, R. W. Brown, M. R. Thompson, and R. Venkatesan. *Magnetic Resonance Imaging: Physical Principles and Sequence Design.* Wiley-Liss, New York, 1999.

[10] E. M. Haacke, R. W. Brown, M. R. Thompson, and R. Venkatesan. *NMR Imaaging of Materials.* Oxford University Press, Oxford, 2000.

[11] W. A. Edelstein, J. M. S. Hutchison, G. Johnson, and T. W. Redpath. Spin warp NMR imaging and applications to human whole-body imaging. *Phys. Med. Biol.,* 25:751, 1980.

[12] P. Mansfield and A. A. Maudsley. Planar spin imaging by NMR. *J. Magn. Reson.,* 27:101, 1977.

[13] J. Hennig, A. Nauerth, and H. Friedburg. RARE imaging: A fast imaging method for clinical MR. *Magn. Reson. Med.,* 3:823, 1986.

[14] D. R. Bailes and D. J. Bryant. NMR imaging. *Contemp. Phys.,* 25:441, 1984.

[15] P. R. Locher. Computer simulation of selective excitation in NMR imaging. *Phil. Trans. Royal Society B,* 289:537, 1980.

[16] J. Frahm and W. Hanicke. Comparitive study of pulse sequences for selective excitation in NMR imaging. *J. Magn. Reson.,* 60:320, 1984.

[17] H. Y. Carr and E. M. Purcell. Effects of diffusion on free prcession in nuclear magnetic resonance experiments. *Phys. Rev.,* 94:630, 1954.

[18] M. Abramowitz and I. A Stegun. *Handbook of Mathematical Functions with Formulas, Graphs, and Mathematical Tables.* Dover, New York, 1972.

[19] K. R. Harris, R. Mills, P. J. Back, and D. S. Webster. Improved spin echo apparatus for measurement of self-diffusion coefficients-diffusion of water in aqueous electrolyte solutions. *J. Magn. Reson.,* 29:473, 1978.

[20] P. P. Zanker, J. Schmidt, J. Schmiedeskamp, R. H. Acosta, and H. W. Spiess. Spin echo formation in the presence of stochastic dynamics. *Phys. Rev. Lett.,* 99:263001, 2007.

[21] H. C. Torrey. Bloch equation with diffusion terms. *Phys. Rev.,* 104:563, 1956.

[22] A. Abragam. *Principles of Nuclear Magnetism.* Clarendon Press, Oxford, 1961.

[23] D. W. McCall, D. C. Douglass, and E. W. Anderson. Self-diffusion studies by means of nuclear magnetic resonance spin echo techniques. *Ber. Bunsenges. Physik. Chem.,* 67:336, 1963.

[24] E. O. Stejskal and J. E. Tanner. Spin diffusion measurements: Spin echoes in the presence of a time-dependent field gradient. *J. Chem. Phys.,* 42:288, 1965.

[25] P. Stilbs and M. E. Moseley. Nuclear spin-echo experiments on standard Fourier transform NMR spectrometers. Application to multicomponent self-diffusion studies. *Chem Scripta,* 13:26, 1979.

[26] P. Stilbs and M. E. Moseley. Multicomponent self-diffusion measurement by the pulsed-gradient spin-echo method on standard Fourier transform NMR spectrometers. *Chem Scripta,* 15:176, 1980.

[27] E. O. Stejskal. Use of spin choes in a pulsed magnetoic field gradient to study anisotropic, restricted diffusion and flow. *J. Chem. Phys.*, 43:3597, 1965.

[28] P. T. Callaghan. Pulsed field gradient nuclear magnetic resonance as a probe of liquid state molecular organisation. *Australian Journal of Physics*, 37:359, 1984.

[29] P. Stilbs. Fourier transform pulsed-gradient spin-echo studies of molecular diffusion. *Progress in Nuclear Magnetic Resonance Spectroscopy*, 19:1, 1987.

[30] W. S. Price. Pulsed-field gradient nuclear magnetic resonance as a tool for studying translational diffusion: Part 1. Basic theory. *Concepts in Magnetic Resonance*, A9:299, 1997.

[31] W. S. Price. Pulsed-field gradient nuclear magnetic resonance as a tool for studying translational diffusion: Part 2. Experimental aspects. *Concepts in Magnetic Resonance*, A10:197, 1998.

[32] W. S. Price. *NMR Studies of Translational Motion*. Cambridge, 2009.

[33] P. T. Callaghan, C. M. Trotter, and K. W. Jolley. A pulsed field gradient system for a Fourier transform nmr spectrometer. *J. Magn. Reson.*, 37:247, 1980.

[34] C. M. Trotter. *PhD thesis*. Massey University, Palmerston North, New Zealand, 1981.

[35] W. S. Price and P. W. Kuchel. Effect of nonrectangular field gradient pulses in the Stejskal and Tanner (diffusion) pulse sequence. *J. Magn. Reson.*, 94:133, 1991.

[36] J. E. Tanner. Use of the stimulated echo in NMR diffusion studies. *J. Chem. Phys.*, 52:2523, 1970.

[37] W. D. Williams, E. F. W. Seymour, and R. M. Cotts. A pulsed-gradient multiple-spin-echo NMR technique for measuring diffusion in the presence of background magnetic field gradients. *J. Magn. Reson.*, 31:271, 1978.

[38] A. Germanus, H. Pfeifer, W. Heink, and J. Kärger. On the application of the spin-locking technique for NMR self-diffusion measrements by pulsed field gradients. *Annalen der Physik*, 495:161, 1983.

[39] R. F. Karlicek and I. J. Lowe. A modified pulsed gradient technique for measuring diffusion in the presence of large background gradients. *J. Magn. Reson.*, 37:75, 1980.

[40] R. M. Cotts, M. J. R. Hoch, T. Sun, and J. T. Markert. Pulsed field gradient stimulated echo methods for improved NMR diffusion measurements in heterogeneous systems. *J. Magn. Reson.*, 83:252, 1989.

[41] P. Z. Sun, J. G. Seland, and D. Cory. Background gradient suppression in pulsed gradient stimulated echo measurements. *J. Magn. Reson.*, 161:168, 2003.

[42] J. G. Seland, G. H. Sorland, K. Zick, and B. Hafskjold. Diffusion measurement at long observation times in the presence of spatially variable internal magnetic field gradients. *J. Magn. Reson.*, 146:14, 2000.

[43] P. Galvosas, F. Stallmach, and J. Kärger. Background gradient suppression in stimulated echo NMR diffusion studies using magic pulsed field gradient ratios. *J. Magn. Reson.*, 166:164, 2004.

[44] C. Buckley, K. G. Hollingsworth, A. J. Sederman, D. J. Holland, M. L. Johns, and L. F. Gladden. Applications of fast diffusion measurement using Difftrain. *J. Magn. Reson.*, 161:168, 2003.

[45] A. Haase, J. Frahm, D. Matthei, and K. D. Merbold. FLASH imaging: rapid NMR imaging using low flip angle pulses,. *J. Magn. Reson.*, 67:258, 1986.

[46] Y-Q Song. Categories of coherence pathways in the CPMG sequence. *J. Magn. Reson.*, 157:82, 2002.

[47] E. E. Sigmund, H. Cho, and Y-Q Song. Multiple-modulation-multiple-echo magnetic resonance. *Concepts in Magnetic Resonance*, 30A:358, 2007.

[48] M. D. Hürlimann. Diffusion and relaxation effects in general stray field NMR experiments. *J. Magn. Reson.*, 148:367, 2001.

[49] J. Hennig. Echoes—how to generate, recognize, use or avoid them in MR-imaging sequences. *Concepts in Magnetic Resonance*, 3:179, 1991.

[50] G. A. Morris and R. Freeman. Selective excitation in Fourier-transform nuclear magnetic-resonance. *J. Magn. Reson.*, 29:433, 1978.

[51] K. Szutkowski and I. Furó. Effective and accurate single-shot nmr diffusion experiments based on magnetization grating. *J. Magn. Reson.*, 195:123, 2008.

[52] L. Zha and I. J. Lowe. Optimized ultra-fast imaging sequence (OUFIS). *Magn. Reson. Med.*, 33:377, 1995.

[53] P. J. Basser, J. Mattiello, and D. LeBihan. MR diffusion tensor spectroscopy and imaging. *Biophysical Journal*, 66:259, 1994.

[54] P. T. Callaghan, C. D. Eccles, and Y. Xia. NMR microscopy of dynamic displacements: k-space and q-space imaging. *Journal of Physics E*, 21:820, 1988.

[55] J. Kärger and W. Heink. The propagator representation of molecular-transport in microporous crystallites. *J. Magn. Reson.*, 51:1, 1983.

[56] A. Caprihan, L. Z Wang, and E. Fukushima. A multiple-narrow-pulse approximation for restricted diffusion in a time-varying field gradient. *J. Magn. Reson.*, 118:94, 1996.

[57] P. T. Callaghan. A simple matrix formalism for the spin echo analysis of restricted diffusion under generalised gradient waveforms. *J. Magn. Reson.*, 129:74, 1997.

[58] J. Stepisnik. Analysis of NMR self-diffusion measurements by a density matrix calculation. *Physica*, 104B:350, 1981.

[59] J. Stepišnik. Measuring and imaging of flow by NMR. *Progr. NMR Spectrosc.*, 17:187, 1985.

[60] N. Bloembergen, E. M. Purcell, and R. V. Pound. Relaxation effects in nuclear magnetic resonance absorption. *Phys. Rev.*, 73:679, 1948.

[61] P. T. Callaghan and J. Stepišnik. Frequency domain analysis of spin motion using modulated gradient NMR. *J. Magn. Reson.*, A117:118, 1995.

[62] D. I. Hoult. Rotating frame zeugmatography. *J. Magn. Reson.*, 33:183, 1979.

[63] D. Canet. Radiofrequency field gradient experiments. *Progr. NMR Spectrosc.*, 30:101, 1997.

[64] F. Humbert, M. Valtier, A. Retournard, and D. Canet. Diffusion measurements using radiofrequency field gradient: Artifacts, remedies, practical hints. *J. Magn. Reson.*, 134:245, 1998.

[65] C. Counsell, M. H. Levitt, and R. R. Ernst. Analytic theory of composite pulses. *J. Magn. Reson.*, 63:133, 1985.

[66] D. Bourgeois and M. Decorps. A B1 gradient method for the detection of slow coherent motion. *J. Magn. Reson.*, 91:128, 1991.

6
Restricted diffusion

In 1963, Donald Woessner [1] carried out a series of steady gradient spin-echo experiments in samples comprising small solvent molecules in the presence of a solid matrix, in each case varying the gradient amplitude at fixed echo time, τ, so that for each value of τ, an apparent diffusion coefficient was obtained from the normalised echo amplitude relation

$$E = \exp(-\tfrac{2}{3}\gamma^2 g^2 D_{\mathrm{app}}\tau^3) \qquad (6.1)$$

Woessner found that D_{app} was a function of the echo time, decreasing as τ increased. In particular for liquid water molecules in the presence of a solid silica suspension or imbibed in a porous sandstone matrix, the rate of decay of D_{app} with respect to τ depended on the nature of the sample microstructure, while the value of D_{app} appeared to extrapolate back to the molecular self-diffusion D_0, as $\tau \to 0$. What Woessner had shown was that gradient spin-echo NMR was an effective tool for measuring restricted diffusion, and that the separate dependencies of echo amplitude on echo time and applied gradient allowed access to more subtle features of the translational dynamics.

Over the next two decades, the use of PGSE NMR to detect the influence of geometric restrictions to free diffusion largely centred on the time-dependent apparent diffusion effect, the relevant dimensionless parameter being the ratio of the diffusion length associated with the diffusion encoding time Δ to the characteristic structural length, l_s, the mean distance between diffusive barriers. For $(D_0\Delta)^{1/2}/l_s \ll 1$, unrestricted self-diffusion was observed, while for $(D_0\Delta)^{1/2}/l_s \gg 1$, $D_{\mathrm{app}} \to 0$ in the case of molecules completely enclosed by pores, or, where an interconnected pathway allowed molecules to diffuse around and between the barriers, D_{app} settled to some reduced asymptotic value.

The first indications that a richer seam of information could be mined from the PGSE NMR experiment were apparent in the deviation of the echo decays from the simple Stejskal–Tanner expression, often found in restricted diffusion experiments. The functional dependence of $E(g, \Delta)$ commonly resembled a linear superposition of Gaussians, for example in the case of samples comprising an isotropic distribution of liquid-crystalline domains in which the local diffusion is restricted in only one or two dimensions.

In this chapter we examine in detail the effects of restricted diffusion on the PGSE NMR signal. The phenomenology is complicated, or perhaps enriched, by the role of surface relaxation of spin magnetisation as molecules collide with boundaries, and by the influence on diffusing spins of magnetic field inhomogeneities resulting from susceptibility differences between the fluid and the solid matrix. And the influence of all these effects is highly dependent on the nature of the gradient waveform. We will find that the subject matter is sufficiently complicated that no general mathematical solution can be found to satisfy all conditions. Rather, the art of the experimenter is to devise the gradient waveform best suited to elucidating the structural and dynamical parameters of interest, and to choose the theoretical framework which most aptly describes that choice.

We begin by defining what we mean by a diffusion coefficient, in the contexts both of the NMR measurement and the physics of the molecular ensemble. Having done so, we then focus our attention on the diffusive behaviour of liquid molecules in a porous medium, independently of the NMR measurement. Next, the effect of bounding surfaces on spin relaxation is considered, followed by a discussion of the effect of susceptibility inhomogeneity on both the NMR free-induction decay and spin echo, as a result of molecular migration in the resulting inhomogeneous magnetic field. Finally we turn to the PGSE NMR experiment, in which external pulsed gradients are applied, allowing for relaxation and inhomogeneous local fields where appropriate and tractable.

6.1 Apparent and effective diffusion coefficients

Woessner's measured diffusion coefficient, D_{app}, was the value of the diffusion parameter needed to fit the data for his steady gradient spin-echo experiment. In magnetic resonance measurements of diffusion using magnetic field gradients and echoes, the term *apparent diffusion coefficient*, or ADC as it is known in the MRI literature, is a parameter intimately related to the particular echo method used to measure diffusion, and to the representation of that method by some suitable echo-attenuation expression parametrised by some D value. These expressions are largely based on a Fickian assumption of unrestricted self-diffusion in which the distribution of displacements is Gaussian and the mean-squared displacement is proportional to time. In that case, D_{app} properly represents the molecular self-diffusion coefficient. But where diffusion is restricted by surfaces or boundaries, these simple expressions no longer apply, so that while one might fit some parameter D_{app}, its value may depend on the strength of the gradient or the echo time used in that particular measurement. Consequently, its precise meaning may be obscure [2].

From a molecular standpoint, we may unequivocally declare what we mean by a diffusion coefficient, even where restricted diffusion applies. To do so, we simply return to the Einstein definition of eqn 1.44, in which D relates to the time dependence of mean-squared displacement. That relationship provides a working definition of a D value, even where the distribution is no longer Fickian. That value can be termed an *effective diffusion coefficient*, D_{eff}, and in the next section we show how that definition applies where diffusion is restricted.

Given the uncertainty surrounding D_{app}, how might we use gradient echo methods to yield parameters that relate directly to well-understood molecular quantities? First, it helps if we can model the system under study in a manner that is reflected in more appropriate echo-attenuation expressions than those based on a naive application of Fick's law! But even where we are unable to know in advance the relevant physics for our system, the magnetic resonance measurement can provide unambiguous molecular insight. First, in the case of the narrow gradient pulse version of the PGSE NMR experiment, the low q limit returns the mean-squared displacement corresponding to a migration time Δ, so that in this case $D_{\mathrm{app}} = D_{\mathrm{eff}}$. Second, in the steady gradient CPMG experiment described in Section 5.7.2, eqn 5.146 tells us that the measurement returns $D(\omega)$, where $\omega = 2\pi/T$, T being the period repeat interval. This $D(\omega)$ is precisely the spectral component of the molecular velocity autocorrelation function, as defined by eqn 5.132.

6.2 Time-dependent mean-squared displacement

The simplest probe of restricted diffusion involves a measurement of the effective diffusion coefficient, $D_{\mathrm{eff}}(t)$, as a function of the diffusion time, where, using the Einstein definition,

$$D_{\mathrm{eff}}(t) = \frac{\langle (\mathbf{r}(t) - \mathbf{r}(0))^2 \rangle}{6t} \tag{6.2}$$

In the limit as $t \to 0$, the diffusion coefficient for most of the molecules in a fluid with bounding walls will be the free molecular diffusion coefficient D_0. However, for a small fraction of molecules, close to the walls, diffusion is inhibited. How big that fraction is, and how much the diffusion is inhibited, we consider in the following sections. For the moment, however, we will take a coarse view in which our description will be dominated by molecules in the bulk.

Suppose the mean distance between walls is given by some length scale d. d might be a typical pore diameter in the case of a porous medium. Fig. 6.1 illustrates the effect of collision with the boundaries as the diffusion time is increased. At times sufficiently small that few molecules collide with the boundaries, $\langle (\mathbf{r}(t) - \mathbf{r}(0))^2 \rangle$ increases linearly with time, with a slope corresponding to the free molecular self-diffusion coefficient, D_0. At longer times, wall collisions cause the diffusion coefficient to reduce, and for times $t \gg d^2/6D_0t$, the diffusion coefficient becomes zero in the case of isolated pores, with $\langle (\mathbf{r}(t) - \mathbf{r}(0))^2 \rangle$ limiting at a value on the order of d^2. By contrast, for interconnected pores, the diffusion coefficient reaches an asymptotic value $D_{\mathrm{eff}}(\infty)$, reduced from the free diffusion coefficient by tortuosity effects, but nonetheless finite.

6.2.1 Tortuosity and the long-time limit

Tortuosity is a measure of the pore connectivity in an interconnected porous medium. The tortuosity, α, is defined by the relation [3]

$$\lim_{t \to \infty} \frac{D_{\mathrm{eff}}(t)}{D_0} = \frac{D_{\mathrm{eff}}(\infty)}{D_0} = \frac{1}{\alpha} \tag{6.3}$$

In simple model systems, for example randomised bead packs, α may be calculated as $\phi^{-1/2}$ [4], where ϕ is the volume fraction of the pore space. However, in more complex porous structures, such a simple relationship breaks down and α is traditionally measured from the 'formation factor'

$$\frac{1}{F} = \frac{\sigma_p}{\sigma_0} \tag{6.4}$$

where σ_0 and σ_p are the respective electrical conductivities for an electrolyte solution in the bulk and in the case where the solution fills the porous medium. Then $\alpha = F\phi$ [5] and

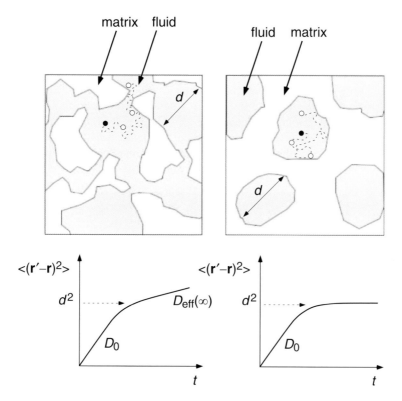

Fig. 6.1 Schematic representation of time-dependent diffusion in an interconnected porous medium (left) and in a system of isolated pores (right).

$$\frac{D_{\text{eff}}(\infty)}{D_0} = \frac{1}{F\phi},$$ (6.5)

The approach to asymptotic conditions can be sufficiently gradual that the determination of $D_{\text{eff}}(\infty)$ may be difficult in practice. There is broad agreement [6, 7] that this approach may be represented as

$$\frac{D_{\text{eff}}(t)}{D_0} = \frac{1}{\alpha} + \frac{\beta_1}{t} - \frac{\beta_2}{t^{3/2}}$$ (6.6)

Presumably, there exists some macroscopic length scale, $L_{macro} \sim REV^{1/3}$ (see Section 2.3.2), such that molecules must diffuse a distance L_{macro} in order to reach the asymptotic limit. On that basis Mair *et al.* [8] have proposed $\beta_1 \sim L_{macro}^2/D_0 t$.

The narrow gradient pulse PGSE NMR experiment is ideally suited to the measurement of $D_{\text{eff}}(t)$, the low q-response yielding the molecular mean-squared displacements directly. At finite q, a much more complex behaviour of the echo attenuation is found. These effects, along with their dependence on the actual gradient waveform, will be discussed in later sections. For the moment, we turn our attention from the the long-time behaviour to the short-time response of $D_{\text{eff}}(t)$.

6.2.2 The short-time limit for restricted diffusion

The indication that a layer of molecules near the boundaries of a porous medium experience inhibited diffusion is dramatically demonstrated in NMR microimaging experiments. Because MRI experiments are generally performed using a spin echo, the effect of diffusion under the influence of the imaging gradients is to cause an attenuation of the image, this attenuation becoming severe when the diffusion distance over the echo time is comparable with the dimension of an imaging voxel [9]. Figure 6.2 shows a high-resolution proton NMR image of water in a 1.9 mm inner-diameter capillary tube, in which a crescent of enhancement is seen due to the effect of the read gradient attenuating the NMR signal from molecules diffusing in the horizontal direction.

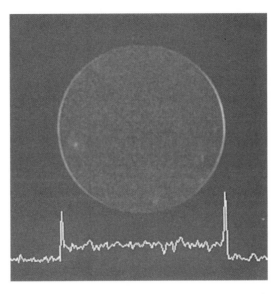

Fig. 6.2 Proton NMR microimage of a 1.9 mm inner diameter water filled capillary, showing edge enhancement where molecular diffusion along the read gradients direction is restricted by the walls. (Reproduced by permission from reference [9].)

The thickness of the restricted layer is on the order of the rms diffusion distance, $(D_0 t)^{1/2}$, when the characteristic measurement time is t. Thus the proportion of fluid experiencing the restriction on a wall of surface area S for an enclosing pore will be approximately $S(D_0 t)^{1/2}/V_p$, where V_p is the pore volume. By this simple scaling argument, we might expect that the short-time behaviour of the diffusion will follow

$$\frac{D_{\text{eff}}(t)}{D_0} = 1 - \frac{(D_0 t)^{1/2} S}{V_p} \tag{6.7}$$

An exact solution to this problem was found in a very elegant manner by Mitra and co-workers [10, 11], who obtained a result

$$\frac{D_{\text{eff}}(t)}{D_0} = 1 - \frac{4\sqrt{D_0 t}}{9\sqrt{\pi}} \frac{S}{V_p} + \mathcal{O}(D_0 t) \tag{6.8}$$

which, remarkably, holds for any pore shape, provided that the pore walls are smooth. Their derivation is based on a careful consideration of the behaviour of the diffusion probability distribution in the vicinity of the wall.

When diffusing near an impermeable boundary, molecules colliding with the wall are reflected. To understand this process quantitatively it is helpful to return to the conditional probability, $P(\mathbf{r}|\mathbf{r}', t)$, where $\mathbf{r}$ and $\mathbf{r}'$ represent the respective start and finish coordinates[1] of the diffusing molecule, and t is the time over which we will measure diffusion. Diffusing molecules will be aware of the wall only if their starting distance from the wall is not much greater than $2D_0 t$. We will treat the wall as smooth, in other words flat over the length $2D_0 t$. Since we are interested in the limiting case $t \to 0$, this flatness condition will always apply, unless the wall geometry is fractal. Suppose we represent the displacement normal to the wall by x. That means the range of x is from 0 to ∞, the latter meaning any distance much larger than the diffusion length $2D_0 t$. y and z, by contrast, will range from $-\infty$ to ∞.

The effect of the reflective boundary condition is such that we can represent the conditional probability for molecules starting at (x, y, z) by including in addition an identical image displaced beyond the wall at $(-x, y, z)$, as shown in Fig. 6.3. Hence the apparent diffusion coefficient measured over time t will be simply obtained.

In the following derivation, we assume pore volume V_p and hence a uniform pore density function V_p^{-1}. The integration of y and z over the range $-\infty$ to ∞ will be equivalent to integrating over any surface area ΔS where the extent of the area in the y and z directions exceeds $2D_0 t$. Given $t \to 0$ then we can conveniently choose ΔS sufficiently small that no pore surface curvature effects need be considered. The elements, ΔS and $\Delta V_p = \Delta S \int_0^\infty dx$ can then be summed to encompass the entire pore surface area, S, and volume, V_p.

We start by evaluating the mean-squared displacement over time t,

$$\langle (\mathbf{r}' - \mathbf{r})^2 \rangle = \int \int \rho(\mathbf{r}) P(\mathbf{r}|\mathbf{r}', t)(\mathbf{r}' - \mathbf{r})^2 d\mathbf{r}' d\mathbf{r}. \tag{6.9}$$

Including the image term in $P(\mathbf{r}|\mathbf{r}', t)$, we have

$$\langle (\mathbf{r}' - \mathbf{r})^2 \rangle = \frac{1}{V_p} \int_{-\infty}^{\infty} dz \int_{-\infty}^{\infty} dz' \int_{-\infty}^{\infty} dy \int_{-\infty}^{\infty} dy' \int_{0}^{\infty} dx \int_{0}^{\infty} dx'$$

$$\times \frac{1}{(4\pi D_0 t)^{3/2}} \exp(-\frac{(z' - z)^2}{4D_0 t}) \exp(-\frac{(y' - y)^2}{4D_0 t})$$

$$\times [\exp(-\frac{(x' - x)^2}{4D_0 t}) + \exp(-\frac{(x' + x)^2}{4D_0 t})]$$

$$\times [(x' - x)^2 + (y' - y)^2 + (z' - z)^2]$$

$$\tag{6.10}$$

---

[1] In other words $\mathbf{r}(0)$ and $\mathbf{r}(t)$ of our previous notation.

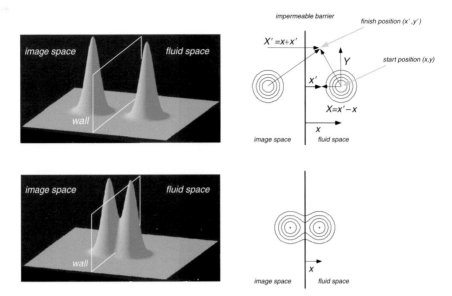

Fig. 6.3 The effect of diffusion in the vicinity of a smooth impermeable barrier in which diffusing particles are reflected at the wall. The process can be represented by adding to the Gaussian conditional probability for diffusion an image Gaussian in the space beyond the wall. The upper diagrams show a 2-D representation of the behaviour when the starting displacement from the wall, x is much larger than $2D_0 t$, while the lower diagram shows the case $x \sim 2D_0 t$. Note the definition of coordinates, where X and Y represent the displacement from origin (x, y) to position (x', y') at time t.

and hence

$$
\begin{aligned}
\langle (\mathbf{r}' - \mathbf{r})^2 \rangle = \frac{1}{V_p} & \int_{-\infty}^{\infty} dz \int_{-\infty}^{\infty} dZ \int_{-\infty}^{\infty} dy \int_{-\infty}^{\infty} dY \int_{0}^{\infty} dx \int_{-x}^{\infty} dX \\
& \times \frac{1}{(4\pi D_0 t)^{3/2}} [\exp(-\frac{X^2}{4D_0 t}) \exp(-\frac{Y^2}{4D_0 t}) \exp(-\frac{Z^2}{4D_0 t}) \\
& \times (X^2 + Y^2 + Z^2) \\
& + [\exp(-\frac{(X+2x)^2}{4D_0 t}) \exp(-\frac{Y^2}{4D_0 t}) \exp(-\frac{Z^2}{4D_0 t}) \exp(-\frac{X^2}{4D_0 t}) \\
& \times (X^2 + Y^2 + Z^2)] \\
= 6 D_0 t & - \frac{S}{V_p} \frac{1}{(4\pi D_0 t)^{1/2}} \int_{0}^{\infty} dx \int_{-\infty}^{-x} dX (X^2 + 4D_0 t) \\
& + \frac{S}{V_p} \frac{1}{(4\pi D_0 t)^{1/2}} \int_{0}^{\infty} dx \int_{x}^{\infty} dX' ((X' - 2x)^2 + 4D_0 t) \exp(-\frac{(X')^2}{4D_0 t})
\end{aligned}
$$

$$(6.11)$$

where $X' = (x + x') = X + 2x$. The remaining part of the derivation requires careful successive integration by parts, leading finally to

$$\langle(\mathbf{r}'-\mathbf{r})^2\rangle = 6D_0t - 4\frac{(2D_0t)^2}{(4\pi D_0t)^{1/2}}\frac{S}{V_p} + \frac{1}{(4\pi D_0t)^{1/2}}\frac{S}{V_p}\int_0^\infty 4x^2\int_x^\infty \exp(-\frac{X^2}{4D_0t}) \quad (6.12)$$

Again integrating by parts, the last double integral is easily shown to be equivalent to $8(2D_0t)^2/3$, whence

$$\langle(\mathbf{r}'-\mathbf{r})^2\rangle = 6D_0t - \frac{4(2D_0t)^2}{3(4\pi D_0t)^{1/2}}\frac{S}{V_p} \quad (6.13)$$

The effective diffusion coefficient, $D_{\text{eff}}(t)$, is given by $\langle(\mathbf{r}'-\mathbf{r})^2\rangle/6t$ and, normalised to the molecular self-diffusion coefficient, yields

$$\lim_{t\to 0}\frac{D_{\text{eff}}(t)}{D_0} = 1 - \frac{4\sqrt{D_0t}}{9\sqrt{\pi}}\frac{S}{V_p} \quad (6.14)$$

6.2.3 Interpolation of short- and long-time limits for restricted diffusion

Having expressions for both the short- and long-time limits for $D_{\text{eff}}(t)/D_0$, the Pade approximant [12] can be used to interpolate. A Pade approximant is the 'best' approximation of a function by a rational function of given order. In the present case, a two-point approximant (agreement as $t\to 0$ and $t\to\infty$) gives [13]

$$\frac{D_{\text{eff}}(t)}{D_0} = 1 - (1 - \frac{1}{\alpha})\frac{c\sqrt{t} + (1 - 1/\alpha)t/\theta}{(1 - 1/\alpha) + c\sqrt{t} + (1 - 1/\alpha)t/\theta} \quad (6.15)$$

where $c = (4/9\sqrt{\pi})(S/V_p)(\sqrt{D_0})$ and θ is a parameter with dimensions of time. Gas and water diffusion experiments [8, 13] in monodisperse bead packs of bead diameter, b, suggest $\sqrt{D_0\theta}\sim 0.14b$. However, for general porous media comprising a wide distribution of pore sizes, θ is a fitting parameter for which no clear geometric relationship has been identified.

Figure 6.4 shows the results of PGSE NMR experiments carried out by Mair *et al.* [8] to measure diffusion for hyperpolarised ^{129}Xe gas at 6.5 bar in a porous medium of pore volume fraction 0.38, comprising randomly packed spherical glass beads of $b = 4\,\text{mm}$ diameter. $D_0 = 8.1\times 10^{-7}$, and $\sqrt{D_0t} = 0.13b$. The data are plotted as a function of normalised diffusion length $b^{-1}\sqrt{D_0t}$. The tortuosity limit $1/\alpha$ is 0.62, equivalent to the random monodisperse bead pack value $\sqrt{\phi}$ [4], while $bS/V_p = 9.8$, again using the random monodisperse bead pack value, $bS/V_p = 6(1 - \phi)\phi$ [13]. The solid line is a Pade approximant fit using eqn 6.15, while the dashed line shows the $\sqrt{D_0t}$-dependence at short times. The fitted value of θ is 0.34 s.

6.2.4 $D_{\text{eff}}(t)$ and calculation of the apparent diffusion coefficient for any gradient waveform

$D_{\text{eff}}(t)$ can be easily measured using the narrow gradient pulse PGSE NMR experiment, where, in the low-q limit, $D_{\text{eff}}(t) = D_{\text{app}}(t)$. For other effective gradient waveforms, a general expression for $D_{\text{app}}(t)$, based on the Gaussian phase approximation and an effective gradient that is piecewise constant in time, has been derived by Zielinski and Sen [14]. Utilising their expression simply requires one to evaluate the integrals

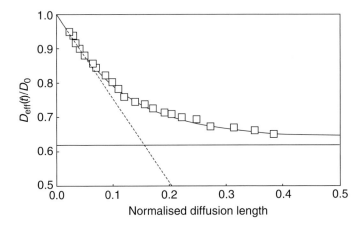

Fig. 6.4 NMR measurements of ^{129}Xe gas (6.5 bar) diffusion in randomly packed spherical glass beads of 4 mm diameter. The solid line is a Pade approximant fit, while the dashed line shows the $\sqrt{D_0 t}$-dependence at short times. (Adapted from Mair *et al.* [8].)

$\int_{t'}^{t''} t D_{\text{eff}}(t) dt$ and $\int_{t'}^{t''} t^2 D_{\text{eff}}(t) dt$, for which the Pade approximant form provides a suitable candidate.

6.2.5   The definition of the asymptotic limit

The discussion of dispersion in Chapter 2 involved a definition (eqn 2.54) of a time-dependent dispersion coefficient expressed in terms of a time derivative of mean-squared displacement. Let us label this case D_{deriv}. By contrast, the effective diffusion coefficient, D_{eff}, defined in eqn 6.2, involves a simple division by time. Clearly the asymptotic behaviours resulting from these two definitions will be very different, as illustrated in Fig. 6.5, where a simple exponential correlation function is used to describe the transition of the rms displacement, $< Z^2(t) >$, from a short-time diffusion coefficient to a long-time value five times smaller.

   The question as to which definition of time-dependent diffusion is more appropriate is a matter of standpoint. Of course, just as we define electrical resistance as V/I rather than dV/dI, so we might argue that the basic Einstein definition, involving a simple division of the mean-squared displacement by time, suggests that D_{eff} is the correct choice. It is also the choice that relates directly to the measurement of diffusion by PGSE NMR, since the Stejskal–Tanner equation contains this implicit definitional assumption. However, equally it could be argued that the initial decay of the echo attenuation yields $< Z^2(t) >$ directly, and so, if one is prepared to pay the signal-to-noise ratio price of differentiating experimental data, $D_{\text{deriv}}(t)$ is accessible. Further, the great advantage of $D_{\text{deriv}}(t)$ is that it may be directly identified with the time integral of the velocity autocorrelation function, as in eqn 2.56. This connection to the velocity helps explain why $D_{\text{deriv}}(t)$ is the preferred definition in the physics of dispersion, where particle migration is driven by the flow.

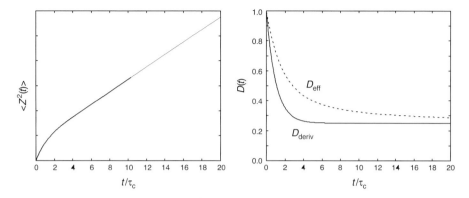

Fig. 6.5 The transition of $\langle Z^2(t) \rangle$, $D_{\text{eff}}(t)$, and $D_{\text{deriv}}(t)$, from an initial diffusion coefficient $D = 1$ to a final value of $D = 0.2$, illustrating the much more rapid approach to the asymptotic limit of $D_{\text{deriv}}(t)$.

So long as one is clear about the choice made and consistent in its use, either $D_{\text{deriv}}(t)$ or $D_{\text{eff}}(t)$ provide a suitable means of tracking the time dependence of diffusive migration.

6.3 Spin relaxation in microscopically inhomogeneous media

In this section we consider how spin relaxation may be influenced by sample structure and geometry. For the experimentalist wishing to interpret echo-attenuation data, some knowledge of relaxation rates is essential. In PGSE NMR experiments, this knowledge will enable the diffusion observation time, Δ, to be made sufficiently short that relaxation weighting effects can be avoided. Where longer values of Δ are needed, this knowledge might permit interpretation of the echo-attenuation signal using a model that correctly assigns relative weights to signals arising from different local geometries.

The interplay between diffusion and relaxation may be treated in a number of different ways. One simple approach is to assign different relaxation rates to different physical regions of a sample, such that diffusion has the effect of causing molecules to migrate between these different relaxation domains or 'phases'. This perspective, which contains no details of sample geometry, may be termed 'exchange between sites', and is parametrised by assigning relative site occupancies, site residence times, and local site relaxation rates. Exchange theory is the subject of Section 6.3.1.

Another approach is to consider a geometric description in which we distinguish spin relaxation in the bulk liquid phase from that which occurs at the interface with a surrounding solid matrix. For molecules experiencing restricted diffusion in a porous structure, geometry plays an important role in influencing spin relaxation. One relaxation mechanism for spins in pores involves the presence of strong relaxation 'sinks' at the pore surface. These sinks may be due to the presence of paramagnetic centres at the surface or due to local hindering of rotational reorientation allowing an enhanced intra- and inter-molecular dipolar interaction as the liquid molecules are temporarily immobilised. Clearly the ratio of the pore surface to pore volume will vary according to the pore size, so that *a priori* we might expect the overall relaxation behaviour to be similarly size-dependent. Section 6.3.2 deals with effects deriving from surface collisions.

Near the pore surface, local magnetic field gradients may exist, due to the heterogeneous diamagnetic susceptibility associated with the boundaries. Here, molecular diffusion will result in spin dephasing, an effect which is discussed in Section 6.4.

6.3.1   Exchange between sites

In the exchange model for multi-site relaxation, molecules hop or 'exchange' between sites with different local relaxation rates. The problem is easy to simulate on a computer using Monte Carlo methods. However, the model is amenable to mathematical treatment and is outlined as follows.

Limiting cases for exchange

Exchange may be classified as slow or fast depending on the relative speed of exchange and relaxation. Suppose that the relaxation process is described by an exponential decay of the magnetisation, $M(t)/M_0 = \exp(-t/T)$, where T is the relaxation time such as T_1 ($M(t)$ being the longitudinal magnetisation) or T_2 (with $M(t)$ being the transverse magnetisation). We begin by assigning some characteristic 'exchange time', τ_e, for migration between sites. This set of sites or 'phases' have occupancies P_i, which are taken to be time-independent. Of course $\sum_i P_i = 1$.

Labelling the relaxation time in the ith phase as $T^{(i)}$, [2] then in the slow-exchange case ($T^{(i)} \ll \tau_e$) the total magnetisation relaxation is multi-exponential and given by

$$M(t)/M_0 = \sum_i P_i \exp(-t/T^{(i)}) \tag{6.16}$$

while in fast exchange ($T^{(i)} \gg \tau_e$) a common relaxation is observed for the whole system, with

$$\frac{1}{T} = \sum_i P_i \frac{1}{T^{(i)}} \tag{6.17}$$

For intermediate exchange rates non-exponential relaxation is observed, but the simple subdivision of eqn 6.16 no longer applies.

Intermediate exchange and the Zimmerman–Brittin treatment

Zimmerman and Brittin [15] solved the problem of exchange between sites of differing relaxation rates by defining a probability $P_{ij}(t)$ that a spin initially in the i phase will be found in the j phase a time t later. The Chapman–Kolmogorov equations [16] then require

$$\frac{dP_{ij}(t)}{dt} = -\frac{1}{\tau_j} P_{ij}(t) + \sum_k P_{ik}(t) \frac{1}{\tau_k} p_{kj} \tag{6.18}$$

where $\frac{1}{\tau_k}$ is the probability per second that a spin in the kth phase leaves the kth phase, equivalent to the inverse of the mean k phase residence time for the spin, τ_k.

---

[2]Note that the superscript labels the site. The relaxation process may be T_1 or T_2 depending on the experiment, in which case, for site i we would have $T_1^{(i)}$ or $T_2^{(i)}$.

p_{kj} is the conditional probability that if a spin leaves the kth phase it will transfer to the jth phase.

Clearly $p_{kk} = 0$ and $\sum_j p_{kj} = 1$. Zimmerman and Brittin then introduce the matrices $\underline{P}(t) = [P_{ij}(t)]$ and $\underline{\underline{D}} = [\delta_{ij}\frac{1}{\tau_j} - p_{ij}\frac{1}{\tau_i}]$, and rewrite eqn 6.18 as a matrix equation

$$\dot{\underline{P}}(t) = -\underline{P}\,\underline{\underline{D}} \tag{6.19}$$

with initial condition $\underline{P} = \underline{1}$, leading to solution

$$\underline{P}(t) = \exp(-\underline{D}t) \tag{6.20}$$

The next step is to define the probability $I_{ij}(t)$ of a spin, which started in the i phase, being in the j phase at a later time t, and, in addition, being 'alive'. Hence the differential equation governing I_{ij} not only involves hopping between phases but also the possibility of spin relaxation at rate $1/T^{(j)}$. This leads to an equation similar to eqn 6.18,

$$\frac{dI_{ij}(t)}{dt} = -\frac{1}{\tau_j}I_{ij}(t) - \frac{1}{T^{(j)}}I_{ij}(t) + \sum_k I_{ik}(t)\frac{1}{\tau_k}p_{kj} \tag{6.21}$$

with matrix form

$$\dot{\underline{I}}(t) = -\underline{I}\,\underline{F} \tag{6.22}$$

where $\underline{F} = \underline{D} + \underline{E}$ and $\underline{E} = [\delta_{ij}1/T^{(j)}]$. Of course the total signal is simply obtained from the sum

$$\bar{I}(t) = \sum_{i,j} I_{ij}(t)P_i \tag{6.23}$$

In matrix notation

$$\bar{I}(t) = \tilde{\underline{\phi}}_0\,\underline{I}\,\underline{\phi}_0 \tag{6.24}$$

where $\tilde{\underline{\phi}}_0 = [1,1,1\ldots]^T$ and $\underline{\phi}_0 = [P_1, P_2, P_3 \ldots..]$.

The two-site problem for relaxation

The Chapman–Kolmogorov analysis is particularly simple in the case of exchange between two sites. Since the hopping particles have no choice as to their destination, $p_{12} = 1$ and

$$\underline{D} = \begin{bmatrix} 1/\tau_1 & -1/\tau_1 \\ -1/\tau_2 & 1/\tau_2 \end{bmatrix} \tag{6.25}$$

and

$$\underline{D}^2 = (\frac{1}{\tau_1} + \frac{1}{\tau_2})\underline{D} \tag{6.26}$$

from which all powers of $\underline{D}$ may be obtained, resulting in

$$\underline{P} = \exp(-\underline{D}t)$$
$$= \underline{1} + \frac{e^{-\lambda t} - 1}{\lambda}\underline{D} \tag{6.27}$$

where

$$\lambda = 1/\tau_1 + 1/\tau_2 \tag{6.28}$$

This defines an effective exchange time $\tau_e = \lambda^{-1}$.

Of course the steady-state occupancies, P_1 and P_2, are defined by the ratio of the rates $1/\tau_1$ and $1/\tau_2$. Given that constraint, eqn 6.27 is equivalent to

$$
\begin{aligned}
P_{11} &= P_1 + P_2 \exp(-\lambda t) \\
P_{12} &= P_2 - P_2 \exp(-\lambda t) \\
P_{21} &= P_1 - P_1 \exp(-\lambda t) \\
P_{22} &= P_2 + P_1 \exp(-\lambda t).
\end{aligned}
\tag{6.29}
$$

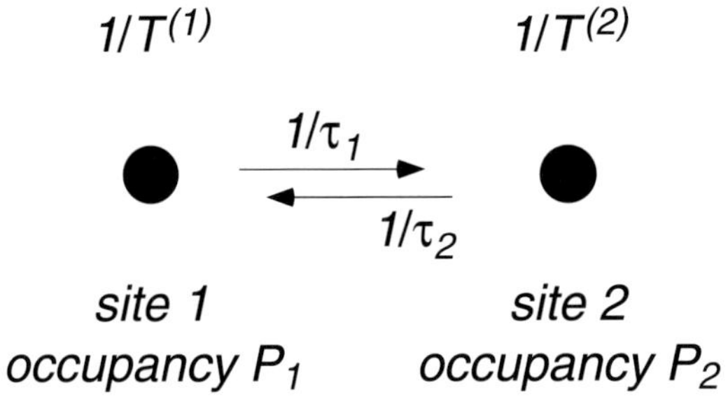

Fig. 6.6 Schematic representation of two-site exchange with local relaxation rates $1/T^{(1)}$ and $1/T^{(2)}$.

The Zimmerman–Brittin solution to the relaxation problem for two-site exchange is equally simple, with the matrix $\underline{\underline{F}}$ given by

$$
\underline{\underline{F}} = \begin{bmatrix} 1/\tau_1 + 1/T^{(1)} & -1/\tau_1 \\ -1/\tau_2 & 1/\tau_2 + 1/T^{(2)} \end{bmatrix}
\tag{6.30}
$$

with eigenvalues

$$
\lambda_1, \lambda_2 = \frac{1}{2} \left\{ \frac{1}{T^{(1)}} + \frac{1}{T^{(2)}} + \frac{1}{\tau_1} + \frac{1}{\tau_2} \mp \left[\left(\frac{1}{T^{(2)}} - \frac{1}{T^{(1)}} + \frac{1}{\tau_2} - \frac{1}{\tau_1} \right)^2 + \frac{4}{\tau_1 \tau_2} \right]^{1/2} \right\}
\tag{6.31}
$$

and solution

$$
\bar{I}(t) = a_1 \exp(-\lambda_1 t) - a_2 \exp(-\lambda_2 t)
\tag{6.32}
$$

where

$$
\begin{aligned}
a_1 &= \frac{1}{\lambda_2 - \lambda_1} [\lambda_2 - 1/T^{(av)}] \\
a_2 &= \frac{1}{\lambda_2 - \lambda_1} [\lambda_1 - 1/T^{(av)}]
\end{aligned}
\tag{6.33}
$$

and

$$
1/T^{(av)} = P_1/T^{(1)} + P_2/T^{(2)}.
\tag{6.34}
$$

Strictly speaking, the requirement for slow exchange is that $1/\tau_1 \ll 1/T^{(1)}$ and $1/\tau_2 \ll 1/T^{(2)}$, while for fast exchange, $1/\tau_e = 1/\tau_1 + 1/\tau_2 \gg 1/T^{(1)}$ *and* $1/T^{(2)}$

suffices. Of course, the fast-exchange description is too simplistic when τ_e becomes comparable with the correlation time governing the relaxation process and relaxation and exchange processes become coupled. A more general treatment of such exchange has been discussed in detail by Wennerstrom [17].

Bound water and exchange

One simple example of two-site exchange concerns the migration of water molecules between 'bound' and 'free' states. The rotational correlation time for free water is around 10^{-12} s [18]. 'Bound water' is an ill-defined term, generally used to describe water molecules closely associated with larger molecules or with solid surfaces, and having much slower tumbling rates, with rotational correlation times typically 10^{-8} s [19]. One indication that bound water is structurally modified comes from the depression of the freezing point, an effect apparent in an NMR experiment by the persistence of a liquid-like signal at temperatures below 0°C [20–22]. The question of 'structure' in bound water is a contentious topic that need not concern us [23]. Nonetheless it is clear that at least two phases of water—bound and free—exist in biological tissue, as well as in mineral systems, where there are solid surfaces with which the water can interact.

The slowing of rotational motion in the bound phase leads to a reduction in both T_1 and T_2 until the correlation time for dipolar fluctuation is of order the Larmor period of around 10^{-9} s. This is the characteristic T_1 minimum apparent in Fig. 4.15, and for a proton pair undergoing isotropic motion, the relaxation values at this minimum ($\omega_0 \tau_c \sim 1$) are of order 10–100 ms for $\omega_0 \sim 100$ MHz. At slower tumbling rates, T_2 continues to fall while T_1 increases. The slowing of reorientational motion in the 'bound' water molecule will inevitably lead to altered proton relaxation for the entire water system, in accordance with the model of Zimmerman and Brittin [15]. The reader is referred to the excellent reviews on the subject of water relaxation [24–27].

Note that the mean time taken for a proton to 'sample' the differing water phases, the exchange correlation time, τ_e, may be due to molecular diffusion or via chemical exchange of protons between adjacent water molecules. While chemical exchange can be important for short-range processes, it is quite slow, taking on the order of milliseconds per jump, so that the dominant process determining the rate of translation between spatially separated regions is molecular self-diffusion.

Equation 6.17 leads to a fast-exchange result for bound and free water,

$$\frac{1}{T_1} = \left(\frac{1}{T_1}\right)_f + P_b\left\{\left(\frac{1}{T_1}\right)_b - \left(\frac{1}{T_1}\right)_f\right\} \tag{6.35}$$

The second term in eqn 6.18 is generally dominant when the bound water correlation time is small [28, 29], leading to a simple proportionality between the T_1 relaxation rate and the bound fraction, P_b [30]. This is a very useful relationship since it indicates that over a narrow composition range, T_1 should be approximately proportional to water content.

It is clear that in biological samples the transition between the bound and free water is not sharp, so that a two-phase description is simplistic. This means that there is seldom a single characteristic exchange time, but rather a continuous spectrum [18, 31].

A variety of correlation time distribution models, applicable to biological tissues, is available [32–34]. Because of the role of diffusion in determining transfer between phases, the exchange process will also be strongly influenced by geometry. In the next section we discuss spin relaxation in the context of diffusion between the bulk and the surrounding surface on which 'relaxation sinks' are present. This model can be used to explain relaxation of water in plant cell and animal cells, but is particularly important in porous mineral structures such as clays and sandstones.

6.3.2 Relaxation sinks and normal modes: wall relaxation and the Brownstein–Tarr relations

We now return to the problem of surface relaxation for molecules experiencing restricted diffusion in a porous structure. The problem has been treated in detail by Brownstein and Tarr [35], who adopt the classical 'magnetisation diffusion' approach of Bloch and Torrey in assigning a magnetisation density, $M(\mathbf{r}, t)$, that obeys the Fick's Law differential equation

$$D\nabla^2 M(\mathbf{r}, t) = \frac{\partial M(\mathbf{r}, t)}{\partial t}. \tag{6.36}$$

The treatment is exactly as in Chapter 1, where the diffusion of molecular probability density was described. Now $M(\mathbf{r}, t)$ describes the combined probability of finding the molecule at $(\mathbf{r}, t)$ such that its spin has not relaxed. $M(\mathbf{r}, t)$ could refer to the longitudinal magnetisation, in the case of T_1 relaxation, or transverse magnetisation, in the case of T_2. Just as we saw in Chapter 1, for the case of collision with partially absorbing walls, for walls that partially relax the magnetisation,

$$D\hat{\mathbf{n}} \cdot \nabla M(\mathbf{r}, t) + \bar{\rho} M(\mathbf{r}, t))|_S = 0 \tag{6.37}$$

where $\bar{\rho}$ is the surface relaxivity parameter. The effect of bulk relaxation can be ignored for the moment as it simply adds an additional fixed relaxation decay to the problem. Equation 6.36 applies within the volume of a pore and reflects the transport of magnetisation via the diffusion of the molecules. Equation 6.37 is the boundary condition on the pore surface, taking into account the sink.

The general solution to eqn 6.36 can be written [36]

$$M(\mathbf{r}, t) = \sum_{n=0}^{\infty} a_n u_n(\mathbf{r}) \exp(-t/T_n) \tag{6.38}$$

where u_n and T_n are eigenfunctions and eigenvalues of the Helmholtz equation

$$u_n/T_n + D\nabla^2 u_n = 0 \tag{6.39}$$

with boundary condition

$$D\hat{\mathbf{n}} \cdot \nabla u_n(\mathbf{r}) + \bar{\rho} u_n(\mathbf{r})|_S = 0 \tag{6.40}$$

The solution takes the form of a sum of normal modes that will depend on the geometry and on the sink strength, $\bar{\rho}$. The eigenfunctions $u_n(\mathbf{r})$ may be normalised with respect to an integration over the pore volume V by $(1/V) \int d\mathbf{r} u_n^2 = 1$.

This generalised solution will prove to be useful in discussing 2-D relaxation experiments in Chapter 9. For the moment, however, we will return to the original notation of Brownstein and Tarr, in which the relaxation of the signal $M(t)$ is obtained by equating $M(t)$ with $\int_V M(\mathbf{r}, t)d\mathbf{r}$, and allowing the initial condition $M(\mathbf{r}, 0) = M(0)/V$. The form of the normal modes solution was written by these authors as

$$M(t) = M(0) \sum_{n=0}^{\infty} I_n \exp(-t/T_n) \tag{6.41}$$

The connection with eqn 6.38 is obvious, with $I_n = a_n \int_V d\mathbf{r} u_n$.

Equation 6.41 can describe either the T_1 or T_2 relaxation process, depending on the value of $\bar{\rho}$ chosen. The parameters that determine I_n and T_n are the molecular self-diffusion coefficient, D, the pore size, a, and the average sink strength, $\bar{\rho}$, over the pore surface. This latter parameter is somewhat empirical and a variety of methods are employed in its estimation. Of course, the interesting feature of the normal modes is their dependence on pore size. Solutions are as follows:

(a) Planar geometry (bounded by $z = -a$ and $z = a$)[3]

$$I_n = \frac{4\sin^2(\xi_n)}{2\xi_n^2 + \xi_n \sin(2\xi_n)}$$
$$T_n = \frac{a^2}{D\xi_n^2} \tag{6.42}$$

where the ξ_n are the positive roots of $\xi_n tan(\xi_n) = \bar{\rho}a/D$.

(b) Cylindrical geometry (bounded by $r = a$)

$$I_n = \frac{4J_1^2(\xi_n)}{\xi_n^2[J_0^2(\xi_n) + J_1^2(\xi_n)]}$$
$$T_n = \frac{a^2}{D\xi_n^2} \tag{6.43}$$

where the ξ_n are the positive roots of $\xi_n J_1(\xi_n)/J_0(\xi_n) = \bar{\rho}a/D$.

(c) Spherical geometry (bounded by $r = a$)

$$I_n = \frac{12[sin(\xi_n) - \xi_n cos(\xi_n)]^2}{\xi_n^2[2\xi_n - sin(2\xi_n)]}$$
$$T_n = \frac{a^2}{D\xi_n^2} \tag{6.44}$$

where the ξ_n are the positive roots of $1 - \xi_n cot(\xi_n) = \bar{\rho}a/D$.

Figure 6.7 shows the dependence of the mode amplitudes on the parameter $\bar{\rho}a/D$. The fast-exchange limit corresponds to $\bar{\rho}a/D \ll 1$. Here the relaxation is single mode,

---

[3]In their original paper, Brownstein and Tarr formulate the rectangular geometry with only one planar surface 'active'. Hence their definition of the boundaries as being $z = 0$ to $z = a$.

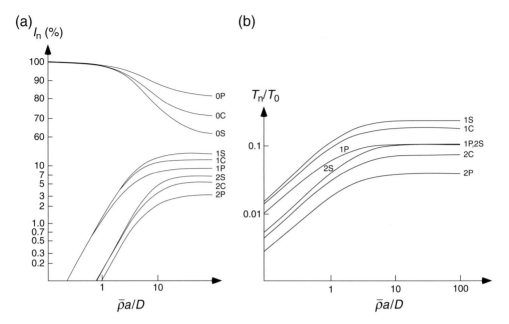

Fig. 6.7 (a) Relative amplitude (I_n) of the nth mode as a function of the dimensionless sink parameter, $\bar{\rho}a/D$, for three geometries: planar (nP), cylindrical (nC), and spherical (nS). (b) Corresponding relaxation times of different modes, normalised to T_0. (Adapted from K. R. Brownstein and C. E. Tarr [35]).

dominated by the slowest relaxation rate, $T_0^{-1} \sim \bar{\rho}(S/V)$, where S/V is the pore surface-to-volume ratio.

The intermediate ($\bar{\rho}a/D \sim 1$) and slow ($\bar{\rho}a/D \gg 1$) regimes feature multi-exponential relaxation, although in the former case the decay is overwhelmingly dominated by the lowest mode. The slow regime is characterised by $\xi_0 \sim 1$ (for example $\xi_0 = \pi$ for the spherical case). Hence $T_0 \sim a^2/D$ for $\bar{\rho}a/D \gg 1$. The physical interpretation is that the relaxation time corresponds to the time taken to migrate across the pore, each spin that hits the wall being lost to relaxation in this slow limit.

Examination of eqns 6.42, 6.43, and 6.44 shows that in the slow regime, the higher mode decay rates, T_n^{-1}, are almost independent of $\bar{\rho}$ and of order $a^2/n^2\pi^2D$. This is helpful because it gives us an estimate of the range of pore sizes that can be probed. Relaxation times can be measured in the range 1 ms to 2 s leading to a size range of 1–30 μm. Unfortunately, the higher modes represent a small faction of the total signal and are difficult to identify independently. Because of the initial slope is also sensitive to the mode amplitudes, it depends strongly on $\bar{\rho}$. In slow and intermediate regimes both the initial decay rate, T_i^{-1}, and the ratio of T_i to the longest relaxation time, T_0, have characteristic values,

$$T_i^{-1} = \bar{\rho}\frac{S}{V} \tag{6.45}$$

Table 6.1 Dominant mode relaxation times in the fast-diffusion and diffusion-limited regimes according to the Brownstein-Tarr model

	Fast-diffusion $(\bar{\rho}a/D) \ll 1$	Diffusion-limited $(\bar{\rho}a/D) \gg 1$
T_0	$a/\bar{\rho}$	a^2/D

and

$$\frac{T_0}{T_i} = \alpha \frac{\bar{\rho}a}{D} \tag{6.46}$$

where α is, respectively, 0.41, 0.35, and 0.31 for the three geometries considered. The second relation is helpful in estimating the dimensionless parameter $\bar{\rho}a/D$ from the ratio T_i/T_0.

Despite the difficulty in obtaining reliable independent estimates for $\bar{\rho}$, the Brownstein–Tarr model has been widely and successfully used to obtain pore-size distributions from multi-exponential relaxation data [37–39]. In this respect it provides a useful complement to q-space imaging. Conversely, the Brownstein–Tarr model tells us how the PGSE components will be weighted according to local geometry. Because PGSE NMR will tend to emphasise the longest relaxation time components, the behaviour of the dominant $n = 0$ term is of interest. Since V/S is of order a, T_0 is given to within a factor of order unity by Table 6.1.

In discussing the problem of diffusion in fractal volumes, we noted that surface sink relaxation could provide a dimensional measure in the case of surface fractal behaviour. This problem has been treated by de Gennes [40] in the diffusion-limited regime.

6.4 Diffusion in local inhomogeneous fields

Heterogeneous media, when placed in a the polarising magnetic field of an NMR apparatus, will exhibit local spatial variations in field due to susceptibility inhomogeneity. We will label these local fields $\Delta\mathbf{B}_0(\mathbf{r})$, and note that they are superposed on the polarising field $B_0\hat{\mathbf{k}}$. For the molecules of a liquid occupying the pore space of such a medium, the field variations result in differential Larmor precession and hence a phase-spreading, which is manifest as an NMR signal attenuation. Unlike the Brownstein–Tarr (surface-sink) relaxation process associated with wall collisions, relaxation due to susceptibility inhomogeneity arises from through-space interactions that are modulated as the molecule diffuses. Whether surface-sink or susceptibility-inhomogeneity effects are more important will depend on the particular material being studied.

We have seen that the Brownstein–Tarr model leads to a specific signature in the form of the relaxation normal modes. Furthermore the surface-sink mechanism influences both T_1 and T_2 relaxation. In this respect relaxation due to susceptibility effects is quite different. T_1 relaxation requires spin Hamiltonian fluctuations at the Larmor frequency. The spin Zeeman interactions due to susceptibility variations in the sample modulate much more slowly than the Larmor frequency as the particle diffuses and so influence T_2 alone. The existence of susceptibility effects is therefore indicated by $T_2 \ll T_1$. One nice feature of this difference is that we are able to separate the influences of surface-sink and susceptibility-variation effects by separately observing T_1 and T_2.

6.4.1   Calculating the local field

In a porous medium, typical diamagnetic susceptibility variations, $\Delta\chi$, are on the order 10^{-6}, leading to local magnetic field variations, $\Delta B \sim \chi B_0$, on the order of a few parts per million of the polarising field. This fact greatly simplifies the calculation of local fields, $\Delta\mathbf{B}(\mathbf{r})$, since the local magnetisation induced by the field is dominated by B_0 alone, allowing a simple linear superposition of fields calculated by summing local magnetisation components. In particular, in a region of the material where the diamagnetic susceptibility is $\chi(\mathbf{r})$, the local magnetisation is

$$\mathbf{M}(\mathbf{r}) = \frac{1}{\mu_0}\chi(\mathbf{r})B_0\hat{\mathbf{k}} \qquad (6.47)$$

with the local magnetic dipole moment associated with an element of volume dV being $\mathbf{M}(\mathbf{r})dV$. In consequence, the local field offset at any position $\mathbf{r}'$ is simply given by the dipolar field superposition

$$\Delta\mathbf{B}(\mathbf{r}') = \frac{\mu_0}{4\pi}\int_V \Big[3\frac{\mathbf{M}(\mathbf{r})\cdot(\mathbf{r}'-\mathbf{r})}{|\mathbf{r}'-\mathbf{r}|^5}(\mathbf{r}'-\mathbf{r}) - \frac{\mathbf{M}(\mathbf{r})}{|\mathbf{r}'-\mathbf{r}|^3}\Big]d\mathbf{r} \qquad (6.48)$$

The summation process is illustrated in Fig. 6.8, for a simple two-phase medium consisting of a solid with susceptibility χ_1 and a liquid with susceptibility χ_2.

Because the magnitude of local field variations, $|\Delta\mathbf{B}(\mathbf{r})|$, are significantly smaller than the magnitude of the polarising field, B_0, from an NMR perspective we need only consider components of that local field parallel to the polarising field. $\Delta\mathbf{B}(\mathbf{r})\cdot\hat{\mathbf{k}}$ will henceforth be labelled $\Delta B(\mathbf{r})$.

A particularly simple example of a porous medium is that of a random monodisperse sphere pack. In this case each sphere generates a local field beyond its boundary equivalent to that of a dipole placed at the sphere centre. Thus the field in the liquid may be obtained by simply summing the contributions from each sphere. An example of the resulting local field map is shown in Fig. 6.9.

6.4.2   Effect of molecular diffusion and the Anderson–Weiss treatment

Of course, spin translational motion will cause the local field to fluctuate. In analogy with the Eulerian velocity field, $\mathbf{v}(\mathbf{r})$, and the Lagrangian distribution of time-dependent velocities, $\mathbf{v}(t)$, we could define, for the spin ensemble, a distribution of fluctuating fields $\Delta B(t)$ and associated Larmor frequencies, $\Delta\omega_0(t)$. The crucial factor in determining the effect on the transverse magnetisation is the rate at which this frequency fluctuates. This rate is characterised by defining the correlation time,

$$\tau_c = \int_0^\infty \frac{\langle\Delta\omega_0(t+\tau)\Delta\omega_0(t)\rangle}{\langle\Delta\omega_0^2\rangle}d\tau \qquad (6.49)$$

where $\langle\Delta\omega_0^2\rangle$ is the mean-square frequency fluctuation. $\langle\Delta\omega_0^2\rangle$ is the second moment (M_2) of the linewidth that prevails in the case of a stationary interaction

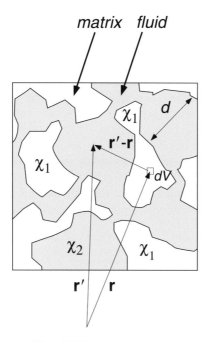

Fig. 6.8 Geometry of local field integration associated with eqn 6.48.

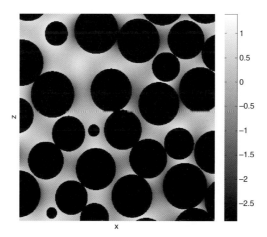

Fig. 6.9 Simulated internal magnetic field (in dimensionless units) for a slice through a sphere pack in which the diamagnetic susceptibility of the spheres and surrounding fluid differ. (Reproduced with permission from Burcaw *et al.* [41].)

$(\tau_c \to \infty)$. τ_c depends on the molecular motion and on the correlation length of the field offset, $\Delta\omega_0(\mathbf{r})$. For the moment we ignore these details. We can therefore define 'rapid fluctuations' and 'slow fluctuations' according to $\tau_c^{-1} \gg (\langle\Delta\omega_0^2\rangle)^{1/2}$ or $\tau_c^{-1} \ll (\langle\Delta\omega_0^2\rangle)^{1/2}$, respectively.

The effect of fluctuating local fields on the nuclear transverse magnetisation is well known and has been treated in detail by Abragam [42]; the application of this theory to relaxation induced by diffusion in a spatially varying local field has been made by Packer [43] and Hazelwood *et al.* [44]. Slow fluctuations result in inhomogeneous line broadening. This means that the differing local fields cause differing local precession rates, thus leading to a decay of the transverse magnetisation of the form $\exp(-\frac{1}{2}M_2 t^2)$. The essential feature of inhomogeneous broadening is that this decay can be refocused in a spin echo. By contrast, rapid fluctuations result in homogeneous broadening, a true relaxation mechanism common to all spins in the system. Decay due to homogeneous broadening is irreversible because of its stochastic nature. For rapid fluctuations this relaxation is described by a T_2 value of $(M_2 \tau_c)^{-1}$.

The general behaviour of both the spin-echo signal and the transverse magnetisation can be treated at any timescale using the Anderson–Weiss theory and the fluctuating field correlation function,

$$g_\omega(\tau) = \frac{\langle \Delta\omega_0(t+\tau)\Delta\omega_0(t)\rangle}{\langle \Delta\omega_0^2\rangle} \tag{6.50}$$

From the assumption that the distribution of $\Delta\omega_0$ is Gaussian, it may be shown that the normalised FID signal, $S(t)$,[4] and the normalised Hahn echo amplitude, $E(2\tau)$, are given respectively by

$$S(t) = \exp(-\langle \Delta\omega_0^2\rangle \int_0^t (t-t')g_\omega(t')dt')$$

$$E(2\tau) = \exp(-\langle \omega_0^2\rangle\{4\int_0^\tau (\tau-t')g_\omega(t')dt' - \int_0^{2\tau}(2\tau-t')g_\omega(t')dt'\})$$

$$\tag{6.51}$$

For stochastic processes it is common to assume a correlation function $g_\omega(\tau) = exp(-|\tau|/\tau_c)$. This leads to

$$S(t) = \exp(-\langle \Delta\omega_0^2\rangle\tau_c^2\{\exp(-t/\tau_c) - 1 + t/\tau_c\})$$

$$E(2\tau) = \exp(-\langle \omega_0^2\rangle\tau_c^2\{4\exp(-\tau/\tau_c) - \exp(-2\tau/\tau_c) + 2\tau/\tau_c - 3\})$$

$$\tag{6.52}$$

Figure 6.10 shows the transition from slow to fast regimes in the decay of the FID and spin echo, as represented by eqns 6.52. Although the concept of a relaxation time can only be defined precisely in the fast limit, it is convenient to use the symbols T_2^* and $T_2^\dagger$ to represent the time taken for the FID and echo amplitudes to decay to e^{-1}.

In Chapter 5 we dealt with the problem of diffusion in a steady, uniform, field gradient and generated an expression for magnetisation decay which depended exponentially on the square of the gradient. This expression is not generally applicable in the case of a randomly varying field offset. This can be immediately recognised by

---

[4]Here we mean the signal $M(t)$ normalised to its initial value $M(0)$.

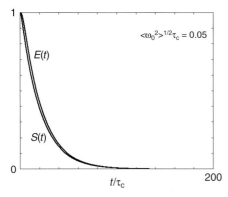
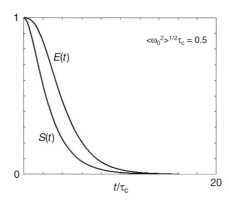

Fig. 6.10 Dependence of FID ($S(t)$) and echo amplitude ($E(t) = E(2\tau)$) on time t for (left) fast motion, where $\langle \Delta\omega_0^2\rangle\tau_c^2 = 0.05$ and (right) intermediate motion, where $\langle \Delta\omega_0^2\rangle\tau_c^2 = 0.5$. Note that $S(t)$ and $E(t)$ converge to a common exponential decay in the fast-motion limit, where relaxation effects dominate.

considering the spin phase, $\int \gamma \Delta B dt$. In the uniform gradient there is no upper bound on the phase as the particle diffuses, since $\Delta B = \mathbf{G} \cdot \mathbf{r}$ and $\mathbf{r}$ is unbounded. This means that the process can be viewed as an unrestricted random phase walk. In the case of the fluctuating local field offset, there is clearly an upper bound on ΔB, which oscillates about 0. To this extent the latter process may be viewed as a restricted phase walk, and the treatment required is quite different.

However the echo expression in eqn 6.52 does have a correspondence with the steady gradient expression for echo attenuation, provided that the diffusion time τ is much shorter than that required to diffuse a characteristic length l_c of the fluctuating field. In the limit $\tau \ll \tau_c$ eqn 6.52 becomes

$$
\begin{aligned}
E(2\tau) &= \exp(\;\tfrac{2}{3}\langle\omega_0^2\rangle\tau_c^{-1}\tau^3) \\
&\approx \exp(-\tfrac{2}{3}\gamma^2\langle G_0^2\rangle D\tau^3)
\end{aligned}
\tag{6.53}
$$

where $\gamma^2\langle G_0^2\rangle = \langle\omega_0^2\rangle/l_c^2$ and we have set $l_c^2 \approx D\tau_c$. The resemblance to eqn 5.45, is obvious.

6.4.3 Measuring relaxation in porous media.

NMR spin relaxation is used extensively in petrophysical applications to measure porosity, pore size distributions, and even permeability. In porous rocks of petrophysical interest, pore sizes range from sub-micron to millimetre. However it is the small pores that provide the dominant interest, since these determine the permeability of the rock and the degree to which ingressed hydrocarbon can be extracted. For such pores, on the order of 10-micron size or smaller, the fast diffusion limit applies and so T_1 and T_2 may be directly related to the pore surface-to-volume ratio. For water in large pores, bulk relaxation behaviour dominates, and from a practical perspective one may write the summed effect as

$$\frac{1}{T_2} = \frac{1}{T_{2bulk}} + \bar{\rho}_2 \frac{S}{V} + \frac{\gamma^2 \langle G_0^2 \rangle D(2\tau)^2}{12}$$

$$\frac{1}{T_1} = \frac{1}{T_{1bulk}} + \bar{\rho}_1 \frac{S}{V} \tag{6.54}$$

where $\bar{\rho}_2$ and $\bar{\rho}_1$ are, respectively, the surface relaxivities for T_2 and T_1 relaxation and the diffusion term in the T_2 expression allows for attenuation due to local inhomogeneous fields, and may be rendered insignificant by choosing the echo time $TE = 2\tau$ sufficiently short.

The distribution of pore sizes leads to a distribution in relaxation rates. Multi-exponential relaxation data may be inverted to yield relaxation-time distributions, using the inverse Laplace technique described in Chapter 9. Of course, converting a distribution of T_2 into a pore-size distribution requires some knowledge of $\bar{\rho}_2$, a parameter which will vary depending on the nature of the rock formation. However, other laboratory methods, such as mercury porosimetry, can be used for calibration purposes, thus enabling petroleum engineers to interpret NMR logging data obtained using tools that are lowered down the well at the drill site.

There exists a considerable body of engineering experience in interpreting NMR relaxation data obtained in this manner. However, amongst the most useful concepts is the idea of a 'T_2 cutoff'. This limit corresponds to pore sizes for which water cannot be removed from the rock by centrifugation, the so-called BVI or 'bulk volume irreducible' by which this bound water is labelled. Not only does the cutoff value give some idea of the degree to which fluid may be extracted from the rock, separating as it does the BVI and FFI (free fluid) fractions, when combined with a knowledge of the porosity, it is also an important parameter in helping estimate rock permeability [45]. Total porous medium porosity can be measured by NMR, comparing the initial, unrelaxed signal intensity (extrapolated from a CPMG train) of a water-saturated sample and a bulk water sample of the same volume. The cutoff is determined by carrying out T_2 distribution measurements in both a fully water-saturated sample and in the same sample after centrifugation has removed all but the irreducible water. The point where the cumulative distributions for the saturated and irreducible water samples separate defines the cutoff [45]. A schematic illustrating these ideas is shown in Fig. 6.11.

6.4.4   Decay due to diffusion in the internal field

The discussion of the previous section concerns the effect of molecular migration in the inhomogeneous field in the case of two canonical NMR measurements, the FID following a 90° RF pulse, and the $90° - 180° -$ spin echo. In both cases T_2 relaxation associated with surface collisions causes additional attenuation. By clever design of the NMR pulse sequence, it is possible to access field migration and relaxation effects separately. One such method is the DDIF (decay due to diffusion in the internal field) experiment of Song *et al.* [46]. The RF pulse sequence is shown in Fig. 6.12. t_e is chosen sufficiently short ($t_e \ll \tau_c$) that the molecules diffuse shorter distances than the characteristic length, $\sqrt{2D\tau_c}$, over which variation occurs in the local field, $\Delta B(\mathbf{r})$. By contrast, t_d is sufficiently long that diffusion results in significant changes in $\Delta B(\mathbf{r})$. Of course, there will be a decay caused by T_1 relaxation over the time t_d. This may

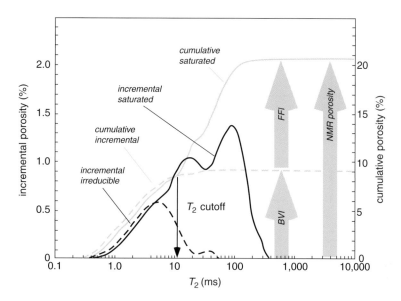

Fig. 6.11 Schematic NMR determination of T_2 cutoff in a porous rock. The solid data lines of the incremental and cumulative T_2 distributions refer to a fully water-saturated rock sample while the dashed data lines refer to measurements in a rock that has been centrifuged to leave only the irreducible water. (Adapted from Coates *et al.* [45])

be removed by normalising the echo signal, $E(t_e, t_d)$, by the reference signal, $R(t_e, t_d)$, obtained using pulse sequence (b) in Fig. 6.12. With this step one obtains the DDIF signal

$$E_n(t_e, t_d) = \int \int \rho(\mathbf{r})e^{i\gamma t_e(\Delta B(\mathbf{r}) - \Delta B(\mathbf{r}'))} P(\mathbf{r}|\mathbf{r}', t_d)d\mathbf{r}d\mathbf{r}' \qquad (6.55)$$

where the signal is taken to be 'on-resonance', so that the precession due to the polarising field, B_0, is neglected.

The decay represented by eqn 6.55 is rather similar to that resulting from molecular diffusion in a steady gradient, but with one major difference: unlike the case for steady gradients, $B_0(\mathbf{r})$ is bounded with the full range of values sampled over the pore length scale. At short times t_d, $E_n(t_e, t_d)$ decays approximately as $1 - \frac{1}{2}\gamma^2 t_e^2 \langle (\Delta B(\mathbf{r}) - \Delta B(\mathbf{r}'))^2 \rangle$, the term in the angular brackets increasing with increasing t_d. But because of the normalisation with $R(t_e, t_d)$, the DDIF signal reaches an asymptotic value for large t_d, as $\Delta B(\mathbf{r}')$ and $\Delta B_0(\mathbf{r})$ become uncorrelated. Expressing the DDIF signal in terms of the distribution of local fields, $f(\Delta B)$,

$$E_n(t_e, t_d) = \int_{\Delta B_{min}}^{\Delta B_{max}} \int_{\Delta B_{min}}^{\Delta B_{max}} f(\Delta B)e^{i\gamma t_e(\Delta B - \Delta B')} P(\Delta B|\Delta B', t_d)d\Delta Bd\Delta B' \quad (6.56)$$

For $t_c \gg \tau_c$, the conditional probability $P(\Delta B|\Delta B', t_d)$ is independent of starting field and reduces to $f(\Delta B')$. Whence

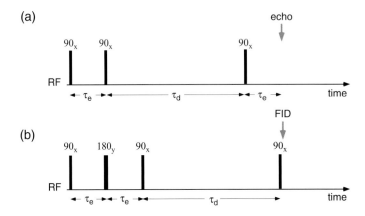

Fig. 6.12 Pulse sequences used for the DDIF method. (a) Encoding due to dephasing in the the local field over two periods t_e separated by t_d. (b) Encoding for T_1 relaxation while dephasing in the local field is cancelled by the spin echo.

$$E_n(t_e, t_d \to \infty) = |\int_{\Delta B_{min}}^{\Delta B_{max}} f(\Delta B) e^{i\gamma t_e \Delta B} d\Delta B|^2$$
$$= |S(t_e)|^2 \tag{6.57}$$

where $S(t)$ is the (relaxation-normalised) FID.

Figure 6.13 shows an example of the DDIF signal obtained from water diffusing in a random glass sphere pack.

The rate of the DDIF decay is determined by the characteristic length, $l_c = \sqrt{2D\tau_c}$, over which the local field fluctuations are manifest. Clearly, this length will depend upon pore size, so that a multi-exponential analysis of the decay can yield a pore-size distribution, much in the same way that multiexponential analysis of relaxation times is used in porous media.

6.4.5   *q*-space analysis of internal field correlations

Equation 6.55 demonstrates the role of the displacement propagator in governing the effects of diffusion in the inhomogeneous field within a porous medium. Access to that propagator is provided in natural manner by the use of pulsed magnetic field gradients. Cho and Song [47] have suggested a clever PGSE NMR approach for examining magnetic field correlations for liquid molecules diffusing in porous media. The method depends on the use of two pulsed gradient sequences, one in which internal field effects are cancelled out and another in which they induce phase shifts. In both sequences relaxation due to T_1 and T_2 plays an identical role, so that by taking a ratio of the signals from each experiment, relaxation effects can be revealed.

The two sequences, which each generate stimulated echoes, are shown in Fig. 6.14. In the first of these (a), the 180° RF pulses refocus the effect of dephasing in the local field $\Delta B(\mathbf{r})$, while the gradient pulses produce an encoding for displacement over the time period of approximately τ_d. Note that the sequence shown effectively consists of a single effective gradient pair, each gradient pulse being split around the 180° RF pulse

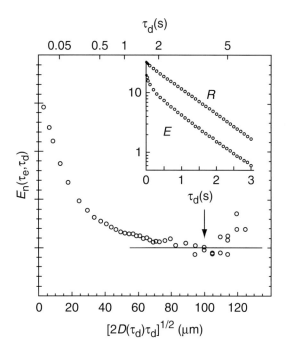

Fig. 6.13 DDIF ratio (E/R) for water diffusing in a uniform-size, randomly packed 100-micron diameter glass bead sample. The ratio is shown as a function of the rms displacement,$(2D(t_d)t_d)^{1/2}$, with the corresponding t_d shown at the top of the graph. The line is the square of the FID. (Figure adapted from Song *et al.* [46].)

with opposite sign in the dephasing and rephasing periods τ_e, so that each contributes the same sign effective gradient to the $q-$encoding, along the lines of the 13-interval Cotts sequence. For this sequence the signal is

$$E(\mathbf{q}, \tau_d) = \int \rho(\mathbf{r}) P(\mathbf{r}|\mathbf{r} + \mathbf{R}, \tau_d) e^{i\mathbf{q}\cdot\mathbf{R}} d\mathbf{r} d\mathbf{R}$$

$$= \int \bar{P}(\mathbf{R}, \tau_d) e^{i\mathbf{q}\cdot\mathbf{R}} d\mathbf{R} \tag{6.58}$$

In the second sequence, (b), the positions of the 180° RF pulses are shifted so that the dephasing (τ_e) and rephasing (τ_e') periods are unequal, leading to additional phase shifts due to the local fields, and the result

$$E(\mathbf{q}, \tau_d) = \int \rho(\mathbf{r}) P(\mathbf{r}|\mathbf{r} + \mathbf{R}, \tau_d) e^{i\mathbf{q}\cdot\mathbf{R}} e^{it\gamma(\Delta B(\mathbf{r}) - \Delta B(\mathbf{r}+\mathbf{R}))} d\mathbf{r} d\mathbf{R}$$

$$= \int \bar{P}(\mathbf{R}, \tau_d) e^{i\mathbf{q}\cdot\mathbf{R}} \langle e^{it\gamma(\Delta B(\mathbf{r}) - \Delta B(\mathbf{r}+\mathbf{R}))} \rangle d\mathbf{R}$$

$$\tag{6.59}$$

where $t = \tau' - \tau$, and where $\langle \dots \rangle$ in eqn 6.59 represents an average over all starting positions $\mathbf{r}$. Taking the Fourier transform with respect to $\mathbf{q}$ of eqns 6.58 and 6.59 gives the probabilities $\bar{P}(\mathbf{R}, \tau_d)$ and $\bar{P}'(\mathbf{R}, \tau_d)$, respectively, such that

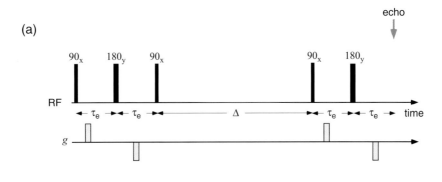

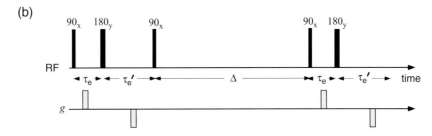

Fig. 6.14 Pulse sequence used by Cho and Song [47] to measure pair correlation function of internal magnetic field: (a) balanced sequence and (b) imbalanced sequence.

$$\bar{P}'/\bar{P} = \langle e^{it\gamma(\Delta B(\mathbf{r}) - \Delta B(\mathbf{r}+\mathbf{R}))}\rangle$$
$$\approx 1 - \gamma^2 t^2 \langle \Delta B^2 \rangle + \gamma^2 t^2 \langle \Delta B(0) \Delta B(\mathbf{R})\rangle \qquad (6.60)$$

where $\langle \Delta B(0) \Delta B(\mathbf{R})\rangle$ is the inhomogeneous field pair correlation function, and $\langle \Delta B^2 \rangle$ is the second moment of the field, easily obtained by examination of the NMR linewidth. Note that the role of τ_d is simply to determine the sensitivity of the signal to the full range of displacements $\mathbf{R}$. Reference [47] contains a number of examples of pair correlation function measurements for water diffusing in glass beadpacks.

6.5   Restricted diffusion for spin echoes with steady gradients

Section 6.2 dealt with the effect of boundaries and restrictions on diffusion. Sections 6.3 and 6.4 outlined the influence on spin relaxation of molecular motion in heterogeneous structures, both from the standpoint of exchange between regions of differing local relaxation rates, and where structural heterogeneities result in locally inhomogeneous magnetic fields. The remaining sections of this chapter deal with the influence of restricted motion on spin-echo NMR experiments in which external gradients are deliberately applied.

We begin by considering the steady gradient spin echo, in part because of its historical interest, but also because it provides a limiting case for the PGSE experiment and one with considerable practical application. In early papers, Robertson [48] and Neuman [49] addressed the problem of the steady gradient spin-echo decay in a bounded region, in both cases utilising the Gaussian phase approximation in order to obtain closed-form expressions for spin echo amplitudes. Before examining examples for specific geometries based on their ideas, it is helpful to outline a more general physical picture of relevant motional regimes.

6.5.1 Characteristic length scales

The most obvious characteristic length scale is the distance diffused by molecules over a given time τ. The diffusion length, l_D, is defined by

$$l_D = (D_0\tau)^{1/2} \tag{6.61}$$

The second length scale is that associated with dephasing in the magnetic field gradient. Equation 5.42 tells us that the mean-squared phase shift experienced by an ensemble of spins borne by molecules in free diffusion, in the presence of a constant magnetic field gradient g, and after a diffusion time τ, is given by

$$\langle \phi^2 \rangle = \tfrac{1}{3}\gamma^2 g^2 \tau_s^2 \xi^2 n^3$$
$$= \tfrac{1}{3}\gamma^2 g^2 D_0\tau^3 \tag{6.62}$$

Note that the accumulation of phase depends in part on the mean-squared displacement $n\xi^2$, but also on the residence time per step, during which a phase $\gamma g\tau_s$ is acquired. This means that the more rapid the diffusion, the greater rms distance a spin-bearing molecule must diffuse in order to acquire a given rms phase shift. This idea is neatly encapsulated by defining a 'dephasing length' [50]

$$l_g = \left(\frac{D_0}{\gamma g}\right)^{1/3} \tag{6.63}$$

l_g is a measure of the molecular diffusion distance corresponding to a spin phase shift on the order of 2π. Note that l_g is related to both the gradient amplitude and the rate of diffusion, but is independent of the diffusion time τ. One gradient-related length scale that does depend on τ is the magnetisation helix wavelength,

$$\lambda = \frac{2\pi}{\gamma g\tau} \tag{6.64}$$

The length scales l_D, l_g, and λ are interdependent since $l_g = (\lambda l_D^2/2\pi)^{1/3}$. Together, eqns 6.63, 6.61, and 6.64 tell us that the echo attenuation for free diffusion in the presence of a steady gradient at echo time 2τ, may be written [51]

$$E = \exp(-\tfrac{2}{3}\gamma^2 g^2 D_0\tau^3)$$
$$= \exp\left(-\frac{2}{3}\left[\frac{l_D}{l_g}\right]^6\right)$$
$$= \exp\left(-\frac{2}{3}4\pi^2\left[\frac{l_D}{\lambda}\right]^2\right) \tag{6.65}$$

Finally, when dealing with restricted diffusion, another length scale emerges, the characteristic distance between boundaries, l_s [51]. This is the distance a molecule may freely diffuse before encountering restrictions to its Brownian motion. Note that for the case of free diffusion, as $l_D \gg l_g$, the echo signal disappears. However, in the case of restricted motion, very different behaviours are apparent, according to the relative sizes of l_D, l_g, and l_s.

6.5.2   Regimes of interest

Figure 6.15 shows a fluid bounded by plane slabs in which molecular diffusion is depicted under a range of comparative length scales. Neglecting the effect of gradients, we can clearly distinguish the diffusive behaviour in the cases $l_D \ll l_s$ and $l_D \gg l_s$. In the former, the diffusion is largely free apart from a small proportion of molecules experiencing wall collisions, the impact on the apparent diffusion coefficient having been discussed in Section 6.1.2. In the latter case the apparent diffusion coefficient reduces asymptotically to zero, while the mean-squared displacement becomes fixed and on the order of l_s^2. However, in a steady gradient spin-echo experiment the behaviour is more subtle, as the length scale associated with dephasing in the gradient starts to play a role. Hürlimann *et al.* [51] have identified three regimes where differing behaviours dominate. While these can be characterised by using the set (l_D, l_s, λ), a better parametrisation is provided by the set (l_D, l_s, l_g).

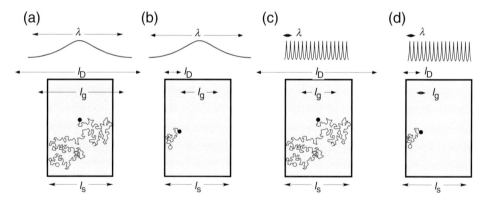

Fig. 6.15 Characteristic length scales for molecules diffusing between two plane parallel slabs, and with the magnetic field gradient applied normal to the planes: (a) corresponds to the motional averaging regime, (b) to the free-diffusion regime, while (c) and (d) correspond to the localisation regime.

Free diffusion regime: $l_D/l_s \ll 1$ and $l_D/l_g \ll 1$ (l_D shortest length)
In the free-diffusion regime only a small proportion of molecules collide with the walls and the gradient is sufficiently weak (l_g sufficiently long) that severe echo attenuation does not result. Here one might expect that the free diffusion result holds, but with the apparent diffusion coefficient modified slightly, as outlined in Section 6.1.2, so as to allow for an impeded layer of thickness l_D near the walls. However the precise

expression for the apparent diffusion coefficient that applies in the case of the steady gradient echo under the definition

$$E(g, \tau) = \exp(-\gamma^2 g^2 D_{\text{app}} \tau^3) \tag{6.66}$$

is slightly different from that of eqn 6.14, which is calculated on the basis of the mean-squared displacement. The appropriate expression for D_{app}, which is consistent with eqn 6.66, has been shown by de Swiet and Sen [50] to be

$$D_{\text{app}} \approx D_0(1 - \alpha(S/V)\sqrt{D_0 \tau}) \tag{6.67}$$

where $\alpha = 32(2\sqrt{2} - 1)/(105\sqrt{\pi})$. This coefficient is very close to the $4/9\sqrt{\pi}$ that applies in eqn 6.14.

Motional narrowing regime: $l_s/l_D \ll 1$ and $l_g/l_s \gg 1$ (l_s shortest length)
This is the regime analysed by Robertson [48], who found, in the case of restricted diffusion between planar boundaries, the steady gradient echo-attenuation expression

$$E(g, \tau) = \exp\left(-\frac{1}{120}\frac{\gamma^2 g^2 l_s^4 2\tau}{D_0}\right)$$

$$= \exp\left(-\frac{1}{60}\left[\frac{l_D}{l_g}\right]^2 \left[\frac{l_s}{l_g}\right]^4\right) \tag{6.68}$$

Note the appearance of the diffusion coefficient in the exponent denominator. For faster diffusion, the phase spreading, and hence the echo attenuation, reduces, and for the simple reason that the molecules are approaching an asymptotic 'averaged' state, in which the molecules appear to reside at their mean position at the box centre between the plane walls. The motional narrowing has occurred because the fluctuation rate for the spin Larmor frequencies exceeds the frequency spread created by the magnetic field gradient across the rectangular pore.

Similar expressions [49] are obtained for different geometries, but with different numerical prefactors in the exponents. Note that these derivations all rely on the use of the Gaussian phase approximation. This approximation becomes questionable when $\gamma^2 g^2 D_0 \tau^3$ is greater than or of order unity, in other words when $l_D \sim l_g$. As we will see in the next section, this breakdown is associated with the appearance of diffraction-like effects in the echo-attenuation function.

Localisation regime: $l_g/l_s \ll 1$ and $l_g/l_D \ll 1$ (l_g shortest length)
This curious regime corresponds to the onset of the edge-enhancement effects seen in Fig. 6.2, where the spins far from the wall have been completely dephased by the gradient pulses and their contribution to the signal disappears. Only spins within a distance l_g of the walls, where restricted diffusion has reduced the spin dephasing effect, contribute significantly to the echo signal [51]. Note that in this regime the phase distribution is not at all Gaussian. This problem in the case of parallel plane boundaries has been addressed in a paper by Stoller *et al.* [52], and again by de Swiet and Sen [50], who find, at long echo times

$$E(g,\tau) = c\frac{D_0^{1/3}}{\gamma^{1/3}g^{1/3}l_s} \exp\left(-a_1\gamma^{2/3}g^{2/3}D_0^{1/3}\tau\right)$$

$$= c\frac{l_g}{l_s}\exp\left(-a_1\left[\frac{l_D}{l_g}\right]^2\right) \tag{6.69}$$

where $a_1 = 1.0188$ and $c = 5.8841$. Note that the exponent, which does not depend on l_s, is expected to apply for all geometries, with only the prefactor c depending on the pore shape.[5]. Experiments in the localisation regime are hard to perform. The

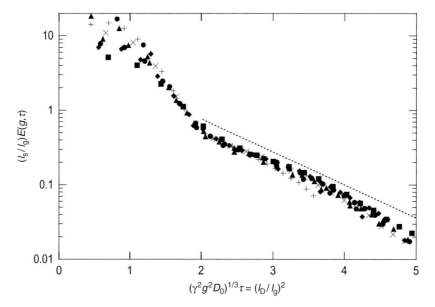

Fig. 6.16 Measured spin-echo amplitude, rescaled according to the predictions of eqn 6.69, as taken from Hürlimann *et al.* [51]. The dashed line corresponds to eqn 6.69.

prefactor l_g/l_s means that signal levels are small and the magnetic field gradients need to be aligned normal to the plane boundaries within a few degrees. However, in a carefully controlled experiment, Hürlimann *et al.* [51] have provided a nice verification of eqn 6.69, as shown in Fig. 6.16.

6.5.3 Alternative non-dimensional parameters

The echo-attenuation expressions for the free diffusion (eqn 6.65), motional narrowing (6.68), and localisation (6.69) regimes can be alternatively expressed in terms of two non-dimensional parameters (p, q) instead of via the three-parameter set (l_D, l_s, l_g). Grebenkov [53] has defined these non-dimensional parameters as[6]

---

[5]a_1 is the first zero of the derivative of the Airy function $Ai(x)$

[6]Note that q should not be confused with the symbol used for the PGSE wavevector.

$$p = \frac{D_0 \tau}{l_s^2}$$

$$= \frac{l_D^2}{l_s^2} \tag{6.70}$$

and

$$q = \gamma g \tau l_s$$

$$= \frac{l_D^2 l_s}{l_g^3} \tag{6.71}$$

whence, eqns 6.65, 6.68, and 6.69 become, respectively,

Free-diffusion regime: $p \ll 1$ **and** $pq^2 \ll 1$

$$E = \exp\left(-\tfrac{2}{3} q^2 p\right) \tag{6.72}$$

Motional-narrowing regime: $p \gg 1$ **and** $p/q \gg 1$
For plane parallel boundaries

$$E = \exp\left(-\tfrac{1}{60} q^2 p^{-1}\right) \tag{6.73}$$

Localisation regime: $p/q \ll 1$ **and** $pq^2 \gg 1$
For plane parallel boundaries at long echo times

$$E = c(p/q)^{1/3} \exp\left(-a_1 (q^2 p)^{1/3}\right) \tag{6.74}$$

The three regimes may be alternatively illustrated in the (q, p) plane [53] or in a plane defined by the ratios $(l_g/l_s, l_D/l_s)$ [51], as shown in Fig. 6.17.

6.6 Pulsed gradient spin-echo NMR for bounded molecules

Section 6.5 dealt with the way in which the confinement of diffusing molecules to a pore or cavity influences the echo attenuation under conditions of a steady magnetic field gradient. In the next chapter we will discover that, when pulsed magnetic field gradients are used, the echo attenuation manifests features akin to that of a diffraction experiment. In the present section, we look beyond restricted diffusion of molecules confined inside a cavity with hard boundaries, and consider environments where spin-bearing molecules or molecular segments exhibit 'softer' restrictions to their motion. For example, a diffusing polymer segment in a cross-linked gel will experience a restoring force due to gel elasticity. This will have the effect of limiting the distribution of translational displacements as Δ becomes large. Another rather unusual form of displacement limitation arises when molecules are confined to diffuse in a curvilinear coordinate system, examples being the motion of solvent molecules along the lamellae in a multi-domain lyotropic liquid crystal or the motion of a random coil polymer molecule in the 'tube' of topological constraints due to neighbouring polymers. In such systems the molecular motion in the laboratory frame exhibits a characteristic dependence on timescale, attenuating as Δ increases. This type of soft bounding might be termed 'dimensionally restricted' diffusion.

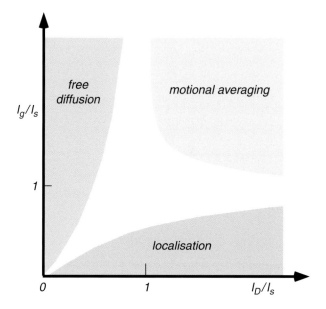

Fig. 6.17 Schematic illustration of the free-diffusion, motional-averaging, and localisation regimes. (Adapted from Hürlimann *et al.* [51].)

6.6.1   Diffusion in a harmonic potential

The problem of diffusion in the presence of an harmonic potential has been considered by Stejskal [54]. In this treatment, the restoring force is balanced by frictional damping due to the viscosity of the surrounding particles. If the ratio of the elastic force constant, k, and the friction coefficient, ζ, is termed β, then the elastic potential is written $V = \frac{1}{2}\beta\zeta r^2$, with associated force $-\beta\zeta\mathbf{r}$. Equating this with the friction force, $-\zeta\mathbf{v}$, gives $\mathbf{v} = \beta\mathbf{r}$. From eqn 2.53, and treating the diffusion as isotropic with scalar diffusion coefficient D, $P(\mathbf{r_0}|\mathbf{r}, t)$ obeys

$$\frac{\partial P}{\partial t} = \beta\nabla \cdot \mathbf{r}P + D\nabla^2 P \tag{6.75}$$

which has the solution

$$P(\mathbf{r_0}|\mathbf{r}, t) = \left[2\pi D\left(1 - e^{-2\beta t}\right)/\beta\right]^{-3/2}$$

$$\times \exp\left[-\frac{\beta\left(\mathbf{r} - \mathbf{r_0}e^{-\beta t}\right)^2}{2D\left(1 - e^{-2\beta t}\right)}\right] \tag{6.76}$$

The equilibrium density can be obtained by using the long-time limit identity of $P(\mathbf{r_0}|\mathbf{r}, t)$ and $\rho(\mathbf{r_0})$, thus

$$\rho(\mathbf{r_0}) = (2\pi D/\beta)^{-3/2} \exp\left(-\beta\mathbf{r_0}^2/2D\right) \tag{6.77}$$

Writing the force constant as $k = \beta\zeta$ and noting that the free particle diffusion coefficient is $k_B T/\zeta$, we can see immediately that eqn 6.79 is the 3-D version of that given in eqn 1.88.

The exponential decay of $\rho(\mathbf{r}_0)$ as the distance from the origin increases is an example of 'soft bounding'. Substitution of $P(\mathbf{r}_0|\mathbf{r},t)$ and $\rho(\mathbf{r}_0)$ into the narrow pulse PGSE expression, eqn 5.78, gives

$$E(\mathbf{q},\Delta) = \exp\left[-4\pi^2 q^2 D \left(1 - e^{-\beta\Delta}\right)/\beta\right] \tag{6.78}$$

The dependence of $E(\mathbf{q},\Delta)$ on $\mathbf{q}$ is Gaussian at all timescales. For $\Delta \ll \beta^{-1}$, the behaviour is the same as for unrestricted diffusion. For $\Delta \gg \beta^{-1}$ $E(\mathbf{q},\Delta)$ is independent of Δ and reduces to $|S(\mathbf{q})|^2$, where $S(\mathbf{q})$ is the 1-D Fourier transform of the equilibrium density $\rho(\mathbf{r}_0)$ given in eqn 6.77,

$$E(\mathbf{q},\infty) = \exp(-q^2 D/\beta) \tag{6.79}$$

The PGSE NMR method has been used to study harmonic-well-restricted diffusion for a polymer gel network [55]. The physics of diffusion in polymer gels has been discussed in detail by de Gennes [56]. Given a friction coefficient ϕ per unit volume and bulk modulus E_b, it can be shown that the elastic constant for longitudinal gel fluctuations of wavevector κ is $E_b\kappa^2$ while the cooperative diffusion coefficient, D_c, is $E_b\phi$. For such a system, β is given by [55] $E_b\kappa^2/F_\zeta$.

6.6.2 Diffusion and exchange between two sites

One commonly encountered morphology is one where the diffusional behaviour may be divided into subregions, with molecules confined to an inhomogeneous distribution of local diffusion rates on a short timescale, but where the ensemble of molecules experience all regions over a sufficiently long timescale, so exhibiting a common, averaged diffusion coefficient. The problem of calculating the PGSE NMR signal for molecules in exchange between two regions may be treated in the same manner as for relaxation [57, 58]. Again, in the short-time limit (slow exchange) the echo attenuation consists of a linear superposition from the sub-regions of weighting P_i and local diffusion coefficient D_i,

$$E(g,\Delta) = \exp(-\gamma^2\delta^2 g^2 \bar{D}\Delta) \tag{6.80}$$

Now the coupled differential equations are

$$\frac{dE_{ij}(g,t)}{dt} = -\frac{1}{\tau_j}E_{ij}(t) - \gamma^2\delta^2 g^2 D_j E_{ij}(g,t) + \sum_k E_{ik}(g,t)\frac{1}{\tau_k}p_{kj} \tag{6.81}$$

For a two-phase system [58]

$$\bar{E}(g,\Delta) = a_1 \exp(-\gamma^2\delta^2 g^2 D_1'\Delta) - a_2 \exp(-\gamma^2\delta^2 g^2 D_2'\Delta)) \tag{6.82}$$

where

$$D_1', D_2' = \frac{1}{2}[D_1 + D_2 + \frac{1}{\gamma^2\delta^2 g^2}(\frac{1}{\tau_1} + \frac{1}{\tau_2}) \tag{6.83}$$

$$\mp \left[\left(D_2 - D_1 + \frac{1}{\gamma^2\delta^2 g^2}(\frac{1}{\tau_2} - \frac{1}{\tau_1})\right)^2 + \frac{4}{\gamma^4\delta^4 g^4 \tau_1\tau_2}\right]^{1/2}] \tag{6.84}$$

and

$$a_1 = \frac{1}{D'_2 - D'_1}[D'_2 - D_{av}]$$

$$a_2 = \frac{1}{D'_2 - D'_1}[D'_1 - D_{(av)}] \tag{6.85}$$

where

$$D_{av} = P_1 D_1 + P_2 D_2 \tag{6.86}$$

In the long timescale limit these equations reduce, as required, to eqn 6.80 with $\bar{D} = D_{av}$.

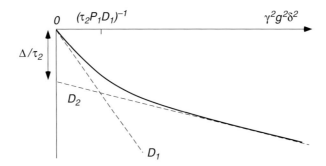

Fig. 6.18 Echo attenuation for two-region diffusion. (Adapted from J. Kärger *et al.* [58].)

One commonly encountered special case concerns the existence of small regions of high mobility ($P_1 \ll P_2$, $D_1 \gg D_2$) for which the echo attenuation simplifies to

$$E(g) = \exp(-\gamma^2 \delta^2 g^2 \left[D_2 + \frac{P_1 D_1}{\gamma^2 \delta^2 g^2 \tau_2 P_1 D_1 + 1} \right] \Delta) \tag{6.87}$$

The exponent term in square brackets represents the effective diffusion coefficient as measured by the PGSE experiment. The limiting slopes and intercepts of the echo-attenuation data allow the determination of D_2, $P_1 D_1$, and τ_2, as illustrated in Fig. 6.18. This approach has been used for describing diffusion in microporous crystallites where the subscripts 1 and 2 refer to the intercrystalline and intracrystalline spaces [59–61].

6.6.3   Dimensional restriction: randomly distributed pipes and sheets

Suppose we try to predict the result of a PGSE NMR experiment in which the molecules have anisotropic diffusion with cylindrical symmetry. In other words, there exists some coordinate frame (x', y', z') in which $D_{x'x'} = D_{y'y'} = D_\perp$, while $D_{z'z'} = D_\parallel$, as shown in Fig. 6.19.

Given that the echo attenuation for spins with a mean-square displacement $\langle Z^2 \rangle$ is $\exp(-\frac{1}{2}q^2\langle Z^2 \rangle)$, for spins in a region where the axis of symmetry is inclined at polar angle θ to the gradient direction, labelled by z [62]

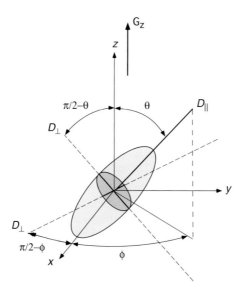

Fig. 6.19 Schematic showing the polar geometry of a randomly oriented locally anisotropic diffusion domain. Cylindrical symmetry is assumed. Two convenient directions in the $D_\perp$ plane are shown, one of which is normal to the z-axis, and the other inclined at $\pi/2 - \theta$.

$$\langle Z^2 \rangle = 2D_{\parallel} \Delta_r \cos^2 \theta + 2D_\perp \Delta_r \sin^2 \theta \qquad (6.88)$$

where the reduced time has been used in order to give the exact echo-attenuation factor. Hence the net $E(\mathbf{q}, \Delta)$ is the sum obtained by averaging element orientations over all angles, weighted by the sphere area element, $\sin \theta d\theta$. Hence [62]

$$E(\mathbf{q}, \Delta_r) = \int_0^\pi \exp\left[-q^2 \Delta_r \left(D_{\parallel} \cos^2 \theta + D_\perp \sin^2 \theta\right)\right] \sin \theta d\theta \bigg/ \int_0^\pi \sin \theta d\theta$$

$$= \exp(-q^2 D_\perp \Delta_r) \int_0^1 \exp\left[-q^2 \Delta_r \left(D_{\parallel} - D_\perp\right) x^2\right] dx$$

$$(6.89)$$

Equation 6.89 is very helpful. Suppose that we are dealing with the the problem of water diffusing between the lamellar sheets of a lyotropic liquid crystal. Then the symmetry is 'two-dimensional' or 'lamellar', and $D_\perp \gg D_{\parallel}$ and, assigning $D_\perp = D$, the echo attenuation varies as

$$E_{2D} = \exp(-q^2 D \Delta_r) \int_0^1 \exp\left[q^2 D \Delta_r x^2\right] dx \qquad (6.90)$$

If, on the other hand, the diffusing molecules are confined to narrow pipes, then the symmetry is 'one-dimensional' or 'capillary' and $D_\perp \ll D_{\parallel}$ with

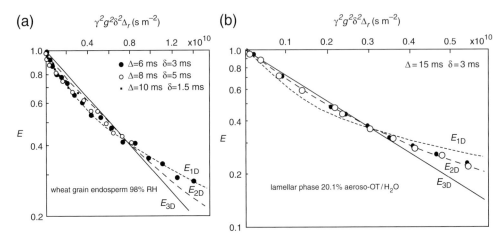

Fig. 6.20 $q^2\Delta_r$-dependence of water proton echo attenuation for (a) wheat grain endosperm tissue equilibrated at 90% relative humidity (RH). The theoretical curves labelled E_{1D}, E_{2D}, and E_{3D} refer to one-, two-, and 3-D diffusion (eqns 6.91, 6.90, and 5.87, respectively) and data obtained at different echo times Δ are consistent with E_{1D}. (Data taken from reference [62].) (b) 20.1% Aerosol OT/water. The data lie on a common curve characteristic of 2-D diffusion, consistent with each molecule residing in a single lamellar domain. Open and closed circles refer to samples after three days and three months' equilibration (Data taken from reference [63].)

$$E_{1D} = \exp(-q^2 D\Delta_r) \int_0^1 \exp\left[q^2 D\Delta_r (1 - x^2)\right] dx$$

$$= \int_0^1 \exp\left[-q^2 D\Delta_r x^2\right] dx \tag{6.91}$$

where in this case we assign $D_{\parallel} = D$.

Equations 6.90 and 6.91 may be contrasted with the echo attenuation behaviour for unrestricted diffusion, $E_{3D} = \exp(-q^2 \Delta_r D)$. The dependence of E_{1D}, E_{2D}, and E_{3D} on $q^2 \Delta_r = \gamma^2 g^2 \delta^2 \Delta_r$ gives a characteristic signature in the curvature of the $\log(E(q, \Delta_r))$ vs $\gamma^2 g^2 \delta^2 \Delta_r$. Figure 6.20 shows the echo-attenuation plot for water protons in the polydomain lamellar phase of Aerosol OT/water [63], along with the best fits using E_{1D}, E_{2D}, and E_{3D}, with the 2-D model representing the data well. Similar agreement using E_{2D} has been found for PGSE experiments on water in bilayers of the polycrystalline smectic liquid crystal, sodium 4-(1'-heptylnonyl)benzenesulphonate [64].

Of course, the echo-attenuation expressions for self-diffusion in a randomly oriented array of dimensionally restricted pores contain an approximation, namely that displacements in the restricted dimensions are ignored. More exact expressions, in which these displacements are accounted for, have been derived by Mitra and Sen [65] using the appropriate eigenmode expansions for the solution to the diffusion equation for liquid molecules in pipes of radius a and between parallel sheets of separation a. Their echo-attenuation relationships are

$$E_{2D} = \exp(-q^2 D\Delta) \int_0^1 \exp\left[q^2 \left(D\Delta - \langle r_\perp(\Delta)^2\rangle/2\right) x^2\right] dx \qquad (6.92)$$

with

$$\langle r_\perp(\Delta)^2\rangle/2 = \frac{a^2}{3}\left[1 - \exp(-3D\Delta/a^2)\right] \qquad (6.93)$$

while for an array of randomly oriented pipes,

$$E_{1D} = \exp(-q^2 D\Delta) \int_0^1 \exp\left[q^2 \left(D\Delta - \langle r_\perp(\Delta)^2\rangle/4\right)(1 - x^2)\right] dx \qquad (6.94)$$

with

$$\langle r_\perp(\Delta)^2\rangle/4 = \frac{a^2}{4}\left[1 - \exp(-4D\Delta/a^2)\right] \qquad (6.95)$$

D being, in each case, the free molecular self-diffusion coefficient. Note that Δ_r is now replaced by Δ, since the narrow gradient pulse condition is required to derive these equations.

6.6.4 Curvilinear diffusion

Motion without branches

Figure 6.21 shows an echo-attenuation plot [63] for water protons in a sample of Aerosol OT/water, a lamellar phase lyotropic liquid crystal system with random domain orientation. The PGSE experiment reveals an interesting behaviour as D is increased. At short times the data exhibit a dependence on g and Δ, as given by the 2-D poly-domain expression, E_{2D}, of eqn 6.90. At longer times the diffusion rate is apparently slower, and the curvature of $E(q)$ vs q^2 reduces, indicating that the diffusion behaviour is tending towards 3-D. For Δ sufficiently short that the water molecules reside in a single local domain with characteristic director orientation, the behaviour E_{2D} is expected. As Δ becomes sufficiently large for the molecules to move to new domains of differing orientation, their motion becomes more 3-D in character. The curious feature of this experiment is that the apparent diffusion coefficient reduces with time.

In the experiment described, we measure molecular displacements in the laboratory frame, but the diffusion, at a microscopic level, is occurring in a curvilinear coordinate system. A similar problem exists in considering the reptational diffusion of a polymer molecule in the curvilinear tube formed by impeding neighbours. It was shown by de Gennes [66] and Edwards [67] that such a motion leads to a dependence of laboratory-frame mean-square displacement on $t^{1/2}$ instead of the usual t-dependence of unrestricted diffusion. A detailed discussion of this problem is found elsewhere [68], but a simple picture can be gained as follows.

Consider molecular motion confined to one- or 2-D local elements. These elements have rms length λ, and are interconnected such that the diffusing particle migrates from one randomly oriented element to another, tracing out a random walk in the laboratory frame as illustrated in Fig. 6.22, with end-to-end distance R. A key factor in this motion is the lack of branch points and the consequent confinement of the particles to specific, albeit tortuous, paths. Suppose we define a total rms curvilinear

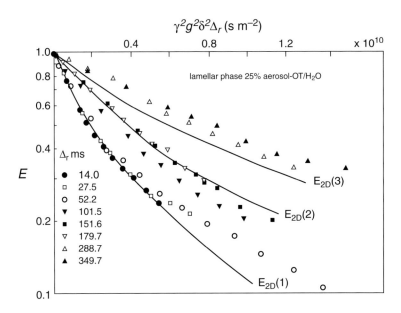

Fig. 6.21 Echo-attenuation plot for a 25% sample of Aerosol OT/water showing dependence of diffusive behaviour as Δ_r is varied over a wide range. For spins residing in one domain only, the data should be coincident. The theoretical curves, $E_{2D}(N)$ correspond to molecules diffusing successively in N domains. (Data taken from reference [63].)

path length Λ, comprising N elements with diffusion time t and local self-diffusion coefficient, D. For 1-D motion we have

$$\Lambda^2 = 2Dt = N^2\lambda^2 \tag{6.96}$$

In the laboratory frame

$$\overline{R^2} = N\lambda^2 = 2DN^{-1}t = \lambda(2Dt)^{1/2} \tag{6.97}$$

The case of two-dimension local diffusion is treated identically, but with a coefficient of 4 rather than 2 appearing in eqn 6.96. Given that we expect $\overline{R^2} = 6D_{\text{eff}}t$, we have for 1-D curvilinear diffusion

$$D_{\text{eff}} = \frac{1}{3}DN^{-1} \tag{6.98}$$

and for the 2-D case

$$D_{\text{eff}} = \frac{2}{3}DN^{-1} \tag{6.99}$$

The distribution of laboratory displacements in such curvilinear diffusion is Gaussian provided N is large, but the dependence of $\overline{R^2}$ as $t^{1/2}$ rather than t is distinctly non-Brownian. In the PGSE experiment the echo signal $E(q)$ will exhibit the usual Gaussian dependence on q but with an effective diffusion coefficient which decreases as $\Delta^{-1/2}$. The data shown in Fig. 6.21 correspond to an experiment in which N is small, so that the Gaussian behaviour is not reached as N increases. Expressions for

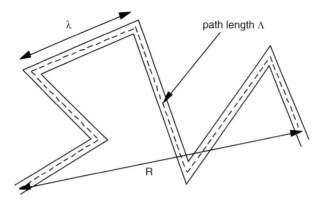

Fig. 6.22 One-dimensional curvilinear diffusion path without branching points, made up of local elements of rms length λ. The path length along the curvilinear path is Λ, while the direct end-to-end distance is R.

$E_{1D}(q)$ and $E_{2D}(q)$, appropriate where N is finite, are suggested in reference [63]. They consist of replacing D in both eqns 6.90 and 6.91 by DN^{-1}. In the limits $N = 1$ and $N \to \infty$, these equations reduce to the correct form. For finite N they do, in the relevant case of $E_{2D}(q)$, provide a good representation of the data shown in Fig. 6.21.

In the polymer problem, the curvilinear confinement exists only for the time, τ_d, taken for polymer to disengage from the tube [66, 68, 70, 71]. This means that the $\overline{R^2} \sim \Lambda^{1/2}$ behaviour can only be observed if Δ is less than τ_d and where q is sufficiently large for the PGSE experiment to detect molecular displacements smaller than the rms dimensions of the polymer, the so-called semi-local motion. In principle this is just a matter of making g or δ large enough. Because of T_2 relaxation, the magnitude of δ is constrained. Given the largest available polymers, the measurement of semi-local motion requires gradient strengths in excess of $10\,\mathrm{Tm^{-1}}$. Figure 6.23 shows an example of mean-squared displacements measured for entangled high molecular weight polymers in semi-dilute solution, at times $t > \tau_d$, where $Z^2 \sim t$, and $t < \tau_d$, where $Z^2 \sim t^{1/2}$. At even shorter NMR observation times, $t = \Delta$, the Rouse time, is reached, where a transition to $Z^2 \sim t^{1/4}$ is observed. By selecting different molar masses, the Rouse and reptation times can be 'tuned' to bring the relevant dynamical regime into the PGSE NMR 'window'.

Motion with branches

Tube disengagement is an example of branching in the paths available for the molecule. Suppose that instead of confinement to a fixed curvilinear path, the diffusing molecule has a choice of paths after moving N_0 elements for a mean-square distance R_0^2 in the lab frame over a time τ. This yields a timescale-independent, laboratory-frame self-diffusion coefficient, D_{eff}, of $R_0^2/6\tau$. Equation 6.97 then yields a particularly simple result,

$$
D_{\mathrm{eff}} = \begin{cases} \dfrac{D}{3N_0} & \text{for 1-D local motion,} \\[2ex] \dfrac{2D}{3N_0} & \text{for 2-D local motion.} \end{cases} \tag{6.100}
$$

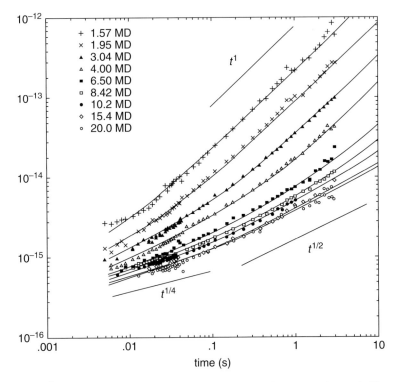

Fig. 6.23 $log(Z^2)$ vs $log(t)$, for nine different molar mass polystyrenes in semi-dilute solution. Asymptotic scaling exponents are shown in the straight-line tangents. Clear $t^{1/4}$ to $t^{1/2}$ and $t^{1/2}$ to t^1 transitions are apparent as the changing molar masses sweep the Rouse and tube disengagement times across the NMR window. From reference [69].

The 1-D expression leads to the well-known $D_{\text{eff}} \sim M^{-2}$ 'reptation law' for polymers, where M is the molar mass.[7] However eqn 6.100 is generally applicable to a wide class of problems, including the diffusion of small molecules in liquid crystals or emulsions. One specific limit of interest is the case $N = 1$, which corresponds to unidirectional motion between branch points. This is equivalent to the curvilinear diffusion problem, but where branching options exist on entering the next domain. Where the motion is locally 1-D or 2-D, D_{eff} is, respectively, $\frac{1}{3}D$ or $\frac{2}{3}D$. Examples of such behaviour have been found for diffusion in microemulsions [72].

6.6.5   Anisotropic diffusion in oriented liquid crystals

One of the problems associated with trying to measure anisotropic diffusion in liquid crystal environments concerns the effect of significant line-broadening caused by dipolar and quadrupolar interactions. When cubic symmetry is broken by the orientational order associated with a lyotropic or thermotropic liquid crystalline phase, these

---

[7]$M \sim N_0$. The 1-D diffusion coefficient D is inversely proportional to the total polymer friction along the curvilinear path, i.e. $D \sim N_0^{-1}$.

interactions are no longer averaged to zero, no matter how fast the probe molecule is diffusing in the mesophase environment. The problem is most severe in thermotropic liquid crystals where one wishes to measure the diffusion of the nematogen. However there does exist a solution, albeit one that requires great experimenter skill and care. What is required is to 'narrow the line' using multiple pulse RF methods, as outlined in Section 4.6.3. The art of the method is to find a way of inserting magnetic field gradient pulses into the mix, so as to produce the desired sensitivity to diffusion.

The first example of the application of PGSE methods to examine diffusion anisotropy in thermotropic liquid crystals was by the Ljubljana group of Blinc *et al.* [73, 74], who used multiple pulse line narrowing techniques to extend the transverse coherence of the liquid crystal spin magnetisation over a timescale sufficient to measure diffusion. These experiments required the gradient pulses to be only a few microseconds long and applied in specific 'windows' of the multiple pulse sequence [73, 75–77].

The 'Achilles heel' when measuring diffusion using multiple pulse line narrowing is this need to apply pulsed gradients in the narrow time windows (on the order of 10 microseconds) of the RF pulse sequence. In consequence, the available gradient strength is limited, and slow diffusion is difficult to access. One brute-force alternative is to use multiple-pulse decoupling in conjunction with a large, although static, magnetic field gradient, an approach used to measure slow diffusion in plastic crystals [78]. The presence of the gradient ensures that the RF pulses are slice-selective by default, a point discussed again in Section 12.2.6. However, there are deeper problems to be addressed when gradients are present during the RF pulse transmission. Clearly, different parts of the sample experience different magnetic field strengths, leading to a position-dependent resonance frequency offset across the sample space. This causes a corresponding variation of the decoupling efficiency, with resulting signal-to-noise ratio loss, and, in addition, the applied magnetic field gradient is scaled by the scaling factor of the decoupling sequence that reduces Zeeman spin operator terms such as chemical shift [79] (see Section 4.6.3). The latter effect means that the effective gradient experienced by spins becomes inhomogeneous across the sample, leading to an inhomogeneous attenuation of the signal by diffusion.

Dvinskikh and Furó [79, 80] have proposed an elegant means of reducing the influence of the variation of the decoupling effect by using a controlled slice selection, thus cancelling the signal from regions with large resonance offset so that the measurement is representative of that part of the sample where the Zeeman scaling factor is well defined. Their pulse sequence, based on MREV-8, is shown in Fig. 6.24(a). They choose to perform the slice selection before the PGSE sequence by applying a soft inversion pulse, this pulse being absent in every second scan so that when two subsequent FIDs are subtracted, signal from longitudinal magnetisation in the selected slice is summed, while that arising from outside the region is cancelled. By adjusting the width of the excited slice, they can control the Zeeman scaling factor to any required accuracy. Here the gradient amplitude and duration are kept constant, and the diffusion time Δ varied. Dvinskikh and Furó applied the method to ^{19}F spins in caesium perfluoroctanoate/D_2O (50 wt %), using the known chemical shifts of the ^{19}F spins to confirm that the Zeeman scaling factor agreed with the theoretical MREV-8 value of $\frac{\sqrt{2}}{3}$. Figure 6.24(b) shows the PGSE NMR echo attenuation (as a function of Δ)

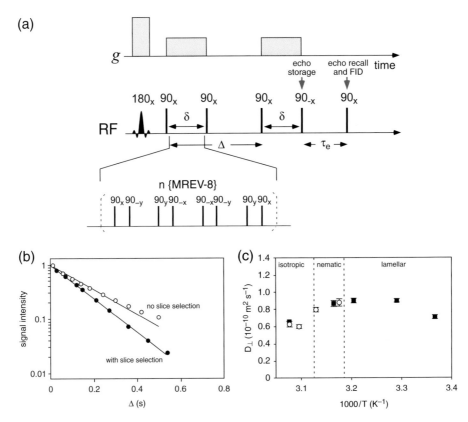

Fig. 6.24 (a) Pulse sequence for stimulated-echo PGSE NMR combined with MREV-8 in which a longitudinal eddy current delay (LED) period, τ_e is included. Slice selection is performed using a prior 180° inversion RF pulse, which is absent every second scan, so that subtraction of alternate signals leads to signal from the desired slice. (b) Application of the pulse sequence to ^{19}F spins in caesium perfluoroctanoate/D_2O, showing the echo attenuation with and without slice selection. (c) Measured $D_\perp$ for the perfluoroctanoate across the phase diagram. (Adapted from Dvinskikh and Furó [79].).

for measurements obtained without and with slice selection. In the latter case, a more uniform gradient results, consistent with the known Zeeman scaling factor.

6.6.6 Diffusion in fractal geometries

Natural structures often exhibit fractal character [81–85] in which self-similarity and non-integer dimensionality play a role. A nice example of fractal behaviour is found in the geography of coastlines [82]. The appearance of a coastline is similarly indented and tortuous, whether viewed from an orbiting satellite or from a distance of a few metres. This self-similarity (sometimes called dilation symmetry) is consistent with a shape that is scale independent. One of the consequences of this is that the total coastline length depends on the smallest step, ϵ, used in its measurement. If the total number of ϵ-intervals in the measurement is $N(\epsilon)$ and the total length is $L(\epsilon)$ then

$$N(\epsilon) \sim F\epsilon^{-d} \tag{6.101}$$

and

$$L(\epsilon) \sim F\epsilon^{1-d} \tag{6.102}$$

where d is called the similarity dimension. A non-fractal coastline would have $d = 1$, whereas fractal behaviour is exhibited when $d > 1$.

Porous materials exhibit analogous behaviour in the volume and surface organisation. Here d tells us the dependence of the 'measure' on the length dimension. In Euclidean behaviour we expect that the total pore volume within a radius R would vary as R^3. 'Volume fractal' behaviour is typified by $V \sim R^d$, where $1 \leq d \leq 3$. 'Surface fractal' behaviour is typified by $S \sim R^d$, where $1 \leq d \leq 2$.

The Brownian random walk provides a very good example of fractal behaviour, a wandering plane-filling 'coastline' in which $d = 2$. Consider a particular walk of end-to-end rms distance r. As ϵ decreases $N \sim F\epsilon^{-d}$. F can be determined as r^d from the result $N = 1$ when $\epsilon = r$. Thus $r \sim N^{1/d}$ and so yields the familiar random walk result, $r \sim N^{1/2}$, when $d = 2$. In fact this same dimensionality applied when the random walk is performed in 3-D space. Note that the fractal dimension associated with the random walk is different to that describing, for example, volume fractal behaviour, and is assigned the symbol d_w. An interesting example is the self-avoiding random 'walk' that defines the random coil polymer conformation in the solution phase [40], for which the chain end-to-end distance scales as $r \sim N^{3/5}$, giving $d = 5/3$.

Self-diffusion is very sensitive to the dimensional character of the path taken via the dependence of rms distance on time, this parameter being directly proportional to N. The fractal character may be defined by the relation [86–88]

$$\overline{r^2} = 6Dt^{2/(2+\theta)} \tag{6.103}$$

in which θ determines the effective dimensionality of the random walk, reducing to zero in the case of Brownian motion. A timescale-dependent self-diffusion experiment can be used to measure θ, and this quantity may be related to the volume fractal dimension [86–88], an effect that can be exploited in the PGSE experiment. From the relation $E = \exp(-\frac{1}{2}\gamma^2 g^2 \delta^2 \overline{Z^2})$, the fractal dimension enters [37, 89–91] through the term $\overline{Z^2}$ as $\Delta^{2/(2+\theta)}$, giving

$$E(q, \Delta) = \exp(-\gamma^2 g^2 \delta^2 D^{2/(2+\theta)}) \tag{6.104}$$

Note that the fractal dimension relates to the migration space of the spins. Where the material exhibits fractal surfaces but the molecules diffuse in a non-fractal (Euclidean) volume, the PGSE experiment exhibits the usual Brownian behaviour. In such an system the fractal behaviour is only apparent when the experimental parameter is surface-dependent, an example being the relaxation of spins due to the existence of surface magnetisation 'sinks' [92–94].

6.7 Frequency-domain measurements and restricted motion

In Chapter 5, the concept of frequency-domain modulated gradient NMR measurement of translational motion was introduced. Using a periodic effective gradient waveform, the induced time-dependent signal attenuation, $\exp(-\beta(t))$, can be related to the spectrum, $\underline{\underline{D}}(\omega)$, of the velocity autocorrelation function, $\langle \mathbf{u}(\tau)\mathbf{u}(0)\rangle$, where $\mathbf{u}$ has zero time average, such that the total velocity is $\mathbf{v} = \mathbf{u} + \langle \mathbf{v}\rangle$. The mean flow, $\langle \mathbf{v}\rangle$, introduces a time-dependent phase shift, $\exp(-i\alpha(t))$. These relationships are outlined in detail in section 5.7.2.

The question of interest in the present context is how the effect of restricted diffusion imprints the velocity autocorrelation spectrum and may be thereby observed in the echo attenuation. The particular details of the gradient pulse train used need not concern us here, although the use of a CPMG train under steady or pulsed gradient conditions is the most convenient to use, leading to the simple result $\beta(t) = \frac{1}{2}NT\gamma^2 g^2 T^2 D_{zz}(2\pi/T)$, where $t = NT$ and z is the gradient axis. The challenge is therefore to find the appropriate velocity autocorrelation spectrum, $D_{zz}(\omega)$.

In seeking $D_{zz}(\omega)$, it is helpful to note two important relations that follow from eqn 1.58. First

$$D_{zz}(\omega = 0) = \lim_{t\to\infty}\int_0^t < u_z(\tau)u_z(0) > d\tau.\tag{6.105}$$

and

$$\langle u_z(t)u_z(0)\rangle = \frac{1}{2}\frac{\partial^2\langle Z_u(t)^2\rangle}{\partial t^2}\tag{6.106}$$

the total displacement, $Z(t)$, being composed of $Z_u(t) + \langle v_z\rangle t$. Where $Z_u(t)$ has zero time average and zero ensemble average, it is follows that[8]

$$\begin{aligned}\langle v_z(t)v_z(0)\rangle &= \frac{1}{2}\frac{\partial^2\langle Z(t)^2\rangle}{\partial t^2}\\ &= \frac{1}{2}\frac{\partial^2\langle Z_u(t)^2\rangle}{\partial t^2} + \langle v_z\rangle^2\\ &= \langle u_z(t)u_z(0)\rangle + \langle v_z\rangle^2\end{aligned}\tag{6.107}$$

It is clear from eqn 6.107 that any estimation of the autocorrelation function of the mean-zero part of the velocity from the total time-dependent displacement, $Z(t)$, can be easily made by subtracting the flow contribution, as implied by the definition of $\beta(t)$ in eqn 5.128. In handling restricted diffusion, however, we only need focus on the stochastic contribution to displacement, $Z_u(t)$.

A knowledge of $\langle Z_u(t)^2\rangle$ therefore provides a starting point for calculating $\langle u_z(t)u_z(0)\rangle$, and hence its Fourier spectrum $D(\omega)$. Another route to $D(\omega)$ is from the initial standpoint of the time-dependent dispersion or diffusion coefficient $D(t)$ as defined by eqn 2.56, since

$$\langle u_z(t)u_z(0)\rangle = \frac{\partial}{\partial t}D(t)\tag{6.108}$$

This particular route turns out to be very convenient from a phenomenological standpoint, since one may often have a priori knowledge of long- or short-time limiting values of $D(t)$, as well as a knowledge of the characteristic times at which crossovers in diffusive behaviour occur, for example those times sufficient for molecules to diffuse known length scales.

---

[8]For the stationary ensemble $\langle Z(t)^2\rangle = \langle Z_u(t)^2\rangle + \langle v_z\rangle^2 t^2$

0.7.1   Spectral densities for free and restricted diffusion

Before considering restricted diffusion, it is helpful to examine the consequences of eqns 6.105 and 6.106 for free diffusion. The integration of eqn 6.105 requires a description of the mean-squared displacement at all times, including the ballistic regime of molecular collisions. From eqn 1.73, and denoting $t_\zeta = \gamma$, we may write

$$
\begin{aligned}
< u_z(t)u_z(0) > &= \frac{1}{2}\frac{\partial^2 \langle Z_u(t)^2\rangle}{\partial t^2} \\
&= \frac{k_B T}{m}\exp(-\gamma t) \\
&= D_0\gamma\exp(-\gamma t)
\end{aligned}
\tag{6.109}
$$

The diffusion spectrum is the Lorentzian

$$
D_{zz}(\omega) = D_0\frac{1}{1+\omega^2/\gamma^2}
\tag{6.110}
$$

For free diffusion, the diffusion coefficient is a constant at all times in excess of γ^{-1}, and hence at all frequencies much lower than γ. The required result $D_{zz}(0) = D_0$ follows directly from eqns 6.105 and 6.110.

Phenomenological model for restricted diffusion

Restriction to diffusion leads to a reduction in the zero-frequency part of the $D_{zz}(\omega)$ spectrum. Given that the ballistic contribution to $< u_z(t)u_z(0) >$ is always present on the timescale of molecular collisions, the effect of boundary restrictions is to introduce velocity anti-correlation, a negative contribution to $< u_z(t)u_z(0) >$. The physical reason for anti-correlation is obvious. Collisions introduce velocity reversals. In consequence, the schematic form of the VACF where molecules are confined in a pore of size a will be via introduction of an additional negative term, such that

$$
\langle u_z(t)u_z(0)\rangle = D_0\gamma\exp(-\gamma t) - D_0\tau_c^{-1}\exp(-t/\tau_c)
\tag{6.111}
$$

where τ_c is a time comparable with that required to traverse the pore, namely $\tau_c \sim a^2/D_0$. The corresponding spectrum is

$$
D_{zz}(\omega) = D_0\frac{1}{1+\omega^2/\gamma^2} - D_0\frac{1}{1+\omega^2\tau_c^2}
\tag{6.112}
$$

For an interconnected porous medium where molecules experience wall collisions within pores, but on a longer timescale, experience an effective tortuosity-reduced diffusion coefficient $D_{\text{eff}}(\infty)$, we may write

$$
\langle u_z(t)u_z(0)\rangle = D_0\gamma\exp(-\gamma t) - (D_0 - D_{\text{eff}}(\infty))\tau_c^{-1}\exp(-t/\tau_c)
\tag{6.113}
$$

and in consequence

$$
D_{zz}(\omega) = D_0\frac{1}{1+\omega^2/\gamma^2} - (D_0 - D_{\text{eff}}(\infty))\frac{1}{1+\omega^2\tau_c^2}
\tag{6.114}
$$

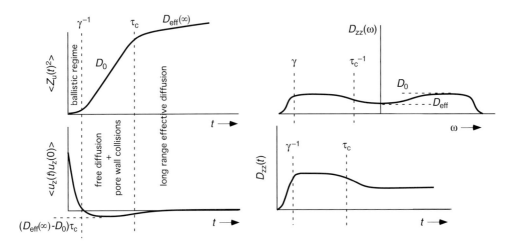

Fig. 6.25 Schematic of Z_u^2 and $\langle u_z(t)u_z(0)\rangle$ vs t, along with the associated diffusion spectrum $D_{zz}(\omega)$ and time-dependent diffusion coefficient $D_{zz}(t)$, in the case of an interconnected porous medium. For an enclosing pore, $D_{\text{eff}}(\infty) = 0$. Note that this simple picture applies only when $\tau_c \gg \gamma^{-1}$.

and, by eqn 2.56,

$$D_{zz}(t) = D_0 \left(1 - \exp(-\gamma t)\right) - (D_0 - D_{\text{eff}}(\infty)) \left(1 - \exp(-t/\tau_c)\right) \tag{6.115}$$

This picture is illustrated schematically in Fig. 6.25. Note that the size of the negative term in the velocity autocorrelation function is much smaller than that associated with ballistic collisions, by the ratio γ^{-1}/τ_c. However, because of the integration to obtain $D_{zz}(0)$ over the correlation time associated with wall collision effects, its effect is significant.

Plane parallel boundaries
In the case of plane parallel boundaries separated by a, where diffusion is measured normal to the walls, there exists an exact expression [95] for the velocity autocorrelation function. In the limit $\tau_c \gg \gamma^{-1}$,

$$\langle u_z(t)u_z(0)\rangle = D_0\gamma \exp(-\gamma t) - \frac{4D_0^2}{a^2} \sum_{k=-\infty}^{\infty} \exp(-(2k+1)^2 \pi^2 D_0 t/a^2) \tag{6.116}$$

The zero-frequency (long-time) diffusion coefficient is indeed zero because of the identity $\sum_{k=-\infty}^{\infty} \pi^{-2}(2k+1)^{-2} = \pi^2/4$.

Undulating lamellae
Diffusion normal to plane parallel boundaries has been studied using FDMG NMR [96] in the case of a lamellar phase lyotropic liquid crystal, in the limit $t \gg \tau_c = d^2/2D_0$ where d is the spacing between the layers. Water molecules diffusing in the interlamellar layer will, for times in excess of τ_c, be confined to 2-D diffusion, $D_\perp$. However

where undulations in the lamellae are present, such transverse diffusion converts $\perp$ displacements into displacements normal to the bilayer. Furthermore, undulation fluctuations will carry water molecules, providing an additional mechanism for movement in the direction of the layer normal. In this study, the diffusion spectrum measured in this normal direction gave access to the characteristic bilayer undulation rates and amplitudes.

Multilamellar vesicles

One of the particular advantages of the FDMG NMR method, especially using the Carr–Purcell train variant, is the opportunity to refocus all coherent phase shifts due to flow. This has particular application in the case of rheo-NMR, where a non-uniform flow field has the effect of introducing phase-spreading and echo train attenuation. Hence the FDMG NMR method has the potential to remove these velocity shear effects, revealing decay due to diffusion alone. One example of the application of this approach is in the measurement of the size of multi-lamellar onion vesicles (MLVs) under shear.

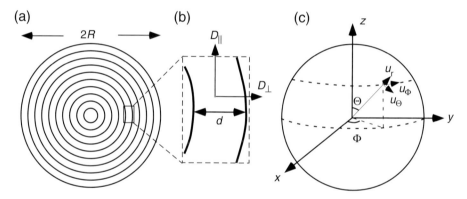

Fig. 6.26 A multilamellar vesicle. (a) The interlayer spacing value $d \sim 10^{-8}$m leads to a time, τ_c, for diffusion perpendicular to the membranes, $\sim 10^{-8}$ s, which is unobservable by FDMG NMR. (b) The much larger onion size $R \sim 10^{-6}$ m leads diffusion parallel to the membranes with decay time, $\tau_d \sim 10^{-3}$ s. The gradient is assumed to be applied along the z-direction, as in (c).

MLVs are formed from certain lyotropic lamellar phases under shear. Providing the shear rate is sufficiently high, spherical vesicles may be formed by shear-induced buckling of the membranes [97] and onion formation has been confirmed experimentally using various techniques and chemical systems [97–99]. Onion size estimation is based on measuring the restricted diffusion of inter-lamellar water molecules using the FDMG NMR sequence [100]. Figure 6.26 shows the coordinate system for the MLV onion, with the multi-bilayer structure, a chosen shell of radius r in which a local section of water layer is shown. Diffusion may be described respectively perpendicular and parallel to the bilayer normals as $D_\perp(t)$ and $D_\parallel(t)$. In the NMR experiment the magnetic field gradient used to encode for motion is applied along some pre-determined z-axis. For diffusion along z, the displacements are azimuth-independent and given by

$$D_{zz}(t, \Theta) = D_{\perp}(t) \cos^2 \Theta(t) + D_{\parallel}(t) \sin^2 \Theta(t) \tag{6.117}$$

The velocity autocorrelation function follows from proposing a time-dependent diffusion coefficient in accordance with eqn 2.56. At the shortest time ($t \ll \tau_c = d^2/2D_0$), few water molecules will have a chance to collide with the bilayer, and so $D_{\parallel}(t)$ and $D_{\perp}(t)$ reduce to the free diffusion coefficient D_0 of the solvent. At longer times, molecules encounter the bilayers in their displacements perpendicular to the membranes and to encompass this phenomenologically, we may write

$$D_{\perp}(t, \Theta) = D_{perm} + (D_0 - D_{perm}) \exp(-t/\tau_c) \tag{6.118}$$

where D_{perm} is the diffusion coefficient corresponding to permeation of the water molecules through the various defects present on the membranes. While the limits of eqn 6.118 are correct, details of the crossover are approximate. Nonetheless the essential physics is represented in terms of a mathematical function that is easily handled in subsequent analytic manipulations. This simple device applies in the higher timescale hierarchies as well.

Of course, one needs to sum contributions from all (Θ, Φ) segments in each of the shells of radius r. Provided that diffusion distances are sufficiently short that a unique angle Θ can be ascribed to each water molecule, we may write:

$$
\begin{aligned}
\langle D_{zz}(t) \rangle &= \frac{\int_0^{\pi} D_{zz}(t, \Theta) P(\Theta) d\Theta}{\int_0^{\pi} P(\Theta) d\Theta} \\
&= \frac{\int_0^{\pi} \left(D_{\perp}(t) \cos^2 \Theta(t) + D_{\parallel}(t) \sin^2 \Theta(t) \right) \sin \Theta d\Theta}{\int_0^{\pi} \sin \Theta d\Theta} \\
&= \frac{1}{3} \left[D_{perm} + (D_0 - D_{perm}) \exp(-t/\tau_c) + 2D_0 \right]
\end{aligned}
\tag{6.119}
$$

where $P(\Theta)$ is the polar angle distribution.

Allowing for longer times in which water molecules explore differing Θ coordinates, the expression becomes more complex. However, a simpler approach is possible [100], again using limiting-case expressions. For $t \gg \tau_d(r) = r^2/2D_0$, $D_{zz}(t)$ must reach the limit D_{perm}, since only permeation through the bilayers can permit diffusion on a greater distance than r. Using the same phenomenological approach

$$\langle D_{zz}(t) \rangle = D_{perm} + \frac{1}{3}(D_0 - D_{perm})(\exp(-t/\tau_c) + 2\exp(-t/\tau_d(r))) \tag{6.120}$$

Finally, the integration over shell size $0 < r < R$ must be performed where R is the onion size. This demands some allowance for permeation effects between layers. An analysis of the relative timescales for permeation and diffusion around the entire spherical shell suggests that it is reasonable to assume a single correlation time, $\tau_d(r)$, for each shell over the time to diffuse the dimensions of the onion. Hence one may average NMR signals acquired from layers with different r values rather than average $\tau_d(r)$ itself across the distribution of layers. Based on a calculation of the diffusion spectrum for a single shell and making radial averaging the final step, a closed-form expression results [100].

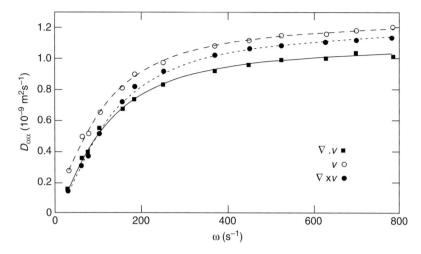

Fig. 6.27 Diffusion spectrum of an onion phase: diffusion is measured along the flow-gradient ($\nabla \mathbf{v}$), flow ($\mathbf{v}$), and vorticity ($\nabla \times \mathbf{v}$) directions. The solid lines are fits to the analytic $D(\omega)$, and give independent D_{perm} ($\sim 10^{-10} m^2 s^{-1}$) and onion radius values ($\sim 7\mu m$) for each direction. (Adapted from reference [100].)

6.7.2 Spectral analysis of flow

Frequency-domain modulated gradient NMR methods based on CPMG echo trains are well suited to the study of the dispersion spectrum for complex flow. In particular, for Péclet numbers in excess of unity, the displacements are generally larger that experienced in purely diffusive motion, and this proves particularly useful given the tendency of the CPMG sequence to reduce sensitivity to diffusive effects. Furthermore, the frequencies that can be accessed by FDMG NMR (from hertz to kilohertz) may be well tailored to characteristic flow times. One example of the application of these methods has been in the study of transverse 'meandering' motion for flow through a beadpack [101]. Figure 6.28 shows a schematic representation of the phenomenon.

Applying the magnetic field gradients in the direction transverse to the mean flow, for water flowing in a monodisperse latex bead pack (diameter 50 microns) and at Péclet numbers ranging from 10 to 5000, these authors observe spectral peaks at a frequency corresponding to the inverse time for flow around a bead, an effect they attribute to coherent meandering flow around the bead. These spectra agree very well with those calculated from lattice Boltzmann simulations [102] in which negative velocity autocorrelation function transients are seen. Similar agreement was found using a range of bead packs comprising different bead diameters, lending credence to the interpretation.

Representing the transverse component of meandering flow by a simple oscillatory motion with frequency $2/\tau_v$, where the characteristic flow time $\tau_v = d/v$, d being the bead diameter and v the local fluid velocity of mean $\langle v \rangle$, a simple velocity autocorrelation function was derived in which both meandering flow and dispersion were accommodated,

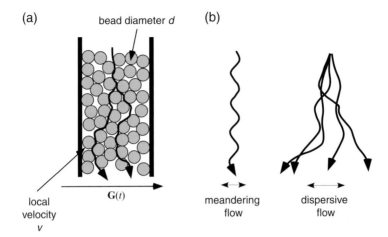

Fig. 6.28 Meandering flow superposed on dispersion for water flowing through a bead pack. The time-dependent gradient is applied transverse to the mean flow direction. (Adapted from reference [101].)

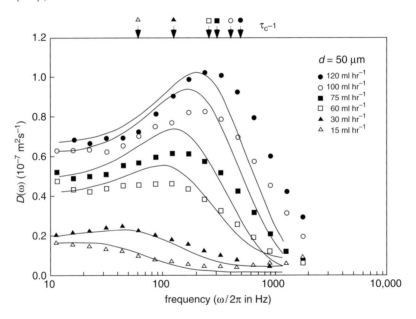

Fig. 6.29 The effective transverse dispersion coefficients, $D(\omega)$, measured at each frequency $f = 1/T$ for the FDMG NMR sequence of Fig. 5.32(b), over the range of flow rates as shown, for a 50-micron bead pack. The theoretical fits are obtained using the velocity autocorrelation functions of Maier *et al.* [102]. (Adapted from reference [101].)

$$\langle u(t)u(0)\rangle = D_0\gamma \exp(-t\gamma) + \int_0^\infty (bv^2)P(v)\exp(-t/\tau_v)dv$$

$$+\frac{1}{2}\int_0^\infty a^2v^2 P(v)\cos(2t/\tau_v)dv \qquad (6.121)$$

a and b being coefficients of order unity, and $P(v)$ the local velocity distribution, taken to be Gaussian with standard deviation σ. This leads to a diffusion spectrum

$$D(\omega) = D_0 + b \int_0^\infty P(v) \frac{dv}{1 + \omega^2 \tau_v^2} dv +$$
$$+ \frac{1}{4} a^2 (2\pi\sigma^2)^{-1/2} (d/2)^3 \exp(-(\omega d/2 - \langle v \rangle)^2 / 2\sigma^2) \qquad (6.122)$$

This simple model provides a good representation of the essential features of Fig. 6.29 and, in particular, the presence of a peak at a frequency corresponding to the inverse time for flow around a bead. The use of such frequency-domain methods for the analysis of translational motion clearly has considerable potential for identifying coherent flow phenomena in otherwise complex, semi-random flow structures.

References

[1] D. E. Woessner. NMR spin-echo self-diffusion measurements on fluids undergoing restricted diffusion. *J. Phys. Chem.*, 67:1365, 1963.

[2] D. S. Grebenkov. Use, misuse, and abuse of apparent diffusion coefficients. *Concepts in Magnetic Resonance*, 36A:24, 2010.

[3] J. Bear. *Dynamic of Fluids in Porous Media*. Elsevier, New York, 1972.

[4] P. N. Sen, L. M. Schwartz, P. P. Mitra, and B. I. Halperin. Surface relaxation and the long time diffusion coefficient in porous media: periodic geometries. *Phys. Rev. B*, 49:215, 1994.

[5] P-Z. Wong. *AIP conference proceedings 154*, chapter Physics and Chemistry of Porous Media II. AIP, New York, 1990.

[6] J. W. Haus and K. W. Kehr. Diffusion in regular and disordered lattices. *Phys. Rep.*, 150:263, 1987.

[7] T. de Swiet and P. N. Sen. Time dependent diffusion coefficient in a disordered medium. *J. Chem. Phys.*, 104:206, 1996.

[8] R. W. Mair, P. N. Sen, M. D. Hürlimann, S. Patz, D. G. Cory, and R. L. Walsworth. The narrow pulse approximation and long length scale determination in xenon gas diffusion NMR studies of model porous media. *J. Magn. Reson.*, 156:202, 2002.

[9] P.T. Callaghan, A. Coy, L. C. Forde, and C. J. Rofe. Diffusive relaxation and edge enhancement in NMR microscopy. *J. Magn. Reson. A*, 101:347, 1993.

[10] P.P Mitra, P.N. Sen, L.M. Schwartz, and P. Le Doussal. Diffusion propagator as a probe of the structure of porous media. *Phys. Rev. Lett.*, 68:3555, 1992.

[11] P. P. Mitra, P. N. Sen, and L. M. Schwartz. Short time behaviour of the diffusion coefficient as a geometrical probe of porous media. *Phys. Rev. B*, 47:8565, 1993.

[12] George A. Baker Jr and Peter Graves-Morris. *Pade Approximants*. Cambridge University Press, New York, 1996.

[13] L. L. Latour, P. P. Mitra, R. L. Kleinberg, and C. H. Sotak. Time-dependent diffusion coefficient of fluids in porous media as a probe of surface-to-volume ratio. *J. Magn. Reson. A*, 101:342, 1993.

[14] L.J. Zielinski and P. N. Sen. Effects of finite-width pulses in the pulsed-field gradient measurement of the diffusion coefficient in connected porous media. *J. Magn. Reson.*, 165:153, 2003.

[15] J. R. Zimmerman and W. E. Brittin. Nuclear magnetic resonance studies in multiple phase systems: Lifetime of a water molecule in an adsorbing phase on silica gel. *J. Phys. Chem.*, 61:1328, 1957.

[16] W. Feller. *Probability Theory and Its Application.* Wiley, New York, 1950.

[17] H. Wennerstrom. Nuclear magnetic relaxation induced by chemical exchange. *Mol. Phys.*, 24:69, 1972.

[18] Paul A. Bottomley, Thomas H. Foster, Raymond E. Argersinger, and Leah M. Pfeifer. A review of normal tissue hydrogen NMR relaxation times and relaxation mechanisms from 1–100 MHz: Dependence on tissue type, NMR frequency, temperature, species, excision, and age. *Med. Phys.*, 11:425, 1984.

[19] E. E. Burnell, M. E. Clark, J. A. M. Hinke, and N. R. Chapman. Water in barnacle muscle. III. NMR studies of fresh fibers and membrane-damaged fibers equilibrated with selected solutes. *Biophys. J.*, 33:1, 1981.

[20] P. S. Belton, R. R. Jackson, and K. J. Packer. Pulsed NMR studies of water in striated muscle: I. Transverse nuclear spin relaxation times and freezing effects. *Biochim. Biophys. Acta.*, 286:16, 1972.

[21] B. M. Fung and T. W. McGaughy. The state of water in muscle as studied by pulsed NMR. *Biochim. Biophys. Acta.*, 343:663, 1974.

[22] S. N. Rustgi, H. Peemoeller, R. T. Thompson, D. W. Kydon, and M. M. Pintar. A study of molecular dynamics and freezing phase transition in tissues by proton spin relaxation. *Biophys. J.*, 22:439, 1978.

[23] G. B. Kolata. Water structure and ion binding: A role in cell physiology? *Science*, 192:1220, 1976.

[24] F. Franks. *Water: a Comprehensive Treatise, Vols 1–4.* Plenum, New York, 1972.

[25] W. Derbyshire. Heterogeneous systems. *Chem. Soc., Spec. Period. Rep. Nuclear Magnetic Resonance*, 5:264, 1978.

[26] W. Derbyshire. Heterogeneous systems. *Chem. Soc., Spec. Period. Rep. Nuclear Magnetic Resonance*, 7:193, 1978.

[27] K. J. Packer, D. A. T. Dick, and D. R. Wilke. The dynamics of water in heterogeneous systems. *Phil. Trans. R. Soc. London Ser. B*, 278:59, 1977.

[28] R. R. Knipsel, R. T. Thomson, and M. M. Pintar. Dispersion of proton spin-lattice relaxation in tissues. *J. Magn. Reson.*, 14:44, 1974.

[29] Seymour H. Koenig and Walter E. Schillinger. Nuclear magnetic relaxation dispersion in protein solutions: I. Apotransferrin. *J. Biol. Chem.*, 244:3283, 1969.

[30] O. K. Daszkiewicz, J. W. Hennel, B. Lubas, and T. W. Szczepkowski. Proton magnetic relaxation and protein hydration. *Nature*, 200:1006, 1963.

[31] P. J. Lillford, A. M. Clark, and D. V. Jones. Water in polymers. In *ACS Symposium Series*, volume 127, page 177, 1980.

[32] G. Held, F. Noack, V. Pollack, and B. Melton. Proton spin relaxation and mobility of water in muscle tissue. *Z. Naturforsch.*, 28c:59, 1973.

[33] J. M. Escayne, D. Canet, and J. Robert. Frequency dependence of water proton longitudinal nuclear magnetic relaxation times in mouse tissue at 20 C. *Biochim. Biophys. Acta.*, 721:305, 1982.

[34] S. H. Koenig, R. D. Brown, D. Adams, D. Emerson, and C. G. Harrison. Magnetic field dependence of 1/T1 of protons in tissue. *Invest. Radiol.*, 19:76, 1984.

[35] K. R. Brownstein and C. E. Tarr. Importance of classical diffusion in NMR studies of water in biological cells. *Phys. Rev. A*, 19:2446, 1979.

[36] Y-Q Song, L. Zielinski, and S. Ryu. Two-dimensional NMR of diffusion systems. *Phys. Rev. Lett.*, 100:248002, 2008.

[37] J. R. Banavar and L. M. Schwartz. *Molecular Dynamics in Restricted Geometries*, chapter Probing porous media with nuclear magnetic resonance, page 273. Wiley, New York, 1989.

[38] K.P. Whittall and A.L. Mackay. Quantitative interpretation of NMR relaxation data. *J. Magn. Reson.*, 84:134, 1989.

[39] M. Lipsicas, J.R. Banavar, and J. Willemsen. Surface relaxation and pore sizes in rocks—a nuclear magnetic resonance analysis. *Appl. Phys. Lett.*, 48:1544, 1986.

[40] P. G. de Gennes. Transfert d'excitation dans un milieu aléatoire. *C. R. Acad. Sci.*, 295:1061, 1982.

[41] L. M. Burcaw, M. W. Hunter, and P. T. Callaghan. Propagator-resolved 2D exchange in porous media in the inhomogeneous magnetic field. *J. Magn. Reson.*, 205:209, 2010.

[42] A. Abragam. *Principles of Nuclear Magnetism.* Clarendon Press, Oxford, 1961.

[43] K. J. Packer. The effects of diffusion through locally inhomogeneous magnetic fields on transverse nuclear spin relaxation in heterogeneous systems. proton transverse relaxation in striated muscle tissue. *J. Magn. Reson.*, 9:438, 1973.

[44] C. F. Hazelwood, D. C. Chang, B. L. Nichols, and D. E. Woessner. Propagator-resolved 2D exchange in porous media in the inhomogeneous magnetic field. *Biophys. J.*, 14:583, 1974.

[45] G. R. Coates, L. Xiao, and M. G. Prammer. *NMR Logging. Principles and Applications.* Halliburton Energy Services, Houston, 1999.

[46] Y-Q Song, S. Ryo, and P.N. Sen. Determining multiple length scales in rocks. *Nature*, 406:178, 2000.

[47] H. Cho and Y-Q Song. NMR measurement of the magnetic field correlation function in porous media. *Phys. Rev. Lett.*, 100:025501, 2008.

[48] B. Robertson. Spin-echo decay of spins diffusing in a bounded region. *Phys. Rev.*, 151:273, 1966.

[49] C. H. Neuman. Spin-echo of spins diffusing in a bounded medium. *J. Chem. Phys.*, 60:4508, 1974.

[50] T. M. de Swiet and P. N. Sen. Decay of nuclear magnetization by bounded diffusion in a constant gradient. *J. Chem. Phys.*, 100:3597, 1994.

[51] M. D. Hürlimann, K. G. Helmer, T. M. de Swiet, and P. N. Sen. Spin echoes in a constant gradient in the presence of simple restriction. *J. Magn. Reson. A*, 113:260, 1995.

[52] S. D. Stoller, W. Happer, and F. J. Dyson. Transverse spin relqaxation in inhomogeneous magnetic fields. *Phys. Rev. A*, 44:7459, 1991.

[53] D. S. Grebenkov. NMR survey of restricted Brownian motion. *Rev. Mod. Phys.*, 79:1077, 2007.

[54] E. O. Stejskal. Use of spin echoes in a pulsed magnetic field gradient to study anisotropic, restricted diffusion and flow. *J. Chem. Phys.*, 43:3597, 1965.

[55] P. T. Callaghan and D. N. Pinder. Dynamics of entangled polystyrene solutions studied by pulsed field gradient nuclear magnetic resonance. *Macromolecules*, 13:1085, 1980.

[56] P. G. de Gennes. Dynamics of entangled polymer solutions I. Inclusion of hydrodynamic interactions, II. the Rouse model. *Macromolecules*, 9:587, 1976.

[57] J. Kärger, M. Kočiřik, and A. Zikánová. Molecular transport through assemblages of microporous particles. *J. Colloid and Interface Sci.*, 84:240, 1981.

[58] J. Kärger, H. Pfeifer, and W. Heink. Principles and applications of self-diffusion measurement by nuclear magnetic resonance. In *Advances in Magnetic Resonance, Vol. 12*. Academic Press, 1988.

[59] J. Kärger. Diffusionsuntersuchung von Wasser an 13X- sowie an 4A- und 5A-zeoliten mit hilfe der Methode der gepulsten Feldgradienten. *Z. Phys. Chem. (Leipzig)*, 248:27, 1971.

[60] H. Lechert. NMR investigations on the structure and sorption problems of Faujasite-type Zeolites. *Catal. Rev.*, 14:1, 1976.

[61] R. M. Barrer. *Zeolites and Clay Minerals as Sorbents and Molecular Sieves*. Academic Press, London, 1978.

[62] P. T. Callaghan, K. W. Jolley, and J. Lelievre. Diffusion of water in the endosperm tissue of wheat grains as studied by pulsed field gradient nuclear magnetic resonance. *Biophys. J.*, 28:133, 1979.

[63] P. T. Callaghan and O. Söderman. Examination of the lamellar phase of aerosol OT/water using pulsed field gradient nuclear magnetic resonance. *J. Phys. Chem.*, 87:1737, 1983.

[64] F. D. Blum, A. S. Padmanabhan, and R. Mohebbi. Self-diffusion of water in polycrystalline smectic liquid crystals. *Langmuir*, 1:127, 1985.

[65] P. P. Mitra and P. N. Sen. Effects of microgeometry and surface relaxation on NMR pulsed-field-gradient experiments: simple pore geometries. *Phys. Rev. B*, 45:143, 1992.

[66] P.G. de Gennes. Reptation of a polymer chain in the presence of fixed obstacles. *J. Chem. Phys.*, 55:572, 1971.

[67] M. Doi and S. F. Edwards. Dynamics of concentrated polymer systems. Part 1.—Brownian motion in the equilibrium state. *J. Chem. Soc., Faraday Trans. 2*, 74:1789, 1978.

[68] M. Doi and S. F. Edwards. *The Theory of Polymer Dynamics*. Oxford University Press, Oxford and New York, 1987.

[69] M. E. Komlosh and P. T. Callaghan. Segmental motion of entangled random coil polymers studied by pulsed gradient spin echo nuclear magnetic resonance. *J. Chem. Phys.*, 109:10053, 1998.

[70] P. G. de Gennes. *Scaling Concepts in Polymer Physics*. Cornell University Press, Ithaca and London, 1979.

[71] W. W. Graessley. Entangled linear, branched and network polymer systems— molecular theories. *Adv. Polymer Sci.*, 47:67, 1982.

[72] M. Clarkson, D. Beaglehole, and P. T. Callaghan. Molecular diffusion in a microemulsion. *Phys. Rev. Lett.*, 54:1722, 1985.

[73] R. Blinc, J. Pirs, and I. Zupancic. Measurement of self-diffusion in liquid crystals by a multiple-pulse NMR method. *Phys. Rev. Lett.*, 30:546, 1973.

[74] R. Blinc, M. Burgar, M. Luzar, J. Pirs, I. Zupancic, and S. Zumer. Anisotropy of self-diffusion in smectic-A and smectic-C phases. *Phys. Rev. Lett.*, 33:1192, 1974.

[75] L. Miljkovic, L. Thompson, M. M. Pintar, R. Blinc, and I. Zupancic. Self-diffusion in isotropic phase of para-alkoxybenzoic acid homologous series. *Chem. Phys. Lett.*, 38:15, 1976.

[76] M. Silva Crawford, B. C. Gerstein, A. L. Kuo, and C. G. Wade. Diffusion in rigid bilayer membranes. Use of combined multiple pulse and multiple pulse gradient techniques in nuclear magnetic resonance. *J. Am. Chem. Soc.*, 102:3728, 1980.

[77] M. I. Hrovat and C. G. Wade. NMR pulsed gradient diffusion measurements. 2. Residual gradients and lineshape distortions. *J. Magn. Reson.*, 45:67, 1981.

[78] I. Chang, G. Hinze, G. Diezemann, F. Fujara, and H. Sillescu. Self-diffusion coefcient in plastic crystals by multiple-pulse NMR in large static field gradients. *Phys. Rev. Lett.*, 76:2523, 1996.

[79] S. V. Dvinskikh and I. Furó. Combining PGSE NMR with homonuclear dipolar decoupling. *J. Magn. Reson.*, 144:142, 2000.

[80] S. V. Dvinskikh and I. Furó. Anisotropic self-diffusion in the nematic phase of a thermotropic liquid crystal by H-1-spin-echo nuclear magnetic resonance. *J. Chem. Phys.*, 115:1946, 2001.

[81] B. B. Mandelbrot. *Fractals: Form, Chance and Dimension*. Freeman, San Francisco, 1977.

[82] B. B. Mandelbrot. *The Fractal Geometry of Nature*. Freeman, San Francisco, 1982.

[83] H. O. Peitgen and P. H. Richter. *The Beauty of Fractals*. Springer-Verlag, Berlin, 1986.

[84] L. Pietronero and E. Tosatti, editors. *Fractals in Physics*. North-Holland, Amsterdam, 1986.

[85] P. Fischer and W. R. Smith, editors. *Chaos, Fractals and Dynamics*. M. Dekker, New York, 1985.

[86] S. Alexander and R. Orbach. Density of states on fractals: fractons. *J. Physique Lett.*, 43:625, 1982.

[87] R. Rammel and G. Toulouse. Random walks on fractal structures and percolation clusters. *J. Physique Lett*, 44:13, 1983.

[88] Y. Gefen, A. Aharony, and S. Alexander. Anomalous diffusion on percolating clusters. *Phys. Rev. Lett.*, 50:77, 1982.

[89] J. R. Banavar, M. Lipsicas, and J. F. Willemsen. Determination of the random-walk dimension of fractals by means of NMR. *Phys. Rev. B*, 32:6066, 1985.

[90] G. Jug. Theory of NMR field-gradient spectroscopy for anomalous diffusion in fractal networks. *Chem. Phys. Lett.*, 131:94, 1986.

[91] J. Kärger and G. Vojta. On the use of NMR pulsed field-gradient spectroscopy for the study of anomalous diffusion in fractal networks. *Chem. Phys. Lett.*, 141:411, 1987.

[92] E. Courtens and R. Vacher. Structure and dynamics of fractal aerogels. *Z. Phys. B. Condensed Matter*, 68:355, 1987.

[93] K. S. Mendelson. Nuclear magnetic relaxation in fractal pores. *Phys. Rev. B*, 34:6503, 1986.

[94] A. Blinc, G. Lahajnar, R. Blinc, A. Zidanšek, and A. Sepe. Proton NMR study of the state of water in fibrin gels and blood clots. *Bulletin of Magnetic Resonance*, 11:370, 1989.

[95] I. Oppenheim and P. Mazure. Brownian motion in systems of finite size. *Physica*, 30:1833, 1964.

[96] A. Lutti and P. T. Callaghan. Undulations and fluctuations in a lamellar phase lyotropic liquid crystal and their suppression by weak shear flow. *Phys. Rev. E*, 73:011710, 2006.

[97] A. Léon, D. Bonn, J. Meunier, A. Al-Kahwaji, O. Greffier, and H. Kellay. Coupling between flow and structure for a lamellar surfactant phase. *Phys. Rev. Lett.*, 84:1335, 2000.

[98] L. Courbin, J. P. Delville, J. Rouch, and P. Panizza. Instability of a lamellar phase under shear flow: Formation of multilamellar vesicles. *Phys. Rev. Lett.*, 89:148305, 2002.

[99] O. Diat, D. Roux, and F. Nallet. 'Layering' effect in a sheared lyotropic lamellar phase. *Phys. Rev. E*, 51:3296, 1995.

[100] O. Diat, D. Roux, and F. Nallet. Measurement of multilamellar onion dimensions under shear using frequency domain pulsed gradient NMR. *J. Magn. Reson.*, 187:251, 2007.

[101] S. L. Codd and P. T. Callaghan. Flow coherence in a bead pack observed using frequency domain modulated gradient nuclear magnetic resonance. *Physics of Fluids*, 13:421, 2001.

[102] R. Maier, D. M. Kroll, H. T. Davis, and R. Bernard. Simulation of flow in bead packs using the Lattice-Boltzmann method. *Physics of Fluids*, 10:60, 1998.

7
Restricted displacements and diffraction phenomena

In 1978, Hayward, Tomlinson, and Packer [1] carried out a PGSE NMR experiment on water flowing at low Reynold's number through a pipe. Under Poiseuille flow, the probability distribution of velocities in the pipe is uniform, so that the averaged propagator for displacements is a hat function, as shown in Fig. 2.8. The corresponding echo attenuation, the Fourier spectrum of this propagator, therefore has the oscillatory character of a 'sinc' function. This pioneering experiment demonstrated for the first time that PGSE NMR echo-attenuation behaviour could encompass more than a superposition of Gaussians, and indeed could include manifestations of phase coherence arising from the spin displacement distribution.

Of course, laminar flow implies intrinsic correlations in molecular displacement. By contrast, self-diffusion is, by conventional wisdom, characterised by entirely uncorrelated molecular motion. From such a standpoint, one would not expect to observe coherence phenomena in the attenuation expression for a PGSE NMR measurement on diffusing molecules. Nonetheless, in 1991, observation of such coherence is precisely what was seen [2]. Using a molecular ensemble in which diffusion provided the only means of spin transport, this experiment revealed oscillatory echo-attenuation functions highly reminiscent of diffraction patterns. The crucial factor governing phase coherence in the spin system was the presence of reflecting walls. Remarkably, it was restricted diffusion that lay behind the effect. Despite the fact that the echo resulted from spin-bearing molecules diffusing under locally Gaussian displacement distributions, the boundary collisions were found to imprint in $E(g, \Delta)$, a spatio-spectral structure factor relating to the pore geometry, thus resulting in an intimate relationship between structure and dynamics. With this insight came recognition of earlier conjecture [3] that the PGSE NMR was akin to a scattering experiment in which the scattering wavevector q was determined by the time integral of the gradient pulse. So began the field of research known as 'diffusive diffraction' or 'q-space imaging'.

The 1991 diffraction pattern is shown in Fig. 7.1.[1] This pattern results from a PGSE NMR experiment carried out on water diffusing in a porous matrix comprising close-packed polymer spheres of diameter ~ 16 microns. As the diffusive observation

---

[1]Note that we consistently use the angular frequency representation of q, whereas a more direct correspondence with inverse structural features results from the use of the cyclic spatial frequency, $\breve{q} = q/2\pi$. Consequently 'diffractograms' are generally plotted in terms of $\breve{q}$, a practice followed here.

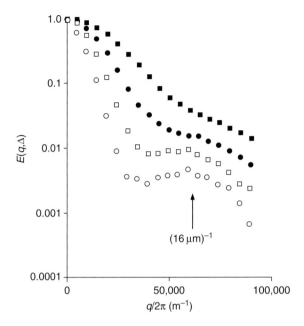

Fig. 7.1 Diffusive diffraction pattern observed for water molecules in a random packing of spherical polymer beads. The data are plotted for increasing values of Δ, 20, 40, 70, and 110 ms. (Adapted from reference [2].)

time, Δ, is increased, a distinct diffraction peak is observed at a reciprocal space value close to the bead diameter value and mean spacing between pores.

In this chapter we outline the physics underlying diffraction effects in PGSE NMR measurements on molecules undergoing restricted motion, providing examples of its application. Of course, a key factor underpinning our description is the narrow gradient pulse condition, since this allows direct access to the propagators for spin displacement. Nonetheless, the condition is not a constraint. We know how to extend our description in the case where gradient pulses are finite. Interpretation of the 'diffractograms' demands a little more care, as coherence effects are dampened and peak positions shifted, but the essential physics remains. Only when the gradient pulse width becomes so long that steady gradient conditions apply do we lose diffraction effects significantly. Such steady gradient experiments in restricted motion were considered in the previous chapter. In this chapter, pulsed gradients take centre stage.

7.1   PGSE NMR 'diffraction' in pores

While the steady gradient spin-echo experiment is capable of revealing the role of restrictions to motion caused by collision of molecules at surfaces bounding the fluid, any detail concerning the shape of those surfaces is buried in numerical prefactors appearing in the exponent of near-Gaussian decays. But in the narrow gradient pulse version of PGSE NMR, where the role of the wavevector $\mathbf{q}$ makes possible the direct sampling of the probability distribution of molecular displacements, a more transparent structural analysis arises. Furthermore, under conditions in which sufficient echo time is allowed for molecules to be able to diffuse through all the pore space, the echo signal bears a Fourier power spectrum relationship to the sample structure factor [2, 4, 5].

7.1.1 A historical precursor: rectangular boundaries

We start by examining a very simple example in which molecules in the liquid state are confined within a rectangular box with distance a between the boundaries. Suppose that a narrow gradient pulse PGSE experiment is performed with the gradient applied parallel to one side (z) of the box of length a, i.e. $\mathbf{g} = g\hat{\mathbf{k}}$. This problem was solved exactly by Tanner and Stejskal [6] under the assumption of perfectly reflecting boundaries. For the conditional probability they obtained the result

$$P_s(z|z', \Delta) = 1 + 2 \sum_{n=1}^{\infty} \exp\left(-\frac{n^2\pi^2 D_0 \Delta}{a^2}\right) \cos\left(\frac{n\pi z'}{a}\right) \cos\left(\frac{n\pi z}{a}\right) \qquad (7.1)$$

with corresponding echo-attenuation function

$$E(g, \Delta) = \frac{2[1 - \cos(\gamma g\delta a)]}{(\gamma g\delta a)^2}$$

$$+ 4(\gamma g\delta a)^2 \sum_{n=1}^{\infty} \exp\left(-\frac{n^2\pi^2 D_0 \Delta}{a^2}\right) \times \frac{1 - (-1)^n \cos(\gamma g\delta a)}{(\gamma g\delta a)^2 - (n\pi)^2}$$

$$(7.2)$$

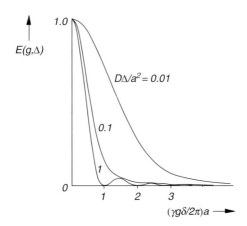

Fig. 7.2 Echo attenuation for diffusion in a box as represented by eqn 7.2. $E(g, \Delta)$ is shown for diffusion times, Δ, both small and comparable with the mean time taken to diffuse across the box, a^2/D_0. For larger values of Δ the behaviour is essentially time-independent and equivalent to that shown for $D_0\Delta/a^2 = 1$.

The short timescale limit ($\Delta \ll a^2/2D_0$) of eqn 7.2 corresponds to the free-diffusion regime of Section 6.5.2, for which $E(g, \Delta) = \exp(-\gamma^2 g^2 \delta^2 D_0 \Delta)$, while in the long timescale limit ($\Delta \gg a^2/2D_0$)

$$E(g, \Delta) = \frac{2\left(1 - \cos(\gamma g \delta a)\right)}{(\gamma g \delta a)^2} \qquad (7.3)$$

This result is particularly interesting. First, it is independent of both Δ and D_0. Expanding eqn 7.2 to fourth order in $\gamma g \delta a$ we find

$$E(g, \Delta) \approx 1 - \tfrac{1}{12}(\gamma g \delta a)^2$$
$$\approx \exp(-\gamma^2 g^2 \delta^2 D_{app}\Delta) \qquad (7.4)$$

where $D_{app} = a^2/12\Delta$. Hence the apparent diffusion coefficient decreases as the observation time, Δ, increases. Its magnitude is simply the mean-square distance travelled by the molecules, $a^2/6$, divided by twice the observation time. Note that where the PGSE gradient amplitude is weak ($\gamma g \delta a \ll 1$), we might expect to reproduce the motional narrowing ($l_D/l_s \gg 1$ and $l_g/l_s \gg 1$) regime of Section 6.5.2. But the result is very different from the motional narrowing result, eqn 6.68, found in the case of the steady gradient spin echo. The reason is simply that in the case of infinitesimally narrow gradient pulses, there is no spectral spread to be averaged by the spin diffusion, since the gradient is turned off during the diffusion period Δ. Instead, this idealised narrow pulse experiment measures directly the mean-squared displacement of the molecules.

The second interesting aspect of eqn 7.3 concerns its non-Gaussian character. This idea that the apparent diffusion coefficient, as measured by PGSE, is described by a term of order the mean-squared displacement divided by the timescale, is only true where the gradient applied is weak, meaning that the magnitude of $q = \gamma g \delta$ is much less than the reciprocal of the barrier spacing. The finite gradient form of eqn 7.3 has a very peculiar feature associated with the condition $qa = 2\pi$. Here, the echo signal is completely zero, as if the spins had diffused an infinite distance. The exact behaviour represented by eqn 7.2 is shown in Fig. 7.2. The meaning of the pathological behaviour of $E(q)$ in the case $qa = 2\pi$ will be apparent once we have developed a suitable formalism for the long time limit signal.

7.1.2   Diffusive diffraction in an enclosed pore

Consider the behaviour of molecules enclosed in a pore of volume V described by the magnetisation density function $M(\mathbf{r}, t)$, where at time $t = 0$ this magnetisation is uniform and equal to the pore density function, $\rho(\mathbf{r}) = V^{-1}$, inside the pore and zero outside. For the moment, we will presume that the pore surface is perfectly reflecting, or in other words, that there is no special relaxation at the walls and that all loss of magnetisation due to relaxation occurs equally for all molecules. This condition ensures that relaxation effects may be uniformly normalised.

In calculating the NMR signal we need to sum all contributions by integrating over the pore space, as indicated by $\int_V ...d\mathbf{r}$. In fact the V subscript is optional provided that we define the pore density function, $\rho(\mathbf{r})$, to be zero outside the pore. Hence we will often omit this subscript and allow that the integral may be taken over all space.

Memory loss and the pore density assumption

There will be some length scale l_s that characterises the pore size. On a timescale long compared with that taken to diffuse a distance l_s, molecules will have diffused backwards and forwards across the box many times and have lost all 'memory' of their initial positions. In mathematical terms this means that the conditional probability $P(\mathbf{r}|\mathbf{r}', \Delta \to \infty)$ is simply related to the pore geometry.

The loss of memory of starting positions as $D_0\Delta/l_s^2 \to \infty$ is commonly expressed by stating that the probability of finding a molecule at any final coordinates $\mathbf{r}'$ within the pore may be written as $\rho(\mathbf{r}')$. We will soon see that this assumption is not precisely correct, but for the moment, it will serve our purpose.

In Chapter 5 it was shown that $E(\mathbf{q}, \Delta)$ has a Fourier relationship with $\bar{P}(\mathbf{R}, \Delta)$, the probability distribution for displacement $\mathbf{R}$ over time Δ, also known as the 'displacement propagator'. If $P_s(\mathbf{r}|\mathbf{r}', \Delta \to \infty) = \rho(\mathbf{r}')$, then the average propagator has a very special character, namely

$$\bar{P}(\mathbf{R}, \infty) = \int \rho(\mathbf{r} + \mathbf{R})\rho(\mathbf{r})d\mathbf{r}. \tag{7.5}$$

This expression is exactly the auto-correlation function of the molecular density! Figure 7.3 shows $\bar{P}(\mathbf{R}, \infty)$ for the rectangular box example previously considered. Properly normalised, it has the form

$$\bar{P}(Z, \infty) = \frac{a + Z}{a^2} \qquad -a \leq Z \leq 0$$
$$\frac{a - Z}{a^2} \qquad 0 \leq Z \leq a \tag{7.6}$$

We are now in a position to calculate the rectangular box mean-square displacement.

$$\begin{aligned}\langle Z^2 \rangle &= \int_{-\infty}^{\infty} Z^2 \bar{P}(Z, \infty)dZ \\ &= \frac{1}{a^2}\int_{-a}^{0} Z^2(a + Z)dZ + \frac{1}{a^2}\int_{0}^{a} Z^2(a - Z)dZ \\ &= \tfrac{1}{6}a^2\end{aligned} \tag{7.7}$$

Note the use of the symbol $\langle .... \rangle$ to represent the ensemble average, in this case implicitly over all starting pore positions and all possible particle displacements.

The diffraction relationship

The idea that the long timescale averaged propagator is simply the auto-correlation function of the molecular density, $\rho(\mathbf{r})$, is a very powerful tool. In particular, since the PGSE experiment (in the narrow pulse approximation) gives us a signal $E(\mathbf{q}, \Delta)$ that is simply the Fourier transform of $\bar{P}(\mathbf{R}, \Delta)$, we obtain another important insight. It is well known in the theory of light scattering that the Fourier transform of a time auto-correlation function is simply the frequency power spectrum, a result sometimes known as the Wiener–Khintchine theorem [7]. We shall express this idea in our own

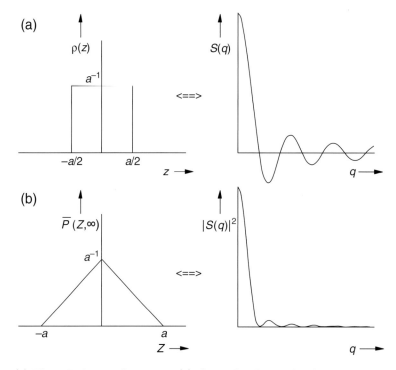

Fig. 7.3 (a) The 1-D density function, $\rho(z)$, for molecules confined to a rectangular box of width a, along with its corresponding Fourier transform, $S(q)$. (b) The infinite time-averaged propagator, $\bar{P}(Z, \infty)$, for the rectangular box and its corresponding transform. Because $\bar{P}(Z, \infty)$ is the auto-correlation function of $\rho(z)$, $\mathcal{F}\{\bar{P}(Z, \infty)\}$ is simply $|S(q)|^2$. In the optical analogy, $|S(q)|^2$ is the diffraction pattern from a single slit.

terms by returning to first principles. In the long-time limit, where $P_s(\mathbf{r}|\mathbf{r}', \Delta \to \infty)$ reduces to $\rho(\mathbf{r}')$, eqn 5.81 tells us

$$
\begin{aligned}
E(\mathbf{q}, \infty) &= \iint \rho(\mathbf{r})\rho(\mathbf{r}') \exp(i\mathbf{q} \cdot (\mathbf{r}' - \mathbf{r}))d\mathbf{r}d\mathbf{r}' \\
&= \int \rho(\mathbf{r}) \exp(-i\mathbf{q} \cdot \mathbf{r})d\mathbf{r} \int \rho(\mathbf{r}') \exp(i\mathbf{q} \cdot \mathbf{r}')d\mathbf{r}' \\
&= S^*(\mathbf{q})S(\mathbf{q}) \\
&= |S(\mathbf{q})|^2
\end{aligned}
\tag{7.8}
$$

$S(\mathbf{q})$ is known as the 'structure factor' and is sometimes denoted $\tilde{\rho}(\mathbf{q})$.

Eqn 7.8 states that in the very long time limit, when the molecules move over the entire sample, the PGSE signal is precisely the power spectrum of the structure factor, an optical analogy pointed out by Cory and Garroway [4]. This is a remarkable result because it means that the PGSE experiment in the long-time limit is an imaging experiment in its own right, returning not the structure factor as in k-space imaging, but the modulus squared of the structure factor. The loss of phase information by this squaring process is a severe disadvantage but it is exactly the price paid in X-ray diffraction experiments. There, the intensity structure factor is interpreted by comparing it with that calculated from a starting model in real space. But in one crucial aspect, PGSE or q-space static imaging has a major advantage over the k-space method in NMR microscopy. Because the signal is sampled in the absence of a gradient, there is no fundamental resolution limit! The only limit is imposed by the magnitude of the pulsed gradient that can be applied, subject to the condition that the two gradient pulses must be well-matched in area.

The rectangular box and single-slit diffraction

Equation 7.8 can be used to give a nice description of the long timescale behaviour in the rectangular box. The structure factor, $S(q)$, for the rectangular box is simply the Fourier transform of a hat function, as shown in Fig. 7.3. Thus

$$E(\mathbf{q}, \infty) = |sinc(\tfrac{1}{2}qa)|^2$$
$$= \frac{4\sin^2(\tfrac{1}{2}qa)}{(qa)^2}$$
$$= \frac{2(1 - cos(qa))}{(qa)^2} \tag{7.9}$$

This is exactly the result of eqn 7.3, an equivalence which can be seen by comparison of Fig. 7.3(b) and the long timescale limit of Fig. 7.2. Its optical analogue is the diffraction pattern of a single slit where the origin of the node when $qa = 2\pi$ is phase cancellation arising from different 'slit elements' across the uniform density, $\rho(z)$, of width a. For single-slit diffraction, each element at z with resulting phase shift $\exp(iqz)$ has an equivalent corresponding element at $(a/2 + z)$, with phase shift $\exp(iq(a/2 + z))$ or $-\exp(iqz)$. For diffusive diffraction, where the density $P(Z)$ is not uniform but given by the autocorrelation function of $\rho(z)$, the same values of q cause cancellation in the sum of elements weighted by their respective phase shifts.

Diffraction from a sphere

It is useful to apply the structure factor idea to other structures in which bounded diffusion occurs. The long timescale behaviour for particles diffusing in a sphere of radius a can be obtained by calculating the 1-D Fourier transform of the normalised density. This latter function is simply $(3/4a^3)(z^2 - a^2)$ for $-a \le z \le a$ and zero elsewhere. Thus

$$S(q) = \int_{-a}^{a} \frac{3}{4a^3}(z^2 - a^2)\exp(iqz)dz$$
$$= \frac{3[qa\cos(qa) - \sin(qa)]}{(qa)^2} \tag{7.10}$$

The PGSE signal, $E(q, \infty)$, is $|S(q)|^2$ and has a node when $qa \approx 3/4$. The behaviour when $qa \ll 2\pi$ is

$$E(q) \approx 1 - \frac{1}{5}(qa)^2$$
$$\approx \exp(-\tfrac{1}{5}\gamma^2 g^2 \delta^2 a^2) \tag{7.11}$$

In fact eqn 7.11 gives quite a good description of $E(q, \infty)$ over a decade of attenuation ($qa \leq \pi$). Comparison with eqn 5.90 shows that the weak gradient apparent diffusion coefficient is $a^2/5\Delta$. As in the case of the rectangular box, this is the mean-square distance diffused by the molecules, $(2/5)a^2$, divided by 2Δ.

7.1.3 The pore density assumption revisited

Suppose we investigate the long-time limit via the molecular probability density $p(\mathbf{r}, t)$, where we allow that $p(\mathbf{r}, 0) = \rho(\mathbf{r})$, and, because of the long term equilibration within the pore, $p(\mathbf{r}, \infty) = \rho(\mathbf{r})$. Again, we will, for the moment, neglect wall relaxation effects and assume that the boundaries are perfectly reflecting. Using the eigenmode expansion

$$p(\mathbf{r}, t) = \sum_n A_n u_n(\mathbf{r}) \exp(-Dk_n^2 t) \tag{7.12}$$

it follows that

$$\rho(\mathbf{r}) = \sum_n A_n u_n(\mathbf{r}) \tag{7.13}$$

Because of the orthogonality of the $\{u_m\}$, $A_m = \int_V \rho(\mathbf{r}) u_m^*(\mathbf{r}) d\mathbf{r}$. It is easy to show that for perfectly reflecting walls, $k_m \neq 0$ except for $m = 0$, so that we have

$$p(\mathbf{r}, \infty) = \left(\int_V \rho(\mathbf{r}') u_0^*(\mathbf{r}') d\mathbf{r}' \right) u_0(\mathbf{r}) \tag{7.14}$$

And since $p(\mathbf{r}, \infty) = \rho(\mathbf{r})$, we are able to identify the lowest eigenmode as proportional to the pore density function via

$$u_0(\mathbf{r}) = \frac{\rho(\mathbf{r})}{\int_V \rho(\mathbf{r}') u_0^*(\mathbf{r}') d\mathbf{r}'}$$

$$\therefore \int_V \rho(\mathbf{r}) u_0(\mathbf{r}) d\mathbf{r} \times \int_V \rho(\mathbf{r}') u_0^*(\mathbf{r}') d\mathbf{r}' = \int_V \rho(\mathbf{r})^2 d\mathbf{r} \tag{7.15}$$

Next, we turn our attention to the conditional probability. Given

$$P(\mathbf{r}|\mathbf{r}', \Delta) = \sum_n u_n^*(\mathbf{r}) u_n(\mathbf{r}') \exp(-Dk_n^2 t) \tag{7.16}$$

it is clear that pore equilibration leads to

$$P(\mathbf{r}|\mathbf{r}', \Delta \to \infty) = u_0^*(\mathbf{r}) u_0(\mathbf{r}')$$

$$= \frac{\rho(\mathbf{r})\rho(\mathbf{r}')}{\int_V \rho(\mathbf{r})^2 d\mathbf{r}} \tag{7.17}$$

In consequence, the simple diffusive diffraction relationship represented by eqn 7.8 must be rewritten[2]

---

[2]The author is grateful to Charles Epstein, of the University of Pennsylvania, for pointing this out.

$$E(\mathbf{q}, \infty) \propto \iint \rho(\mathbf{r})^2 \rho(\mathbf{r}') \exp(i\mathbf{q} \cdot (\mathbf{r}' - \mathbf{r})) d\mathbf{r} d\mathbf{r}' \tag{7.18}$$

and so the echo attenuation does not have the simple $|S(\mathbf{q})|^2$ form as commonly assumed. Does it really matter? If the pore density is uniform, in other words if $\rho(\mathbf{r}) = 1/V$, then $\rho(\mathbf{r})^2 \propto \rho(\mathbf{r})$ and of course eqn 7.18 is indistinguishable from eqn 7.8. In every experimental case examined so far, this simple identity holds. But could one envisage a problem where $\rho(\mathbf{r})$ is non-uniform? Of course, it is possible to begin a diffusive diffraction experiment with a non-uniform magnetisation, for example by allowing wall relaxation in advance of the application of the first magnetic field gradient pulse. This initial distribution would have to be accounted for in any calculation. But where the intrinsic spin density varies, for example due to an inhomogeneous distribution of permeable barriers, as one might find when imaging biological tissue, the problem will be further complicated by heterogeneous diffusion rates.

Certainly for liquids imbibed in any porous matrix where the fluid phase is homogeneous, the simple diffusive diffraction relation, eqn 7.8, applies, provided asymptotic pore equilibration conditions are reached. In the next section we see how to treat the PGSE NMR experiment under restricted diffusion conditions at all timescales and under conditions of wall relaxation.

7.1.4 Diffusive diffraction in a matrix of enclosing pores

Suppose that instead of a single pore, our sample consists of a matrix of N pores, labelled by i, as shown in figure 7.4, where each pore centre is given by a vector $\mathbf{r}_{0i}$ while each pore density function is $\rho_{0i}(\mathbf{r} - \mathbf{r}_{0i})$. The zero subscript distinguishes the local pore density function from the overall sample density. For convenience we define a local pore coordinate $\mathbf{r}_p = \mathbf{r} - \mathbf{r}_{0i}$.

The normalised spin echo signal in the narrow gradient pulse approximation can be written

$$
\begin{aligned}
E(\mathbf{q}, \infty) &= \frac{1}{N} \sum_{i=1}^{N} \int \rho_{0i}(\mathbf{r} - \mathbf{r}_{0i}) \exp(-i\mathbf{q} \cdot \mathbf{r}) d\mathbf{r} \int \rho_{0i}(\mathbf{r}' - \mathbf{r}_{0i}) \exp(i\mathbf{q} \cdot \mathbf{r}') \, d\mathbf{r}' \\
&= \frac{1}{N} \sum_{i=1}^{N} \int \rho_{0i}(\mathbf{r}_p) \exp(-i\mathbf{q} \cdot (\mathbf{r}_{0i} + \mathbf{r}_p)) d\mathbf{r}_p \int \rho_{0i}(\mathbf{r}'_p) \exp(i\mathbf{q} \cdot (\mathbf{r}_{0i} + \mathbf{r}'_p)) \, d\mathbf{r}'_p \\
&= \frac{1}{N} \sum_{i=1}^{N} \int \rho_{0i}(\mathbf{r}_p) \exp(-i\mathbf{q} \cdot \mathbf{r}_p) d\mathbf{r}_p \int \rho_{0i}(\mathbf{r}'_p) \exp(i\mathbf{q} \cdot \mathbf{r}'_p) d\mathbf{r}'_p \\
&= \frac{1}{N} \sum_{i=1}^{N} S^{*}_{0i}(\mathbf{q}) S_{0i}(\mathbf{q}) \\
&= \overline{|S_0(\mathbf{q})|^2} \tag{7.19}
\end{aligned}
$$

Hence the diffusive diffraction experiment returns the averaged squared structure factor for the pores. Fourier transformation of this echo signal yields to averaged pore correlation function, since

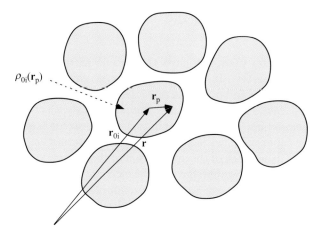

Fig. 7.4 A porous medium consisting of a matrix of similar pores, characterised by centre coordinates $\mathbf{r}_{0i}$, local pore coordinates $\mathbf{r}_p$ and pore densities $\rho_{0i}(\mathbf{r}_p)$.

$$\mathcal{F}\{\overline{|S_0(\mathbf{q})|^2}\} = \overline{\int \rho_0(\mathbf{r}_p)\rho_0(\mathbf{r}_p + \mathbf{R})d\mathbf{r}_p} \qquad (7.20)$$

At first sight the result seems unremarkable in comparison with eqns 7.5 and 7.8. But note that the cancellation of the pore centre coordinates leads to a result in which the structure revealed is that of an 'average pore' rather than the whole sample itself. The Fourier transform of the signal gives a kind of averaged pore image, albeit an image of the pore correlation function rather than the pore itself. But the crucial point is that the field of view (FOV) for that image will have dimensions comparable with the pore size rather than the sample size. In standard MRI experiments, the available signal must be distributed between the various voxels of the image, thus limiting the number of voxels available, since the voxel signal-to-noise ratio decreases as the number of voxels increase. Suppose the sample allows as many N_V voxels, with acceptable signal-to-noise ratio. Then the image resolution will be $FOV/N_V^{1/n}$ for an n-dimensional image. In the diffusive diffraction experiment the potential resolution is therefore very much better, allowing for voxel sizes on the order of microns or even smaller. That is the real power of the diffusive diffraction method in the elucidation of sample structure.

This result begs the question as to whether there exists the equivalent of an imaging experiment in a matrix of enclosed pores, in which the field of view remains on the order of the pore size, rather than the sample size, while the image returned is that of the averaged pore rather than the averaged pore correlation function. Indeed there is such an experiment.

7.1.5 'Long-narrow' PGSE NMR: direct imaging of the structure factor by diffusive diffraction

Laun *et al.* [8] have proposed an ingenious adaptation of the PGSE NMR experiment that provides a solution to the problem of the loss of phase information in the standard narrow gradient pulse diffusive diffraction experiment for which $\overline{|S_0(\mathbf{q})|^2}$ is returned. Their variant, shown in Fig 7.5, exchanges the narrow first effective gradient pulse with an equal area long pulse, and of duration T such that $T \gg l_s^2/D$. The pulse sequence might be labelled 'long-narrow' PGSE NMR.

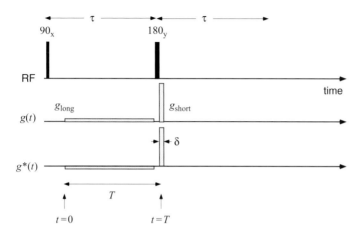

Fig. 7.5 Modified two-pulse PGSE NMR experiment ('long-narrow' PGSE NMR) due to Laun *et al* [8] in which the first narrow gradient pulse is replaced by a weak long pulse of duration T and of equal area. Hence $\mathbf{q} = \gamma T \mathbf{g}_{long} = \gamma \delta \mathbf{g}_{short}$.

Eqn 7.19 tells us that the two-pulse narrow gradient experiment returns the average value of individual pore terms $S_0(\mathbf{q})^* S_0(\mathbf{q})$. The same sort of multipore sample average will be returned by the experiment shown in Fig 7.5. To understand the meaning of the new experiment, we therefore only need to focus on a single pore.

$$E(\mathbf{q}, T) = \int \rho_0\big(\mathbf{r}(0) - \mathbf{r}_0\big) \exp\Big(-i\mathbf{q} \cdot \frac{1}{T}\int_0^T \mathbf{r}(t)dt\Big) d\mathbf{r}(0)$$
$$\times \int \rho_0\big(\mathbf{r}(T) - \mathbf{r}_0\big) \exp\big(i\mathbf{q} \cdot \mathbf{r}(T)\big) d\mathbf{r}(T) \qquad (7.21)$$

Note that the final coordinate sampled by the second, narrow, gradient pulse is $\mathbf{r}' = \mathbf{r}(T)$, the position of a molecule at the end of the first, long pulse. To simplify our notation, let us replace $\int \rho_0(\mathbf{r} - \mathbf{r}_0)....d\mathbf{r}$ by $\langle.... \rangle$. Then

$$E(\mathbf{q}, T) = \Big\langle \exp\Big(-i\mathbf{q} \cdot \frac{1}{T}\int_0^T \mathbf{r}(t)dt\Big) \exp\big(i\mathbf{q} \cdot \mathbf{r}(T)\big)\Big\rangle$$
$$= \Big\langle \exp\Big(-i\mathbf{q} \cdot \frac{1}{T}\int_0^T [\mathbf{r}_0 + \mathbf{r}_p(t)]dt\Big) \exp\big(i\mathbf{q} \cdot [\mathbf{r}_0 + \mathbf{r}_p(T)]\big)\Big\rangle$$
$$= \Big\langle \exp\Big(-i\mathbf{q} \cdot \frac{1}{T}\int_0^T \mathbf{r}_p(t)dt\Big) \exp\big(i\mathbf{q} \cdot \mathbf{r}_p(T)\big)\Big\rangle \qquad (7.22)$$

Laun *et al.* note that for T long compared with the time to diffuse back and forth across a pore, the time integral of $\mathbf{r}_p(t)$ is approximately T multiplied by the time averaged (pore centre) coordinate, and in the present notation, where we choose the coordinate origin at the pore centre, that value is zero. Assuming that the ensemble average may be shifted to the exponent, Laun *et al.* find $E(\mathbf{q}, \infty) = \langle \exp(i\mathbf{q} \cdot \mathbf{r}_p(T)) \rangle = S_0(\mathbf{q})$, so that this new experiment apparently returns the pore structure factor directly. The great advantage of this experiment is that an image of the averaged pore density function, $\rho(\mathbf{r}_p)$, can be directly computed via a Fourier transformation.

Careful analysis suggests that this structure factor could be multiplied by a further attenuation factor (and hence the image blurred) due to motion occurring during the first long gradient pulse. We need to perform the correct ensemble average of the phase factors to find any additional attenuation factor multiplying $S_0(\mathbf{q})$.

Before obtaining this factor, it is interesting to evaluate the echo attenuation obtained using pulse sequence of Fig 7.5 in the case of free diffusion. By making the substitution $\mathbf{r}_p(t) = \mathbf{r}_p(0) + \mathbf{R}(t)$ and noting $\mathbf{r}_p(T) = \mathbf{r}_p(0) + \mathbf{R}(T)$, eqn 7.22 becomes

$$E(\mathbf{q}, T) = \left\langle \exp\left(-i\mathbf{q} \cdot \frac{1}{T} \int_0^T \mathbf{R}(t)dt \right) \exp\left(i\mathbf{q} \cdot \mathbf{R}(T) \right) \right\rangle$$

$$= \left\langle \exp\left(-iq\frac{1}{T} \int_0^T Z(t)dt \right) \exp\left(iqZ(T) \right) \right\rangle \tag{7.23}$$

To evaluate eqn 7.23 we need to know whether $\int_0^T Z(t)dt$ and $Z(T)$ are correlated or uncorrelated. For free diffusion $\int_0^T Z(t)dt$ and $Z(T)$ are indeed correlated[3] and $\left\langle \int_0^T Z(t)dt Z(T) \right\rangle = DT^2$. The easiest way to handle the ensemble average problem in the case of the weak steady gradient is to use the Gaussian phase approximation, in other words to write the solution as $\exp(-\frac{1}{2}q^2\sigma_\chi^2)$ where σ_χ^2 is the mean-squared average of the term in square brackets, i.e.

$$\sigma_\chi^2 = \left\langle \left(\frac{1}{T} \int_0^T Z(t)dt - Z(T) \right)^2 \right\rangle$$

$$= \left\langle \left(\frac{1}{T} \int_0^T Z(t)dt \right)^2 \right\rangle - \left\langle 2(Z(T)\frac{1}{T} \int_0^T Z(t)dt \right)^2 + \left\langle Z(T)^2 \right\rangle \tag{7.24}$$

We met $\left\langle \left(\int_0^T Z(t)dt \right)^2 \right\rangle$ before in the Carr-Purcell evaluation of steady gradient effects in eqn 5.43. It has the value $\frac{2}{3}DT^3$. Given $\langle Z(T)^2 \rangle = 2DT$, $\exp\left(-\frac{1}{2}q^2\sigma_\chi^2 \right) = \exp\left(-\frac{1}{3}q^2DT \right)$ for free diffusion.

The interpretation of eqn 7.24 when the diffusion is restricted in a pore of size l_s is quite subtle. Suppose a particle starts at coordinate $z_p(0)$. Then for large T, it will have diffused back and forth across the pore many times such that it may, with equal probability, finish up anywhere in the pore so that $\langle z_p(T) \rangle = 0$ and $\langle \int_0^T Z dt \rangle = -z_p(0)$. Thus the cancellation of starting pore positions which led to eqn 7.24 is no

---

[3] The problem may be evaluated by noting $(\partial/\partial T)(\int_0^T Z(t)dt)^2 = 2Z(T)\int_0^T Z(t)dt$ and $(\partial/\partial T)Z(T)\int_0^T Z(t)dt = Z^2(T) + \ddot{Z}\int_0^T Z(t)dt$, the ensemble average of the second term being zero as discussed in section 1.3.7.

longer helpful and we need to return to our absolute pore position notation of eqn 7.22. Noting that the final coordinate $z_p(T)$ is completely uncorrelated with prior particle positions,

$$E(q,T) = \left\langle \exp\left(-iq\frac{1}{T}\int_0^T z_p(t)dt\right)\exp\left(iqz_p(T)\right)\right\rangle$$

$$= \left\langle \exp\left(-iq\frac{1}{T}\int_0^T z_p(t)dt\right)\right\rangle S_0(q) \qquad (7.25)$$

And so the pore structure factor is indeed measured, although multiplied by the additional attenuation factor $\left\langle \exp\left(-iq\frac{1}{T}\int_0^T z_p(t)dt\right)\right\rangle$. Again, we use the Gaussian phase approximation, and thus calculate $\sigma_\chi^2 = \left\langle\left(\frac{1}{T}\int_0^T z_p(t)dt\right)^2\right\rangle$. This may be evaluated from a knowledge of the conditional probability $P(z_p(0)|z_p(T),T)$. Let us do this calculation for a simple rectangular pore of width $l_s = 2a$ for which z_p ranges from $-a$ to a. The conditional probability is given in eqn 7.27 and after some algebra, we find $\sigma_\chi^2 = 4(a^2/(\frac{\pi}{2})^6)(a^2/DT)$ which vanishes as $T \gg a^2/D$.

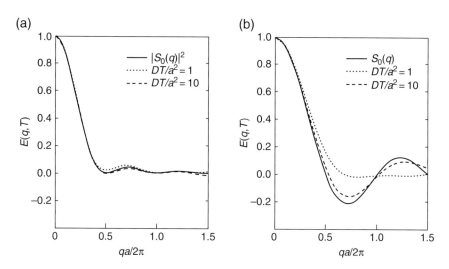

Fig. 7.6 Comparison of simulated echo attenuations, $E(q,T)$, for spins diffusing between rectangular barriers, in the case of the (a) narrow pulse and (b) 'long-narrow' versions of PGSE NMR. 1000 particles are used and the two diffusion times shown correspond to DT/a^2 values of 1 and 10.

The blurring of the image therefore disappears provided we make the diffusion time sufficiently greater than the time to diffuse across a pore. This conclusion is verified in the simulation shown in Fig 7.6 where the results of narrow PGSE pair and 'long-narrow' PGSE NMR for two different diffusion times are compared.

Despite the elegance of long-narrow PGSE NMR, it implies two experimental challenges. First, in stepping the value of q, one must do so with two very differently shaped pulses while scrupulously maintaining the equal area condition. Second, the approach to the asymptotic $S_0(q)$ form as T increases appears slower than that of the standard narrow pulse PGSE method approaching $|S_0(q)|^2$. These challenges, if met, may be rewarded by the greater information content of $S_0(q)$.

7.2 Finite time diffraction in planar, cylindrical and spherical pores

7.2.1 Finite Δ: the exact treatment

If our interpretation of diffusive diffraction effects were to be restricted to the condition $\Delta \gg l_s^2/D$ and to structures with perfectly reflecting walls, then we would indeed be severely constrained. Wall relaxation effects are a fact of life. Furthermore, the upper limit imposed by spin relaxation to observation time Δ, may, in large pores, make the '$\Delta \to \infty$' condition impossible, while in interconnected porous media, multiple length scales will certainly make this condition difficult to achieve. Yet diffraction effects are still manifest when Δ is finite, as is apparent in Fig. 7.1.

What is needed is a theory more sophisticated than that provided by eqn 7.8 alone, notwithstanding the useful physical insight that this simple limiting case provides. Here we will outline theories which account for finite diffusion times and wall relaxation.

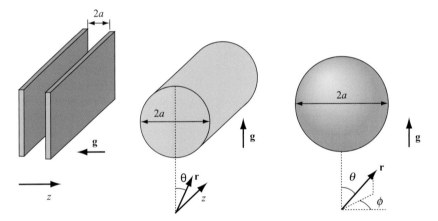

Fig. 7.7 Plane parallel pore, cylindrical pore, and spherical pore, showing relevant dimensions and coordinate frames.

In order to understand the key physics behind the diffusive diffraction phenomenon, it is helpful to show what happens for molecules diffusing in three simple standard geometries, between plane parallel boundaries, and in cylindrical and spherical pores. These are exactly the geometries considered by Brownstein and Tarr in the context of relaxation phenomena, and as part of the analysis presented here, wall relaxation effects will be allowed for. Note that in each case considered, a single pore length scale is relevant,[4] and while these ideal pore shapes may be special cases, they will provide a unique opportunity to calculate closed-form analytic expressions, since each of these boundary conditions leads to well-known eigenmodes of the diffusion equation.

---

[4]For the plane boundaries and for the cylindrical pore, we shall only be interested in diffusion normal to the pore surface.

A narrow pulse approximation expression for $E(q, \Delta)$ in the case of rectangular pores with relaxing walls was was first published in 1992 by Mitra and Sen [9], although a more complicated, independently derived expression was later published by Snaar and Van As [10] in 1993. The cylindrical pore case was first obtained in 1995 [11], while a formula for $E(q, \Delta)$ in the case of spherical pores was published in reference [9] and corrected in reference [11].

The derivations of the echo-attenuation expressions in each case start with finding the spatial part of the conditional probability, $P(\mathbf{r}|\mathbf{r}', t)$, $u_n(\mathbf{r})$. That is achieved by finding the eigen-solutions to the Helmholtz equation, $\nabla^2 u_n = \lambda_n u_n$, taking advantage of any symmetry, and allowing for the appropriate boundary conditions. To satisfy Fick's second law, the temporal part of each term in the eigenmode expansion is simply $\exp(D_0 \lambda_n t)$, and, as explained in Chapter 1, a finite solution as $t \to \infty$ requires a negative exponent, namely $\lambda_n = -k_n^2$.

7.2.2 Parallel plane pore

This is a 1-D problem in which the gradient is applied along the z-direction normal to a pair of bounding planes, and these relaxing planes are separated by a distance $2a$ and placed at $z = \pm a$.[5] The Helmholtz equation eigenmodes for planar geometry are

$$u_n(z) = \left\{ \begin{array}{c} \cos(\xi_n z) \\ \sin(\zeta_n z) \end{array} \right\} \tag{7.26}$$

The eigenfunction expansion for the propagator is given by

$$\begin{aligned} P(z|z', \Delta) = a^{-1} \sum_{n=0}^{\infty} &\exp(-\frac{\xi_n^2 D \Delta}{a^2})(1 + \sin(2\xi_n)/2\xi_n)^{-1} \\ &\times \cos(\xi_n z/a) \cos(\xi_n z'/a) \\ +a^{-1} \sum_{m=0}^{\infty} &\exp(-\frac{\zeta_m^2 D \Delta}{a^2})(1 \quad \sin(2\zeta_m)/2\zeta_m) \quad ^1 \\ &\times \sin(\zeta_m z/a) \sin(\zeta_m z'/a) \end{aligned} \tag{7.27}$$

where the eigenvalues ξ_n and ζ_m are determined by the equations

$$\xi_n \tan(\xi_n) = \frac{\bar{\rho} a}{D} \tag{7.28}$$

and

$$\zeta_m \cot(\zeta_n) = -\frac{\bar{\rho} a}{D} \tag{7.29}$$

The above expression for the conditional probability leads directly to the echo attenuation

---

[5]Note our preference in denoting the spacing between the plates as $2a$ rather than a, which arises from a desire to more directly compare with the case of the cylinder and sphere, where a is taken to be the radius.

$$E(q, \Delta) = \sum_{n=0}^{\infty} \exp(-\frac{\xi_n^2 D\Delta}{a^2})2(1 + \sin(2\xi_n)/2\xi_n)^{-1}$$

$$\times \frac{[(qa)\sin(qa)\cos\xi_n - \xi_n\cos(qa)\sin\xi_n]^2}{[(qa)^2 - \xi_n^2]^2}$$

$$+ \sum_{m=0}^{\infty} \exp(-\frac{\zeta_m^2 D\Delta}{a^2})2(1 - \sin(2\zeta_m)/2\zeta_m)^{-1}$$

$$\times \frac{[(qa)\cos(qa)\sin\zeta_m - \zeta_m\sin(qa)\cos\zeta_m]^2}{[(qa)^2 - \zeta_m^2]^2}$$

$$(7.30)$$

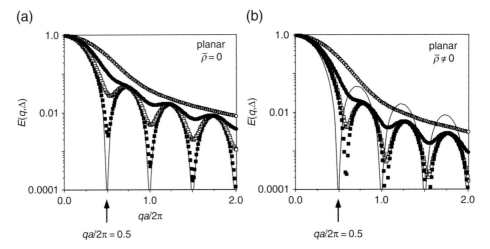

Fig. 7.8 (a) Echo attenuation, $E(q, \Delta)$, for spins trapped between parallel plane barriers separated by $2a$, in which the gradient is applied normal to the planes, and for perfectly reflecting walls ($\bar{\rho}a/D = 0$). Four successive time intervals are displayed. In multiples of a^2/D, Δ is, respectively, 0.2 (open circles), 0.5 (solid circles), 1.0 (open squares), and 2.0 (solid squares). Also shown is the theoretical curve for the infinite time diffraction pattern. Note that the first diffraction minimum occurs near $qa \approx \pi$. (b) as for (a) but for partially absorbing walls ($\bar{\rho}a/D = 2$).

The validity of eqn 7.30 may be checked in two ways. First, it reproduces the standard planar pore relaxation expression of Brownstein and Tarr when $q = 0$. Second, setting $\bar{\rho} = 0$ and noting our different definition of the dimension a, eqn 7.30 reproduces the well-known narrow gradient pulse echo attenuation result of Tanner and Stejskal [6] for rectangular pores with perfectly reflecting walls.

Figure 7.8 shows echo attenuations calculated using eqn 7.30 for a range of echo times both with and without wall relaxation. At times short compared with that required to diffuse between the walls, the echo attenuation is approximately Gaussian. At longer times, as wall collisions start to dominate, distinct diffraction effects are

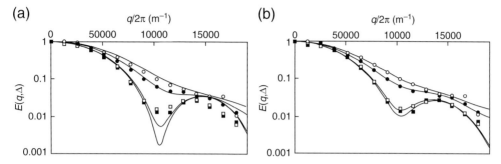

Fig. 7.9 (a) Echo attenuation, $E(q, \Delta)$, obtained in a narrow gradient pulse PGSE NMR experiment using a stack of pentane-filled rectangular capillaries with $2a = 100 \mu m$, at different fixed times, Δ, of 200 ms (open circles), 300 ms (closed circles), 700 ms (open squares), and 900 ms (closed squares), and corresponding to the range $0.4 a^2/D$ to $1.8 a^2/D$. Superposed are theoretical curves obtained from eqn 7.30, in which the known system parameters are used, including $\bar{\rho} = 0.00005$ corresponding to the known T_1. (b) As for (a) but where a distribution of a values is allowed and the system parameters are fitted. (Adapted from reference [12].)

observed. Note the characteristic echo minimum at $qa = \pi$, as well as the fact that the first diffraction maximum occurs at quite a significant degree of echo attenuation where $E(q, \Delta) < 0.1$. The effect of relaxation, apparent in Fig. 7.8(b), is to shift the diffraction minimum to higher q values, suggesting a reduction in the apparent pore size.

Figure 7.9 shows the results of narrow gradient pulse PGSE NMR experiments [12] carried out on pentane molecules diffusing between the walls of a stack of rectangular microcapillaries of spacing $2a = 100 \, \mu m$, both with theoretical curves based on the known system parameters and where the parameters are adjusted to allow the best fit for each Δ-value dataset. In the latter case, a is within a few per cent of the manufacturer's specification, while the alignment/spacing standard deviation value is less than 5%. D_0 values, which arise predominantly from the fit to the low q part of the data, are all close to the known self-diffusion coefficient of pentane at 28°C, $5 \times 10^{-9} m^2 s^{-1}$. Note that in the case of the plane parallel pore, the size of the coherence effects is remarkable, with more than a doubling of the echo amplitude with increasing gradient from the minimum near $qa/2\pi = 0.5$ to the maximum near $qa/2\pi = 0.75$.

7.2.3   Cylindrical pore

This is a 2-D problem handled in cylindrical polar coordinates in which the longitudinal z-axis is a symmetry axis for the system. The relevant coordinates are (r, θ) and the gradient is applied along the polar axis direction (i.e. across a diameter). The relaxing boundary is at a radial distance $r = a$ from the cylinder's central axis.

The Helmholtz eigenmodes are

$$u_n(r,\theta) = \left\{ \begin{array}{c} J_n(\beta_{nk}r) \\ N_n(\beta_{nk}r) \end{array} \right\} \left\{ \begin{array}{c} \cos(n\theta) \\ \sin(n\theta) \end{array} \right\} \tag{7.31}$$

where J_n and N_n are Bessel functions of the first and second kind. Since $N_n(\beta_{nk}r)$ diverges as $r \to 0$, only the $J_n(\beta_{nk}r)$ solution is permitted for a cylindrical pore. The eigenfunction expansion for the propagator is given by

$$P(\mathbf{r}|\mathbf{r}',\Delta) = \sum_{n,k}^{\infty} \exp(-\frac{\beta_{nk}^2 D\Delta}{a^2}) J_n(\beta_{nk}r/a) J_n(\beta_{nk}r'/a)$$

$$\times A_{nk}^2 \cos(n\theta)\cos(n\theta') \tag{7.32}$$

where for $n \neq 0$

$$A_{nk}^2 = \frac{(2/\pi a^2)(\beta_{nk}^2/J_n^2(\beta_{nk}))}{(\bar{\rho}a/D)^2 + \beta_{nk}^2 - n^2} \tag{7.33}$$

and

$$A_{0k}^2 = \frac{(1/\pi a^2)(\beta_{0k}^2/J_0^2(\beta_{0k}))}{(\bar{\rho}a/D)^2 + \beta_{0k}^2} \tag{7.34}$$

The J_n are standard (cylindrical) Bessel functions, while the eigenvalues β_{nk} are determined by the equations

$$\beta_{nk} \frac{J_n'(\beta_{nk})}{J_n(\beta_{nk})} = -\frac{\bar{\rho}a}{D} \tag{7.35}$$

From eqn 7.32, the echo-attenuation expression follows

$$E(q,\Delta) = \sum_k^{\infty} 4\exp(-\frac{\beta_{0k}^2 D\Delta}{a^2}) \frac{\beta_{0k}^2}{(\bar{\rho}a/D)^2 + \beta_{0k}^2}$$

$$\times \frac{[(qa)J_0'(qa) + (\bar{\rho}a/D)J_0(qa)]^2}{[(qa)^2 - \beta_{0k}^2]^2}$$

$$+ \sum_{nk}^{\infty} 8\exp(-\frac{\beta_{nk}^2 D\Delta}{a^2}) \frac{\beta_{nk}^2}{(\bar{\rho}a/D)^2 + \beta_{nk}^2 - n^2}$$

$$\times \frac{[(qa)J_n'(qa) + (\bar{\rho}a/D)J_n(qa)]^2}{[(qa)^2 - \beta_{nk}^2]^2} \tag{7.36}$$

As required by setting $q = 0$, eqn 7.36 reproduces the Brownstein–Tarr result for relaxation within cylindrical pores. The spin-echo-attenuation result for perfectly reflecting cylindrical pores was independently derived by Linse and Söderman [14] and eqn 7.36 reduces to their result when $\bar{\rho} = 0$.

Figure 7.10 shows echo attenuations calculated using eqn 7.36 for a range of echo times both with and without wall relaxation. The phenomena are similar to those

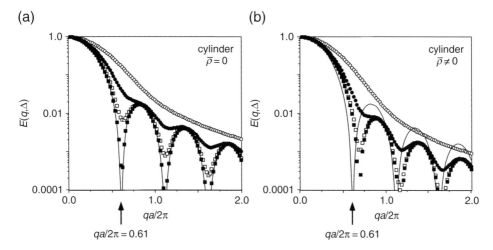

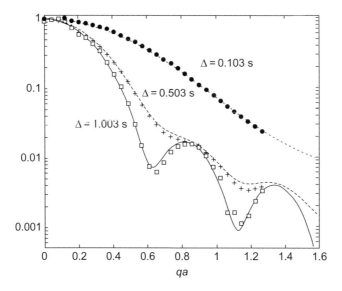

Fig. 7.10 (a) Echo attenuation, $E(q, \Delta)$, for spins trapped in a cylindrical pore of radius a in which the gradient is applied across a diameter and for perfectly reflecting walls ($\bar{\rho}a/D = 0$). Four successive time intervals are displayed. In multiples of a^2/D, Δ is, respectively, 0.2 (open circles), 0.5 (solid circles), 1.0 (open squares), and 2.0 (solid squares). Also shown is the theoretical curve for the infinite time diffraction pattern. Note that the first diffraction minimum occurs near $qa \approx 0.61$. (b) As for (a) but for partially absorbing walls ($\bar{\rho}a/D = 2$).

Fig. 7.11 Experimental data (points) and prediction of eqn 7.36 for pulsed field gradients applied orthogonal to the axis of a water-filled, 50 micron i.d. capillary, for three different diffusion times corresponding to $D\Delta/a^2 = 0.098$, 0.48, and 0.96. (Adapted from Gibbs [13].)

observed in the plane parallel pore, except that the echo minimum is at $qa/2\pi \approx 0.61$, and the first diffraction maximum occurs at an even greater degree of echo attenuation. Again, the effect of relaxation is to shift the diffraction minimum to higher q.

Equation 7.36 has been tested experimentally by S.J. Gibbs [13]. The theoretical lines, which correspond to $\bar{\rho} = 0$, are not fits, but predictions based on independently determined parameters. The agreement is very good, with residual deviations from the theory possibly due to non-orthogonality of the cylinder axis and the pulsed gradient direction.

7.2.4 Spherical pore

For the spherical case, the gradient of magnitude q is applied along the polar axis of the spherical polar coordinate frame. The relaxing boundary is at a radial distance $r = a$ from the sphere centre. The Helmholtz eigenmodes are

$$u_n(r, \theta, \phi) = \left\{ \begin{array}{c} j_n(\alpha_{nk}r) \\ h_n(\alpha_{nk}r) \end{array} \right\} \left\{ \begin{array}{c} P_n^l(\theta) \\ Q_n^l(\theta) \end{array} \right\} \left\{ \begin{array}{c} \cos(l\phi) \\ \sin(l\phi) \end{array} \right\} \tag{7.37}$$

where j_n and h_n are spherical Bessel functions of the first and second kind, and the P_n^l are associated Legendre functions. Since $h_n(\alpha_{nk}r)$ diverges as $r \to 0$, the $j_n(\alpha_{nk}r)$ solution is required for a spherical pore. Given azimuthal symmetry around the polar axis defined by that applied magnetic field gradient in the PGSE NMR experiment, only the $\cos(l\phi)$ solution with $l = 0$ is permitted, and P_n^l reduces to the Legendre polynomial P_n.

The eigenfunction expansion for the propagator is given by

$$P(\mathbf{r}|\mathbf{r}', \Delta) = \frac{1}{2\pi a^3} \sum_{n,k} \exp(-\frac{\alpha_{nk}^2 D\Delta}{a^2}) \frac{2n+1}{j_n^2(\alpha_{nk}) - j_{n-1}(\alpha_{nk})j_{n+1}(\alpha_{nk})}$$
$$\times j_n(\alpha_{nk}r/a)j_n(\alpha_{nk}r'/a)P_n(\cos \theta)P_n(\cos \theta') \tag{7.38}$$

where the j_n are spherical Bessel functions. The eigenvalues are determined by

$$\alpha_{nk} \frac{j_n'(\alpha_{nk})}{j_n(\alpha_{nk})} = -\frac{\bar{\rho}a}{D} \tag{7.39}$$

Noting that $\int_0^\pi P_n(\cos \theta) \exp(i2\pi qr \cos \theta)d(\cos \theta)$ is $2(-i)^n j_n(qr)$, and using the usual recurrence relations on spherical Bessel functions [15], it can be shown, after some algebra, that

$$E(q, \Delta) = \frac{3}{2} \sum_{n,k} \exp(-\frac{\alpha_{nk}^2 D\Delta}{a^2}) \frac{2n+1}{j_n^2(\alpha_{nk}) - j_{n-1}(\alpha_{nk})j_{n+1}(\alpha_{nk})}$$
$$\times \frac{\{\alpha_{nk}j_n(qa)[j_{n-1}(\alpha_{nk}) - j_{n+1}(\alpha_{nk})]}{[(qa)^2 - \alpha_{nk}^2]^2} \tag{7.40}$$

which reduces to the simpler form,

$$E(q,\Delta) = \sum_{n,k} 6\exp(-\frac{\alpha_{nk}^2 D\Delta}{a^2})\frac{(2n+1)\alpha_{nk}^2}{(\bar{\rho}a/D - \frac{1}{2})^2 + \alpha_{nk}^2 - (n+\frac{1}{2})^2}$$

$$\times \frac{\{(qa)j_n'(qa) + (\bar{\rho}a/D)j_n(qa)\}^2}{[(qa)^2 - \alpha_{nk}^2]^2} \tag{7.41}$$

Setting $q = 0$, eqn 7.41 reproduces the Brownstein–Tarr spherical pore result. Setting $\bar{\rho} = 0$ it encompasses the perfectly reflecting wall case derived by Balinov *et al.* [16].

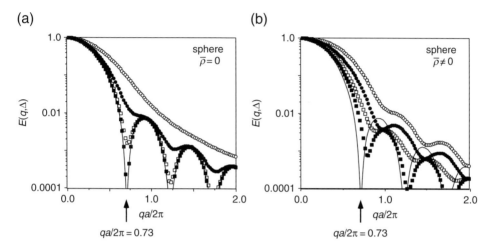

(a)

(b)

Fig. 7.12 (a) Echo attenuation, $E(q,\Delta)$, for spins trapped in a spherical pore of radius a and for perfectly reflecting walls ($\bar{\rho}a/D = 0$). Four successive time intervals are displayed. In multiples of a^2/D, Δ is, respectively, 0.2 (open circles), 0.5 (solid circles), 1.0 (open squares), and 2.0 (solid squares). Also shown is the theoretical curve for the infinite time diffraction pattern. Note that the first diffraction minimum occurs near $qa \approx 0.61$. (b) As for (a) but for partially absorbing walls ($\bar{\rho}a/D = 2$).

Figure 7.12 shows echo attenuations calculated using eqn 7.41 for a range of echo times both with and without wall relaxation. The phenomena are similar to those observed in the plane parallel and cylindrical pores, except that the echo minimum is at $qa/2\pi \approx 0.73$, and the first diffraction maximum occurs at $E(q,\Delta) < 0.01$. Figure 7.13 shows the absolute effect of relaxation on the unnormalised echo attenuation, while Fig. 7.12(b) shows the effect on the normalised attenuations.

In each of the geometries considered, the effect of wall relaxation is to shift the minima of the diffraction patterns to lower q values, thus making the pore size appear smaller. A similar effect is found even for non-relaxing walls when the duration of the gradient pulses becomes significant compared with the time to diffuse across the pore.

7.2.5 Finite width gradient pulses and relaxation effects

The apparent pore 'shrinking' evident in Figs 7.8, 7.10, and 7.12, as wall relaxation effects are included, is easy to understand. The 'spin-killing' in the vicinity of the wall

(a)   (b)

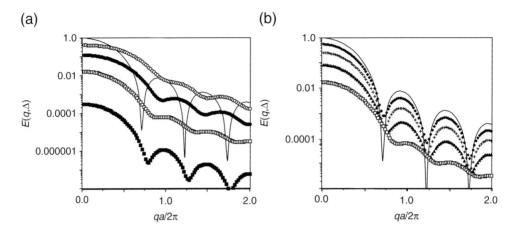

Fig. 7.13 (a) As for Fig. 7.12 (spherical pore case with $\bar{\rho}a/D = 2$) but with absolute rather than normalised echo amplitudes. Note that the diffraction minimum is strongly right shifted but moves closer to $qa/2\pi \approx 0.73$ at longer observation times. (b) Absolute echo attenuation for spins trapped in spherical pores of radius a at a fixed time $\Delta = a^2/D$ and for four different values of $\bar{\rho}a/D = 2$, namely 0.2 (solid diamonds, 0.5 (crosses), 1.0 (solid triangles), and 2.0 (open squares). Also shown for comparison is the theoretical curve for the infinite time diffraction pattern where $\bar{\rho} = 0$.

has the effect of reducing the local magnetisation density, causing a corona of reduced spin probability density in the vicinity of the wall. The immediate effect is to cause the apparent pore size, as viewed in the diffraction pattern, to reduce.

Careful inspection of Fig. 7.12(b) suggests a time-dependent shift in the position of the diffraction minimum for the spherical pore, an effect also observed more subtly in the cylindrical case. The reason is also easily understood. For the sphere, the relevant spin motion is parallel to the magnetic field gradient, which also defines the polar axis. At short times, reflection and relaxation of spins proximal to the wall will be dominated by the high proportion of molecules located at equatorial latitudes (near $\theta = 90°$), where the shorter chord lengths parallel to the polar axis make wall collisions more likely and the proportionate 'shrinking' of the apparent sphere diameter greater. At longer times, the repetitive relaxation suffered by such spins reduces their influence on the diffraction pattern, and the role of spins away from the equator increases. These shift phenomena indicate the need to account for both relaxation and finite observation time effects in analysing PGSE NMR data obtained for spherical pores.

Another 'pore shrinkage effect' results from the use of a finite duration gradient pulse in the PGSE NMR experiment. Impulse propagator matrix theory [18] can be used to predict the echo attenuation, and in the case of the plane parallel pore the results are shown in Fig. 7.14. These predictions agree precisely with independent Monte Carlo simulations [12, 14] and numerical solutions [17] to the Bloch–Torrey equation. It is apparent in Fig. 7.14 that the position of the first minimum in $E(q, \Delta)$ moves to higher values of q as the pulse duration is increased. This shift, which has been noted by a number of authors [12, 17, 19, 20], is, like the wall relaxation effect,

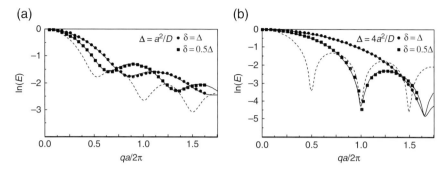

Fig. 7.14 Comparison of the impulse-propagator theory with numerical solutions to the magnetisation diffusion equation [17] for finite pulse-width PGSE attenuation in the case of restricted diffusion between parallel planes of spacing $2a$. No parameters have been adjusted and the comparison is absolute. The solid line shows the result calculated using the impulse propagator matrix theory [18], while the dashed curve shows the result predicted by the narrow gradient pulse approximation. (These data are adapted from reference [11] but with the plane spacing changed to $2a$ rather than the value of a used in that reference.)

associated with a narrowing in the effective well size by virtue of collisions of molecules with the boundaries. This narrowing was nicely explained by Mitra and Halperin [19] by invoking the 'smoothing' effect over the particle trajectory of the finite encoding time δ. Figure 7.15 illustrates how this results in the effective position of the particle during encoding being shifted out from the wall.

The probability distribution shown in Fig. 7.15(b) indicates a form of motional averaging that increases progressively as δ increases. Remarkably, Fig. 7.14 shows that the position of the first diffraction minimum continues to move to higher q as the diffusion length $D\Delta/a^2$ increases, with no asymptotic limit being evident.

7.3   Interconnected pores

While the simple 'diffusive-diffraction' long-time limit, $D\Delta/l_s^2 \to \infty$, may be achievable for an enclosing pore in which one length scale l_s describes the pore size, for interconnected pores, when multiple length scales exist, the diffractive limit breaks down, and the description is necessarily more complex. Of course, the eigenmode solution to the diffusion equation adopted in the previous section incorporates the finite displacements of molecules over an adjustable timescale and has diffusive-diffraction as a limiting case for long times. The same eigenmode approach can, in principle, be adopted for any enclosing geometry, including an array of interconnected pores, and where that geometry has some natural periodicity the general solution is quite practicable. For disordered structures, alternative strategies are required.

One means of making the problem tractable is to adopt a separation of timescales in which the diffractive limit applies at the length scale of the pore, while interpore diffusion spreads the diffraction pattern across a lattice of pores. This is known as the

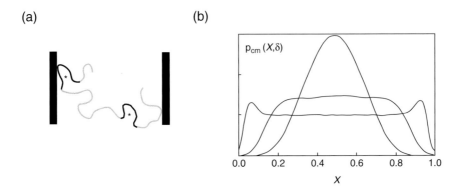

Fig. 7.15 (a) Trajectory of molecule diffusing between parallel plates with reflecting walls. The dark part of the trajectory corresponds to the periods δ when the gradient pulse is applied. The asterisks show the effective centre of mass of the particle during that pulse, and illustrate how collision with the wall necessarily shifts the centre of mass away from the wall. (b) Centre of mass distribution obtained by simulating molecules diffusing in a plane parallel pore ($2a = 1.0$) with reflecting walls. Note the narrowing as $\sqrt{D_0\delta} = 0.1$, 0.5, and 1. (Adapted from Mitra and Halperin [19].)

'pore equilibration assumption'. Within that assumption, two models are available to calculate the lattice contribution, one adopting a simple spreading Gaussian envelope of pore occupancy, and the other allowing, through a pore-hopping process, for the inter-pore diffusion to arise implicitly. Both give simple analytic expressions for the echo attenuation over a range of times longer than that required to diffuse across a single pore.

Note that material structures comprising an array of pores can be depicted as the convolution of a lattice with a pore density function. The Fourier space of such a material thus comprises a pore structure factor multiplied by a reciprocal lattice. This reciprocal lattice idea will be important in treating the problem of diffusion in an interconnected pore space.

7.3.1 Eigenmodes of the interconnected pore space

In the case of a porous medium with periodic microstructure, it is possible to adopt some of the methods used to describe the electronic properties of crystalline solids.[6] In particular the energy eigenfunctions (with energy eigenvalue $E_n = \hbar\omega_n$) for an electron in a periodic potential may be described by 'Bloch waves', products of a plane wave and functions that are periodic with the lattice potential, whence

$$\Psi_{n\mathbf{p}}(\mathbf{r}, t) = \phi_{n\mathbf{p}}(\mathbf{r}) \exp(i\mathbf{p} \cdot \mathbf{r} - i\omega_n t). \tag{7.42}$$

[6]The diffusion equation is similar to the Schrödinger equation for a moving particle but with an imaginary time. Consequently, solutions to the Schrodinger equation can form a nice starting point.

p is a wavevector in the first Brillouin zone[7] in reciprocal space and $\phi_{n\mathbf{p}}(\mathbf{r})$ is a function with the lattice periodicity. The eigenmode index n is said to label the 'band'. In this manner, Mitra *et al.* [9] suggested that the eigenmodes of the diffusion equation may be written in the form, $\phi_{n\mathbf{p}}(\mathbf{r})\exp(i\mathbf{p}\cdot\mathbf{r}-\lambda_{n\mathbf{p}}t)$, and using the Bloch–Floquet theorem, expanded $\phi_{n\mathbf{p}}(\mathbf{r})$ in a Fourier series of the reciprocal lattice vectors, **g**, as

$$\phi_{n\mathbf{p}}(\mathbf{r}) = \sum_{\mathbf{g}} \tilde{\phi}_{n\mathbf{p}}(\mathbf{g})\exp(i\mathbf{g}\cdot\mathbf{r}) \tag{7.43}$$

where

$$\tilde{\phi}_{n\mathbf{p}}(\mathbf{g}) = \frac{1}{V_a}\int_{V_a} d\mathbf{r}\phi_{n\mathbf{p}}(\mathbf{r})\exp(-i\mathbf{g}\cdot\mathbf{r}) \tag{7.44}$$

and V_a is the volume of the unit cell.

It follows directly from eqns 1.47 and 1.52 that the periodic part of the diffusion eigenmode satisfies

$$\lambda_{n\mathbf{p}}\phi_{n\mathbf{p}} + D(\nabla + i\mathbf{p})^2\phi_{n\mathbf{p}} = 0 \tag{7.45}$$

and[8]

$$D\hat{\mathbf{n}}\cdot\nabla\phi_{n\mathbf{p}}(\mathbf{r}) + \bar{\rho}\phi_{n\mathbf{p}}(\mathbf{r}) = 0 \tag{7.46}$$

The diffusion propagator in the periodic medium can therefore be written

$$P(\mathbf{r}|\mathbf{r}',t) = \sum_{n\mathbf{p}} e^{-\lambda_{n\mathbf{p}}t}\phi_{n\mathbf{p}}(\mathbf{r})\phi_{n\mathbf{p}}(\mathbf{r}')e^{i\mathbf{p}\cdot(\mathbf{r}-\mathbf{r}')} \tag{7.47}$$

In principle, the echo-attenuation expression may be simply calculated by evaluating the expression

$$\begin{aligned} E(\mathbf{q},t) &= \int_{V_p} d\mathbf{r}\rho(\mathbf{r})\int_{V_p} d\mathbf{r}'P(\mathbf{r}|\mathbf{r}',t)\exp(i\mathbf{q}\cdot(\mathbf{r}'-\mathbf{r})) \\ &= \frac{1}{V_p}\int_{V_p} d\mathbf{r}\int_{V_p} d\mathbf{r}'P(\mathbf{r}|\mathbf{r}',t)\exp(i\mathbf{q}\cdot(\mathbf{r}'-\mathbf{r})) \end{aligned} \tag{7.48}$$

where V_p is the volume of the pore space and the fluid density is taken to be uniform.[9]

Because $\phi_{n\mathbf{p}}(\mathbf{r})$ is a function restricted to the pore space, solving the diffusion equation for any pore shape is not trivial. However, Bergman and Dunn [21–23] tackled this problem in an ingenious manner by treating it as the solution to a chemical-potential diffusion equation, which must be satisfied everywhere. They do this by defining a spatially dependent diffusion coefficient, $D(\mathbf{r})$, which vanishes in the solid matrix and is equal to the free diffusion value, D, outside, as well as by defining the characteristic or indicator function of the pore space, $\theta(\mathbf{r})$, which is 1 for **r** in the pore

---

[7]The first Brillouin zone is simply a Wigner–Seitz primitive cell of the reciprocal lattice of the periodic structure.

[8]$\mathbf{p} = 0$ at the boundary.

[9]Note that we here write **q** in angular frequency units, so as to match the convention for the momentum vectors of solid state physics. The **q** used here is equivalent to the $2\pi\mathbf{q}$ that applies when cyclic frequencies are used.

space and zero in the matrix. Restriction of $\phi_{n\mathbf{p}}(\mathbf{r})$ to the pore space implies that the relevant Fourier spectrum in terms of lattice vectors is

$$\theta(\mathbf{r})\phi_{n\mathbf{p}}(\mathbf{r}) = \sum_{\mathbf{g}} \tilde{\psi}_{n\mathbf{p}}(\mathbf{g})\exp(i\mathbf{g}\cdot\mathbf{r}) \tag{7.49}$$

where

$$\tilde{\psi}_{n\mathbf{p}}(\mathbf{g}) = \frac{1}{V_a}\int_{V_a} d\mathbf{r}\theta_p(\mathbf{r})\phi_{n\mathbf{p}}(\mathbf{r})\exp(-i\mathbf{g}\cdot\mathbf{r}) \tag{7.50}$$

The echo-attenuation function then becomes

$$E(\mathbf{q},t) = \frac{1}{V_p}\sum_{n,\mathbf{p}}\int_{V_p} d\mathbf{r}\sum_{\mathbf{g}}\tilde{\psi}_{n\mathbf{p}}(\mathbf{g})\exp(i(\mathbf{g}+\mathbf{p}-\mathbf{q})\cdot\mathbf{r})$$
$$\times \int_{V_p} d\mathbf{r}'\sum_{\mathbf{g}}\tilde{\psi}_{n\mathbf{p}}^*(\mathbf{g})\exp(-i(\mathbf{g}+\mathbf{p}-\mathbf{q})\cdot\mathbf{r}')e^{-\lambda_{n\mathbf{p}}t}$$

$$\tag{7.51}$$

Since $\int d\mathbf{r}\exp(i\mathbf{k}\cdot\mathbf{r})$ is the Dirac delta function $\delta(\mathbf{k})$, the integrals require $\mathbf{p} = \mathbf{q} - \mathbf{g}$, which restricts $\mathbf{p}$ in the sum. Further, since $\mathbf{p}$ is in the first Brillouin zone, the only reciprocal space vector permitted is that closest to $\mathbf{q}$, and which is labelled $\mathbf{g_q}$. $\mathbf{g_q}$ is, in effect, the unique reciprocal lattice vector that returns $\mathbf{q}$ to the first Brillouin zone. In consequence, the echo-attenuation expression reduces to

$$E(\mathbf{q},t) = \frac{1}{\phi}\sum_{n} e^{-\lambda_{n\mathbf{p}}t}|\tilde{\psi}_{n\mathbf{p}}(\mathbf{g_q})|^2 \tag{7.52}$$

where $\mathbf{p} = \mathbf{q} - \mathbf{g_q}$.

Figure 7.16 shows a set of normalised echo-attenuation calculations, $E(\mathbf{q},t)/E(0,t)$, based on eqn 7.52, and calculated over a range of echo times, for a cubic array of overlapping spheres ($\phi = 0.202$) where the wall relaxivity is taken to be both zero and finite [24]. The gradient wavevector is aligned with the (100) direction, and the data exhibit the expected echo maximum corresponding to the Bragg peak at $qa/2\pi = 1$. The results agree remarkably well with simulations, and the diffraction patterns exhibit very little dependence on relaxation. In particular, we note that unlike the case of the isolated pore, where wall relaxation makes the pore appear smaller and hence the diffraction peak moves to larger q value, in the case of a periodic lattice the reciprocal lattice spacings are unaffected by relaxation and so the Bragg peak positions remain fixed.

Of course the periodic lattice model, while tractable analytically, is unlikely to be representative of most porous media where structural disorder, orientational and translational, will tend to dominate the behaviour. In the next section we outline the pore equilibration model whereby a tractable theory may be developed. However, the periodic lattice-Bloch eigenmode approach has an especially useful contribution to make in calculating the limiting behaviour of the effective diffusion as $t \to \infty$, yielding a result that may well have applicability in disordered systems. Of course the asymptotic limit will be associated with the echo-attenuation data as $q \to 0$, and

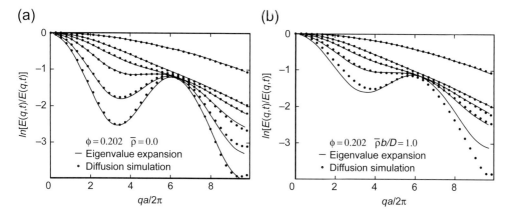

Fig. 7.16 Normalised echo attenuations obtained over a range of echo times for a porous medium consisting of a cubic array of overlapping spheres in (a) for $\bar{\rho}b/D = 0$ and (b) $\bar{\rho}b/D = 1$, where b is the side length of the cubic unit cell. In each case the eigenmode expansion based on Bloch modes agrees well with simulations. (Adapted from Bergman *et al.* [24].)

hence for $\mathbf{g_q} = 0$. For $t \gg b^2/D_0$ the lowest $(n = 0)$ mode dominates, and reference to eqn 7.45 suggests that D_{eff} will depend on the curvature of the lowest energy band at the centre of the Brillouin zone [25]. Using this approach for a dilute array of spheres, Sen *et al.* [25] obtained $D_{\text{eff}}/D_0 = 2/(3 - \phi)$, a result consistent with the known formation factor $1/F = 2\phi/(3 - \phi)$ [26].

7.3.2 Pore equilibration model

Let us suppose that the porous medium may be described by an interconnected series of pores, as shown in Fig. 7.17. Each pore centred at $\mathbf{r}_{0i}$ has a normalised local fluid density function,[10] $\rho_{0i}(\mathbf{r} - \mathbf{r}_{0i})$, so that the density function, $\rho(\mathbf{r})$, of the array can be represented by a superposition of local structures,

$$\rho(\mathbf{r}) = \frac{1}{N} \sum_{i=1}^{N} \rho_{0i}(\mathbf{r} - \mathbf{r}_{0i}) \tag{7.53}$$

Now we will use the pore equilibration assumption to write the conditional probability that describes the diffusive displacement of any molecule starting in the ith pore. To do so, we require that the time taken to diffuse across a pore is much shorter than the time taken to diffuse between pores. This assumption permits us to assign some probability C_{ij} for diffusion to the jth pore, but allocates the molecule equal probability of being anywhere within the jth pore once it arrives. Hence

$$P_i(\mathbf{r}|\mathbf{r}', \Delta) = \sum_{j=1}^{N} C_{ij}\rho_{0j}(\mathbf{r} - \mathbf{r}_{0j}) \tag{7.54}$$

---

[10]This function could include associated local pore throats, but in effect we regard these throats primarily as obstacles which slow diffusion between pores.

The validity of pore equilibration depends on the assumption that in the time needed to diffuse to distant pores, the labelled particles will diffuse back and forth several times within those pores, which they partially occupy. In effect this means that as time advances, the occupancy of pores distant from the starting pore will increase, but the local density distributions will remain the same as in equilibrium, albeit with a smaller amplitude. In other words we shall be concerned with timescales that are longer than a^2/D_0, but finite on the scale b^2/D_p, where a is a measure of the pore size, b is the average pore spacing, and D_p is the interpore diffusion coefficient. Thus, pore equilibration requires $D_0 b^2 / D_p a^2 \gg 1$.

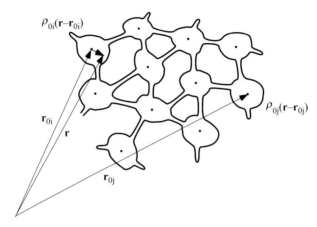

Fig. 7.17 Schematic representation of a general array of interconnected pores in which the local pore density functions are distributed on a lattice of points at locations $\mathbf{r}_{0i}$.

Pore equilibration: the Gaussian envelope

From eqn 7.54, we may write the echo attenuation when the gradient pulses are applied along the z-axis as

$$E(q, \Delta) = \frac{1}{N} \sum_{i=1}^{N} \sum_{j=1}^{N} C_{ij} \int \int \rho_{0i}(z - z_{0i}) \rho_{0j}(z - z_{0j})$$
$$\times \exp(iq(z' - z_{0j})) \exp(-iq(z - z_{0i})) dz' dz$$
$$\times \exp(iq(z_{0j} - z_{0i}))$$
$$= \frac{1}{N} \sum_{i=1}^{N} \sum_{j=1}^{N} C_{ij} \exp(iq(z_{0j} - z_{0i})) S_{0i}^*(q) S_{0j}(q)$$

$$(7.55)$$

If the pores are identical, then the product $S_{0i}^*(q) S_{0j}(q)$ is $|S_0(q)|^2$. For a structure with variable pore geometry we will define (somewhat roughly) an 'average pore structure factor', $\overline{|S_0(q)|^2}$ given by $\overline{S_{0i}^*(q) S_{0j}(q)}$. Thus

$$E(q, \Delta) = \overline{|S_0(q)|^2} \frac{1}{N} \sum_{i=1}^{N} \sum_{j=1}^{N} C_{ij} \exp\left(iq(z_{0j} - z_{0i})\right)$$

$$= \overline{|S_0(q)|^2} F(q, \Delta)$$

$$(7.56)$$

The connectivity matrix can now be applied to the problem of diffusion between the pores and the calculation of the lattice contribution to the echo attenuation, $F(q, \Delta)$. In the Gaussian envelope model the probability to be in neighbouring pores gradually increases according to a spreading diffusion profile written

$$C_{ij} = \frac{b}{(4\pi D_p \Delta)^{1/2}} \exp\left(-\frac{n_{ij}^2 b^2}{4 D_p \Delta}\right)$$

$$= \sum_i \{\sum_j \delta(Z - n_{ij}b)\} \otimes (4\pi D_p \Delta)^{1/2} \exp(-Z^2/4D_p\Delta)$$

$$(7.57)$$

where n_{ij} is a real number describing the fraction of lattice separations, b, that separate pores i and j. Thus $F(q, \Delta)$ involves a convolution between a 'lattice correlation function', $L(Z) = \sum_i\{\sum_j \delta(Z - n_{ij}b)\}$, and a diffusive envelope, $d(Z, \Delta) = (4\pi D_p \Delta)^{1/2} \exp(-Z^2/4D_p\Delta)$. For a regular periodic lattice with the gradient aligned along a crystallographic axis, n_{ij} will be a simple integer. For an irregular lattice or pore glass, we could replace $\sum_i\{\sum_j \delta(Z - n_{ij}b)\}$ by some general averaged lattice correlation function $\overline{L(Z)}$ and write the equation

$$E(q, \Delta) = \overline{|S_0(q)|^2} \left[\mathcal{F}\{d(Z, \delta)\} \otimes \mathcal{F}\{\overline{L(Z)}\}\right] \qquad (7.58)$$

To evaluate this expression we need some suitable description of $\overline{L(Z)}$.

Figure 7.18 shows three examples of a perfectly regular lattice, a regular lattice with irregular spacing,[11] and a pore glass with irregular spacing. A practical description of the the irregular lattice correlation function, $\overline{L(Z)}$, is given by presuming a mean pore spacing b with standard deviation (on a Gaussian distribution) ζ. A molecule starting in one pore and moving to the next will find its position at a distance b with standard deviation ζ, but the next nearest neighbour at $2b$ will have standard deviation 2ζ and so on, the lattice correlation being gradually lost, as shown in Fig. 7.18. In this sense ζ is a correlation length, which defines the distance over which regularity in the lattice displacements decays.

For molecules migrating from a starting pore to neighbouring pores in a pore glass, the vectors describing the set of possible displacements will have some mean length b but no orientational regularity. Averaged over the entire sample we could consider the first neighbour set to be uniformly distributed on a surrounding spherical shell of radius b. To predict the outcome of the PGSE experiment, we require the distribution of projected distances, Z, resulting from diffusion to the nearest neighbour. A simple

---

[11] The case of regular periodic lattices is covered in detail in the author's earlier book, reference [27].

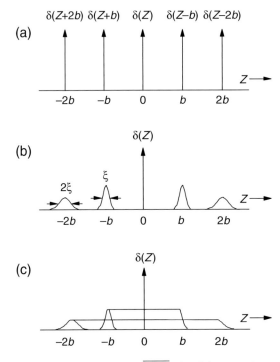

Fig. 7.18 The lattice correlation functions, $\overline{L(Z)}$, for (a) a regular 1-D lattice; (b) an ir-regularly-spaced 1-D lattice; (c) a pore glass. In (b) the Gaussians at each lattice site are successively broader for each pore successively displaced from the starting pore, with $\overline{L(Z)}$ becoming constant for $Z \gg b$. In (c) the pore glass lattice correlation function is a sum of normalised rectangular densities at each successive pore shell, convoluted with successively broader Gaussian pore separation distributions.

analysis shows that this probability density along Z is uniform for $-b < Z < b$, with magnitude $(2b)^{-1}$, and zero outside. For migration to the nth pore shell, the density is $(2nb)^{-1}$ between $-nb$ and nb. Thus, for the pore glass, $\overline{L(Z)}$ is a sum of hat functions of increasing width and decreasing height (conserving probability) associated with successive further neighbour pore shells, convolved with a Gaussian spread of increasing standard deviation $n\zeta$, and described by $(2\pi\zeta^2)^{-1/2}\exp(-Z^2/2n^2\zeta^2)$. Hence, the echo attenuation becomes

$$E(q, \Delta) = \overline{|S_0(q)|^2} \sum_n C_n \frac{\sin(qnb)}{qnb} \exp(-\tfrac{1}{2}q^2n^2\zeta^2) \qquad (7.59)$$

where

$$C_n = 4\pi(nb)^2 \frac{b}{(4\pi D_p\Delta)^{3/2}} \exp(-\frac{n^2b^2}{4D_p\Delta}) \qquad (7.60)$$

Pore equilibration-the pore hopping model

The need to assume a Gaussian envelope for diffusive occupancy of the pore lattice can be avoided using a simple trick [28]. Suppose that we consider a time interval, τ,

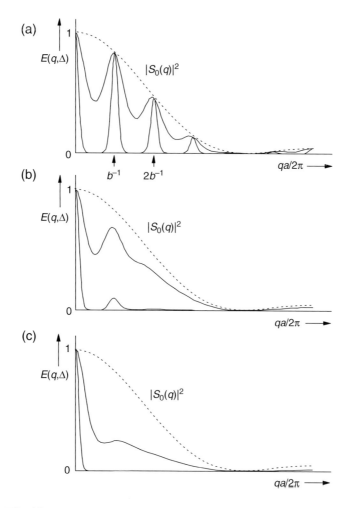

Fig. 7.19 $E(q, \Delta)$ vs q calculated using eqn 7.58: (a) regular lattice, (b) irregular 1-D lattice where $\xi = 0.2b$, and (c) irregular pore glass where $\xi = 0.2b$. In each case the pore is treated as a sphere with radius a approximately $b/3$ and $E(q, \Delta)$ is shown for Δ values of $0.2b^2/D_p$ and $2b^2/D_p$.

sufficiently short that molecules are most unlikely to have diffused further than the nearest pore. The resulting probability density function, $C(Z, \tau)$, involves only the starting pore and the nearest neighbour, and is easily written down exactly. Then, to calculate the pore hopping probability density for a finite time, $\Delta = M\tau$, all we need do is simply convolve $C(Z, \tau)$ M times so that

$$C(Z, \Delta) = C(Z, \tau) \otimes C(Z, \tau) \otimes \ldots]_{M\ times} \tag{7.61}$$

The contribution, $F(q, \Delta)$ of pore hopping to the echo-attenuation function is therefore

$$F(q, \Delta) = \mathcal{F}\{C(Z, \Delta)\}$$
$$= \mathcal{F}\{[C(Z, \tau) \otimes C(Z, \tau) \otimes \ldots]_{M \ times}\}$$
$$= [\mathcal{F}\{C(Z, \tau)\}]^M \tag{7.62}$$

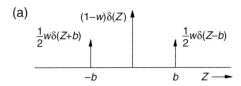

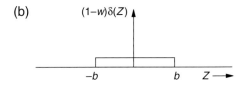

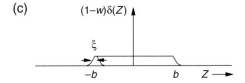

Fig. 7.20 Probability density functions, $C(Z, \tau)$, for hopping in an infinitesimal time interval τ, for (a) regular lattice, (b) pore glass with uniform pore spacing, (c) pore glass with irregular pore spacing. (Adapted from reference [28].)

Figure 7.20 shows $C(Z, \tau)$ for both a regular lattice and a pore glass, where we assume that the hopping probability in time τ is w. They may be written as:
(a) regular lattice

$$C(Z, \tau) = (1 - w)\delta(Z) + \tfrac{1}{2}w[\delta(Z - b) + \delta(Z + b)] \tag{7.63}$$

(b) regularly spaced pore glass

$$C(Z, \tau) = (1 - w)\delta(Z) + \frac{w}{2b}[H(Z - b) - H(Z + b)] \tag{7.64}$$

(c) irregularly spaced pore glass

$$C(Z, \tau) = (1 - w)\delta(Z) + \frac{w}{2b}[H(Z - b) - H(Z + b)]$$
$$\otimes (2\pi\zeta^2)^{-1/2} \exp(-Z^2/2\zeta^2) \tag{7.65}$$

where H represents the Heaviside step function.

One more step is needed before the final echo attenuation expressions can be calculated, and that is to relate w and b to the interpore diffusion coefficient, D_p. To do so, we consider the low q limiting behaviour of the echo attenuation $E(q, \Delta) = |S_0(q)|^2 F(q, \Delta)$, thus emphasising the long-range displacements associated with the pore hopping. Of course the low q limit of $E(q, \Delta)$ becomes $1 - 4\pi^2 q^2 \langle Z^2(\Delta) \rangle$. In consequence, noting $|S_0(q)|^2 \to 1$ as $q \to 0$,

$$\langle Z^2(\Delta) \rangle = -F''(q = 0, M\tau)$$
$$= Mwb^2$$
$$= 2D_p\Delta \tag{7.66}$$

where the double prime denotes the second derivative with respect to q. Hence, we can identify

$$w = \frac{2D_p\Delta}{Mb^2} \tag{7.67}$$

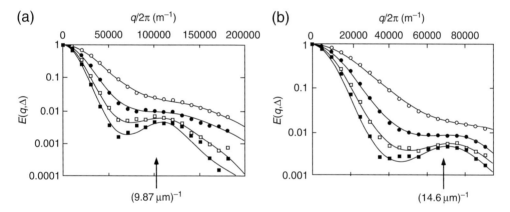

Fig. 7.21 Echo-attenuation functions, $E(q, \Delta)$, resulting from the diffusion of water in a bed of randomly packed monodisperse polystyrene spheres of diameter (a) 9.87 microns and (b) 14.6 microns, where the observation times, Δ, range from 10 ms (open circles), 20 ms (closed circles) 30 ms (open squares) to 40 ms (closed squares). The solid lines are fits using eqn 7.70. (Adapted from reference [29].) Note the coherence peak at $q/2\pi \sim b^{-1}$ where the pore spacing b is comparable to the bead diameter.

Evaluating the Fourier spectrum of $C(Z, \tau)$ in each of the cases given in eqns 7.63, 7.64, and 7.65, and taking the Mth power we find:
(a) regular lattice

$$E(q, \Delta) = |S_0(q)|^2 \exp\left(-\frac{4D_p\Delta}{b^2} \sin^2(\tfrac{1}{2}qb)\right) \tag{7.68}$$

(b) regularly spaced pore glass

$$E(q, \Delta) = |S_0(q)|^2 \exp\left(-\frac{6D_p\Delta}{b^2}\left[1 - \frac{\sin(qb)}{qb}\right]\right) \tag{7.69}$$

(c) irregularly spaced pore glass

$$E(q, \Delta) = |S_0(q)|^2 \exp\left(-\frac{6D_p\Delta}{b^2 + 3\zeta^2}\left[1 - \exp(-\tfrac{1}{2}q^2\zeta^2)\frac{\sin(qb)}{qb}\right]\right)$$

(7.70)

Figure 7.21 shows the results of PGSE NMR echo attenuation experiments [29] carried out on water diffusing in a porous medium consisting of a pore glass made from randomly packed monodisperse polystyrene spheres with nominal diameters 9.87 microns and 14.6 microns. In each case a coherence peak is clearly visible at $q/2\pi \sim b^{-1}$, the pore spacing, b, being on the order of a bead diameter. Note that the amplitude of this peak is at an attenuation value $E(q, \Delta) < 0.01$, thus placing demands on experimental signal-to-noise ratio. In each case the data are fitted using eqn 7.70, assuming a somewhat simplistic spherical pore shape, radius a. The representation of the data is good, with reasonably consistent parameter values. For example, in the case of the 9.87 micron spheres, of $a \sim 3\,\mu$m, $\xi \sim 2\,\mu$m, $D_p \sim 2x10^{-9}$ m^2 s^{-1}, and with b ranging from $6.2\,\mu$m to $10.6\,\mu$m with increasing Δ. That the fitted pore spacing should depend on observation time is perhaps not surprising given the simplicity of the underlying pore equilibration assumption. Equations 7.68, 7.69, and 7.70 are certainly not exact, but they do provide insight and are a conveniently simple representation of the restricted diffusion process in interconnected porous media.

7.3.3 'Long-narrow' PGSE NMR and interconnected pores

While the long-narrow PGSE NMR approach provides an obvious advantage in diffraction studies of enclosing pores, in the case of interconnected pores the method suffers from a significant disadvantage. Over the long gradient pulse, molecules will diffuse from pore to pore, at an asymptotic diffusion rate given by D_p. The final narrow gradient pulse will return the spectrum $S_0(q)$, as a coefficient in the echo attenuation, provided that the duration of that pulse is much shorter than the time to diffuse across a pore. But this structure factor will be multiplied (in the pore equilibration picture) by a factor relating to the inter-pore diffusion, and, because the inter-pore length distances will generally exceed the pore dimension, the damping factor may be so great as to mask the spectrum.

Of course, it could be argued that the narrow-pulse PGSE NMR method also returns a structure factor (in this case the less information-rich $|S_0(q)|^2$) experiencing a similar damping due to inter-pore diffusion, as evident in eqn 7.58. But now we see a crucial difference between the methods. For narrow pulse PGSE NMR, one is sensitive to inter-pore diffraction effects, in addition to pore-scale diffraction. Indeed, as eqn 7.58 makes clear, inter-pore diffraction dominates the echo attenuation, the term $|S_0(q)|^2$ merely providing a weakly decaying envelope. Alas, inter-pore diffraction cannot be observed in long-narrow PGSE NMR because, in trading the short initial gradient pulse for a long weak pulse, we have gained the pore structure factor directly, but lost the interference effects associated with migration across the pore lattice, accepting in return a gaussian decay associated with inter-pore diffusion in the presence of a weak steady gradient.

In the spirit of eqn 7.25, we may write down the echo attenuation for the intercon-nected pore space signal of the long-narrow PGSE NMR method as

$$E(q,T) = \left\langle \exp\left(-iq\frac{1}{T}\int_0^T z_p(t)dt\right)\right\rangle \overline{S_0(q)}$$

$$= \exp(-\tfrac{1}{3}q^2 D_p T)\overline{S_0(q)} \tag{7.71}$$

It may be possible, for sufficiently small inter-pore diffusion time T, that the diffusive prefactor $\exp(-\tfrac{1}{3}q^2 D_p T)$ does not overwhelm the averaged pore structure factor $\overline{S_0(q)}$, making pore size characterisation feasible.[12] In that case, having measured the non-asymptotic D_p value applicable to the chosen value of T, the factor $\exp(-\tfrac{1}{3}q^2 D_p T)$ can be calculated for each q value, and normalised out, leaving $\overline{S_0(q)}$ directly visible. At the time of writing of this book, that experimental possibility remains an, as yet, unproven possibility.

7.4   Applications of q-space diffraction

7.4.1   Emulsions and capsules

While it is possible to manufacture almost ideal plane parallel pores and cylindrical pipes of highly uniform spacing or radius, in the case of synthetic spherical particles, other factors intervene to complicate the q-space diffraction measurement. An obvious application of diffusive diffraction is in the analysis of the diffusive behaviour of mole-cules trapped inside spherical colloidal particles. Of course, the signal response will be complicated by the fact that the internal restricted diffusion will be superposed on the diffusion of the particle itself. What that means in effect is that the propagator for in-ternal motion is convolved with the Gaussian propagator associated with the diffusion of the sphere. Provided that sphere radius a exceeds the length scale for sphere diffusion over the time $\Delta \gg a^2/D_0$, then diffusive diffraction effects will dominate. But even outside this condition the behaviours are inherently separable. Of course, inter-sphere diffusion (migration of molecules between spheres) and sphere radius polydispersity will further complicate or 'smear' diffusive diffraction effects. Nonetheless, such fac-tors are are exactly analytically describable [30], and their convolution to generate the propagator that characterises the resultant molecular translational dynamics may be written down.

Examples of spherical colloidal particle systems include emulsions, core-shell par-ticles, multi-lamellar onion phases, and polyelectrolyte multi-layer capsules. For emul-sions, size polydispersity and inter-droplet diffusion certainly play an important role in determining the PGSE NMR signal response. Haakanson *et al.* have shown that in close-packed emulsions, it is possible to use diffraction effects resulting from inter-droplet migration to determine the local emulsion packing structure [31]. Wassenius *et al.* [32] have investigated the echo signal from molecules inside 'core-shell' latex parti-cles with a liquid core of hexadecane and a solid polystyrene shell. At echo times long compared with that required to diffuse across the sphere, they found the root mean

---

[12]It is worth noting, however, that $\overline{S_0(q)}$ decays, with increasing q, at a much slower rate than $|\overline{S_0(q)}|^2$, making it more resilient to measurement in the face of inter-pore diffusive decay.

square displacement of oil inside the particle core to be constant for all diffusion times and, from that, calculated the mean particle radius. While the diffusive diffraction pattern in the echo decay was almost completely smeared out due to polydispersity and wall relaxation effects, a small 'bump' in $E(q)$ was observed at $qa \approx 0.7$. Yadav and Price [33] have seen both intra-sphere and inter-sphere diffraction effects from PGSE NMR experiments on water-in-oil emulsions, carrying out a detailed fit to the data using the multiple-propagator matrix method to allow for the influence of finite-width gradient pulse effects.

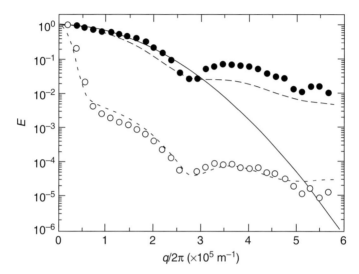

Fig. 7.22 $E(q, \Delta)$ vs q for both water (open circles) and tetramethyl ammonium chloride (TMA; closed circles) for a water/TMA in heptane emulsion. Here $\delta = 0.8ms$ and $\Delta = 0.5s$, while g is incremented from 0 to $17.2\,\text{T/m}$. The solid line is a fit using the Gaussian phase approximation model, while the dashed lines are multiple propagator diffraction fits. Note that inter-droplet effects are indicated for the water, but not the TMA signal. (Redrawn from Yadav and Price [33].)

Their echo-attenuation data, obtained from both water and tetramethyl ammonium chloride, are shown in Fig. 7.22 along with theoretical estimates. A simple Gaussian phase approximation model fails badly, as can be seen in Fig. 7.22, while the diffusive diffraction model works well when a distribution of sphere sizes is incorporated.

7.4.2 Biology and medicine

Ironically, monodispersity is more often observed in nature than achieved in synthesis. A remarkable example of distinctive size monodispersity in colloids is the mammalian blood cell. Kuchel and co workers [34–39] have carried out extensive studies of the size, shape, and permeability of human blood cells using PGSE NMR diffusive diffraction effects. This work has not only shown that erythrocytes orient in a magnetic field, but accurate shape and size determinations have been made using diffusive diffraction, and shape evolution through various forms has been observed on poisoning with sodium fluoride.

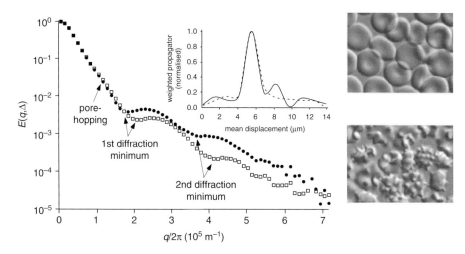

Fig. 7.23 $E(q, \Delta)$ vs q for red blood cell samples showing the influence of shape. The signals were obtained from predominantly discocytes (squares) and spherocytes (closed circles). The insets show the displacement distributions for both graphs. Discocyte and spherocyte microscope images are shown on the right of the figure. (Reproduced with permission from Pages *et al.* [39].)

An example of these data are shown in Fig. 7.23, where distinctly different diffraction patterns are obtained subsequent to introduction of the NaF. Also shown are size distributions based on a second-derivative method described in reference [38]. Note that at low q the inflection labelled 'pore-hopping' is due to diffraction effects related to water moving between cells. Using these methods Kuchel and co-workers have been able to monitor shape changes in erythrocytes over many hours.

7.4.3 Pulsed gradient spin-echo ESR

Given the short T_2 so common in electron paramagnetic resonance,[13] typically on the order of or short than $1\,\mu s$, the duration of gradient pulses in any possible PGSE time-domain ESR experiment will be similarly constrained to sub-microsecond. In addition, the short T_1, again typically on the order of or shorter than $1\ \mu s$, confines Δ to on the order of microseconds. Despite the larger gyromagnetic ratio of the electron spin, a pulsed gradient of $10\,\mathrm{T}\,\mathrm{m}^{-1}$, at the high end of the experimentally feasible, would only allow the measurement of diffusion coefficients faster than $10^{-7}\,\mathrm{m^2}\,\mathrm{s}^{-1}$, much faster than molecules in the liquid state, except at highly elevated temperatures.

---

[13] Also known as electron spin resonance or ESR.

There do, however, exist 'pathological' EPR samples with long relaxation times, one of the most notable being the 1-D organic metal (fluoroanthene)$_2$PF$_6$ [40], for which $T_1 \approx T_2 \approx 10\,\mu s$. Steady gradient diffusion methods have indicated [40, 41] that the upper limit for electron diffusion perpendicular to the conducting channels is at least three orders of magnitude smaller than in the longitudinal direction. In 1994 a PGSE ESR experiment [42] was performed on this system, in which both Δ and q (parallel to the longitudinal axis) were varied. The results were consistent with restricted diffusion in a one-dimensional well [43].

The switching of gradient pulses with duration on the order of $1\,\mu s$ is difficult, even with gradient coils designed specifically for low inductance. The switching time achieved with conventional commercial linear current amplifiers is around $20\,\mu s$ and limited by feedback loop circuitry. Overcoming this limitation is feasible using the 'clipped L-C resonance' idea of Conradi *et al.* [44] in which driven, semi-sinusoidal current pulses are generated. This method allows current pulses of any duration, δ, down to $2\,\mu s$, resulting in peak field gradients up to 1.0 T m^{-1} using a quadrupolar gradient coil of 0.2 T m^{-1}A^{-1} [42].

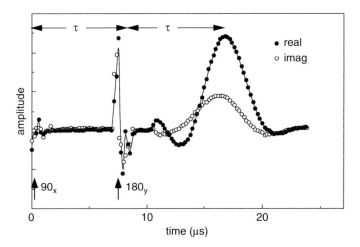

Fig. 7.24 Real and imaginary echo signals from a $90_x - \tau - 180_y$ spin-echo sequence ($\tau = 7.8\,\mu s$) after signal averaging from 1024 transients. The acquisition begins $0.5\,\mu s$ before the first RF pulse although the correct phase cycling for signal co-addition does not begin until $0.5\,\mu s$ following the second pulse. The additional delay before the signal appears is due to the $4\,\mu s$ deadtime. (Reproduced with permission from reference [42].)

Figure 7.24 shows an example of an echo signal obtained from the electrons in the (fluoroanthene)$_2$PF$_6$ sample, obtained at 300 MHz using $B_0 = 10.5$ mT in the stray field region of a 7 T superconducting magnet [42]. The RF pulses were $0.2\,\mu s$ and $0.4\,\mu s$, respectively, for the 90 and 180 degree RF pulses and were generated using a Bruker AMX300 NMR spectrometer. The echo shown was obtained with 1000 transients at $2\tau = 16\,\mu s$ and at a repetition time of 2 ms, limited by the spectrometer data transfer rate rather than the sample T_1. Remarkably, the response of the echo amplitude to q

and Δ is consistent with restricted electron diffusion, in which the size, l, of the 1-D compartments is governed by an exponential distribution

$$p(l) = l\bar{l}^{-2} \exp(-l/\bar{l}) \tag{7.72}$$

associated with random defects. Using this length distribution, the planar restricted diffusion expression of eqn 7.30 provides a good fit to the data, as shown in Fig. 7.25.

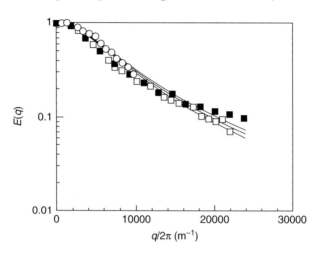

Fig. 7.25 Normalised echo-attenuation data for PGSE ESR from electrons in (fluoroanthene)$_2$PF$_6$ sample, corresponding to different Δ and δ values of (open circles) 14 μs and 2 μs, (closed squares) 16.5 μs and 2.6 μs, and (open squares) 19.5 μs and 2.6 μs, along with three curves representing best fits to the impenetrable relaxing wall model where $D = 1.8 \times 10^{-4}\,\mathrm{m}^2\,\mathrm{s}^{-1}$, $\bar{l} = 47\,\mu$m, and $\bar{p} = 0.01$. (Reproduced by permission from reference [43].)

7.5   Flow diffraction

The diffraction pattern shown in Fig. 7.1 arises because diffusing water molecules migrate from pore to pore in the beadpack, thus embedding in the displacement propagator features characteristic of the inter-pore geometry. In particular there exists a higher probability of displacement at distances corresponding to the pore spacing d, thus leading to the characteristic Bragg peak at $q/2\pi \sim 1/d$.

Of course diffusion merely provides the transport mechanism whereby the pore space is interrogated. It is equally possible to use advection. Such 'flow diffraction' effects were first observed for a system comprising a monodisperse packing of 90.7 micron diameter polystyrene spheres in deionised water. Just as the optimal time for observation of coherence via Brownian motion is $\Delta \sim b^2/D_{\mathrm{eff}}$, so the optimal evolution time for flow coherence is $b/<v>$. Note that for this system the Brownian time is around 4 s, a time at which magnetic relaxation completely destroys the signal. Hence the flow method provides a means of examining much larger pore spacings, in the present case giving an optimal diffraction for a conveniently chosen $\Delta = 30\,\text{ms}$ when $<v> = 3.3\,\text{mm s}^{-1}$. Results for the echo-attenuation function $E(q)$ are shown in Fig. 7.26, where the coherence peak due to displacement correlation of nuclei over the pore spacing is clearly evident as velocity is increased. Note the convenient control in this type of measurement, as increasing the velocity results in an increased length scale in the sampling of the pore spaces displaced from the starting pore.

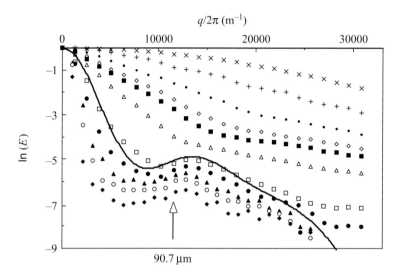

Fig. 7.26 $E(q)$ vs q for water flowing in a 90.7-micron diameter bead pack in which the magnetic field gradient is applied along the axial flow direction. Here $\Delta = 30\,\text{ms}$ and a coherence peak is evident for velocities corresponding to $\Delta > b/<v>$. The data shown range from 0 to 8.7 mm/s. The line is a fit of eqn 7.70 to the 3.3 mm/s data (open squares). (Reproduced from reference [45].)

One set of data, corresponding to a mean flow rate of $<v> = 3.3\,\text{mm/s}$, is fitted in Fig. 7.26, using the diffusive diffraction pore hopping relation, eqn 7.70. In this case D_{eff} corresponds to the longitudinal dispersion coefficient obtained from the low-q limit of the data. The resulting parameters are $b = 82.5\,\mu\text{m}$, $a = 22.4\,\mu\text{m}$, and $\xi = 12.2\,\mu\text{m}$, quite reasonable estimates given the known bead diameter.

7.6   Related issues

7.6.1   Frequency-domain modulated gradient NMR and diffusive diffraction

An interesting question, which has generated a number of articles [46–49], concerns the ability of FDMG NMR experiments to be sensitive to diffusive diffraction effects. Recall that the mode of operation is typically a CPMG echo train, interspersed with gradient pulses or carried out in the presence of a constant magnetic field gradient. In order to clearly define the frequency, the echo train comprises many periods, each involving a small attenuation effect, which gradually accumulates with period number.

The method is effectively a manifestation of the central limit theorem in action, each period representing a low-q perturbation in which the Gaussian approximation applies, the cumulative effect of many Gaussians being ultimately a Gaussian echo train decay. Of course the Gaussian phase case can never embody coherence effects, and no such effects have been seen in these experiments.

The positive side of this ledger is that the cumulant truncation that underpins the interpretation of FDMG NMR is soundly based. And of course, there are good theories available for restricted diffusion under the Gaussian phase approximation, those of Robertson [50] and Neuman [51] being early examples. Kuchel *et al.* [52] have considerably extended this work, carrying out detailed calculations and numerical simulations of restricted diffusive effects in spheres with relaxing walls, under the GPA.

7.6.2 Return to origin probability

The narrow gradient pulse PGSE NMR experiments provide a means of accessing information about the probability that spin-bearing molecules displace a characteristic distance over a clearly defined diffusion time $t = \Delta$. One particularly interesting probability, $P_{RTO}(t)$, concerns the return of particles to origin, since through its time dependence we might gain some insight regarding pore space restrictions and connectivity [53, 54]. The measurement of $P_{RTO}(t)$ follows directly from the definition of the narrow gradient pulse echo attenuation, in eqn 5.81. Integrating each side of this equation we have

$$\int E(\mathbf{q}, t) d\mathbf{q} = \int \rho(\mathbf{r}) \int P(\mathbf{r}|\mathbf{r}', t) \left[\int \exp(i\mathbf{q} \cdot [\mathbf{r}' - \mathbf{r}]) d\mathbf{q} \right] d\mathbf{r}' d\mathbf{r}$$
$$= \int \rho(\mathbf{r}) P(\mathbf{r}|\mathbf{r}, t) d\mathbf{r}$$
$$= \overline{P}(0, t) \tag{7.73}$$

where $P_{RTO}(t) = \overline{P}(0, t)$ is the average propagator in the case $\mathbf{R} = 0$. The return to origin probability is thereby obtained by integrating the normalised echo attenuation over all $\mathbf{q}$.

Of course eqn 7.73 involves a 3-D integration, unnecessary in the case of an isotropic material where one may write [53]

$$\overline{P}(0, t) = 4\pi \int_0^\infty E(q, t) q^2 dq \tag{7.74}$$

In practice, there will be some upper limit q_{max} to the integration, so that one may define the origin only to a spatial precision on the order of q_{max}^{-1}.

For a freely diffusing particle, for which the propagator may be written $(4\pi D_0 t)^{-3/2} e^{-\mathbf{R}^2/4D_0 t}$, $P_{RTO}(t) = (4\pi D_0 t)^{-3/2}$. In a simple enclosing pore space, the rate of decay of $P_{RTO}(t)$ is slower than $t^{-3/2}$. The starting point for this analysis is the eigenmode definition of the conditional probability as $\sum_n u_n(\mathbf{r}) u_n^*(\mathbf{r}') \exp(-k_n^2 t)$, from which we may write

$$\overline{P}(0,t) = \frac{1}{V_p} \sum_n |u_n(\mathbf{r})|^2 \exp(-k_n^2 t)$$

$$= \frac{1}{V_p} \sum_n \exp(-k_n^2 t) \tag{7.75}$$

the pore density being represented by $\rho(\mathbf{r}) = V_p^{-1}$. Consider the limit $t \to 0$. Just as the time-dependent diffusion coefficient, normalised by its free diffusion value D_0, may be shown to diminish at a rate proportional to both $t^{1/2}$ and the pore surface-to-volume ratio, S/V_p, so it may be shown [53] that for small t, $\overline{P}(0,t)$ normalised by its free diffusion limit $(4\pi D_0 t)^{-3/2}$ increases at this same rate. From reference [53]

$$\overline{P}(0,t)(4\pi D_0 t)^{3/2} = 1 + \frac{\sqrt{\pi}}{2} \frac{S}{V_p} \sqrt{D_0 t} + O(D_0 t) \tag{7.76}$$

While diffusion in the pore space is hindered by the reflecting surfaces, reducing the mean-square distance travelled by the particles, these same surfaces have the effect of increasing the return to origin probability. Thus the early-time behaviour of both D_{eff}, which derives from the low q limit of $E(q,t)$, and the return to origin probability, which derives from the q-integral of $E(q,t)$, provide similar insight.

At times comparable with or longer than that required to diffuse, the pore scale, $P_{RTO}(t)$, can provide additional information regarding the connectedness of a porous matrix. Schwartz *et al.* [54] have carried out both simulations and periodic lattice eigenmode calculations to show that $P_{RTO}(t)$, normalised by its free diffusion limit, rises to a maximum at a time $t \sim a^2/D_0$, declining at longer times. These authors argue that, in disordered systems, the existence of the maximum arises from pore connectedness and is controlled by the two competing effects of 'bounce back' and the trapping of diffusing particles in very small pores. While these effects are not well understood, they do have the potential to be used to complement q-space diffraction measurements and the determination of $D_{\text{eff}}(t)$ from the low-q behaviour of the echo attenuation.

References

[1] R. J. Hayward, K. J. Packer, and D. J. Tomlinson. Pulsed field-gradient spin echo N.M.R. studies of flow in fluids. *Molecular Physics*, 23:1083, 1972.

[2] P. T. Callaghan, A. Coy, D. Macgowan, K. J. Packer, and F. O. Zelaya. Diffraction-like effects in NMR diffusion studies of fluids in porous solids. *Nature*, 351:467, 1991.

[3] P. T. Callaghan. Pulsed field gradient nuclear magnetic resonance as a probe of liquid state molecular organisation. *Australian Journal of Physics*, 37:359, 1984.

[4] D. G. Cory and A. N. Garroway. Measurement of translational displacement probabilities by NMR: An indicator of compartmentation. *Magn. Reson. Med.*, 14:435, 1990.

[5] P. T. Callaghan, D. MacGowan, K. J. Packer, and F. O. Zelaya. High resolution q-space imaging in porous structures. *J. Magn. Reson*, 90:177, 1990.

[6] J. E. Tanner and E. O. Stejskal. Restricted self-diffusion of protons in colloidal systems by pulsed-gradient spin echo method. *J. Chem. Phys.*, 49:1768, 1968.

[7] D. C. Champeney. *Fourier transformations and their physical applications*. Academic Press, New York, 1973.

[8] F. B. Laun and T. A. Kuder and W. Semmler and B. Stieltjes. Determination of the defining boundary in magnetic resonance diffusion measurements. *Phys. Rev. Lett.*, 107:048102, 2011.

[9] Partha P. Mitra, Pabitra N. Sen, Lawrence M. Schwartz, and Pierre Le Doussal. Diffusion propagator as a probe of the structure of porous media. *Phys. Rev. Lett.*, 68(24):3555, 1992.

[10] J. E. M. Snaar and H. Van As. NMR self-diffusion measurements in a bounded system with loss of magnetization at the walls. *J. Magn. Reson. A*, A102:318, 1993.

[11] P. T. Callaghan. Pulsed gradient spin echo NMR for planar, cylindrical and spherical pores under conditions of wall relaxation. *J. Magn. Reson. A*, 113:53, 1995.

[12] A. Coy and P. T. Callaghan. Pulsed gradient spin echo nuclear magnetic resonance for molecules diffusing between partially reflecting rectangular barriers. *J. Chem. Phys.*, 101:4599, 1994.

[13] S. J. Gibbs. Observations of diffusive diffraction in a cylindrical pore by PFG NMR. *J. Magn. Reson*, 124:223, 1997.

[14] P. Linse and O. Söderman. The validity of the short-gradient-pulse approximation in NMR studies of restricted diffusion. Simulations of molecules diffusing between planes, in cylinders and spheres. *J. Magn. Reson. A*, 116:77, 1995.

[15] G. B. Arfken and H. J. Weber. *Mathematical Methods for Physicists*. Cambridge, New York, 2005.

[16] B. Balinov, B. Jönsson, P. Linse, and O. Söderman. The NMR self-diffusion method applied to restricted diffusion. Simulation of echo attenuation from molecules in spheres and between planes. *J. Magn. Reson. A*, 104:17, 1993.

[17] M. H. Blees. The effect of finite duration of gradient pulses on the Pulsed-Field-Gradient NMR method for studying restricted diffusion. *J. Magn. Reson. A*, 109:203, 1994.

[18] P. T. Callaghan. A simple matrix formalism for spin echo analysis of restricted diffusion under generalized gradient waveforms. *J. Magn. Reson.*, 129:74, 1997.

[19] P. P. Mitra and B. I. Halperin. The effect of finite gradient-pulse widths in pulsed gradient diffusion measurements. *J. Magn. Reson. A*, 113:94, 1995.

[20] A. Caprihan, L. Z. Wang, and E.Fukushima. A multiple-narrow-pulse approximation for restricted diffusion in a time-varying field gradient. *J. Magn. Reson.*, 118:94, 1996.

[21] D.J. Bergman and K-J. Dunn. Theory of diffusion in a porous medium with applications to pulsed field gradient NMR. *Phys. Rev. B*, 50:9153, 1994.

[22] D. J. Bergman and K-J. Dunn. Self-diffusion of nuclear spins in a porous medium with a periodic microstructure. *J. Chem. Phys.*, 102:3041, 1994.

[23] D. J. Bergman and K-J. Dunn. Self-diffusion in a periodic porous medium with interface absorption. *Phys. Rev. E.*, 51:3401, 1995.

[24] D. J. Bergman, K-J. Dunn, L. M. Schwartz, and P. P. Mitra. Self-diffusion in a periodic porous medium: a comparison of different approaches. *Phys. Rev. E.*, 51:3393, 1995.

[25] P. N. Sen, L. M. Schwartz, P. P. Mitra, and B. I. Halperin. Surface relaxation and the long time diffusion coefficient in porous media: Periodic geometries. *Phys. Rev. B*, 49:215, 1994.

[26] P. N. Sen, C. Scala, and M. H. Cohen. A self-similar model for sedimentary rocks with application to the dielectric constant of fused glass beads. *Geophysics*, 46:781, 1981.

[27] P. T. Callaghan. *Principles of Nuclear Magnetic Resonance Microscopy*. Oxford, New York, 1991.

[28] P. T. Callaghan, A. Coy, T. P. J. Halpin, D. MacGowan, K. J. Packer, and F. O. Zelaya. Diffusion in porous systems and influence of pore morphology in pulsed gradient spin echo nuclear magnetic resonance studies. *J. Chem. Phys.*, 97:651, 1991.

[29] A. Coy and P. T. Callaghan. Pulsed gradient spin echo NMR diffusive diffraction experiments on water surrounding close-packed polymer spheres. *J. Colloid Interface Sci.*, 168:373, 1994.

[30] Pang-Chih Jiang, Tsyr-Yan Yu, Wann-Cherng Perng, and Lian-Pin Hwang. Pore-to-pore hopping model for the interpretation of the pulsed gradient spin echo attenuation of water diffusion in cell suspension systems. *Biophys. J.*, 80:2493, 2001.

[31] B. Håkansson, R. Pons, and O. Söderman. Structure determination of a highly concentrated W/O emulsion using Pulsed-Field-Gradient spin-echo nuclear magnetic resonance diffusion diffractograms. *Langmuir*, 15:988, 1999.

[32] Helena Wassenius, Magnus Nyden, and Brian Vincent. NMR diffusion studies of translational properties of oil inside coreshell latex particles. *J. Colloid Interface Sci.*, 264:538, 2003.

[33] Nirbhay N. Yadav and William S. Price. Impediments to the accurate structural characterisation of a highly concentrated emulsion studied using NMR diffusion diffraction. *J. Colloid Interface Sci.*, 348:163, 2009.

[34] P. W. Kuchel, A. Coy, and P. Stilbs. NMR diffusive diffraction of water revealing alignment of erythrocytes in a magnetic field and their dimesnions and membrane transport characteristics. *Magn. Reson. Med.*, 37:637, 1997.

[35] A. M. Torres, R. J. Michniewicz, B. E. Chapman, G. A. R. Young, and P. W. Kuchel. Characterisatiuon of erythrocyte shapes and sizes by NMR diffusive diffraction of water: correlation with electron micrographs. *Magn. Reson. Imaging*, 16:423, 1998.

[36] G. H. Benga, B. E. Chapman, G. C. Cox, and P. W. Kuchel. Comparative NMR studies of diffusional water permeability of red blood cells from different species: XIV. Little penguin, eudyptula minor. *Cell Biology International*, 27:921, 2003.

[37] D. G. Regan and P. W. Kuchel. Simulations of NMR-detected diffusion in suspensions of red cells: the effects of variation in membrane permeability and observation time. *Eu. Biophys. J.*, 32:671, 2003.

[38] P. W. Kuchel, T. R. Eykyn, and D. G. Regan. Measurement of compartment size in q-space experiments: Fourier transform of the second derivative. *Magn. Reson. Med.*, 52:907, 2004.

[39] G. Pages, D. Szekely, and P. W. Kuchel. Erythrocyte-shape evolution recorded with fast-measurement NMR diffusion-diffraction. *J. Magn. Reson. Imaging*, 28:1409, 2008.

[40] G. G Maresch, A. Grupp, M. Mehring, J. U. van Schutz, and H. C. Wolf. Direct observation of one-dimensional electron spin transport in the organic conductor (FA)2AsF6 by the electron spin echo field gradient technique. *J. Phys. (Paris)*, 46:461, 1983.

[41] R. Ruf, N. Kaplan, and E. Dormann. Restricted diffusion of the conduction electrons in quasi-one-dimensional organic conductors. *Phys. Rev. Lett.*, 74:2122, 1995.

[42] P. T. Callaghan, A. Coy, E. Dormann, R. Ruf, and N. Kaplan. Pulsed gradient spin echo ESR. *J. Magn. Reson. A*, 111:127, 1994.

[43] N. Kaplan, E. Dormann, R. Ruf, A. Coy, and P. T. Callaghan. Restricted electronmotion in a one-diemnsional organic conductor: Pulsed gradient spin echo ESR in (fluoranthene)2PF6. *Phys. Rev. B*, 52:16385, 1995.

[44] M. Conradi, A. N. Garroway, D. G. Cory, and J. Miller. Generation of short intense gradient pulses. *J. Magn. Reson.*, 94:370, 1991.

[45] J. D. Seymour and P. T. Callaghan. 'Flow-diffraction' structural characterization and measurement of hydrodynamic dispersion in porous media by PGSE NMR. *J. Magn. Reson. A*, 122:90, 1996.

[46] J. Stepisnik. Spin echo attenuation of restricted diffusion as a discord of spin phase structure. *J. Magn. Reson.*, 131:339, 1998.

[47] J. Stepisnik. Validity limits of Gaussian approximation in cumulant expansion for diffusion attenuation of spin echo. *Physica B*, 270:110, 1999.

[48] J. Stepisnik. A new view of the spin echo diffusive diffraction in porous structures. *Europhys. Lett.*, 60:453, 2002.

[49] J. Stepisnik. Averaged propagator of restricted motion from the Gaussian approximation of spin echo. *Physica B*, 344:214, 2004.

[50] B. Robertson. Spin-echo decay of spins diffusing in a bounded region. *Phys. Rev.*, 151:273, 1966.

[51] C. H. Neuman. Spin-echo of spins diffusing in a bounded medium. *J. Chem. Phys.*, 60:4508, 1974.

[52] P. W. Kuchel, A. J. Lennon, and C. Durrant. Analytical solutions and simulations for spin-echo measurements of diffusion of spins in a sphere with surface and bulk relaxation. *J. Magn. Reson. B*, 112:1, 1996.

[53] P. P. Mitra, L. L. Latour, R. L. Kleinber, and C. H. Sotak. Pulsed field gradient measurements of restricted diffusion and the return-to-origin probability. *J. Magn. Reson. A*, 114:47, 1995.

[54] L. M. Schwartz, M. D. Hürlimann, K-J. Dunn, P. P. Mitra, and D. J. Bergman. Restricted diffusion and the return to the origin probability at intermediate and long times. *Phys. Rev. E*, 55:4225, 1996.

8
Double wavevector encoding

It is a remarkable feature of nuclear spin magnetisation that coherences created by RF pulses can persist for times on the order of seconds, thus making possible controlled evolution under the various terms of the spin Hamiltonian. Therein lies the basis of multi-dimensional NMR, in which evolution prior to the detection of the NMR signal imparts amplitude and phase information, which may be analysed in terms of those precursor time intervals. This chapter and the next address multiple encoding as it relates to spin translational motion. In particular, the use of magnetic field gradient pulses allow one to derive information about diffusion, flow, or molecular displacement distributions directly. If two successive PGSE pairs are used, then depending on the directions of these pulses, successive displacements of the same molecule may be correlated by orientation, or simply by directional sign. In the latter case an 'opposed' pair of identical amplitude gradient pulses will leave a phase shift that depends only on the change of velocity between the separate encodings, This is the so-called 'compensated double PGSE' NMR experiment. Alternatively, the amplitudes of each pair might be independently varied and the resulting signal Fourier transformed in each independent dimension, to produce a 2-D plot of correlated displacement distributions, commonly labelled a velocity exchange or VEXSY experiment. Such multidimensional PGSE NMR methods will be discussed in detail in Chapter 9.

First, however, we focus our attention on the double PGSE NMR experiment. Simply doubling the $\mathbf{q}$ encoding step gives us access to quite new information, though care is needed to ensure that the chosen pulse sequence and phase cycle select the appropriately encoded magnetisation coherence. Double PGSE NMR has particularly powerful application in the study of complex flow, in revealing the velocity autocorrelation function, and, in a targeted superposition of experiments, the non-local dispersion tensor. But we will also see that double encoding produces interesting outcomes even when the translational dynamics are governed by Brownian motion. In the case where molecules experience restricted diffusion in locally anisotropic pores or domains, the echo attenuation will depend on the relative orientation of the gradient pulse pairs, thus characterising local anisotropy despite a globally random orientation distribution for the domains. As we will see, a critical factor concerns the length of time allowed between the PGSE NMR pulse pairs, the so-called 'mixing time', τ_m. If τ_m is short compared with the time to diffuse between boundaries, then diffusive diffraction behaviour is dramatically different from that experienced with single wavevector encoding—more pronounced, and more information rich.

8.1 Double PGSE NMR

8.1.1 The double scattering process

Figure 8.1(a) shows a schematic double PGSE NMR experiment with independent gradient pairs. The delay between the gradient pulse pairs is the mixing time, τ_m. In Chapter 5, the analogy between the PGSE NMR measurement and inelastic neutron scattering was outlined. In the latter, neutrons undergo a momentum transfer in which a small amount of energy is exchanged due to the motion of the scattering nucleus. Extending this analogy to the representation of double PGSE NMR as a double scattering process requires an extra leap of imagination. Nonetheless, the coherent nature of the NMR evolution, in which the same nucleus experiences two independent sets of phase shifts from successive PGSE pairs, $\mathbf{q}$ and $\mathbf{q}'$, demands that we picture the scattering analogy as shown in Fig. 8.1.

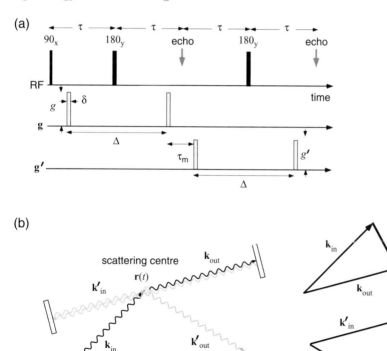

Fig. 8.1 (a) Schematic double PGSE sequence with independent gradients $\mathbf{g}'$ and $\mathbf{g}'$. The delay between the gradient pulse pairs is the mixing time, τ_m. (b) A somewhat implausible neutron scattering representation of double PGSE encoding, in which the scattered neutron is first reflected by mirrors (which define the initial emergence direction and hence $\mathbf{q}$) so as to return for a second scattering event, such that the final scattering vector is $\mathbf{q}'$, and defined by the direction of the detected neutron. The second scattering waves are shown in grey.

Here, the first scattered neutron is rerouted via mirrors to undergo a second scattering event, where judicious choice of path lengths could in principle be used to define τ_m. We know $\mathbf{q}$ by our choice of mirror arrangement, and we know $\mathbf{q}'$ by virtue of our detection of the emerging neutron. Of course, this picture is fanciful, especially given that the second scattering event must involve the original scattering centre, but it represents what we must do in order to make the inelastic neutron scattering experiment match the performance of double PGSE NMR. The very implausibility of such a scattering picture underscores the remarkable flexibility of PGSE NMR when it comes to tailored experimental design.

8.1.2 Compensated and uncompensated sequences

Spin-echo and stimulated-echo versions of the double PGSE NMR pulse sequence are shown in Fig. 8.2, along with their effective gradient equivalents. In these examples the second gradient pulse pair is chosen to be of exactly the same amplitude as the first, and applied along the same direction, or, as in the case of (c), in the same direction but with opposite sense. These particular versions of double PGSE NMR fall into a special class known as 'compensating', as in Fig. 8.2(a) and (b), or 'uncompensating', as in Fig. 8.2(c). While the uncompensated version has only the zeroth ($\int_0^t g^*(t')dt'$) moment zero at the echo formation, the compensated version has both zeroth and first moments ($\int_0^t t'g^*(t')dt'$) zero, and so is insensitive to mean flow, but sensitive to acceleration or velocity fluctuations. At first sight, the uncompensated double PGSE NMR sequence would appear to have little value, but, as we shall see later, it can provide a valuable reference experiment.

The constraining of the $\mathbf{q}$ and $\mathbf{q}'$ vectors to be collinear, albeit with the same or opposite sense, is particularly useful in a wide class of double scattering experiments, especially where flow compensation is desired. However, as we will see in later sections, allowing $\mathbf{q}$ and $\mathbf{q}'$ to assume a range of relative orientations can result in some quite unique insights.

8.2 Double PGSE NMR and dispersion

8.2.1 Propagator and ensemble descriptions

For the moment, we focus our attention on the compensated double PGSE method. For simplicity, we will analyse the effect of this sequence under the assumption of the narrow gradient pulse approximation, although we may, by simulation or by use of the multiple propagator method, extend our analysis to account for finite pulse-width effects. Let us allow that at the first gradient pulse, the spin-bearing molecule is at position $\mathbf{r} = \mathbf{r}(0)$, while at application of the second pulse a time Δ later, its position is $\mathbf{r}' = \mathbf{r}(\Delta)$. After the 'mixing time' and at the application of the third gradient pulse it is $\mathbf{r}'' = \mathbf{r}(\Delta + \tau_m)$, and then at the final gradient pulse it is $\mathbf{r}''' = \mathbf{r}(2\Delta + \tau_m)$. The accumulated spin phase is therefore $\exp\left(i\mathbf{q} \cdot [\mathbf{r}(2\Delta + \tau_m) - \mathbf{r}(\Delta + \tau_m) - \mathbf{r}(\Delta) + \mathbf{r}(0)]\right)$. Thus the normalised echo signal may be written, as the ensemble average

$$E_D(\mathbf{q}) = \langle \exp\left(i\mathbf{q} \cdot [\mathbf{r}(2\Delta + \tau_m) - \mathbf{r}(\Delta + \tau_m) - \mathbf{r}(\Delta) + \mathbf{r}(0)]\right)\rangle \qquad (8.1)$$

or, using using the conditional probabilities,

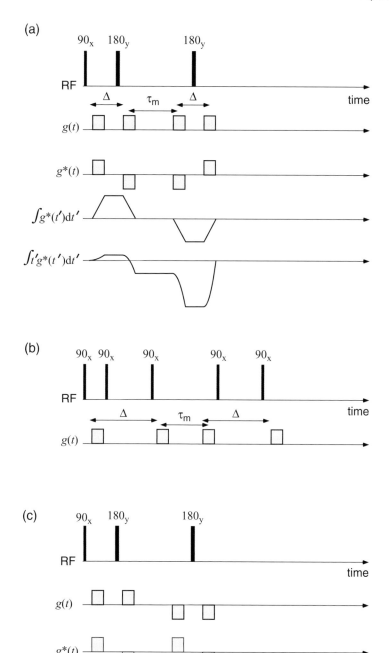

Fig. 8.2 Schematic double PGSE sequences in which the mixing time is indicated by τ_m. (a) Compensated double PGSE using spin echoes, (b) compensated double PGSE using stimulated echoes, (c) uncompensated double PGSE sequence.

$$E_D(\mathbf{q}) = \int d\mathbf{r}\rho(\mathbf{r}) \int d\mathbf{r}' P(\mathbf{r}|\mathbf{r}', \Delta) \int d\mathbf{r}'' P(\mathbf{r}'|\mathbf{r}'', \tau_m)$$
$$\times \int d\mathbf{r}''' P(\mathbf{r}''|\mathbf{r}''', \Delta) \exp\left(i\mathbf{q} \cdot [\mathbf{r}''' - \mathbf{r}'' - \mathbf{r}' + \mathbf{r}]\right) \qquad (8.2)$$

The propagator description turns out to be especially helpful when we look at the effect of restricted diffusion in double PGSE NMR sections. For the moment, however, we will take a look at how the method may be used to study fluctuating flows.

8.2.2   Dispersion measurement and the low-q limit

Single PGSE measurement of dispersion
Fluid dispersion is most easily described in terms of the ensemble of Lagrangian velocities, $\mathbf{v}(t)$, in which the mean flow velocity is $\mathbf{V} = \langle \mathbf{v}(t) \rangle$, while the fluctuation about the mean is $\mathbf{u}(t)$. In relating this to the PGSE NMR experiment, it becomes natural to substitute the displacement $\mathbf{r}(\Delta) - \mathbf{r}(0)$ by $\int_0^\Delta \mathbf{v}(t)dt$. For a single PGSE NMR sequence,

$$E_S(\mathbf{q}) = \left\langle \exp\left(i\mathbf{q} \cdot \int_0^\Delta \mathbf{v}(t)dt\right) \right\rangle$$
$$= \exp(i\mathbf{q} \cdot \mathbf{V}\Delta) \left\langle \exp\left(i\mathbf{q} \cdot \int_0^\Delta \mathbf{u}(t)dt\right) \right\rangle \qquad (8.3)$$

whence

$$|E_S(\mathbf{q})| \approx 1 - \tfrac{1}{2}q^2 \int_0^\Delta \int_0^\Delta \langle \mathbf{u}(t)\mathbf{u}(t') \rangle dt dt' \qquad (8.4)$$

Note that these relationships allow us to define an effective dispersion coefficient according to the second definition of Section 6.1, namely

$$D^*_{eff} = -\frac{1}{\Delta} \lim_{q \to 0} \frac{\partial^2}{\partial q^2} \ln|E_S(q)| \qquad (8.5)$$

where $q = |\mathbf{q}|$ and the dispersion is measured for the velocity component of $\mathbf{u}(t)$ in the direction of $\mathbf{q}$ [1]. Such an effective dispersion coefficient is simply obtained in practice by finding the slope of $\ln|E_S(q)|$ *vs* q^2 in the low-q limit. A typical flow geometry might be that through a cylindrical porous medium in which axial symmetry prevails. In that case $\parallel$ and $\perp$ directions may be defined and

$$D^*_{\parallel,\perp} = \frac{1}{2\Delta} \int_0^\Delta \int_0^\Delta \langle u_{\parallel,\perp}(t)u_{\parallel,\perp}(t') \rangle dt dt' \qquad (8.6)$$

Such PGSE NMR methods have been extensively used [1–4] to study dispersion in porous media.

Double PGSE measurement of dispersion

For the compensated double PGSE NMR sequence shown in Fig. 8.2(a) and (b) the mean flow term $\mathbf{V}$ is automatically cancelled [5, 6], as illustrated in Fig. 8.3, and

$$E_D(\mathbf{q}) = \left\langle \exp\left(i\mathbf{q} \cdot \int_0^{\Delta} \mathbf{u}(t)dt \right) \right.$$

$$\left. \times \exp\left(-i\mathbf{q} \cdot \int_{\tau_m+\Delta}^{\tau_m+2\Delta} \mathbf{u}(t)dt \right) \right\rangle \tag{8.7}$$

Now, in the low-q limit

$$E_D(\mathbf{q}) \approx 1 - \tfrac{1}{2}q^2 \left[\int_0^{\Delta} \int_0^{\Delta} \langle \mathbf{u}(t)\mathbf{u}(t')\rangle dt dt' \right.$$

$$- 2 \int_0^{\Delta} \int_{\tau_m}^{\tau_m+\Delta} \langle \mathbf{u}(t)\mathbf{u}(t')\rangle dt dt'$$

$$\left. + \int_{\tau_m+\Delta}^{\tau_m+2\Delta} \int_{\tau_m+\Delta}^{\tau_m+2\Delta} \langle \mathbf{u}(t)\mathbf{u}(t')\rangle dt dt' \right] \tag{8.8}$$

and, allowing for the stationarity of $\mathbf{u}(t)$,

$$E_D(\mathbf{q}) \approx 1 - \tfrac{1}{2}q^2 \left[2\int_0^{\Delta} \int_0^{\Delta} \langle \mathbf{u}(t)\mathbf{u}(t')\rangle dt dt' \right.$$

$$\left. - 2 \int_0^{\Delta} \int_{\tau_m}^{\tau_m+\Delta} \langle \mathbf{u}(t)\mathbf{u}(t')\rangle dt dt' \right] \tag{8.9}$$

Equation 8.9 is particularly interesting. First, consider the case where the mixing time τ_m is much longer than any characteristic correlation time τ_v for the fluctuating $\mathbf{u}(t)$. Then the displacements during the two successive encoding intervals are completely uncorrelated and the second integral in the sum is zero. In that case the double PGSE NMR experiment will yield twice the stochastic part of the exponent for single PGSE NMR, and $E_D = |E_S|^2$. In this sense, any use of the low-q limit to define an effective dispersion coefficient using the double PGSE method, whatever the value of τ_m, implies an additional factor of two in our analysis of the initial slope.

Second, because there are two pairs of gradient pulses, each fulfilling the echo condition, the double PGSE sequence may be run with different directions for the first and second pulse pairs. Let us label these α and β. In that case we would rewrite eqn 8.8 as

$$E_D(q_\alpha, q_\beta) \approx 1 - \tfrac{1}{2} \left[q_\alpha^2 \int_0^{\Delta} \int_0^{\Delta} \langle u_\alpha(t)u_\alpha(t')\rangle dt dt' \right.$$

$$+ q_\beta^2 \int_{\tau_m+\Delta}^{\tau_m+2\Delta} \int_{\tau_m+\Delta}^{\tau_m+2\Delta} \langle u_\beta(t)u_\beta(t')\rangle dt dt'$$

$$\left. - 2q_\alpha q_\beta \int_0^{\Delta} \int_{\tau_m}^{\tau_m+\Delta} \langle u_\alpha(t)u_\beta(t')\rangle dt dt' \right] \tag{8.10}$$

and by this means it is possible to correlate independent velocity directions at different times.

Measuring the velocity autocorrelation function

Finally, suppose we are able to make Δ sufficiently short ($\Delta \ll \tau_v$) that the fluctuating part of the velocity is effectively constant over the encoding interval Δ. In that case eqns 8.9 and 8.10 tell us that the double PGSE NMR method gives us a nice means of measuring the velocity autocorrelation function of $\mathbf{u}(t)$.

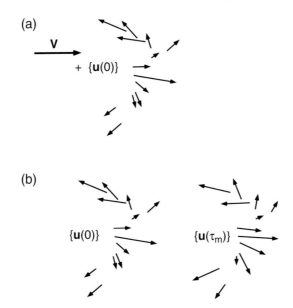

Fig. 8.3 The sensitivity of single and double PGSE to velocity distributions in the case $\Delta \ll \tau_v$. (a) Single PGSE, where the mean flow $\mathbf{V}$ imparts a phase shift $\exp(i\mathbf{q} \cdot \mathbf{V}\Delta)$ and the remaining velocity distribution contributes a phase factor $\langle \exp(i\mathbf{q} \cdot \mathbf{u}\Delta)\rangle$. (b) Double PGSE, where the mean flow term disappears and the phase factor arises from contributions at times 0 and τ_m as $\langle \exp(i\mathbf{q} \cdot [\mathbf{u}(0) - \mathbf{u}(\tau_m)]\Delta)\rangle$.

For $\Delta \ll \tau_v$, as $q^2 \to 0$, and with $\mathbf{q}$ in the α direction, we have

$$E_D(q_\alpha) \approx 1 - \tfrac{1}{2}q_\alpha^2 \left[2\Delta^2 \langle u_\alpha^2 \rangle dt dt' \right.$$

$$\left. - 2\Delta^2 \langle u_\alpha(0) u_\alpha(\tau_m)\rangle dt dt' \right] \tag{8.11}$$

Clearly, analysis of the low-q^2 slope gives

$$-\frac{1}{\Delta} \lim_{q_\alpha \to 0} \frac{\partial^2}{\partial q_\alpha^2} \ln|E_D(q_\alpha)| = \langle u_\alpha^2 \rangle \Delta - \langle u_\alpha(0) u_\alpha(\tau_m)\rangle \Delta \tag{8.12}$$

where $\langle \mathbf{u}(0)\mathbf{u}(\tau_m)\rangle$ is the velocity autocorrelation function, VACF. This function may be obtained by measuring the effective dispersion coefficient for the double PGSE NMR experiment as a function of τ_m, so that the normalised VACF is

$$\frac{\langle u_\alpha(0) u_\alpha(\tau_m)\rangle}{\langle u_\alpha^2 \rangle} = \frac{D_{eff}(\infty) - D_{eff}(\tau_m)}{D_{eff}(\infty)} \tag{8.13}$$

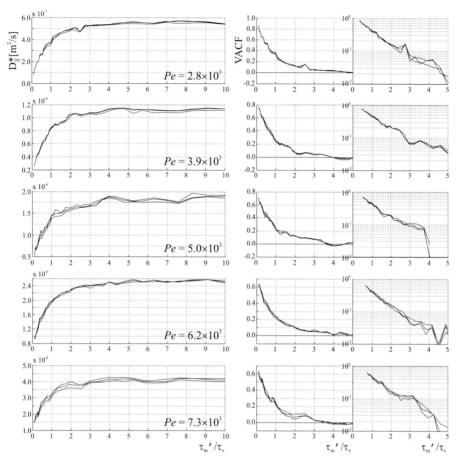

Fig. 8.4 τ_m dependence of apparent longitudinal dispersion for double PGSE and velocity autocorrelation functions (VACF) as a function of Péclet number. Note that the effective mixing time, $\tau_m' = \tau_m + \Delta$, is used. The semi-logarithmic plots of VACF reveal near single exponential behaviour. Note that the three curves in each graph were obtained by analysing the echo signals to three different levels of attenuation, exp(-0.3), exp(-0.4), and exp(-0.5). (Reproduced with permission from reference [7].)

Figure 8.4 shows the result of just such an experiment, performed for water flowing through a bead pack comprising 0.5 mm diameter latex spheres, and for a number of flow rates corresponding to Péclet numbers ranging from 2800 to 7300. In each case the times are shown normalised to the calculated flow correlation time $\tau_v = d/V$. The VACFs obtained appear single exponential and in each case exhibit a time constant close to τ_v

The extension to the case of a pair of gradient pulses of different direction is possible, given eqn 8.10. Here $\langle u_\alpha(0)u_\beta(\tau_m)\rangle$ may be obtained by from the 2-D function $E_D(q_\alpha, q_\beta)$, using the derivative $\partial^2/\partial q_\alpha \partial q_\beta$.

8.2.3   Reversible and irreversible dispersion

The double PGSE NMR experiment is particularly useful when it comes to distinguishing ensemble mean and temporal mean behaviour. For example, the steady velocity shear produced by a plane Couette cell has zero ensemble velocity average, despite there being a steady flow for each streamline. Of course, Taylor dispersion [8] will introduce fluctuations given sufficient time for molecules to laterally diffuse, but for the moment, let us assume that the 'static' distribution of velocities fluctuates much more slowly that the NMR timescale. The point is that the single PGSE NMR experiment will return a dispersion effect from this static component while the double PGSE NMR experiment will not.

Static and fluctuating parts
Suppose we represent these ideas mathematically by decomposing $\mathbf{u}(t)$ into a 'static' distribution $\delta\mathbf{u}$, with zero ensemble average, and a fluctuating component with zero time and ensemble average, $\mathbf{u}_f(t)$. In our double PGSE NMR experiment, the phase shifts arising from $\delta\mathbf{u}$ will be reversed in the second encoding and so any net contribution will be absent. Assuming that $\delta\mathbf{u}$ and $\mathbf{u}_f(t)$ are completely uncorrelated

$$D^*_{\text{single}} = \frac{1}{2}\langle\delta u^2\rangle\Delta + \frac{1}{2\Delta}\int_0^\Delta\int_0^\Delta\langle u_f(t)u_f(t')\rangle dt dt' \tag{8.14}$$

and

$$D^*_{\text{double}} = \frac{1}{2\Delta}\int_0^\Delta\int_0^\Delta\langle u_f(t)u_f(t')\rangle dt dt'$$
$$- \frac{1}{2\Delta}\int_0^\Delta\int_{\tau_m}^{\tau_m+\Delta}\langle u_f(t)u_f(t')\rangle dt dt' \tag{8.15}$$

Following reference [7], D^*_{single} and D^*_{double} can be evaluated in the case where $u_f(t)$ is describable by an Ornstein–Uhlenbeck process, such that the autocorrelation function for $u_f(t)$ is given by the exponential decay

$$\langle u(t)u(0)\rangle = \langle u^2\rangle\exp(-t/\tau_c) \tag{8.16}$$

Allowing for the stationarity of the ensemble, the integrals are simple to perform and yield

$$D^*_{\text{single}} = \frac{1}{2}\langle\delta u^2\rangle\Delta + \langle u_f^2\rangle\tau_c\left[1 + \frac{\tau_c}{\Delta}\left(e^{-\frac{\Delta}{\tau_c}} - 1\right)\right] \tag{8.17}$$

and

$$D^*_{\text{double}} = \langle u_f^2\rangle\tau_c\left[1 + \frac{\tau_c}{\Delta}\left(e^{-\frac{\Delta}{\tau_c}} - 1\right)\right]$$
$$- \frac{1}{2}\langle u^2\rangle\tau_c\frac{\tau_c}{\Delta}\exp\left(-\frac{\tau_m}{\tau_c}\right)\left[1 - e^{-\frac{\Delta}{\tau_c}}\right]\left[e^{\frac{\Delta}{\tau_c}} - 1\right] \tag{8.18}$$

Limiting cases

The various limiting cases resulting from eqn 8.18 are outlined in detail in reference [7]. Again, considering the limit where the velocity-encoding time, Δ, is much less than the fluctuation correlation time, τ_c.

$$D^*_{\text{single}} \approx \frac{1}{2}\langle \delta u^2 \rangle \Delta + \frac{1}{2}\langle u_f^2 \rangle \Delta \tag{8.19}$$

and

$$D^*_{\text{double}} \approx \frac{1}{2}\langle u_f^2 \rangle \Delta \left[1 - \exp\left(-\frac{\tau_m}{\tau_c} \right) \right] \tag{8.20}$$

Hence comparison of the single and double PGSE NMR experiments not only has the ability to reveal very slowly fluctuating or 'static' velocity spreads, it also provides a nice means by which temporal fluctuations may be accurately determined. As we will see in later sections, the double PGSE experiment has the capacity to significantly enhance the information content obtainable from the sole use of a single pair of gradient pulses.

It is worth noting that the double PGSE experiment, with its sequential reversal in the sign of velocity encoding, is not quite the same as a flow reversal experiment in which the apparent displacements are measured after an equal period of reversed flow. The distinction between these two types of measurement goes to the heart of the meaning of 'irreversibility' in complex flows, and distinctions which arise from the relevant length and timescales of each experiment.

8.3   Phase cycling for double PGSE NMR

Implementing double PGSE NMR is no trivial matter. It is customary to use spin-echo or stimulated-echo methods, so as to refocus unwanted background field inhomogeneity or protect the spins from T_2 relaxation. The extra RF pulses resulting from the doubling process can lead to complications. First, it is necessary to ensure that the phases chosen for the RF pulses do indeed produce the phase superposition required, for example as compensated or uncompensated PGSE versions. Indeed, where stimulated echoes are used, superpositions beyond these two simple cases are possible. Second, phase cycling to remove signal artifacts is inherently longer, since with every RF pulse added to a sequence comes the potential to introduce additional unwanted signal. The reason for this is that NMR pulse sequences are generally based on a presumption of perfect 90° or 180° RF pulses, which have the effect of manipulating the magnetisation generated by the first RF pulse in the train, but which themselves do not introduce additional coherence. Of course, in practice, these pulses are never perfect, if for no other reason than that the RF field is never perfectly homogeneous over the entire sample. For this reason, it is customary to phase-alternate these pulses, at the same time maintaining constant receiver phase locked to the first 90° excitation pulse, so as to always keep the desired signal positive whilst unwanted signals are cancelled. Each additional pulse means a doubling of the phase cycle, and so multiple RF pulse schemes necessarily involve much longer cycles.

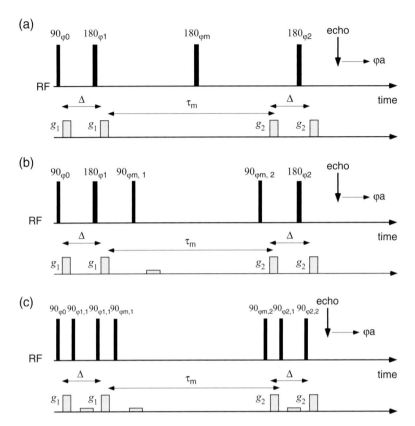

Fig. **8.5** Schematic RF and gradient pulse sequences for double PGSE NMR experiments in which the gradient pulse area $(g\delta)$ is stepped. (a) $90° - g_1 - 180° - g_1 - \tau_m/2 - 180° - \tau_m/2 - g_2 - 180° - g_2 - acq$ version is shown. $g_1 = \pm g_2$ depending on whether compensated or uncompensated phase addition is required. The corresponding phase cycle is given in Table 8.1. (b) As for (a), but for the $90° - g_1 - 180° - g_1 - 90° - \tau_m - 90° - g_2 - 180° - g_2 - acq$ version in which z-storage is used for the mixing period. The corresponding phase cycle is given in Table 8.2. (c) as for (a), but for the $90° - g_1 - 90° - 90° - g_1 - 90° - \tau_m - 90° - g_2 - 90° - 90° - g_2 - acq$ version, in which z-storage is used for the encoding and mixing periods. The corresponding phase cycle is given in Table 8.3. Note that homospoil gradient pulses are included where stimulated echoes or z-storage are used. (Adapted from reference [9].)

8.3.1   Characterising the coherences

Let us deal first with the issue of wanted or unwanted coherence, a subject examined in detail in reference [9]. To do so we will take the pulse sequences of Fig. 8.5 as practical contenders for double PGSE NMR. In Fig. 8.5 a spin-echo scheme is shown, in which the mixing time also contains a 180° RF pulse so as to remove local field inhomogeneity dephasing. It is immediately obvious that this sequence is uncompensated, unless g_1

Table 8.1 Phase cycle table for double PGSE NMR based on three 180° refocusing pulses (Fig. 8.5(a)). Other combinations of RF pulse phases are possible but they generate no coherences beyond the two examples given.

$$90^\circ_{\varphi_0} - g_1 - 180^\circ_{\varphi_1} - g_1 - 180^\circ_{\varphi_m} - g_2 - 180^\circ_{\varphi_2} - g_2 - acq_{\varphi_a}$$

RF pulses				$g_1 = g_2$				$g_1 = -g_2$			
φ_0	φ_1	φ_m	φ_2	U	C	M	φ_a	U	C	M	φ_a
x	$\pm x$	$\pm x$	$\pm x$	$-$	0	0	x	0	$-$	0	x
x	$\pm y$	$\pm y$	$\pm y$	$+$	0	0	x	0	$+$	0	x

and g_2 are of opposite sign. Figures 8.5(b) and (c) show combinations where z-storage and stimulated echoes are used. In each case the RF pulse phase is labelled by a subscript φ, which may be taken to be x or y in the rotating frame, the first sequence in the train being taken to be always x, thus inducing a magnetisation along the y-axis of the rotating frame, which as discussed in Section 5.8, we take for convenience as the real axis of our transverse Argand plane. As discussed in Section 5.8 the effect of the very first gradient pulse is to induce a phase factor $\exp(i\phi_{1,1})$ and, using the rules outlined in that section, we must work out the cumulative effect of the subsequent RF and gradient pulses. In the case of the two pairs of gradient pulses labelled g_1 and g_2, the respective phases induced on a spin isochromat by each of the pulses are $\phi_{1,1}$, $\phi_{1,2}$, $\phi_{2,1}$, and $\phi_{2,2}$, with pair phase differences $\Delta\phi_1 = \phi_{1,2} - \phi_{1,1}$ and $\Delta\phi_2 = \phi_{2,2} - \phi_{2,1}$. The desired phase factor for the compensated (C) double PGSE sequence is $\exp(i\Delta\phi_1 - i\Delta\phi_2)$, while that of the uncompensated (U) is $\exp(i\Delta\phi_1 + i\Delta\phi_2)$.

8.3.2   Coherence selection

The pulse trains of Fig. 8.5 may be analysed through a process that is tedious [9], but with outcomes as summarised in the tables. Note that in the case of the all 180° pulse sequence of Fig. 8.5(a), both pure uncompensated and compensated signals can be obtained, the latter requiring opposite-signed g_1 and g_2 pulse pairs. By contrast, the pulse sequence comprising two spin echoes with a period of z-storage for the mixing period will always result in a superposition of uncompensated (U) and compensated (C) signals. The reason for this is straightforward: since the effect of the first gradient pulse pair is an induced phase difference $\exp(i\Delta\phi_1) = \exp(i[\phi_{1,1} - \phi_{1,2}])$. z-storage involves the maintenance of only one component of the magnetisation generated in the transverse Argand plane, the other component remaining in that place during the mixing time being destroyed by homospoiling or T_2 relaxation. In algebraic terms, we retain only the $\cos(\Delta\phi_1)$ or $\sin(\Delta\phi_1)$ from $\exp(i\Delta\phi_1)$. The result is that both positive and negative phase differences are present to be acted upon by the second gradient pulse pair in the second spin echo. Thus we will find both $\Delta\phi_1 + \Delta\phi_2$ and $-\Delta\phi_1 + \Delta\phi_2$ in the final signal contributions. To obtain pure U or C, a phase cycle involving both x- and y-phase storage pulses is needed. By this means, the appropriate admixture of real and imaginary terms can be obtained after the recall RF pulse following z-storage.

Table 8.2   Phase cycle table for double PGSE NMR based on two 180° refocusing pulses and one 90° pair for z-storage during the mixing time (Fig. 8.5(b)). Again, 180° pulses may have phases of either sign, for the same effect. In this table we include the y phase only for convenience. The table is not exhaustive but gives a sample of outcomes, all of which contain a superposition of U and C terms.

$$90^\circ_{\varphi_0} - g_1 - 180^\circ_{\varphi_1} - g_1 - 90^\circ_{\varphi_{m,1}} - 90^\circ_{\varphi_{m,2}} - g_2 - 180^\circ_{\varphi_2} - g_2 - acq_{\varphi_a}$$

	RF pulses				$g_1 = g_2$				$g_1 = -g_2$		
φ_0	φ_1	$\varphi_{m,1}\ \varphi_{m,2}$	φ_2	U	C	M	φ_a	U	C	M	φ_a
x	$\pm y$	$(-x, x)$	$\pm y$	$+$	$+$	0	x	$+$	$+$	0	x
x	$\pm y$	(x, x)	$\pm y$	$-$	$-$	0	x	$-$	$-$	0	x
x	$\pm y$	$(-y, y)$	$\pm y$	$+$	$-$	0	x	$-$	$+$	0	x
x	$\pm y$	(y, y)	$\pm y$	$-$	$+$	0	x	$+$	$-$	0	x
x	$\pm y$	$(-y, y)$	$\pm y$	$+$	$+$	0	y	$+$	$+$	0	y
x	$\pm y$	(y, y)	$\pm y$	$-$	$-$	0	y	$-$	$-$	0	y

Table 8.3   Phase cycle table for double PGSE NMR based on two stimulated echoes and one 90° pair for z-storage during the mixing time (Fig. 8.5(c)). The table is not exhaustive but gives a sample of outcomes. Note that the mixing term, labelled M, is generally undesired and can be removed by a suitable phase cycle superposition.

$$90^\circ_{\varphi_0} - g_1 - 90^\circ_{\varphi_{1,1}} - 90^\circ_{\varphi_{1,2}} - g_1 - 90^\circ_{\varphi_{m,1}} - 90^\circ_{\varphi_{m,2}} - g_2 - 90^\circ_{\varphi_{2,1}} - 90^\circ_{\varphi_{2,2}} - g_2 - acq_{\varphi_a}$$

	RF pulses				$g_1 = g_2$				$g_1 = -g_2$		
φ_0	φ_1	$\varphi_{m,1}\ \varphi_{m,2}$	φ_2	U	C	M	φ_a	U	C	M	φ_a
x	(x, x)	(x, x)	(x, x)	$-$	$-$	$-$	x	$-$	$-$	$-$	x
x	(y, y)	(y, y)	(y, y)	$+$	$-$	$+$	y	$-$	$+$	$-$	y
x	(x, x)	(x, x)	(x, x)	$+$	$-$	$+$	y	$-$	$+$	$-$	y
x	(y, y)	(x, x)	(y, y)	$-$	$-$	$-$	x	$-$	$-$	$-$	x
x	(x, x)	(x, x)	(y, y)	$+$	$+$	$-$	y	$+$	$+$	$-$	y
x	(y, y)	(x, x)	(x, x)	$+$	$+$	$-$	y	$+$	$+$	$-$	y
x	(y, y)	(y, y)	(x, x)	$-$	$+$	$+$	x	$+$	$-$	$-$	x
x	(x, x)	(y, y)	(y, y)	$-$	$+$	$+$	x	$+$	$-$	$-$	x

The most convenient form of the double PGSE NMR sequence, where spins are to be protected from T_2 relaxation, is that involving two stimulated echoes and one z-storage period, as shown in Fig. 8.5(c). Intriguingly, this sequence generates yet another coherence, neither U nor C in character. Instead it is of the form $\exp(i[\phi_{1,1} + \phi_{1,2}] - i[\phi_{2,1} + \phi_{2,2}])$ and denoted M. Where the magnitudes of g_1 and g_2 are the same, as would customarily be the case, this phase factor approximately represents an encoding for displacement over the mixing time τ_m. It is an unintended and undesired contribution and requires suitable phase cycling to be removed.

Table 8.4 32-step phase cycle table for the double PGSE NMR sequence used to measure the non-local dispersion tensor, and based on two 180° refocusing pulses and one 90° pair for z-storage during the mixing time τ. (Fig. 8.6(b) and (c)). The eight steps shown are repeated for positive and negative first 180° RF pulse phases so as to remove FID effects from this pulse, and repeated a second time, with opposite sign excitation pulse phase and opposite sign acquisition phase, to remove baseline artifacts.

φ_0	φ_1	$\varphi_{m,1}$ $\varphi_{m,2}$	φ_2	φ_a
x	$\pm y$	(x, x)	y	x
x	$\pm y$	$(-x, -x)$	$-y$	x
x	$\pm y$	(y, y)	$-x$	$-x$
x	$\pm y$	$(-y, -y)$	x	$-x$
x	$\pm y$	$(x, -x)$	y	$-x$
x	$\pm y$	$(-x, x)$	$-y$	$-x$
x	$\pm y$	$(y, -y)$	$-x$	x
x	$\pm y$	$(-y, y)$	x	x

Table 8.3 exhibits a wide range of U, C, and M superpositions. By suitable combinations each coherence may be selected as desired. For example, to obtain a pure compensated double PGSE outcome using the stimulated echo, the z-storage sequence of Fig. 8.5(c), the $(x, x)(x, x)(x, x) - x$ and $(y, y)(y, y)(y, y) - y$ combinations serve to eliminate U and M. Where $g_1 = g_2$ the summed cycle is effective, whereas for $g_1 = -g_2$ the difference may be used.

8.4 Non-local dispersion tensor

In Chapter 2, the non-local dispersion tensor was introduced. Defined by the relation

$$\underline{\underline{D}}^{NL}(\mathbf{R}, \tau) = \int P(\mathbf{r})\mathbf{u_E}(\mathbf{r}, 0)P(\mathbf{r}|\mathbf{r} + \mathbf{R}, \tau)\mathbf{u_E}(\mathbf{r} + \mathbf{R}, \tau)d\mathbf{r} \qquad (8.21)$$

$\underline{\underline{D}}^{NL}$ provides insight regarding spatio-temporal correlations in the fluctuating part of the velocity field in dispersive flow. Its measurement provides one of the more technically challenging applications of double PGSE NMR methods [10], requiring the use of a superposition of signals arising from a pulse sequences such as those shown in Fig. 8.6. The elements of the $\underline{\underline{D}}^{NL}$ tensor can be extracted from the response of the spin-echo signal superposition to the independent dimensions of the various gradient pulses, an analysis that requires a careful signal fitting in the low-q domain.

8.4.1 The pulse sequence for velocity and displacement encoding

The pulse sequences of Fig. 8.6 have the gradients $\mathbf{g}$ and $\mathbf{g}_u$ stepped independently [11], where $\mathbf{q}_D = \gamma \mathbf{g} \delta_D$ and $\mathbf{q}_u = \gamma \delta_u \Delta_u \mathbf{g}_u$ are, respectively, conjugate to the dynamic displacement, $\mathbf{R}$, and the local velocity, $\mathbf{v}$. As shown in Figs 8.6(b) and (c), the displacement gradient pulse may be split in two around a 180° RF pulse, thus enabling background inhomogeneity refocusing [12]. Hence $\delta_u = \delta$ and $\delta_D = 2\delta$.

In the case of $\mathbf{q}_D$, it is helpful to use a fully bi-polar displacement encoding to produce a good representation, via Fourier transformation, of propagator information

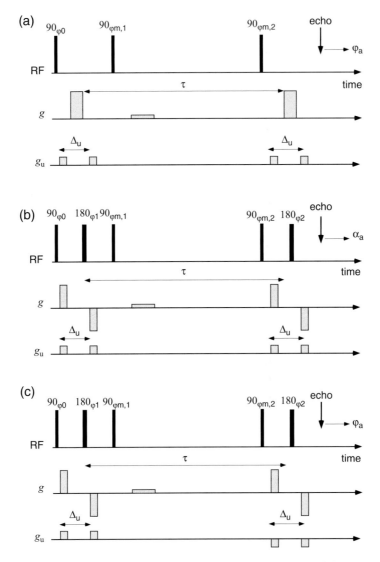

Fig. 8.6 (a) Schematic RF and gradient pulse sequences for non-local dispersion measurement, in which the gradient pulses labelled g refer to displacement encoding, and those labelled g_u are for compensated or uncompensated velocity encoding depending on the sign of the second g_u pair. τ is the mixing time over which displacement is measured. (b) Practical implementation of the compensated sequence based on splitting of the displacement gradient pulse into two, around a $180°$ RF pulse for background inhomogeneity refocusing. The phase cycle is shown in Table 8.4. (c) As for (b), but for the uncompensated sequence. (Adapted from Hunter *et al.* [11].)

in the space $\mathbf{R}$ of $\underline{\underline{D}}^{NL}(\mathbf{R},\tau)$. For the velocity encoding, the low-q_u limit is used, as in the case of direct measurements of the velocity autocorrelation function via double PGSE experiments [7], described in Section 8.1.2. As in those measurements, the time τ, the separation of the double PGSE pulses, corresponds to a 'mixing-time' in the NMR sense, and is the temporal displacement referred to in $\underline{\underline{D}}^{NL}(\mathbf{R},\tau)$.

The NMR signal
The NMR signal is acquired at the final spin echo from the sequence shown in Fig. 8.6(b). When normalised it can be written

$$
\begin{aligned}
E(\mathbf{q}_D,\mathbf{q}_u) = \iint &\exp(i\mathbf{q}_D{\cdot}\mathbf{R})\exp(-i\mathbf{q}_u \cdot (\mathbf{u}_E(\mathbf{r},0)+\langle\mathbf{v}\rangle)) \\
&\times P(\mathbf{r})P(\mathbf{r}|\mathbf{r}+\mathbf{R},\tau)\exp(i\mathbf{q}_u{\cdot}(\mathbf{u}_E(\mathbf{r}+\mathbf{R},\tau)+\langle\mathbf{v}\rangle))d\mathbf{R}d\mathbf{r}
\end{aligned}
\tag{8.22}
$$

The next step in the analysis involves taking the inverse Fourier transform in the $\mathbf{q}_D$ dimension, thus taking us to the domain of the displacement $\mathbf{R}$ and the intermediate signal

$$
\begin{aligned}
S(\mathbf{R},\mathbf{q}_u) = \mathcal{F}_{\mathbf{q}_D}^{-1}\{E(\mathbf{q}_D,\mathbf{q}_u)\} = \int &\exp(-i\mathbf{q}_u{\cdot}\mathbf{u}_E(\mathbf{r},0))P(\mathbf{r})P(\mathbf{r}|\mathbf{r}+\mathbf{R},\tau) \\
&\times \exp(i\mathbf{q}_u{\cdot}\mathbf{u}_E(\mathbf{r}+\mathbf{R},\tau))d\mathbf{r}
\end{aligned}
\tag{8.23}
$$

This notation can be compacted through a convenient change [11] in which the E subscript is dropped, and the integral over starting positions rewritten as an ensemble average. Using this approach the integrand term $P(\mathbf{r})P(\mathbf{r}|\mathbf{r}+\mathbf{R},\tau)$ is replaced by the averaged propagator $\overline{P}(\mathbf{R},\tau)$. Hence

$$
\begin{aligned}
\underline{\underline{D}}^{NL}(\mathbf{R},\tau) &= \int P(\mathbf{r})\mathbf{u}_E(\mathbf{r},0)P(\mathbf{r}|\mathbf{r}+\mathbf{R},\tau)\mathbf{u}_E(\mathbf{r}+\mathbf{R},\tau)d\mathbf{r} \\
&= \langle\mathbf{u}(0)\overline{P}(\mathbf{R},\tau)\mathbf{u}(\tau)\rangle
\end{aligned}
\tag{8.24}
$$

By this means, eqn 8.23 may be rewritten

$$
S(\mathbf{R},\mathbf{q}_u) = \langle\exp(-i\mathbf{q}_u{\cdot}\mathbf{u}(0)\overline{P}(\mathbf{R},\tau)\exp(i\mathbf{q}_u{\cdot}\mathbf{u}(\tau)\rangle
\tag{8.25}
$$

Of course, it is important to treat this new notation as only a shorthand for the full integral expression.

The low-q limit and extracting the elements of D^{NL}
The elements of $\underline{\underline{D}}^{NL}$ can be found by carrying out a succession of measurements using different gradient directions. To illustrate how these elements are extracted, it is helpful to consider, as an example, the case where displacement $(\mathbf{q}_D)$ encoding and velocity $(\mathbf{q}_u)$ encoding are both in the Z direction. The next step in the analysis is to expand the echo-attenuation expression in the low-q_u limit. For the compensated sequence

$$
\begin{aligned}
S_{comp}(Z,q_u) = \langle(1 - iq_u u_z(0) - \tfrac{1}{2}q_u^2 u_z(0)^2 + \dots)\overline{P}(Z,\tau) \\
\times (1 + iq_u u_z(\tau) - \tfrac{1}{2}q_u^2 u_z(\tau)^2 - \dots)\rangle
\end{aligned}
\tag{8.26}
$$

where q_u represents the magnitude of a $\mathbf{q}_u$ vector applied along the z-axis. The equivalent expansion for the compensated pulse sequence gives

$$
S_{comp}(Z, q_u) = \left\langle \left(\left(1 + iq_u \left(u_z(0) - u_z(\tau) \right) - \tfrac{1}{2}q_u^2 \left(u_z(0)^2 + u_z(\tau)^2 \right) \right. \right. \right.
$$
$$
\left. \left. +q_u^2 u_z(0)u_z(\tau) \right) \overline{P}(Z,\tau) \right\rangle
$$
$$
+iO(q_u^3) + O(q_u^4) \tag{8.27}
$$

where truncation errors in q_u^3 and q_u^4 are notated separately, since they arise respectively from imaginary and real parts of the data.

Sequence superposition

The desired non-local dispersion tensor element is present as a partial coefficient of q_u^2. The trick therefore is to extract it cleanly. This involves the use of a second experiment performed with the sign of the second velocity-encoding pulse reversed, as shown Fig. 8.6(b). This version yields

$$
S_{uncomp}(Z, q_u) = \exp\left(i2q_u \langle v_z \rangle \right)
$$
$$
= \times \left\langle \left(\left(1 + iq_u \left(u_z(0) + u_z(\tau) \right) - \tfrac{1}{2}q_u^2 \left(u_z(0)^2 + u_z(\tau)^2 \right) \right. \right. \right.
$$
$$
\left. \left. -q_u^2 u_z(0)u_z(\tau) \right) \overline{P}(Z,\tau) \right\rangle
$$
$$
+iO(q_u^3) + O(q_u^4) \tag{8.28}
$$

The extra phase factor, $\exp\left(i2q_u \langle v_z \rangle \right)$, at the beginning of the expression, arises because the uncompensated version of the double PGSE experiment is sensitive to the mean flow. The phase factor is easily determined and corrected by examining the data $E(0, q_u)$. Performing this phase correction and implementing a difference superposition results in

$$
S_{comp}(Z, q_u) - \exp(-i2q_u \langle v_z \rangle)S_{uncomp}(Z, q_u)
$$
$$
= -i2q_u \langle u_z \overline{P}(Z,\tau) \rangle
$$
$$
+2q_u^2 \langle u_z(0)\overline{P}(Z,\tau)u_z(\tau) \rangle
$$
$$
+iO(q_u^3) + O(q_u^4) \tag{8.29}
$$

A fit of q_u^2 to the real part of the superposition represented by eqn 8.29 provides the desired encoding for the non-local dispersion tensor component $D_{zz}(Z, \tau)$. Alternatively, the tensor may be obtained from the superposition

$$
S_{comp}(Z, q_u) - \exp(-i2q_u \langle v_z \rangle)S_{uncomp}(Z, q_u)
$$
$$
+ S_{comp}(Z, -q_u) - \exp(i2q_u \langle v_z \rangle)S_{uncomp}(Z, -q_u)
$$
$$
= +4q_u^2 \langle u_z(0)\overline{P}(Z,\tau)u_z(\tau) \rangle + O(q_u^4) \tag{8.30}
$$

The complete D^{NL} tensor

The complete non-local tensor contains elements for each combination of velocity and displacement direction, and can be generalised using the dimension subscripts α and β for the initial and final velocity and $\mathbf{q}_u$, respectively, and γ, for the direction of the displacement encoding, denoted X_γ. The non-local dispersion tensor, written in this subscript form is

$$D^{NL}_{\alpha\beta}(X_\gamma, \tau) = \langle u_\alpha(0)\overline{P}(X_\gamma, \tau)u_\beta(\tau)\rangle \tag{8.31}$$

The three choices of direction available for the three different encodings give a total of 27 terms. However, it turns out that for axial flow in a porous medium there are six non-zero independent terms [11].

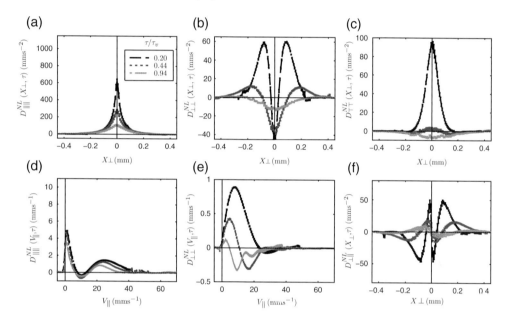

Fig. 8.7 The six independent non-local dispersion tensor components for flow in a random 0.5 mm diameter beadpack of porosity 0.38, measured for three different values of τ. The components in the direction of the main flow are plotted as average velocity, rather than displacement. The error bars show the estimate of the error in the fit to q^2. (Reproduced from Hunter *et al.* [11].)

Omitting the truncation errors, the general expression for the signal superposition of eqn 8.29 becomes

$$
\begin{aligned}
&\exp(iq_{u\alpha}\langle v_\alpha\rangle - iq_{u\beta}\langle v_\beta\rangle)S_{comp}(X_\gamma, q_{u\alpha,\beta}) \\
&- \exp(iq_{u\alpha}\langle v_\alpha\rangle + iq_{u\beta}\langle v_\beta\rangle)S_{uncomp}(X_\gamma, q_{u\alpha,\beta})\}
\end{aligned}
\begin{aligned}
&= iq_{u\alpha}\langle u_\alpha\overline{P}(X_\gamma, \tau)\rangle \\
&\quad + iq_{u\beta}\langle \overline{P}(X_\gamma, \tau)u_\gamma\rangle \\
&\quad + 2q_{u\alpha}q_{u\beta} \\
&\qquad \times \langle u_\alpha(0)\overline{P}(X_\gamma, \tau)u_\beta(\tau)\rangle
\end{aligned} \tag{8.32}
$$

where $q_{u\alpha,\beta}$ means the first and second pairs of velocity-encoding gradients are along α and β, respectively. As before, the extraction of D^{NL} involves either fitting the real part to to $q_{u\alpha}q_{u\beta}$ or using a superposition of $S_{comp}(X_\gamma, -q_{u\alpha,\beta})$ and $S_{un}(X_\gamma, -q_{u\alpha,\beta})$.[1]

8.4.2 Non-local dispersion for porous media flow

Figure 8.7 shows the results obtained using the the double PGSE NMR sequences of Fig. 8.6(b) and (c) for distilled water flowing at a tube velocity of about $10\,\text{mm s}^{-1}$ through a randomly distributed 500-micron diameter latex sphere bead pack ($\tau_v \approx 50\,\text{ms}$) of $10\,\text{mm}$ tube diameter. The velocity-encoding time, $\Delta_u = 2\,\text{ms}$, is much smaller than the displacement encoding times of 10, 21.5, and 46.3 ms. Each measurement takes around 8 h in a 400 MHz magnet. The directions α, β, and γ are fixed by the flow to be longitudinal ($\parallel$) or transverse ($\perp$).

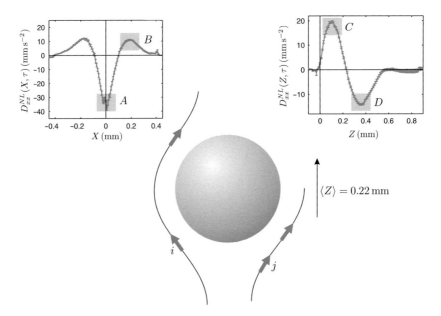

Fig. 8.8 Possible behaviour of particles moving around a bead with experimental measurements. Consider $D_{xx}^{NL}(X,\tau)$ and $D_{xx}^{NL}(Z,\tau)$ at an encoding time such that the mean displacement is approximately half a bead diameter. A guess is made that particles similar to i make up that parts of the graphs A and D, whereas the correlations shown as B and C are from particles similar to j. (Reproduced from Hunter *et al* [11].)

For each of the graphs shown in Fig. 8.7, integration over the displacement X_γ returns one data point for the velocity autocorrelation function at that time. Clearly there is extra information in the non-local dispersion tensor. Just to illustrate the sort of physical insight that is possible, it is instructive to focus attention on the two

---

[1]Where the velocity encoding is along a direction with bulk flow, no phase correction is necessary.

tensor components, $D_{xx}^{NL}(X,\tau)$ and $D_{xx}^{NL}(Z,\tau)$. When integrated over displacement, X or Z, both describe exactly the same temporal correlation function. But look at what they tell us once extracted, as shown in Fig. 8.8, for an encoding time such that the mean displacement is approximately half a bead diameter. They have oscillatory character, with positive lobes (correlations), negative lobes (anti-correlations), and decays to zero (de-correlation). An anti-correlation arises when flows have correlated magnitudes, but opposite directions, exactly as once might expect when considering the transverse velocity components for longitudinal flow in the vicinity of a bead, as seen in the streamline illustrated in the lower part of Fig. 8.8. These anti-correlations are clearly visible in $D_{xx}^{NL}(X,\tau)$ and $D_{xx}^{NL}(Z,\tau)$, at zero displacement along X for $D_{xx}^{NL}(X,\tau)$ and displaced along the flow for $D_{xx}^{NL}(Z,\tau)$. Correlations are visible for small displacements for both tensor elements, while at distances much greater than a bead diameter all correlation disappears.

8.4.3 Non-local dispersion for pipe flow

A nice example of the use of the non-local dispersion tensor to glean physical insight is the case of Taylor dispersion in pipe flow, where diffusion across streamlines causes molecules to be swept apart in the shear field of longitudinal flow. The flow field is shown in Fig. 8.9, and the relevant non-local dispersion is associated with transverse diffusive displacements across stream lines, for which the longitudinal total velocity field is $v_z = \bar{v}(1 - r^2/a^2)$, while the mean-referenced field is $u_z = \bar{v}(1 - 2r^2/a^2)$, $\bar{v}$ being the average longitudinal velocity in the pipe.

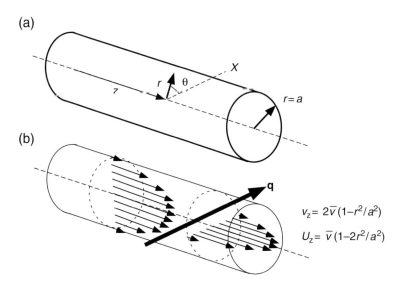

Fig. 8.9 Coordinates for laminar flow in a pipe. The relevant non-local dispersion is associated with transverse diffusive displacements across the stream lines, as probed by wavevector **q**. The direction of **q** defines the displacement axis X used as a reference for cylindrical polar angle θ.

For the pipe, an easy route to D^{NL} is afforded by calculating the Fourier transform of the non-local dispersion tensor with respect to $\mathbf{q}$, the wavevector conjugate to X.

$$E(\mathbf{q}) = \iint d\mathbf{r}d\mathbf{r}' u_z(\mathbf{r}, 0) P(\mathbf{r}) P(\mathbf{r}|\mathbf{r}', \tau) u_z((\mathbf{r}', t) \exp\left(-i\mathbf{q} \cdot (\mathbf{r}' - \mathbf{r}')\right)$$

$$= \frac{1}{\pi a^2} \sum_{n,k} \exp(-\beta_{nk}^2 D_0 \tau/a^2) A_{nk}^2$$

$$\times \int_0^{2\pi} \int_0^{2\pi} \int_0^a \int_0^a \bar{v}(1 - 2r^2/a^2) J_n(\beta_{nk}r/a) \cos(n\theta) J_n(\beta_{nk}r'/a) \cos(n\theta') \bar{v}(1 - 2r'^2/a^2)$$

$$\times \exp(iq(r\cos\theta - r'\cos\theta')) d\theta d\theta' dr dr'$$

$$= \frac{1}{\pi a^2} \sum_{n,k} \exp(-\beta_{nk}^2 D_0 \tau/a^2) A_{nk}^2$$

$$\left| \int_0^{2\pi} \int_0^a \int_0^a \bar{v}(1 - 2r^2/a^2) J_n(\beta_{nk}r/a) \cos(n\theta) \exp(iq(r\cos\theta)) d\theta dr \right|^2$$

$$(8.33)$$

where A_{nk} is given by eqns 7.33 and 7.34, and β_{nk} are the roots of eqn 7.35. Labelling the integral inside the modulus signs I_1, and noting the identity $J_n(x) = (i^{-n}/2\pi) \int_0^{2\pi} \exp(ix\cos\theta) \exp(in\theta) d\theta$,

$$I_1 = \int_0^a \int_0^a \bar{v}(1 - 2r^2/a^2) J_n(\beta_{nk}r/a) \left[\frac{2\pi}{i^n} \frac{1}{2} J_n(qr) + \frac{2\pi i^{4n}}{i^n} \frac{1}{2} J_n(qr) \right] r dr$$

$$= \frac{2\pi}{i^n} \int_0^a \bar{v}(1 - 2r^2/a^2) J_n(\beta_{nk}r/a) J_n(qr) r dr$$

$$(8.34)$$

I_1 may be easily obtained by numerical integration, with eqn 8.33 Fourier transformed to give the non-local dispersion tensor element $D_{zz}^{NL}(X, \tau)$. The results, for a range of values of non-dimensionalised τ values, are shown in Fig. 8.10. Also shown is the integral with respect to X of each of the $D_{zz}^{NL}(X, \tau)$ curves, plotted as a function of τ. This, of course, is the velocity autocorrelation function, which decays, in accordance with Taylor–Aris theory [8, 13, 14], as $\exp(-D_0\mu_1^2\tau/a^2)$, where $\mu_1 = 3.83$ is the first root of the zeroth order Bessel function.[2]

The physical interpretation of the positive and negative correlation and anti-correlation lobes of Fig. 8.10 follows naturally from our knowledge of the Poiseuille flow field. Most obviously, the positive correlation lobe at $X = 0$ decays with time, as more diffusion occurs, and the return to origin probability drops. The negative lobes near $X = \pm a$ arise because of the dominant role of those molecules, the velocities of which change in sign of u_z when displacing by a, the pipe radius. At short τ, these negative lobes are unpopulated, growing as the diffusion time increases and molecules have the opportunity to make significant transverse migrations. The last

---

[2]The decay is, in fact, multiexponential, but the μ_1 term is dominant.

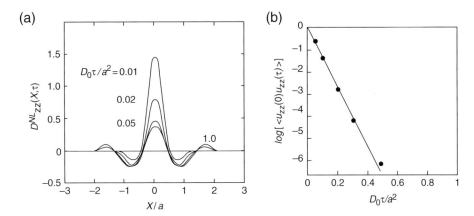

Fig. 8.10 (a) Non-local dispersion tensor element $D_{zz}^{NL}(X,\tau)$ for pipe flow at times τ such that $D_0\tau/a^2$ takes values of 0.01,0.02, 0.05, and 1. (b) Normalised velocity autocorrelation function $\langle u_z(0)u_z(\tau)\rangle$ obtained by integrating the curves in (a). The slope of the line is 3.83^2.

lobes to appear are those positive lobes at displacements near $X = \pm 2a$. These are a direct consequence of the symmetry of the velocity field. Once $D_0\tau/a^2 \gtrsim 1$, the lobe structure persists, since for long τ values all transverse displacements are explored and an asymptotic $D_{zz}^{NL}(X,\tau)$ is reached because of unique starting and end-point velocities for displacements right across the pipe diameter, irrespective of the time taken to make the traverse. Of course, the integral across X of this asymptotic limit is zero, as is immediately apparent in the VACF values shown in Fig. 8.10.

8.5 Restricted diffusion and double wavevector encoding with long mixing time

In studies of fluid advection and dispersion, the advantages of double PGSE NMR are obvious, given the capacity of this method to remove phase shifts due to mean flow. But it also happens that double wavevector encoding provides new insights even when molecular transport is governed by Brownian motion alone and, in particular, when that Brownian motion is restricted by molecular collision with boundaries. A first suggestion of this use was by Cory, Garroway, and Miller [15], who proposed that double encoding could be used to examine effects of local compartment anisotropy in porous media. The first detailed theoretical study of the implications of double wavevector encoding for molecules diffusing in a porous medium was due to P.P. Mitra, who, in a remarkable paper in 1995, showed that where restricted diffusion occurs, even if the pore geometry is isotropic, the echo attenuation paradoxically depends on the relative directions in which the two pairs of gradient pulses are applied. We will examine this paradox in more detail in Section 8.6.

8.5.1 Correlated and uncorrelated phase encoding

One of the remarkable features of NMR is that phase-sensitive signals are acquired from an ensemble of nuclear spins in a coherent manner. In particular, the phase for each spin may undergo a number of controlled evolutions before the signal is measured. In the case of double wavevector encoding, we deal specifically with independent evolution periods associated with each wavevector gradient pulse pair. There are two quite separate senses in which phase evolutions and molecular dynamics may be considered to be correlated or uncorrelated. First, in the most stringent sense, the positions of all spin-bearing molecules may be strictly correlated between the first and second gradient pulse pairs. This would be the case, for example, when the pairs were applied in such rapid succession that molecules had not had a chance to move significantly, so that the position of a molecule at the end of the first pair became the position of the molecule at the beginning of the second. That is the special case dealt with in Section 8.6. But there is a weaker sense of correlation of equal interest, even if molecules have moved significantly over a time interval (the so-called mixing time) that may intervene between the application of the two pulse pairs. The positions of the molecules are now uncorrelated, but provided that they retain the same 'local properties', each molecular-borne spin gains a successive phase shift specific to those individualised properties. For example, if by 'local property' we mean confined to an anisotropic pore or domain, such that any resulting anisotropic translational motion will cause a phase response dependent on the direction of the applied gradient, then we will find that the relative directions of the gradients in the two pulse pairs becomes highly significant.

8.5.2   The propagator description $\tau_m \gg a^2/D$

A clearer understanding of the effect of the double wavevector encoding arises from a use of a propagator description in the narrow gradient pulse approximation. A general version of the pulse sequence is shown in Fig. 8.11. The echo attenuation resulting from this sequence may be written

$$E(\mathbf{q}_1, \mathbf{q}_2) = \iiiint P(\mathbf{r}) \exp(i\mathbf{q}_1 \cdot \mathbf{r}) P(\mathbf{r}|\mathbf{r}', \Delta) \exp(-i\mathbf{q}_1 \cdot \mathbf{r}')$$
$$\times P(\mathbf{r}'|\mathbf{r}'', \tau_m) \exp(-i\mathbf{q}_2 \cdot \mathbf{r}'') P(\mathbf{r}''|\mathbf{r}''', \Delta) \exp(i\mathbf{q}_2 \cdot \mathbf{r}''') d\mathbf{r} d\mathbf{r}' d\mathbf{r}'' d\mathbf{r}''' \quad (8.35)$$

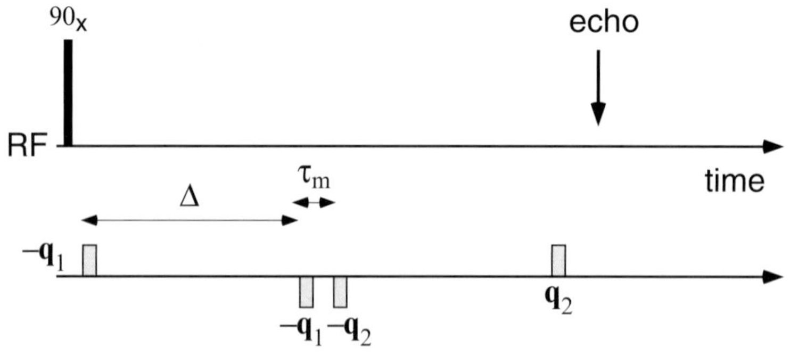

Fig. 8.11 Schematic RF and gradient pulse sequences for a general double wavevector encoding in which the PGSE pairs are represented by wavevectors $\mathbf{q}_1$ and $\mathbf{q}_1$, which may have independent directions.

Let us first deal with the case of free diffusion, where the propagator is always the Gaussian $P(\mathbf{r}|\mathbf{r}', t) = (4\pi Dt)^{-3/2} \exp(-(\mathbf{r}' - \mathbf{r})^2/4Dt)$, and

$$
\begin{aligned}
E(\mathbf{q}_1, \mathbf{q}_2) &= \iiiint P(\mathbf{r}) \exp(-i\mathbf{q}_1 \cdot (\mathbf{r}' - \mathbf{r}))(4\pi D\Delta)^{-3/2} \exp(-(\mathbf{r}' - \mathbf{r})^2/4D\Delta) \\
&\quad \times (4\pi D\tau_m)^{-3/2} \exp(-(\mathbf{r}'' - \mathbf{r}')^2/4D\tau_m) \\
&\quad \times \exp(i\mathbf{q}_2 \cdot (\mathbf{r}''' - \mathbf{r}''))(4\pi D\Delta)^{-3/2} \exp(-(\mathbf{r}''' - \mathbf{r}'')^2/4D\Delta) d\mathbf{r}d\mathbf{r}'d\mathbf{r}''d\mathbf{r}''' \\
&= \exp(-(q_1^2 + q_2^2)D\Delta)
\end{aligned}
$$

(8.36)

This experiment is equivalent to simply multiplying the attenuations obtained from two separate single PGSE experiments carried out with different wavevectors. It contains no new information.

Next we consider what happens when the diffusion is restricted, such that $\Delta \gg a^2/D$ but the mixing time is also sufficiently long, $\tau_m \gg a^2/D$, that restricted diffusion is experienced over τ_m. For each interval, the propagator reduces to $\rho(\mathbf{r})$, so that

$$
\begin{aligned}
E(\mathbf{q}_1, \mathbf{q}_2) &= \iiiint P(\mathbf{r}) \exp(i\mathbf{q}_1 \cdot \mathbf{r})\rho(\mathbf{r}') \exp(-i\mathbf{q}_1 \cdot \mathbf{r}') \\
&\quad \times \rho(\mathbf{r}'') \exp(-i\mathbf{q}_2 \cdot \mathbf{r}'')\rho(\mathbf{r}''') \exp(i\mathbf{q}_2 \cdot \mathbf{r}''') d\mathbf{r}d\mathbf{r}'d\mathbf{r}''d\mathbf{r}''' \\
&= |\int \rho(\mathbf{r}) \exp(i\mathbf{q}_1 \cdot \mathbf{r})d\mathbf{r}|^2 |\int \rho(\mathbf{r}) \exp(i\mathbf{q}_2 \cdot \mathbf{r})d\mathbf{r}|^2 \\
&= |\tilde{\rho}(\mathbf{q}_1)|^2 |\tilde{\rho}(\mathbf{q}_2)|^2
\end{aligned}
$$

(8.37)

where $\tilde{\rho}(\mathbf{q})$ is the pore spectral density. Again, we might be tempted to believe that the experiment is equivalent to simply multiplying the attenuations obtained from two separate single PGSE experiments. However, this is not quite the case. New information is present when the pores are locally anisotropic, because the effect of successive wavevectors of different orientation must be calculated for each pore. The derivation is subtle, and most easily worked out in the low-q limit. We will work to fourth order (q^4) and so expand $\tilde{\rho}(\mathbf{q})$ as

$$
\begin{aligned}
\tilde{\rho}(\mathbf{q}) &= 1 - i \int d\mathbf{r}(\mathbf{q} \cdot \mathbf{r})\rho(\mathbf{r}) - \tfrac{1}{2} \int d\mathbf{r}(\mathbf{q} \cdot \mathbf{r})^2 \rho(\mathbf{r}) \\
&\quad - i\tfrac{1}{6} \int d\mathbf{r}(\mathbf{q} \cdot \mathbf{r})^3 \rho(\mathbf{r}) + \tfrac{1}{24} \int d\mathbf{r}(\mathbf{q} \cdot \mathbf{r})^4 \rho(\mathbf{r}) + \mathcal{O}(q^6) + ...
\end{aligned}
$$

(8.38)

The problem is greatly simplified if we allow for inversion symmetry in $\rho(\mathbf{r})$, in which case all the odd powers disappear and

$$|\tilde{\rho}(\mathbf{q}_1)|^2|\tilde{\rho}(\mathbf{q}_2)|^2 = 1 - \int d\mathbf{r}(\mathbf{q}_1 \cdot \mathbf{r})^2 \rho(\mathbf{r}) - \int d\mathbf{r}(\mathbf{q}_2 \cdot \mathbf{r})^2 \rho(\mathbf{r})$$

$$+ \tfrac{1}{12} \int d\mathbf{r}(\mathbf{q}_1 \cdot \mathbf{r})^4 \rho(\mathbf{r}) + \tfrac{1}{12} \int d\mathbf{r}(\mathbf{q}_2 \cdot \mathbf{r})^4 \rho(\mathbf{r})$$

$$+ \tfrac{1}{4} \left[\int d\mathbf{r}(\mathbf{q}_1 \cdot \mathbf{r})^2 \rho(\mathbf{r}) \right]^2 + \tfrac{1}{4} \left[\int d\mathbf{r}(\mathbf{q}_2 \cdot \mathbf{r})^2 \rho(\mathbf{r}) \right]^2$$

$$+ \int d\mathbf{r}(\mathbf{q}_1 \cdot \mathbf{r})^2 \rho(\mathbf{r}) \int d\mathbf{r}(\mathbf{q}_2 \cdot \mathbf{r})^2 \rho(\mathbf{r}) + \mathcal{O}(q^6) + ...$$

$$(8.39)$$

Now let us address the case where there exists an ensemble of pores, each labelled by j. Then eqn 8.37 must be rewritten

$$E(\mathbf{q}_1, \mathbf{q}_2) = \frac{1}{N} \sum_j^N |\tilde{\rho}_j(\mathbf{q}_1)|^2 |\tilde{\rho}_j(\mathbf{q}_2)|^2 \qquad (8.40)$$

We will see in the next section that the anisotropy to which the double wavevector experiment is sensitive provides a characteristic signal that survives the effect of the ensemble sum.

8.5.3   Local anisotropy, global isotropy

Equation 8.39 involves a correlation of local diffusion along differing relative directions, even though the final average may be taken over all pore orientations. Thus the measurement has an inherent capacity to reveal local pore or diffusion anisotropy, even when the global polydomain structure is isotropic. The way in which this may be analysed has followed two alternative approaches. In the first, due to Cheng and Cory [16], the problem is treated by developing eqn 8.39 as a case of restricted diffusion in isolated pores the shape of which is described by an ellipsoid of revolution, as shown in Fig. 8.12(a). Figure 8.12(b) and (c) present an alternative description in which diffusion is long range but the local diffusion coefficient is anisotropic, as might be found in polydomain lyotropic liquid crystals.

We will deal first with the restricted pore diffusion approach. Suppose we assign $\mathbf{q}_1$ and $\mathbf{q}_2$ the same magnitude q, but with different polar angles with respect to $\mathbf{r}$, θ_1, and θ_2. In eqn 8.39, each term $\mathbf{q}_1 \cdot \mathbf{r}$ is replaced with $qr \cos \theta_1$. Next we allow for all pore orientations by averaging over the ensemble as indicated in eqn 8.40, writing $\frac{1}{N} \sum_j^N = \langle ... \rangle$. Each pore might have some characteristic 'director' expressing its orientation. By way of example, for a pore comprising an ellipsoid of revolution, this director might be the long axis. Clearly, for an isotropic distribution of pore directors, where all possible relative orientations of pore and gradient wavevector are equally allowed, any axis is a suitable polar axis with which to calculate the orientational average of all pore responses. Hence integrals involving a single value of θ are identical, and we may rewrite eqn 8.39 as

$$E(\mathbf{q}_1, \mathbf{q}_2) = 1 - 2q^2 \left\langle \int d\mathbf{r} \cos^2 \theta_1 r^2 \rho(\mathbf{r}) \right\rangle + \tfrac{1}{6} q^4 \left\langle \int d\mathbf{r} \cos^4 \theta_1 r^4 \rho(\mathbf{r}) \right\rangle$$

$$+ \tfrac{1}{2} q^4 \left\langle \left[\int d\mathbf{r} \cos^2 \theta_1 r^2 \rho(\mathbf{r}) \right]^2 \right\rangle$$

$$+ q^4 \left\langle \iint d\mathbf{r} d\mathbf{r}' \cos^2 \theta_1 r^2 \rho(\mathbf{r}) \cos^2 \theta_2 r'^2 \rho(\mathbf{r}') \right\rangle + \mathcal{O}(q^6) + \ldots..$$

$$(8.41)$$

For pores where $\rho(\mathbf{r})$ is anisotropic, the third term in q^4 contains an integral that depends on the relative orientation of the gradients and the shape of the pore. In particular, that integral has an angular weighting $[\cos^2(\theta_1 + \theta_2) + \cos^2(\theta_1 - \theta_2) + 2\cos(\theta_1 - \theta_2)\cos(\theta_1 + \theta_2)]$. The isotropic averaging[3] causes the the third trigonometric term in the square brackets to be zero. However, the second term in the square brackets is non-zero for anisotropic pores and has a coefficient $\cos^2 \Theta$, where $\Theta = (\theta_1 - \theta_2)$, the fixed angle between the two gradient pulses. Thus, it is the this $\cos^2 \Theta$ term that determines the echo-attenuation dependence on the relative orientation of the pulses. Clearly, the greatest difference arises for $\Theta = 0$ and $\Theta = 90°$. The best means of revealing the pore anisotropy will be to carry out two experiments, with the two pairs of gradient pulses collinear or orthogonal. The corresponding echo attenuations are labelled $E_{zz}(q)$ and $E_{zx}(q)$.

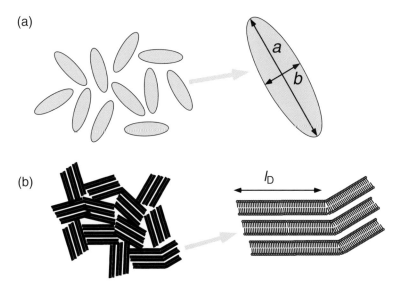

Fig. 8.12 (a) Array of ellipsoidal pores in which diffusion is restricted. The pores are randomly oriented but have a characteristic local anisotropy. (b) Polydomain lamellar lyotropic liquid crystals in which the domains are randomly oriented but have a characteristic local diffusion anisotropy.

---

[3]i.e. allowing θ_1 and θ_2 to independently vary over all polar angles.

Restricted diffusion in ellipsoidal pores

The analysis of Cheng and Cory for ellipsoidal pores leads to the result

$$E_{zz}(q) = 1 - c_2 q^2 + c_4 q^4 + \mathcal{O}(q^6) + \dots$$
$$E_{zx}(q) = 1 - d_2 q^2 + d_4 q^4 + \mathcal{O}(q^6) + \dots \quad (8.42)$$

where

$$c_2 = d_2 = \frac{2}{5}\left(\frac{2}{3}a^2 + \frac{1}{3}b^2\right)$$
$$c_4 - d_4 = \frac{2}{375}\left(a^2 - b^2\right)^2 \quad (8.43)$$

The local anisotropy effect occurs at fourth order in q and, of course, disappears for a sphere where $a = b$. This effect has been nicely demonstrated in a double wavevector restricted diffusion experiment on yeast cells, as shown in Fig. 8.13. The data are fitted to a polynomial, and the value c_2 and difference in the quadratic terms $c_4 - d_4$ are used to fix the ellipsoidal semi-axes at 13 and 2.2 microns for a and b.

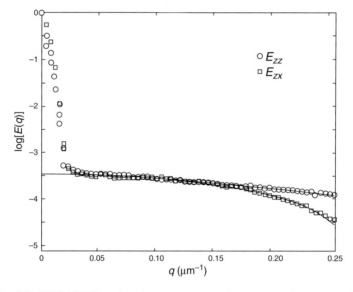

Fig. 8.13 Double PGSE NMR spin-echo-attenuation data curves for yeast cells. The open circles and squares correspond to parallel and perpendicular scattering vectors, respectively. The solid lines are fitted to a power series with seven terms. In these experiments the gradient pulse length was 1.0 ms, $\Delta = 40$ ms, and $\tau_m = 40$ ms. (Adapted from Cheng and Cory [16].)

General expression for anisotropic pores

Lawrenz *et al.* [17] have derived a general fourth order echo-attenuation expression for anisotropic pores in the long mixing time limit. Defining the moment

$R_{jk} = \int_{pore} \rho(\mathbf{r}) r_i r_j d\mathbf{r}$, they find that the double wavevector echo attenuation when $q_1 = q_2 = q$ and the angle between the gradients is Θ, is given by

$$E(q, \Theta) = 1 - \frac{2}{3} \langle r^2 \rangle q^2$$

$$+ \frac{1}{10} \left[\langle r^4 \rangle + \sum_k R_{kk}^2 + \sum_{k \neq l} R_{kk} R_{ll} \right] q^4$$

$$+ \frac{1}{30} \left[\sum_k R_{kk}^2 - \sum_{k \neq l} R_{kk} R_{ll} \right] \cos(2\Theta) q^4 + O(q^6) + \dots$$

$$(8.44)$$

This equation reproduces the result for ellipsoids of revolution given in eqn 8.43. But it also tells us that the angular effect at long mixing times first appears in the q^4 term, and always with an angular dependence $\cos 2\Theta$. Hence the biggest difference in echo attenuations will be between $\Theta = 0$ and $\Theta = \pi/2$.

Locally anisotropic diffusion with ellipsoidal symmetry

In the example shown in Fig. 8.12(b) a different approach is required. Here the use of a locally anisotropic diffusion coefficient allows the calculation of the echo attenuation for the separate cases of E_{zz} and E_{zx}. The method of analysis [18] follows that of Section 6.6.3 and, as for eqn 6.89, assumes that the diffusing molecules remain in their starting domains. In other words, $l_D^2 \gg D\Delta$. For polydomain lamellar liquid crystals the symmetry is such that $D_\perp \gg D_\parallel$ and $D_\perp \approx D_0$, the free molecular self-diffusion coefficient. However, for any $D_\perp$, $D_\parallel$, one obtains the relations

$$E_{zz}(q) = \int_0^1 \exp(-q^2 \Delta [2D_\parallel x^2 + 2D_\perp (1 - x^2)]) dx$$

$$E_{zx}(q) = (2\pi)^{-1} \int_0^{2\pi} \int_0^1 \exp(-q^2 \Delta [D_\parallel x^2 + D_\parallel (1 - x^2) \cos^2 \phi + D_\perp (1 - x^2)$$

$$+ D_\perp \sin^2 \phi + D_\perp x^2 \cos^2 \phi]) d\phi dx$$

$$(8.45)$$

Expanding these expressions as power series to q^4 gives

$$E_{zz}(q) = 1 - q^2 \Delta [\frac{2}{3} D_\parallel + \frac{4}{3} D_\perp]$$
$$+ \frac{1}{2} q^4 \Delta^2 [\frac{12}{15} D_\parallel^2 + \frac{16}{15} D_\perp D_\parallel + \frac{32}{15} D_\perp^2] + \mathcal{O}(q^6) + \dots$$
$$E_{zx}(q) = 1 - q^2 \Delta [\frac{2}{3} D_\parallel + \frac{4}{3} D_\perp]$$
$$+ \frac{1}{2} q^4 \Delta^2 [\frac{8}{15} D_\parallel^2 + \frac{24}{15} D_\perp D_\parallel + \frac{28}{15} D_\perp^2] + \mathcal{O}(q^6) + \dots$$

$$(8.46)$$

Note that as in the previous example, where restricted diffusion in elliptical pores was assumed, the difference between the $E_{zz}(q)$ and $E_{zx}(q)$ attenuations shows up only to fourth order in q. To use the language of the previous example,

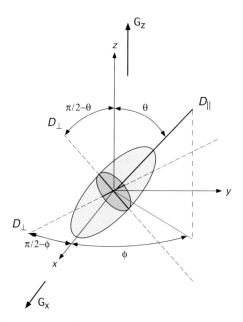

Fig. 8.14 Schematic showing the polar geometry of a randomly oriented locally anisotropic diffusion domain. Cylindrical symmetry is assumed and one of the two $D_\perp$ axes is chosen to be normal to the z direction.

$$c_2 = d_2 = \Delta[\tfrac{2}{3}D_\| + \tfrac{4}{3}D_\perp]$$
$$c_4 - d_4 = \tfrac{2}{15}\Delta^2[D_\| - D_\perp]^2 \qquad (8.47)$$

Again, local isotropy causes the difference in q^4 terms to disappear. Note that for for all relative values of $D_\|$ and $D_\perp$ where these two differ, we always find that the echo attenuation for E_{zx} is greater than for E_{zz}. The reason for this universal effect can be found in the much greater proportion of solid angle available when the polar angle is close to the equator. What eqn 8.47 tells us is that a pulse sequence in which the two gradient pairs have their directions switched between orthogonal orientations will always optimise the proportion of spins with displacements close to the gradient directions. The same universality holds true in the case of restricted diffusion in anisotropic pores, as is evident in the sign independence of $c_4 - d_4$ on the prolate or oblate character of the ellipsoidal pore.

One simple example of a system in which diffusion is locally anisotropic but globally isotropic is provided by a polydomain lyotropic liquid crystal in the lamellar phase. Here the bilayers impede water diffusion normal to the lamellae, while diffusion is relatively free within the intervening aqueous layers. This is an example of 2-D anisotropy in which $D_\| = 0$ and $D_\perp = D$. Figure 8.15 shows results obtained using collinear (E_{zz} and E_{xx}) and orthogonal (E_{zx} and E_{xz}) encoding directions for water diffusing in the polydomain lamellar phase of Aerosol OT/water [19]. This system has been shown by X-ray diffraction experiments [20] to have a lamellar spacing of around 10 nm, at a distance diffused by a water molecule over a time of around 0.025 ms. In the typical PGSE observation times, Δ, of several milliseconds, water molecules will freely diffuse

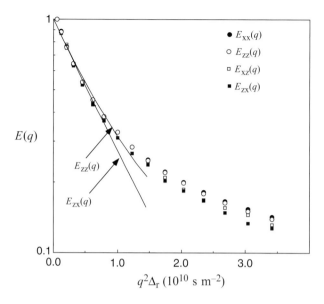

Fig. 8.15 Comparison of echo attention obtained using collinear (E_{zz} and E_{xx}) and orthogonal (E_{zx} and E_{xz}) encoding directions, for water diffusing in the polydomain lamellar phase of Aerosol OT/water. A difference is found in the echo attenuations in the collinear and orthogonal cases, indicating local anisotropy. The theoretical lines are based on eqn 8.45 and fitting the water diffusion coefficient from the initial slope of the data. (Reproduced from reference [18].)

in the $\perp$ directions in the water planes sandwiching the bilayers, the diffusion appearing unrestricted in these directions while effectively zero along the bilayer normals. However, previous experiments suggest domain sizes of around 10 to 15 μm that are traversed only for diffusion observation times of 50 ms or greater, so that values of Δ on the order of 10 ms, as used in the measurements shown in Fig. 8.15, should ensure that water molecules are confined to a single local domain.

The lamellar phase of Aerosol OT is known to be strongly influenced by defect structures and micellar inclusions, which impede free water diffusion along the layers [21]. In the quantitative fit to the data shown in Fig. 8.15, the water diffusion coefficient is 2 x 10^{-10} m^2 s^{-1}, an order of magnitude smaller than for free water. These defects may also explain why the observed ratio between E_{zz} and E_{zx} is less than predicted by eqn 8.47. Nonetheless, the signal difference between collinear and orthogonal double wavevector encoding is apparent.

8.6 Restricted diffusion and double wavevector encoding with infinitesimal mixing

The behaviour of the double wavevector echo attenuation in the case where molecules are restricted to pores becomes quite extraordinary when the mixing time is made to be short, a phenomenon first recognised by Mitra in 1995 [22]. The behaviour is not only paradoxical, but it is also extremely useful, especially when one seeks to elucidate diffusive diffraction effects in disordered systems.

8.6.1 The Mitra paradox

Some physical insight can be obtained by analysing the effect of the double PGSE NMR sequence of Fig. 8.16, where two special cases of are illustrated, both of which have the mixing time τ_m between the pairs much shorter than the characteristic time τ_c to diffuse across a pore. For the purposes of this argument $\tau_m = 0$.

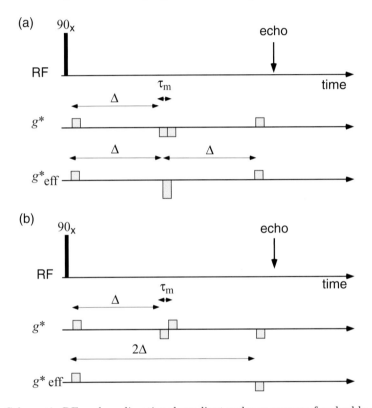

Fig. 8.16 Schematic RF and co-directional gradient pulse sequences for double wavevector encoding using two pairs of gradient pulses applied in quick succession such that the mixing time τ_m is much less that the characteristic time to diffuse across a pore, τ_c. In (a) the pulse pairs have opposite sense, while in (b) they have the same sense. In this latter case the second pulse from the first pair cancels the first from the second pair.

In Fig. 8.16(a), the pairs of gradient pulses are collinear and applied in opposite sense. In (b) they are applied in the identical sense and in consequence the second pulse of the first pair cancels the first pulse of the second pair. The net result is that (a) represents two successive encodings with encoding time Δ, while (b) represents a single encoding pair, but with encoding time 2Δ. Consider first the case of free diffusion, or alternatively $\Delta \ll \tau_c$. In (a), the motions of the molecules in the first and second encoding periods Δ are completely uncorrelated. Both the first and second pulse pairs produce an attenuation $\exp(-q^2 D\Delta)$, such that resulting echo attenuation is $\exp(-2q^2 D\Delta)$. In (b), the single pulse pair results in attenuation $\exp(-q^2 D2\Delta)$. The outcomes of (a) and (b) are identical.

By contrast, consider the case of restricted diffusion, $\Delta \gg \tau_c$. In (b) the equivalent single pulse pair results in attenuation $|S(q)|^2$, but in (a), a rather subtle effect ensues. We might be tempted to argue that the motions of the molecules in the first and second encoding periods Δ are again completely uncorrelated, so that both the first and second pulse pairs produce an attenuation $|S(q)|^2$, resulting in an overall echo attenuation $|S(q)|^4$. The outcomes of (a) and (b) would certainly be quite different. But the $|S(q)|^4$ outcome applies only when the mixing time is long. At infinitesimal mixing, the correlation of molecular positions at the end of the first and start of the second pulse pair requires a more careful analysis. Next we will see how the result may be exactly derived, and for any relative gradient orientations.

8.6.2   Propagator description, $\tau_m \ll a^2/D$

If $\tau_m \ll a^2/D$, the molecules have had insufficient time to change their displacements relative to the local pore geometry, and so we can represent that condition by setting the propagator for the mixing interval to a delta function. Hence eqn 8.37 must be rewritten

$$
\begin{aligned}
E(\mathbf{q}_1, \mathbf{q}_2) = &\iiiint P(\mathbf{r}) \exp(i\mathbf{q}_1 \cdot \mathbf{r}) \rho(\mathbf{r}') \exp(-i\mathbf{q}_1 \cdot \mathbf{r}') \\
&\times \delta(\mathbf{r}'' - \mathbf{r}') \exp(-i\mathbf{q}_2 \cdot \mathbf{r}'') \rho(\mathbf{r}''') \exp(i\mathbf{q}_2 \cdot \mathbf{r}''') d\mathbf{r} d\mathbf{r}' d\mathbf{r}'' d\mathbf{r}''' \\
= &\iiint P(\mathbf{r}) \exp(i\mathbf{q}_1 \cdot \mathbf{r}) \rho(\mathbf{r}') \exp(-i\mathbf{q}_1 \cdot \mathbf{r}') \\
&\times \exp(-i\mathbf{q}_2 \cdot \mathbf{r}') \rho(\mathbf{r}''') \exp(i\mathbf{q}_2 \cdot \mathbf{r}''') d\mathbf{r} d\mathbf{r}' d\mathbf{r}''' \\
= &\int \rho(\mathbf{r}) \exp(i\mathbf{q}_1 \cdot \mathbf{r}) d\mathbf{r} \int \rho(\mathbf{r}''') \exp(i\mathbf{q}_2 \cdot \mathbf{r}''') d\mathbf{r}''' \\
&\times \int \rho(\mathbf{r}') \exp(-i[\mathbf{q}_1 + \mathbf{q}_2] \cdot \mathbf{r}') d\mathbf{r}' \\
= &\tilde{\rho}(\mathbf{q}_1) \tilde{\rho}(\mathbf{q}_2) \tilde{\rho}^*(\mathbf{q}_1 + \mathbf{q}_2)
\end{aligned}
\tag{8.48}
$$

It is interesting to evaluate the prediction of eqn 8.48 along the lines of eqn 8.39. Again, we will assume pores with inversion symmetry, and this time work in the low-q limit and evaluate up to second order (q^2) to obtain

$$
\begin{aligned}
\tilde{\rho}(\mathbf{q}_1) \tilde{\rho}(\mathbf{q}_2) \tilde{\rho}^*(\mathbf{q}_1 + \mathbf{q}_2) = 1 &- \tfrac{1}{2} \int d\mathbf{r} (\mathbf{q}_1 \cdot \mathbf{r})^2 \rho(\mathbf{r}) + \tfrac{1}{2} \int d\mathbf{r} (\mathbf{q}_2 \cdot \mathbf{r})^2 \rho(\mathbf{r}) \\
&- \tfrac{1}{2} \int d\mathbf{r} ((\mathbf{q}_1 + \mathbf{q}_2) \cdot \mathbf{r})^2 \rho(\mathbf{r}) + \mathcal{O}(q^4) + ...
\end{aligned}
\tag{8.49}
$$

Suppose we express eqn 8.49 for the case of isotropic pores where the absolute direction of $\mathbf{q}$ is irrelevant. We will see that their relative direction does, however, turn out to be of considerable importance. For convenience we will assume that the two parts of gradient pulses have equal magnitudes q, but with different directions given the different polar angles, θ_1 and θ_2, which the vector $\mathbf{r}$ makes with respect to $\mathbf{q}_1$ and $\mathbf{q}_2$. With respect to $\mathbf{q}_1 + \mathbf{q}_2$, the polar angle is θ_{12}. Finally, we set the angle between $\mathbf{q}_1$ and $\mathbf{q}_2$ as Θ. Using this notation

$$
\begin{aligned}
E(\mathbf{q}_1, \mathbf{q}_2) &= \tilde{\rho}(\mathbf{q}_1)\tilde{\rho}(\mathbf{q}_2)\tilde{\rho}^*(\mathbf{q}_1 + \mathbf{q}_2) \\
&= \left(1 - \tfrac{1}{2}q^2 \int \cos^2\theta_1 r^2 \rho(\mathbf{r})d\mathbf{r} + \mathcal{O}(q^4) + \ldots\right) \\
&\quad \times \left(1 - \tfrac{1}{2}q^2 \int \cos^2\theta_2 r^2 \rho(\mathbf{r})d\mathbf{r} + \mathcal{O}(q^4) + \ldots\right) \\
&\quad \times \left(1 - \tfrac{1}{2}q^2(2 + 2\cos\Theta) \int \cos^2\theta_{12} r^2 \rho(\mathbf{r})d\mathbf{r} + \mathcal{O}(q^4) + \ldots\right)
\end{aligned}
$$

$$(8.50)$$

where θ_{12} is the angle between the symmetry axis and $\mathbf{q}_1 + \mathbf{q}_2$. Each of the integrals $\int \cos^2\theta r^2 \rho(\mathbf{r})d\mathbf{r}$ are identical and, for the isotropic pore, is simply $\tfrac{1}{3}\langle r^2 \rangle$, where $\langle r^2 \rangle$ is the mean-squared radius of gyration of the pores, $\int r^2 \rho(\mathbf{r})d\mathbf{r}$. Given that, eqn 8.50 becomes

$$
\begin{aligned}
E(\mathbf{q}_1, \mathbf{q}_2) &= 1 - \tfrac{1}{3}q^2 \langle r^2 \rangle (2 + \cos\Theta) + \mathcal{O}(q^4) + \ldots \\
&= 1 - \tfrac{1}{3}q^2 \langle r^2 \rangle \left(1 + 2\cos\frac{\Theta}{2}\right) + \mathcal{O}(q^4) + \ldots
\end{aligned}
$$

$$(8.51)$$

Given our definition in eqn 8.48, where the sign of the $\mathbf{q}_1$ and $\mathbf{q}_2$ vectors is determined by the first and second pulses of the pairs, respectively, the case of Fig. 8.16(a) corresponds to $\cos\Theta = 1$ and Fig. 8.16(b) to $\cos\Theta = -1$. Note that the angular dependence is now in the q^2 term, a point to which we shall return. Unlike the long mixing-time case, the greatest difference in echo attenuation occurs between $\Theta = 0$ and $\Theta = \pi$.

The spherical pore and pore sizing

For a sphere, $\langle r^2 \rangle = \tfrac{3}{5}a^2$. For the case $\cos\Theta = -1$, where the coinciding pulses of the two pairs cancel, $E(\mathbf{q}_1, \mathbf{q}_2) = 1 - \tfrac{1}{5}q^2 a^2$, as found for the single pulse pair in Chapter 7. For $\cos\Theta = 1$, the attenuation is three times greater and $E(\mathbf{q}_1, \mathbf{q}_2) = 1 - \tfrac{3}{5}q^2 a^2$. The fact that the outcome of the experiment depends on the relative orientation of the gradient pulse pairs, despite the fact that no pore anisotropy has been assumed, is indeed remarkable.

One immediate application of the double wavevector encoding is the elucidation of the size of compartments in complex heterogeneous materials such as biological tissue. The appearance of a $\cos\Theta$ angular dependence of the echo attenuation at q^2 indicates restricted diffusion at short mixing time. Most importantly, the coefficient of this term provides an estimate of pore size [22–24]. Figure 8.17 shows the result of a set of double PGSE NMR experiments carried out at different mixing times, in three samples comprising porcine spinal cord, water between acrylate beads, and a radish sample, where the pores sizes are successively increasing. In each case a *cos*Θ angular dependence is seen. At mixing times long compared with the time to diffuse across the pore, the anisotropy attenuates, of course, most rapidly in the sample with smallest pore size. From the amplitude of the $E(q, \pi)/E(q, 0)$ variation, estimates of the pore sizes can be made.

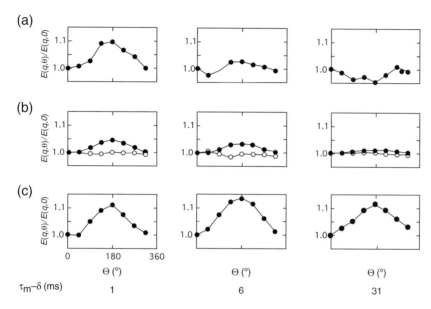

Fig. 8.17 Normalised signal versus Θ for different τ_m values, but at the same gradient pulse strength. (a) Porcine spinal cord ($\delta = 17$ ms), (b) water between acrylate beads ($\delta = 6$ ms) with the bulk water signal (open circles) shown for comparison, and (c) radish sample ($\delta = 6$ ms). The plots in each column share the value $\tau_m - \delta = 1$, 6, and 31 ms (from left to right). Note that the q value is much greater for (a) because of the larger value of δ. The data have been fitted to yield pore sizes (radius a) estimates of 1.9, 4.1, and 7.4 microns, respectively for (a), (b), and (c). Note the absence of Θ dependence for the water sample and the gradual attenuation of the angular effect as the mixing time is increased. (Adapted from Koch and Finsterbusch [23].)

Note that the finite gradient pulse-width value, δ, presents some difficulty in the interpretation, not just because of the narrow gradient pulse approximation associated with the assignment of $\rho(\mathbf{r})$ to the propagator, but also because of uncertainty regarding the nature of the mixing time. However, these effects may be handled using the matrix multiple propagator methods described in Chapter 5. A detailed account of finite time effects using such methods has been given by Özarslan and Basser [25, 26].

Anisotropic pores

How does pore anisotropy reveal itself in the case of short mixing time? Certainly such anisotropy no longer permits the simple symmetry argument that led to eqn 8.51. Instead, eqn 8.49 yields

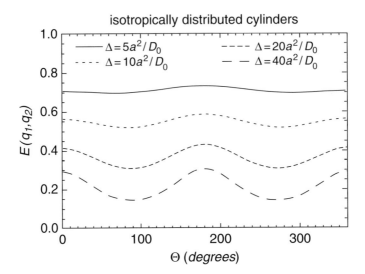

Fig. 8.18 Normalised double PGSE NMR signal from cylinders isotropically distributed in space showing echo attenuation E versus Θ for different Δ values, where $\tau_m = 0.002a^2/D_0$, $(\gamma\delta ga)^2 = 0.1$, and $\delta \to 0$. Note the onset of $\cos^2\Theta$ behaviour as the fourth-order terms appear at deeper attenuation. (Adapted from Özarslan and Basser [26].)

$$E(\mathbf{q}_1, \mathbf{q}_2) = 1 - q^2 \int \cos^2\theta_1 r^2 \rho(\mathbf{r})d\mathbf{r} - q^2 \int \cos^2\theta_2 r^2 \rho(\mathbf{r})d\mathbf{r}$$

$$-q^2(2 + 2\cos\Theta) \int \cos^2\theta_{12} r^2 \rho(\mathbf{r})d\mathbf{r} + \mathcal{O}(q^4) + \ldots \qquad (8.52)$$

But even for anisotropic pores, an isotropic distribution means that any polar axis is suitable for an estimation of the orientational average response. The $1 + 2\cos^2(\Theta/2)$ dependence of the echo attenuation is the same as that observed for isotropic pores, although the coefficient $\langle \int \cos^2\theta_1 r^2 \rho(\mathbf{r})d\mathbf{r} \rangle$ will be different, reflecting not only the size but also the shape of the pore. Nonetheless these is nothing in this coefficient that enables us to separate pore size and pore shape. To measure pore anisotropy, we need to include the q^4 term in the signal. Figure 8.18 shows the transition from $1 + 2\cos^2(\Theta/2)$ to $\cos^2(\Theta)$ behaviour, for an isotropic distribution of long cylinders, as the echo attenuation is increased and the fourth-order term comes into play. Hence the diffusion is restricted normal to the cylinder axis and free along the axis direction.

Figure 8.19, at much weaker attenuation, is dominated by the q^2 term and shows the requirement for very short mixing time in order to reveal the $1 + 2\cos^2(\Theta/2)$ dependence of the echo attenuation. As τ_m is increased to become comparable with the pore diffusion time, the angular dependence in the echo attenuation disappears. The calculation that allows an estimation of the echo attenuation for finite τ_m values [26] is described in the next section.

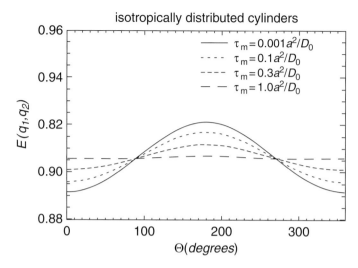

Fig. 8.19 Normalised double PGSE NMR signal from cylinders isotropically distributed in space, showing echo attenuation E versus Θ for different τ_m values, where $\Delta = a^2/D_0$, $(\gamma\delta g a)^2 = 0.1$, and $\delta \to 0$. Only the second-order (q^2) component of the echo attenuation is playing a role and the angular dependence disappears as the mixing time becomes longer. (Adapted from Özarslan and Basser [26].)

8.7 Restricted diffusion and double wavevector encoding with finite mixing time

8.7.1 First-order expression for restricted isotropic geometries with general pulse timings

Using the matrix multiple propagator method, and dividing the gradient waveform into M steps of duration τ and total time $T = M\tau$, Özarslan and Basser [26] have derived a general first-order (q^2) expression for the double wavevector experiment, applicable for all pulse timings. Their calculation involves a multiplication of $M \times M$ matrices [27], as outlined in Chapter 5. To assist the evaluation of the matrix multiplications of $S_k(\mathbf{q}) = V^{-1/2} \int d\mathbf{r} u_k(\mathbf{r}) \exp(i\mathbf{q_k} \cdot \mathbf{r})$ and $A_{kk'} = \int d\mathbf{r} u_k^*(\mathbf{r}) u_{k'}(\mathbf{r}) \exp(i\mathbf{q_k} \cdot \mathbf{r})$, they define terms $_cA(\mathbf{q})$, and $_cS(\mathbf{q})$, which are proportional to $(\mathbf{q} \cdot \mathbf{r})^c$ in the Taylor series expansions. The integrals to be evaluated in order to find the coefficient of q^2 are then

$$_2S^T(\mathbf{q}_i)R\ _0S^*(\mathbf{q}_j) = -\frac{1}{2V} \int_V d\mathbf{r}(\mathbf{q_i} \cdot \mathbf{r})^2 \qquad (8.53)$$

and

$$_1S^T(\mathbf{q}_i)R\ _1S^*(\mathbf{q}_j) = \frac{1}{V} \int_V d\mathbf{r}\mathbf{q_i} \cdot \mathbf{r} \int_V d\mathbf{r}'\mathbf{q_j} \cdot \mathbf{r}'P(\mathbf{r}|\mathbf{r}', t) \qquad (8.54)$$

These terms depend on the shape of the pore. For planar, cylindrical, and spherical pores of plane spacing $2a$ or radii a, the results may easily be obtained in a closed form using the eigenmode expansions given in Chapter 7 as

$$_2S^T(\mathbf{q}_i)R\,_0S^*(\mathbf{q}_j) = -\frac{\frac{1}{2}q_j^2a^2}{2+d} \tag{8.55}$$

and

$$_1S^T(\mathbf{q}_i)R\,_1S^*(\mathbf{q}_j) = 2a^2\mathbf{q}_j \cdot \mathbf{q}_j \sum_{n=1}^{\infty} s_{dn}e^{-\alpha_{dn}^2 D_0(t_j-t_i)/a^2} \tag{8.56}$$

where d is the dimensionality 1, 2, or 3 appropriate to the plane, cylinder, or sphere, and the $\alpha_d n$ are, respectively, the roots σ_n, β_n, and α_n appropriate to the planar, cylinder, or sphere case as outlined in Chapter 7. s_{dn} is given by $[\alpha_{dn}^2(\alpha_{dn}^2 - d+1)]^{-1}$. Allowing $M \rightarrow \infty$ and $\tau \rightarrow 0$, Özarslan and Basser derive the echo-attenuation expression to q^2 in terms of integrals over the effective gradient, $\mathbf{g}^*(t)$, as

$$E \approx 1 - 2\gamma^2a^2 \sum_{n=1}^{\infty} s_{dn} \int dt e^{\alpha_{dn}^2 D_0 t/a^2} \mathbf{g}^*(t) \cdot \mathbf{F}_{dn}(t) \tag{8.57}$$

where

$$\mathbf{F}_{dn}(t) = \int_t^T dt' \mathbf{g}^*(t)e^{-\alpha_{dn}^2 D_0 t'/a^2} \tag{8.58}$$

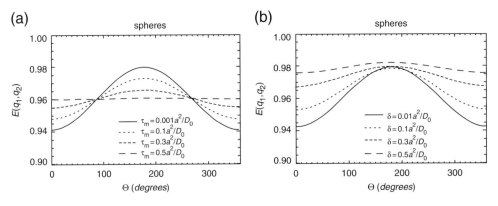

Fig. 8.20 Normalised double PGSE NMR signal from an ensemble of spheres, showing echo attenuation E versus Θ for (a) different τ_m values where $\Delta = a^2/D_0$, $(\gamma\delta ga)^2 = 0.1$, and $\delta \rightarrow 0$ and (b) different δ values where $\Delta = a^2/D_0$, $(\gamma\delta ga)^2 = 0.1$, and $\tau_m = 0.002a^2/D_0$. Only the second-order (q^2) component of the echo attenuation is playing a role, and the angular dependence disappears as the mixing time becomes longer, and as the increasing value of δ causes the experiment to approach the steady gradient case. (Adapted from Özarslan and Basser [26].)

Using these relations, the second-order echo-attenuation term can be calculated, in the case of plane pores, cylinders, or spheres, for any waveform whatsoever. Hence it is possible to examine how the angular dependence is perturbed by increasing the mixing time, or by increasing the gradient pulse duration δ. The results [26] are shown in Fig. 8.20. It is clear that the $1 + 2\cos^2(\Theta/2)$ dependence attenuates sharply as the mixing time becomes comparable with the time to diffuse across the pore, while in the case where δ is varied, the approach to steady gradient conditions also results in a significant weakening of the angular dependence. Clearly, narrow gradient pulse and short mixing time conditions are ideal if the the angular dependence of the q^2 term in the echo attenuation is to be used to measure pore size. However, the finite time expressions at least permit the method to be used under a wider range of parameter space.

8.7.2 Isotropic fluid with unrestricted diffusion and finite width gradient pulse effects

There remains one other cautionary example to be considered in the matter of the angular dependence of differently aligned pulse pairs in double PGSE NMR. This curiosity concerns the case of an isotropic fluid undergoing free diffusion, where no Θ dependence would be expected. It arises when the double PGSE pulse sequence is implemented by superposing the last pulse of the first pair with the first pulse of the second pair, as shown in Fig. 8.21(b). In this case, the echo attenuation for two pulse pairs of magnetic field gradients g_1 and g_2 obeys [26]

$$E = \exp(-\gamma^2\delta^2 D_0[(\Delta - \delta/3)(g_1^2 + g_2^2) - (\delta/3)g_1g_2 \cos\Theta]) \tag{8.59}$$

whereas the 'safer' pulse sequence shown in Fig. 8.21(a) returns the expected result

$$E = \exp(-\gamma^2\delta^2 D_0(\Delta - \delta/3)(g_1^2 + g_2^2) \tag{8.60}$$

The physical reason for this discrepancy is perfectly obvious. For the pulse sequence of Fig. 8.21(b), in the case $\Theta = 0$ we have to account for finite pulse-width effects in a three-pulse experiment, whereas when $\Theta = \pi$, the experiment is closer to a two-gradient pulse experiment.[4] By contrast, the sequence of Fig. 8.21(a) is always a four-gradient pulse experiment and the finite-width pulse sequence effects are proportionately the same for each pulse pair.

Because of this artifact, the pulse sequence of Fig. 8.21(a), with τ_m set as short as possible, should always be used for double PGSE NMR experiments targeted at measuring pore size through the angular dependence of the q^2 term in the echo attenuation. By this means one ensures that any observed angular dependence arises from signal contributions due to molecules experiencing restricted diffusion.

8.8 Diffusive diffraction with double PGSE NMR

Just as $|\tilde{\rho}(\mathbf{q})|^2$ is the PGSE NMR diffraction signal for the long-time limit ($\Delta \gg a^2/D_0$) for spin-bearing molecules diffusing in an enclosing pore, described by the density $\rho(\mathbf{r},$ then $|\tilde{\rho}(\mathbf{q}_1)|^2|\tilde{\rho}(\mathbf{q}_2)|^2$ and $\tilde{\rho}(\mathbf{q}_1)\tilde{\rho}(\mathbf{q}_2)\tilde{\rho}^*(\mathbf{q}_1+\mathbf{q}_2)$ are, respectively, the diffraction signals for double PGSE in the cases $\tau_m \gg a^2/D_0$ and $\tau_m \ll a^2/D_0$. As we have seen, the long mixing time case is of some interest where local anisotropy is to be examined in a system which is globally isotropic. But it is the case $\tau_m \ll a^2/D_0$ which arouses particular interest. In particular, unlike expressions for the long mixing time double

---

[4]It is precisely a two-pulse PGSE NMR experiment when $g_1 = g_2$ and $\Theta = \pi$.

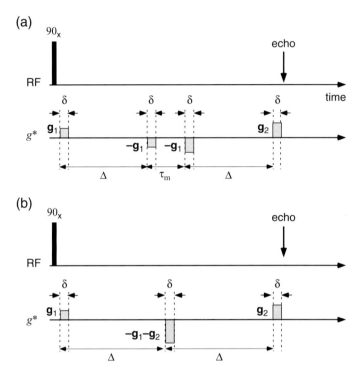

Fig. 8.21 Double PGSE NMR pulse sequences in which (a) the two pairs of gradient pulses are separate blocks separated by each other by an identifiable mixing time τ_m and (b) the last pulse of the first pair and the first pulse of the second pair are superposed so as to reduce τ_m. Sequence (b) can result in artifactual angular dependence due to finite pulse-width effects, even for a freely diffusing isotropic liquid. Pulse sequence (a) should be used to avoid these effects, even where tau_m is to be minimised. (Adapted from Özarslan and Basser [26].)

PGSE or single PGSE pair experiments, the short mixing time double PGSE NMR diffraction signal is not inherently positive. This enables far greater dynamic range in the signal and so can significantly aid the determination of structure.

8.8.1   Diffractograms with signed amplitude

Consider a multiple PGSE experiment in which N pairs of PGSE gradients, each of identical wavevector $\mathbf{q}$, are applied in succession such that in each case the diffusion time Δ is sufficiently long ($\Delta \gg a^2/D_0$) that molecules have diffused many time across the pore. In that case, a very simple form of the echo attenuation may be written down in the limit of long mixing times

$$E_{\infty,\infty}(\mathbf{q}, N) = |\tilde{\rho}(\mathbf{q})|^{2N} \tag{8.61}$$

where we follow reference [25] in using the double ∞ subscript to refer to conditions for Δ and τ_m. The diffraction signal is always positive.

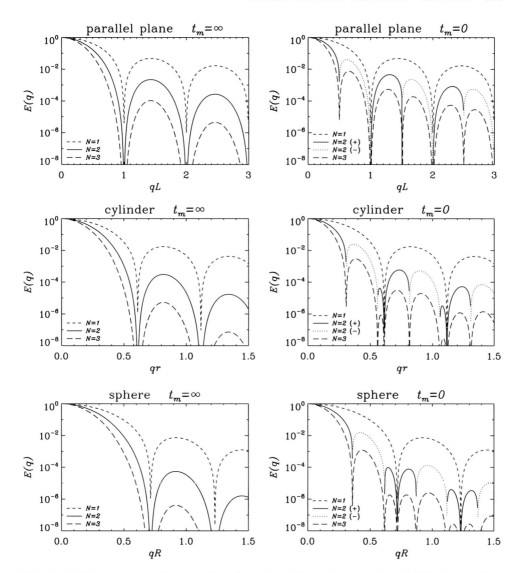

Fig. 8.22 Signal attenuation as a function of q with varying number of diffusion gradient pairs for parallel plane pore with spacing L (top), cylindrical pore with radius r (middle), and spherical pore with radius R (bottom). The left column shows the results obtained in the limit $\tau_m \to \infty$, while the $\tau_m = 0$ case is shown on the right column. In both cases $\delta \to 0$ and $\Delta_m \to \infty$. The continuous and dotted lines both illustrate the curve obtained with $N = 2$, where the former shows the positive sections and the latter shows the negative sections after flipping. (Reproduced from Özarslan and Basser [25].)

When the mixing time is short, signed diffraction signals may be obtained for even N. Following the simple reasoning above, one may write [25] for N odd

$$E_{\infty,0}(\mathbf{q}, N) = |\tilde{\rho}(\mathbf{q})|^2 |\tilde{\rho}(2\mathbf{q})|^{N-1} \tag{8.62}$$

while for N even

$$E_{\infty,0}(\mathbf{q}, N) = \tilde{\rho}(\mathbf{q})^2 \tilde{\rho}^*(2\mathbf{q}) \tilde{\rho}(2\mathbf{q})|^{N-2} \tag{8.63}$$

Figure 8.22 shows the diffractograms for $N = 1, 2, 3$ in the case of planes separated by L, where the gradient is applied normal to the planes, cylinders of radius r, where the gradients are applied normal to the cylinder axis, and spheres of radius R. The right-hand set correspond to short mixing time and for $N = 2$, the dotted line indicates a signal of negative sign. In a sense, such a signed behaviour is merely a curiosity, and it is not obvious that greater information content is present. However, as we shall see in the next section, the importance of the signed oscillatory character of the $N = 2$ short mixing time diffraction signal is that it holds the key to unravelling structures: polydispersity.

8.8.2   Unravelling structure in polydisperse systems

Suppose our sample is made of spherical pores of differing radii. The result of the PGSE NMR experiment is a superposition of diffraction signals from each pore, each weighted by its probability in the distribution. Of course, what we would really like to do is to invert the data to obtain the distribution. Such an inversion is much simpler when the signals associated with each parameter of the distribution belongs to an orthogonal basis function. The real value of oscillatory signals is that they have the potential to provide just such an orthogonal basis.

Let us see how that might work with the three simple geometries of parallel planes, cylinders, and spheres, for which the structure factors, $\tilde{\rho}(\mathbf{q})$ are, respectively, $\sin(qa)/qa$, $J_1(qa)/qa$, and $j_1(qa)/qa$, a being the half-plane spacing, or the cylinder/sphere radius. We could take a pulse sequence similar to that shown in Fig. 8.21(b), where $g_1 = g_2 = g$, such that the superposed central pulse has amplitude $2gx$, where $x = \cos\Theta$. Then, for a distribution of pores sizes $P(a)$, the echo signal at a particular value of g and x would be

$$E_{\infty,0}(\mathbf{q}, x) = \int_0^\infty P(a)\tilde{\rho}(\mathbf{q})^2 \tilde{\rho}^*(2\mathbf{q}x)da \tag{8.64}$$

For convenience, choose the case of cylinders. Then

$$E_{\infty,0}(\mathbf{q}, x) = \int_0^\infty P(a) \left(\frac{J_1(qa)}{qa}\right)^2 \frac{J_1(2qxa)}{2qxa}da \tag{8.65}$$

This equation describes a 2-D experiment in which q and x may be independently varied. Given the Bessel function orthogonality [28],[5] then it follows for a particular q that

[5] $\int_0^\infty J_1(\alpha\rho)J_1(\alpha'\rho)\rho d\rho = \alpha^{-1}\delta(\alpha - \alpha')$.

$$P(a) = a^2 \left(\frac{J_1(qa)}{qa} \right)^{-2} F(a,q) \qquad (8.66)$$

where

$$F(a,q) = \int_0^\infty E_{\infty,0}(\mathbf{q},x)(2qx)^2 J_1(2qxa)d(2qx) \qquad (8.67)$$

the variable x being taken from 0 to 1, and the argument $2q$ being set large enough to sufficiently attenuate the echo when $x=1$. Equations 8.66 and 8.67 are unfriendly to real experiments, where the $(2qx)^2$ factor in eqn 8.67 tends to emphasise the noise at high x values, while the static structure factor squared in eqn 8.66 causes poles in the division at the nodal points. Of course, by working with a sum of echo attenuations obtained at a few judiciously chosen q values, these poles may be avoided and, given good enough signal-to-noise in the experiment, the distribution $P(a)$ may be revealed.

References

[1] J. D. Seymour and P. T. Callaghan. Generalized approach to NMR analysis of flow and dispersion in porous media. *AIChE J.*, 43:2096, 1997.
[2] A. J. Sederman, M. L. Johns, P. Alexander, and L. F. Gladden. Structure-flow correlations in packed beds. *Chem. Eng. Sci.*, 53:2117, 1998.
[3] L. Lebon, L. LeBlond, and J. P. Hulin. Experimental measurement of dispersion processes at short times using a pulsed field gradient NMR technique. *Phys. Fluids*, 9:481, 1966.
[4] B. Manz, P. Alexander, and L. F. Gladden. Correlations between dispersion and structure in porous media probed by nuclear magnetic resonance. *Phys. Fluids*, 11:259, 1999.
[5] P. T. Callaghan and Y. Xia. Velocity and diffusion imaging in NMR microscopy. *J. Magn. Reson.*, 91:326, 1991.
[6] P. T. Callaghan, S. L. Codd, and J. D. Seymour. Spatial coherence phenomena arising from translational spin motion in gradient spin echo experiments. *Concepts in Magnetic Resonance*, 11:181, 1999.
[7] A. A. Khrapitchev and P. T. Callaghan. Reversible and irreversible dispersion in a porous medium. *Phys. Fluids*, 15:2649, 2003.
[8] G. I. Taylor. Dispersion of soluble matter in solvent flowing slowly through a pipe. *Proc. Roy. Soc. London A*, 219:186, 1953.
[9] A. A. Khrapitchev and P. T. Callaghan. Double PGSE NMR with stimulated echoes: phase cycles for the selection of desired encoding. *J. Magn. Reson.*, 152:1, 2001.
[10] M. W. Hunter and P. T. Callaghan. NMR measurement of nonlocal dispersion in complex flows. *Phys. Rev. Lett.*, 89:210602, 2007.
[11] M. W. Hunter, A. N. Jackson, and P. T. Callaghan. Nuclear magnetic resonance measurement and Lattice-Boltzmann simulation of nonlocal dispersion the tensor. *Phys. Fluids*, 22:027101, 2010.
[12] W. D. Williams, E. F. W. Seymour, and R. M. Cotts. A pulsed-gradient multiple-spin-echo NMR technique for measuring diffusion in the presence of background magnetic field gradients. *J. Magn. Reson.*, 31:271, 1978.

[13] R. Aris. On the dispersion of a solute in a fluid flowing through a tube. *Proc. Roy. Soc. A*, 235:67, 1956.

[14] C. Van Den Broeck. A stochastic description of longitudinal dispersion in uniaxial flows. *Physica A*, 112:343, 1982.

[15] D. G. Cory, A. N. Garroway, and J. B. Miller. Applications of spin transport as a probe of local geometry. *Polymer Preprints*, 31:149, 1990.

[16] Y. Cheng and D. G. Cory. Multiple scattering by NMR. *J. Am. Chem. Soc.*, 121:7935, 1999.

[17] M. Lawrenz, M. A. Koch, and J. Finsterbusch. A tensor model and measures of microscopic anisotropy for double-wave-vector diffusion-weighting experiments with long mixing times. *J. Magn. Reson.*, 202:43, 2010.

[18] P. T. Callaghan and M. E. Komlosh. Locally anisotropic motion in a macroscopically isotropic system: displacement correlations measured using double pulsed gradient spin echo NMR. *Magnetic Resonance in Chemistry*, 40:S15, 2002.

[19] J. Rogers and P. A. Winsor. Optically positive, isotropic and negative lamellar liquid crystalline solutions. *Nature*, 216:477, 1967.

[20] K. Fontell. The structure of the lamellar liquid crystalline phase in Aerosol OT–water system. *J. Colloid Interface Sci.*, 44:318, 1973.

[21] P. T. Callaghan and O. Söderman. Examination of the lamellar phase of Aerosol OT/water using pulsed field gradient nuclear magnetic resonance. *J. Phys. Chem.*, 87:1737, 1983.

[22] P. P. Mitra. Multiple wave-vector extensions of the NMR pulsed field gradient spin echo diffusion method. *Phys. Rev. B*, 51:15074, 1995.

[23] M. A. Koch and J. Finsterbusch. Compartment size estimation with double wave vector diffusion-weighted imaging. *Magn. Reson. Med.*, 60:90, 2008.

[24] N. Shemesh, E. Özarslan, P. J. Basser, and Y. Cohen. Measuring small compartmental dimensions with low-q angular double-PGSE NMR: The effect of experimental parameters on signal decay. *J. Magn. Reson.*, 198:15, 2009.

[25] E. Özarslan and P. J. Basser. MR diffusion–'diffraction' phenomenon in multi-pulse-field-gradient experiments. *J. Magn. Reson.*, 188:285, 2007.

[26] E. Özarslan and P. J. Basser. Microscopic anisotropy revealed by NMR double pulsed field gradient experiments with arbitrary timing parameters. *J. Chem. Phys.*, 128:158511, 2008.

[27] P. T. Callaghan. A simple matrix formalism for the spin echo analysis of restricted diffusion under generalised gradient waveforms. *J. Magn. Reson.*, 129:74, 1997.

[28] G. B. Arfken and H. J. Weber. *Mathematical Methods for Physicists*. Cambridge, New York, 2005.

9

Multidimensional PGSE NMR

Much of high-resolution NMR spectroscopy derives its power from the abundance of information present in the spectrum, that distribution of frequencies associated with the evolution of spin phase in the presence of molecular-based interactions, such as the chemical shift, the scalar couplings, and the dipolar and quadrupole interactions. These spectra are usually Fourier conjugates, with respect to time, of oscillatory signals acquired as a function of time or, in the case of multidimensional NMR, arising from spin magnetisation that has previously evolved in a well-defined and prior time domain.

Oscillatory data are naturally analysed using such Fourier transformation. This book deals with the additional spin phase evolution associated with translational displacements under the influence of time-varying magnetic field gradients. These phase evolutions may also induce oscillatory behaviour in the signal, for example where flow is present, and Fourier transformation in this case yields the spectrum of molecular displacements. However, as we have seen, when the molecular ensemble is characterised by diffusive displacements, the pulsed gradient induced phase-spreading results not in an oscillatory signal, but one exhibiting exponential decay with respect to the squared gradient pulse area, q^2. Fourier transformation with respect to q results in a spectrum of displacement Z, a Gaussian the width of which is determined by the molecular diffusion coefficient, D, just as spin–spin relaxation results in a signal which decays with time and the Fourier conjugate with respect to time of which is a frequency spectrum, the width of which is determined by the relaxation time T_2. Of course we could easily analyse those spectra by measuring their widths to find the values of D or T_2.

But suppose the molecular ensemble featured a plethora of sub-ensembles exhibiting different diffusion coefficients or relaxation times. This would result in a multi-exponential signal and spectra comprising a superposition of Gaussians representing each sub-ensemble. Such a superposition would be very hard to analyse since each sub-ensemble would contribute to the overall width and we would need to apply a fitting process involving an unknown number of Gaussians. Of course, the problem is we have simply calculated the wrong 'spectrum'. What we really needed was not the spectrum of displacements or frequencies in each respective case, but instead the spectrum of diffusion coefficients or relaxation times. The type of processing that generates these from multi-exponential decays is known as an inverse Laplace transformation.

As we will see in this chapter, both Fourier transformation and inverse Laplace transformation can be used in conjunction with PGSE NMR as a basis for a multi-dimensional characterisation, whether as separation, correlation, or exchange experiments. And having incorporated inverse Laplace transformation, we may introduce relaxation effects to provide additional dimensions. In a porous medium, molecular migration may also result in changes in local relaxation rates, which depend upon pore size, or in diffusion coefficient, another parameter sensitive to local pore geometry. Such changes will be manifest in exchange experiments, in which we correlate identical parameters measured on the same spin coherences, but at times separated by a mixing period τ_m. Equally we may find it helpful to correlate differing NMR parameters obtained as close as possible in time, in other words as $\tau_m \to 0$. For example, $D - T_2$ maps give us insight regarding the correlation of translational and rotational mobility of molecules. Even the simplest of all NMR experiments, the measurement of the spectrum associated with the FID, yields vital information regarding molecular translational motion. In porous materials, spins in migrating liquid molecules experience local susceptibility-related magnetic fields, which fluctuate as the molecules move about, leading to changes in the NMR spectrum.

This chapter takes us through the various Fourier and Laplace dimensions in which molecular translation plays a role. At the heart of the methods described here is the concept of multiplexing, the encoding of our signal in multiple independent dimensions. In mathematical terms, this simply means that instead of acquiring a signal as a function of one experimental variable, we acquire it as a function of two or more. First, however, we compare the nature of the transformations.

9.1 Fourier or Laplace?

9.1.1 An example from PGSE NMR

Suppose an ensemble of molecules is described by sub-ensembles weighted by probability density $P(D)$, each with diffusion coefficient, D. Then the q-space signal acquired in a narrow gradient pulse PGSE NMR experiment as shown in Fig. 9.1(a) will be the integral over the sub-ensembles

$$E(q) = \int P(D) \exp(-q^2 D\Delta) dD \tag{9.1}$$

with Fourier conjugate $\mathcal{F}\{E\}$ with respect to q given by

$$\bar{P}(Z) = \int P(D)(2\pi D\Delta)^{-1/2} \exp(-Z^2/2D\Delta) dD \tag{9.2}$$

In this transformation, a multi-exponential decay in q^2 has been replaced by a sum of Gaussian propagators. An example is shown in Fig. 9.1(b), where the ensemble consists of the simplest possible non-trivial example with one sub-ensemble comprising two-thirds of the molecules with diffusion coefficient D_1 and another sub-ensemble comprising one-third of the molecules with diffusion coefficient D_2. For convenience we take $D_2 = 4D_1$. In Fig. 9.1(b), the two distinct Gaussians are barely discernable. What we seek is a different type of transformation that returns $P(D)$ directly so as to reveal the distribution of diffusion coefficients shown in Fig. 9.1(c).

9.1.2 Forward and inverse

Consider the following two pairs of equations.

$$f(\omega) = \mathcal{F}^{-1}\{S(t)\} = (2\pi)^{-1/2} \int_{-\infty}^{\infty} \exp(i\omega t) S(t) dt$$

$$S(t) = \mathcal{F}\{f(\omega)\} = (2\pi)^{-1/2} \int_{-\infty}^{\infty} \exp(-i\omega t) f(\omega) d\omega \tag{9.3}$$

and

$$f(R) = \mathcal{L}^{-1}\{S(t)\}$$

$$S(t) = \mathcal{L}\{f(R)\} = \int_{0}^{\infty} f(R) \exp(-Rt) dR \tag{9.4}$$

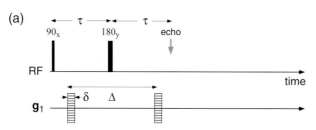

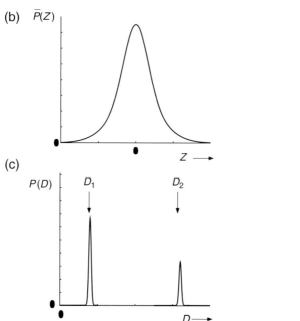

Fig. 9.1 One-dimensional PGSE NMR in which a gradient pulse pair, **g**, encodes the echo. (a) Basic 1-D pulse sequence. (b) The result of Fourier transforming the signal with respect to q when the molecular ensemble exhibits two different diffusion coefficients. A propagator is returned comprising two Gaussians co-centred at zero displacement but with different widths. (c) The result when the signal is inverse Laplace transformed to reveal the two diffusion coefficients directly.

Note that eqn 9.4 may be rewritten to encompass the PGSE NMR result of eqn 9.1,

$$P(D) = \mathcal{L}^{-1}\{E(q^2\Delta)\}$$
$$E(q^2\Delta) = \mathcal{L}\{P(D)\} = \int_0^\infty P(D)\exp(-q^2 D\Delta)dD$$

(9.5)

The first pair represent the inverse and forward Fourier transformations familiar in NMR spectroscopy and MRI. They are symmetric, well-defined, and easily implemented digitally using the FFT algorithm. The second (or third) pair is of particular interest here. The second of the two equations represent the multi-exponential decay in which $f(R)$ is the probability of finding a decay rate R. This integral is known as the Laplace transformation and its outcome, in which $f(R)$ is 'transformed', is the time-domain signal $S(t)$, such as might be found in an NMR relaxation experiment. Its inverse, indicated in the first of the two equations, takes $S(t)$ to $f(R)$ and enables us to calculate the relaxation rate distribution from a knowledge of the NMR signal. This 'inverse Laplace transformation' (ILT) may be desirable, but unlike the Fourier inverse is asymmetric, ill-defined, and not at all easily implemented.

9.2 Inverse Laplace transformations

9.2.1 Analytic and numerical inverse Laplace transformations

There does exist an analytic form to the inverse Laplace transformation of eqn 9.4. In particular, it is the contour integral in the complex plane [1]

$$f(R) = \mathcal{L}^{-1}\{S(t)\} = \frac{1}{2\pi i}\int_{\gamma-i\infty}^{\gamma+i\infty} S(t)\exp(Rt)dt$$

(9.6)

where γ is a vertical contour positioned to the right of any singularities. Numerically, this integral is unhelpful, since the need to multiply by a potentially increasing large exponential $\exp(Rt)$ can lead to exponential divergence of numerical solutions due to noise in $S(t)$ and finite precision errors. The inverse Laplace transform is therefore an ill-posed problem.

The Laplace transform belongs to a class of integrals known as 'Fredholm integrals of the first kind' [1, 2], of the form

$$S(t) = \int_a^b K(t,R)f(R)dR$$

(9.7)

where $K(t,R)$ is known as the kernel. The standard problem is, given a knowledge of the kernel and the signal $S(t)$, how do we find $f(R)$? Because this question is ill-posed, it requires a special approach.

Suppose we discretise the problem and write the signal to be inverted, $S(t)$, as a vector arising from the discrete sampling times, t_i, where $i = 1....N$, with the solution $f(R)$ being discretised in the domain of R_j where $j = 1....M$. If we pre-specify the values of R_j to be used, then we may calculate *a priori* the $N \times M$ kernel matrix $K_{ij} = K(t_i, R_j)$ and

$$S(t_i) = \sum_{j=1}^{M} K(t_i, R_j) f(R_j) \tag{9.8}$$

or

$$\underline{S} = \underline{\underline{K}}\ \underline{f} \tag{9.9}$$

In fact, when we are dealing with measurement, we need to take account of noise. For example, each $S(t_i)$ value will have some noise ϵ_i, so that strictly we should write eqn 9.9 as

$$\underline{S} = \underline{\underline{K}}\ \underline{f} + \underline{\epsilon} \tag{9.10}$$

At first, we might be tempted to think that we could solve for $\underline{f}$ by using the pseudo-inverse $\underline{\underline{K}}^{-1}$ of the matrix $\underline{\underline{K}}$ so that

$$\underline{f} = \underline{\underline{K}}^{-1}\ \underline{S} \tag{9.11}$$

or again, including the noise, by

$$\underline{f} = \underline{\underline{K}}^{-1}\ \underline{S} + \underline{\underline{K}}^{-1}\ \underline{\epsilon} \tag{9.12}$$

The pseudo-inverse is calculated by writing $\underline{\underline{K}}$ as a product of an orthogonal[1] $N \times N$ matrix $\underline{\underline{U}}$, a diagonal $N \times M$ matrix $\underline{\underline{\Sigma}}$, and an orthogonal $M \times M$ matrix $\underline{\underline{V}}$ as

$$\underline{\underline{K}} = \underline{\underline{U}}\ \underline{\underline{\Sigma}}\ \underline{\underline{V}} \tag{9.13}$$

a process known as singular value decomposition. The diagonal values of $\underline{\underline{\Sigma}}$ are known as the singular values of $\underline{\underline{K}}$. The pseudo-inverse of $\underline{\underline{K}}$ is obtained by writing

$$\underline{\underline{K}}^{-1} = \underline{\underline{V}}^{-1}\underline{\underline{\Sigma}}^{-1}\underline{\underline{U}}^{-1} \tag{9.14}$$

where $\underline{\underline{\Sigma}}^{-1}$ is obtained by taking the inverse of all the diagonal elements in $\underline{\underline{\Sigma}}$. While in principle $\underline{\underline{K}}^{-1}$ may be calculated numerically, the matrix $\underline{\underline{K}}$ is 'ill-conditioned'. This is reflected in the large value of the condition number, defined by the product of the norms $||\underline{\underline{K}}|| \cdot ||\underline{\underline{K}}^{-1}||$.[2] The condition number gives the maximum ratio of the fractional error in the solution $\underline{f}$ to the fractional error in $\underline{S}$. A large condition number means that the inversion process tends to have an explosive effect on any errors, $\underline{\epsilon}$, in $\underline{f}$.

There is another sense in which the inverse problem is ill-posed for the Fredholm integral eqn 9.7. Suppose we add to a solution $f(R)$, the oscillatory function $\sin(wR)$, then the Riemann–Lebesque lemma [3] states that for any $K(t, R)$ there exists a frequency w such that

$$\int_a^b K(t, R) \sin(wR) dR = 0 \tag{9.15}$$

This means that for an arbitrarily large amplitude A and for arbitrarily small errors ϵ in $S(t)$, if $f(R)$ is a solution to eqn 9.7, then there exists a frequency w such that

---

[1] An orthogonal matrix $\underline{\underline{U}}$ is one for which $\underline{\underline{U}}^T = \underline{\underline{U}}^{-1}$.

[2] The Frobenius norm of a matrix $\underline{\underline{K}}$ is defined as $||\underline{\underline{K}}|| = \sqrt{Trace(\underline{\underline{K}}\underline{\underline{K}}^*)}$ or $\sqrt{\sum_i \sum_j K_{ij}^2}$. For comparison, the Euclidean norm of a vector, $\underline{x}$ is $||\underline{x}|| = \sqrt{\sum_i x_i^2}$.

$f(R) + A\sin(\omega R)$ is also a solution within that error ϵ. That is a recipe for an infinite solution set! Whichever way we invert this problem, using a measurement of $S(t)$ to obtain an estimate for $f(R)$, there is no unique solution. Another approach is needed in which additional criteria are imposed.

9.2.2   Regularised non-negative least squares in 1-D

A practical approach to the Laplace transform inversion was proposed by Stephen Provencher in 1982 [4]. First, we write down the integral equation in discrete form, allowing for noise in the signal

$$S(t_i) = \sum_{j=1}^{M} \exp(-t_i R_j) f(R_j) + \epsilon_i \tag{9.16}$$

Absolute prior knowledge is then applied by requiring that each spectral amplitude, $f(R_j)$, is positive. Statistical prior knowledge is applied by requiring an optimal, minimum mean-squared error solution to be found. And finally, the principal of parsimony requires that we use the simplest solution, the one that contains the least detail or information not already expected. This last point deals with the Riemann–Lebesque problem, by requiring that our solution be as 'smooth' as possible in the variable R.

Tikhonov regularisation

Provencher implemented parsimony using a method known as Tikhonov regularisation [5]. The idea is as follows: suppose we attempt a least-squares solution $\underline{f}$ to eqn 9.9. Then the residual is $||\underline{K}\underline{f} - \underline{S}||^2$, the quantity we would seek to minimise. Tikhonov regularisation adds a term $\underline{\Gamma}\underline{f}$ in the minimisation process so that the residual to be minimised is $||\underline{K}\underline{f} - \underline{S}||^2 + \alpha^2||\underline{\Gamma}\underline{f}||^2$, where α^2 is known as the regularisation parameter. The operator, Γ, is chosen on the basis of parsimony, that is to favour smoothness with minimum number of peaks. One such smoothness operator might involve the curvature of $f(R)$ so that

$$||\underline{\Gamma}\underline{f}||^2 = \int_{R_{min}}^{R_{max}} [f''(R)]^2 dR \tag{9.17}$$

where (R_{min}, R_{max}) represent the bounds of the solution variable R. Hence a standard way of expressing the solution is

$$V(\alpha) = ||\underline{K}\underline{f} - \underline{S}||^2 + \alpha^2||\underline{\Gamma}\underline{f}||^2$$
$$= minimum \tag{9.18}$$

The way to proceed numerically is as follows [4]. We start with the signal vector $\underline{S}$, the elements of which, S_j, correspond to each sampling value (for example time value) labelled t_i, where $i = 1...N$. Next, we choose *a priori* a range in the solution space of discrete relaxation rates R_j, where $j = 1..M$. The solution vector $\underline{f}$, the 'spectrum of relaxation rates', has values corresponding to each R_j and labelled by f_j with $j = 1...M$. Thus the solution satisfies

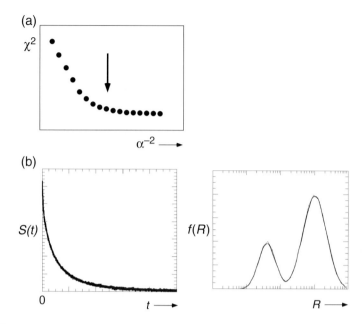

Fig. 9.2 Numerical inverse Laplace transform of time-domain signal $S(t)$ decaying as a multi-exponential, along with spectrum of relaxation rates, $f(R)$. (a) Adjustment of the regularisation parameter α^2 to just minimise χ^2, the arrow showing the chosen value, while (b) and (c) show $S(t)$ and $f(R)$, respectively, with the latter calculated using a logarithmic spacing of R values.

$$V(\alpha) = \sum_i^N \sum_j^M (S_i - \exp(-t_i R_j) f_j)^2 + \alpha^2 \sum_j^M (2f_j - f_{j+1} - f_{j-1})^2$$
$$= minimum \tag{9.19}$$

where the last term contains the numerical form of the second derivative of $f(R)$. With the f_j values constrained to be positive, the least squared minimisation is carried out using a range of α values, and the residual $\chi^2 = \sum_i^N \sum_j^M (S_i - \exp(-t_i R_j) f_j)^2$ calculated for each α^2. By the principle of parsimony, a value of α^2 is chosen (see Fig. 9.2(a)) which minimises χ^2, but only just. Such a solution, when we minimise according to eqn 9.19, has the least degree of curvature needed to well represent the data. The programme developed by Provencher to carry out this 1-D ILT process is known as CONTIN [6], since it allows, in effect, a continuous distribution of relaxation rates to be determined. The method was soon after applied to the analysis of NMR relaxation data [7].

Binning, integration, and resolution
Note that the R_j may be linearly spaced but, more commonly, we might chose them spaced equally in $\log(R)$, so that a wide range of decades can be covered without the need for an excessively large value of M. In operating the algorithm, these pre-assigned R_j are essentially 'bins' in a histogram chosen to represent the data. In that sense they

are discrete and their intensity correctly represents the total probability of finding the relaxation rate in the range defined by adjacent bins. To re-emphasise, the binned intensities are correctly normalised however we choose our binning scale. By contrast, in a continuum representation, one needs a conversion correction (a $d\log(R) = R^{-1}dR$ weighting) in transferring from a logarithmic scale to a linear scale.

Figure 9.2(b) and (c) show an example of noisy, time-domain data undergoing multi-exponential decay, transformed using the Tikhonov regularised least squares process of Provencher. Note the finite width of the peaks in the spectra, characteristic of solutions regularised by curvature according to the parsimony condition. Typically, such broadening means that the method cannot easily resolve relaxation rates less than a factor of three in separation.

Inversion robustness

The quality of the ILT algorithm can best be judged by subjecting the process to data generated from a known spectrum of relaxation times. Prange and Song [8] have demonstrated the degree of consistency available in such a test by using an known bimodal distribution of 100 T_2 values logarithmically spaced between 0.001 and 10 s, as shown in Fig. 9.3(a). From this they generate an echo decay train consisting of 8192 points spaced by of 0.0002 s, starting at 0.0002 s, adding Gaussian noise with standard deviation is 0.025 of the maximum echo value. They then subject this to a non-regularised NNL inversion, resulting in a 'three delta function' solution, and to a regularised inversion, from which a smooth distribution is obtained, closely, but not exactly, resembling the *a priori* spectral distribution. Most importantly, they demonstrate the effect of the noise by re-analysing the data using a different noise realisation based on the same standard deviation and Gaussian distribution. The results are encouragingly similar but, as Prange and Song show in their article, as the noise standard deviation increases, so does the variability of the solution space.

9.2.3 Regularised non-negative least squares in 2-D

Having seen that it is possible to invert multi-exponential data in 1-D, we now turn to 2-D data sets. Because Fourier transforms are linear operations, we may independently transform 2-D data sets along either axis, and in either order, in a well-defined manner. The non-linear inverse Laplace transformation offers no such luxury, and any attempt will founder because of the dependence on the chosen order of transformation directions, and the sensitivity to noise of the outcome on the edges of peaks in the 2-D spectrum. Instead the regularised non-negative least squares approach must be applied to the 2-D data set taken as a whole.

Suppose we acquire a signal in two independent dimensions, both of which exhibit multi-exponential decay. Examples might be $T_1 - T_2$ or $T_2 - D$ experiments. For convenience consider the latter, where the signal $S(t_i, q_j^2)$ is dependent on both time t_i, where $i = 1...N_1$ and wavevector-squared q_j^2, where $j = 1...N_2$ as

$$S(t_i, q_j^2) = \sum_{k=1}^{M_1} \sum_{l=1}^{M_2} \exp(-t_i R_k) \exp(-q_j^2 \Delta D_l) f(R_k, D_l) + \epsilon_{i,j} \qquad (9.20)$$

or

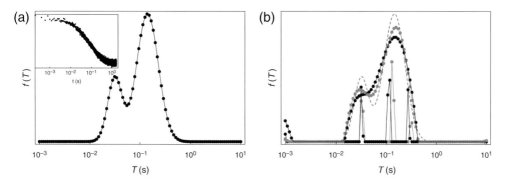

Fig. 9.3 (a) *A priori* T_2 spectrum used to synthesise the echo decay data shown in the inset. The noise standard deviation is 0.025 of the maximum echo value. (b) T_2 spectra obtained by NNLS (these are the 'spiky' results that are scaled to 10% to fit on the plot) along with the smooth regularised T_2 spectral solution. The black and grey dots are two independent solutions for the data shown in (a) but with a different noise realisation at the same standard deviation. The original T_2 spectrum (dashed curve) is shown for comparison. (Reproduced by permission from Prange and Song [8].)

$$\underline{\underline{S}} = \underline{\underline{K_1}} \, \underline{\underline{f}} \, \underline{\underline{K_2}} + \underline{\underline{\epsilon}} \tag{9.21}$$

where $K_{1i,k} = \exp(-t_i R_k)$ and $K_{2l,j} = \exp(-q_j^2 \Delta D_l)$. We seek the 2-D spectrum $\underline{\underline{f}}$, with elements $f(R_k, D_l)$ such that $k = 1...M_1$ and $l = 1...M_2$. Equation 9.21 involves the multiplication of $N_1 \times M_1$, $M_1 \times M_2$, and $M_2 \times N_2$ matrices. Of course, it is possible to re-express this equation making $\underline{\underline{S}}$ and $\underline{f}$ vectors rather than tensors, thus transforming the 2-D character of the equation to a 1-D character as in the original Provencher form of eqn 9.10. All we would have to do is to rewrite the successive rows of matrices $\underline{\underline{S}}$ and $\underline{\underline{f}}$ as a concatenated string of vectors, at the same time appropriately transforming the kernel matrices, $\underline{\underline{K_1}}$ and $\underline{\underline{K_2}}$, to make a single outer product matrix, $\underline{\underline{K_0}} = \underline{\underline{K_1}} \otimes \underline{\underline{K_2}}$, involving appropriately concatenated exponential decays. We could then apply the 1-D regularised non-negative least squares process described earlier. These new vectors would be large, of dimension $N_1 \times N_2$ and $M_1 \times M_2$, respectively. Consequently the demands on computation time would be severe. Worse, the dimension of $\underline{\underline{K_0}}$ is enormous, being $N_1 \times M_1 \times M_2 \times N_2$. To store this matrix requires a great deal more memory than storing the individual $\underline{\underline{K_1}}$ and $\underline{\underline{K_2}}$ matrices. The method is impractical as it stands.

A nice solution to problem has been demonstrated by Venkataramanan *et al.* [9]. Their procedure involves first reducing the dimension of the matrices by singular value decomposition. In this process the matrices are reorganised by unitary transformations so that the singular values along the diagonal are ordered by decreasing size, the matrix dimension being truncated once the singular values become sufficiently small, with typically 30 or so of the largest being retained. That process, in itself, represents a subtle form of regularisation. Then, these smaller dimension matrices are projected as vectors and the 1-D regularised non-negative least squares solution obtained, the solution being finally reorganised into the desired 2-D array.

9.2.4   Testing using known distributions and pearling effects

The ill-defined nature of the inverse Laplace transformation demands that some caution be applied when using it. But like all 'black boxes', the inner workings of which have an element of mystery, the performance of the ILT may be judged on what it does to data the inverse of which is known *a priori*.

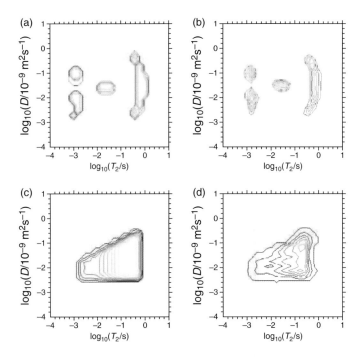

Fig. 9.4 Numerical 2-D inverse Laplace transforms (b and d) of signals generated using known precursor $D - T_2$ distributions (a and c) in which the signal-to-noise ratio is 10^4. Calculations were carried out by Daniel Polders.

The test is straightforward. We generate some chosen distribution of decay rates, calculate the expected signal, add some noise, and then run the result through the transformation to see if we generate that with which we started. Figure 9.4 provides an example of such a procedure in the case of a 2-D ILT. The known distributions, $f(R, D)$, here chosen as $D - T_2$ correlations in which each pixel represents a diffusion and T_2 component, scaled by its intensity, are generated as 100×100 matrices, in (a) as a 'smiley face' and in (c) as a trapezium of ramped intensity. The decaying signals, $S(t, q^2)$, were calculated using eqn 9.20, the gradient strength, G, being ramped from 0 to 1 T m^{-1} in 49 logarithmically spaced steps, with the time, t varied from 2 ms to 2000 ms in steps of 1 ms, corresponding to a PGSE experiment followed by a CPMG train, in which $\tau = 250\,\mu s$ and in which every second echo is collected. The added Gaussian noise is such that the signal-to-noise ratio for $S(t, q^2)$ is 10^4 at the signal origin.

Figure 9.4(d) illustrates a curious effect found in the regularised ILT solution, namely a tendency to convert flat distributions into 'lumpy' distributions. The smoothly ramped trapezoidal distribution used to calculate the input data is represented in the inversion process by a distribution of similar boundary shape, but in which the intensity is no longer smoothly ramped but gathered into local 'islands' of intensity. This effect is commonly termed 'pearling'. Pearling is clearly an artifact of the transformation, but occasionally useful despite that. For example, suppose that one is to quantitatively analyse parts of a continuous distribution. It has the effect of 'lumping' intensity into localised regions, 'bins' of intensity that may be independently integrated.

Sections 9.4 and beyond will present examples of the inverse Laplace transformation in action. But we begin our discussion of multi-dimensional PGSE NMR with a survey of methods confined to Fourier space.

9.3 Multi-dimensional Fourier–Fourier methods

Both k-space and q-space are traversed by the use of magnetic field gradient pulses, with Fourier transformation taking us respectively to the domains of molecular position $\mathbf{r}$ and molecular displacement $\mathbf{R}$. One obvious type of multidimensional Fourier–Fourier application concerns the combination of k-space and q-space encoding to produce an image with voxels endowed with a 'displacement contrast', in its most complete form with the propagator for displacement being measured for each image element. We term this experiment 'dynamic NMR microscopy' and devote the next chapter to its description. In this chapter, however, we shall be concerned only with multidimensional methods where imaging is treated as a separate and independent contrast. Where imaging is not performed, and this will be the usual case here, the experiments to be discussed apply to the whole sample. Should the multidimensional encoding methods discussed in this chapter be concatenated with a further image encoding, an ambitious expansion into further dimensions, then we can consider that the experiments discussed here apply to one image volume element or voxel. Of course there are only so many NMR dimensions we can practically employ, each one requiring a further multiplexing and further dedication of acquisition time. Two or three seems reasonable. We shall, in this chapter, be content to use those up in the service of q-space alone!

9.3.1 VEXSY

Consider an ensemble of molecules that exhibit a propagator $\bar{P}(\mathbf{R}, \Delta)$, and which are subject to the pulse sequence shown in Fig. 9.1 in which two separate pairs of short duration PGSE gradients, $\mathbf{g}_1$ and $\mathbf{g}_2$, independently encode the spin phase, engendering phase shifts $\exp(i\mathbf{q}_1 \cdot \mathbf{R}_1 + i\mathbf{q}_2 \cdot \mathbf{R}_2)$ for molecules that migrate $\mathbf{R}_1$ along the gradient direction over the first pair and $\mathbf{R}_2$ over the second pair. Following the first 90° RF pulse, the transverse magnetisation is phase-encoded according to the displacement $\mathbf{R}_1$, which occurs over the duration Δ between the first PGSE pulse pair. Following a mixing time delay τ_m, spins associated with molecules displaced by $\mathbf{R}_1$ during the first phase-encoding period will have moved to a different part of the displacement spectrum, $\bar{P}(Z, \Delta)$. In other words, there exists a conditional probability distribution $\mathcal{P}(\mathbf{R}_1, \Delta | \mathbf{R}_2, \Delta; \tau_m)$, which is assumed to be common to the entire molecular ensemble. It tells us the likelihood that a molecule that moved by $\mathbf{R}_1$ over the time delay Δ will be displaced by $\mathbf{R}_2$ over Δ, when this latter measurement is made after a mixing delay τ_m. We may thus write down the final, normalised (to zero gradient) phase-encoded echo signal as

$$E(\mathbf{q}_1, \mathbf{q}_2, \tau_m, \Delta) = \iint \bar{P}(\mathbf{R}_1, \Delta) \mathcal{P}(\mathbf{R}_1, \Delta | \mathbf{R}_2, \Delta; \tau_m)$$
$$\times \exp(i\mathbf{q}_1 \cdot \mathbf{R}_1 + i\mathbf{q}_2 \cdot \mathbf{R}_2) d\mathbf{R}_1 d\mathbf{R}_2 \qquad (9.22)$$

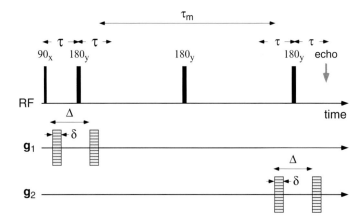

Fig. 9.5 RF and gradient pulse sequence for velocity exchange spectroscopy (VEXSY) in which successive PGSE pulse pairs ($\mathbf{g}_1$, and $\mathbf{g}_2$) are applied separated by a delay time τ_m. To be an effective exchange experiment in which like phenomena are compared at different times, $\mathbf{g}_1$ and $\mathbf{g}_2$ are required to be in the same direction.

The 2-D experiment, known as VEXSY [10], for velocity exchange spectroscopy, is performed by encoding the signal in the two independent $\mathbf{q}$ domains. It is an exchange experiment because the molecular displacements or velocities are probed at different times to see if they have changed. To be an exchange experiment, we need to be comparing the same parameter measured at different times. That means, in effect, that the two gradient pulses should be applied in the same direction, so that following a 2-D inverse Fourier transformation, we compare displacements Z_1 and Z_2 along the same axis, but at different times.[3] Equation 9.22 tells us that the spectrum returned following the 2-D FT with respect to q_1 and q_2 is the joint probability distribution

$$\bar{P}_2(Z_1, \Delta; Z_2, \Delta; \tau_m) = \bar{P}(Z_1, \Delta) \mathcal{P}(Z_1, \Delta | Z_2, \Delta; \tau_m) \qquad (9.23)$$

---

[3]The experiment in which the gradient $\mathbf{g}_1$ and $\mathbf{g}_2$ are not parallel is discussed in Section 9.3.4.

For very short mixing times, such that the molecules have not had the chance to move to a different sub-ensemble, $P(Z_1, \Delta | Z_2, \Delta; \tau_m)$ will be a delta function $\delta(Z_1 - Z_2)$, and the distribution will be diagonal, albeit broadened by diffusion over the time Δ. When the mixing time is very long compared with any characteristic time for the ensemble, $P(Z_1, \Delta | Z_2, \Delta; \tau_m)$ will be independent of Z_1 and return to the *a priori* propagator $\bar{P}(Z_2, \Delta)$. This latter situation always applies for the case of Brownian motion of small molecules where collision times are picoseconds. The VEXSY experiment simply returns a 2-D spectrum that is a product of two Gaussians, as shown in Fig. 9.6

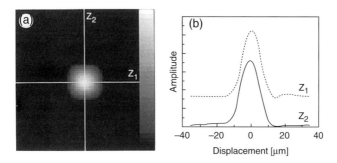

Fig. 9.6 (a) 2-D VEXSY spectrum, $\bar{P}_2(Z_1, \Delta; Z_2, \Delta; \tau_m)$, for unrestricted Brownian motion of water molecules and (b) orthogonal slice profiles shown the independent Gaussian displacement spectrum associated with each encoding. (Reproduced with permission from reference [10].)

For flow in a porous medium, where characteristic ensemble exchange times, τ_v, may be made comparable with the mixing time τ_m, the structure of the VEXSY spectrum is considerably more interesting. Figure 9.7 [11] shows displacement propagators[4] obtained from water flowing in a 0.5-mm diameter (d) bead pack, at a variety of encoding times Δ. In this example, the mean velocity $\langle v \rangle$ is 24 mm s^{-1}, $\tau_v = d/\langle v \rangle = 21$ ms, and $Pe = 5600$. Note that as Δ is made comparable to, and then larger than, τ_v, the distribution of velocities moves towards a Gaussian. Note that it is customary to step q-space over both positive and negative values in order to obtain the best quality representation of the propagator following Fourier transformation.

Figure 9.8 shows the evolution of the joint probability density $\bar{P}_2(Z_1, \Delta; Z_2, \Delta; \tau_m)$ (plotted in velocity units) for the 0.5-mm diameter bead pack, at $Pe = 1900$. A range of mixing times are shown, from $\tau_m \ll \tau_v$ to $\tau_m \gg \tau_v$. In the figure labels, $\tau'_m = \tau_m + \Delta$ is used to provide a more accurate measure of the total mixing time. For negligible mixing time, the velocity distribution is expected not to change between the first and the second encoding interval, leading to intensities along the main diagonal. Some broadening is present because of diffusive motion during Δ. For τ_m comparable with τ_v, significant off-diagonal intensity indicates velocity changes. Figure 9.8(b) shows the conditional probabilities, $P(Z_1, \Delta | Z_2, \Delta; \tau_m)$, obtained by dividing the joint probabilities in (a) by the relevant averaged propagators [11]. The shape of the conditional

---

[4]Note that these are plotted, for convenience, in velocity units rather than displacement units.

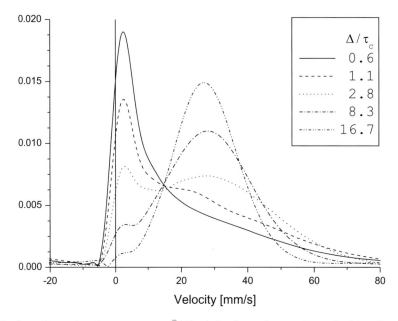

Fig. 9.7 One-dimensional propagators $\bar{P}(Z, \Delta)$ for flow of water in packed beads of 0.5-mm diameter at a volume flow rate of 3.0 l/h, corresponding to an average velocity of 24 mm s^{-1}. The propagators were obtained with variable encoding times Δ. Note that the displacement is plotted in velocity units, conversion between displacement and velocity being simply made via the encoding time Δ. (Reproduced with permission from Khrapitchev *et al.* [11].)

probability describes the probability of finding a velocity v_2 in the second encoding interval given that a velocity v_1 was observed in the first interval. It can be seen that for the shortest mixing time, a strong correlation exists between v_1 and v_2. With increasing mixing time, the fluid elements are allowed to sample a wider range of streamlines and the different components of dispersion lead to a loss of memory, so that the conditional probability of v_2 becomes gradually less dependent on the initial value v_1, the alignment of contour lines changing from the main diagonal toward a flatter slope considerably less than unity.

Of course, the pulse sequence of Fig. 9.5 is idealised. stimulated-echo versions are possible [11], with z-storage during τ_m, as well as sequences in which CPMG trains are inserted to maintain transverse magnetisation during the mixing time, in the latter case, care needing to be taken to sustain both the real and imaginary parts of the signal through the echo train [10]. Note, following our discussion in Chapter 8, the need to use an appropriate phase cycle in order to select the desired magnetisation coherence in the final echo.

Finally, it is worth noting that VEXSY is a pure phase-encoding method and that means that the signal is available with full spectral resolution. That is the upside of a time-costly encoding scheme where separate time steps are needed for each increment of the independent dimensions.

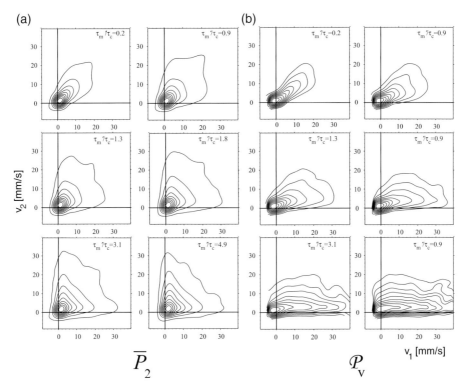

Fig. 9.8 Joint probability density of displacements, $\bar{P}_2(Z_1, \Delta; Z_2, \Delta; \tau_m)$, for water flowing through packed beads of 0.5-mm diameter obtained by the VEXSY experiment, at flow rate of 1.0 l/h, with average velocity $\langle v \rangle$ of 7.9 mm s^{-1}, corresponding to τ_v of 64 ms and $Pe = 1900$. The encoding time is $\Delta = 12$ ms. Velocities measured before and after the mixing interval τ_m are shown along the abscissa and ordinate axes, respectively. Numbers indicate velocities in millimetres per second. Contour lines are in steps of 10% where the peak amplitude is normalised to unity. (b) As in (a), but with the data plotted as the conditional probability $\mathcal{P}(\mathbf{R}_1, \Delta | \mathbf{R}_2, \Delta; \tau_m)$. (Reproduced by permission from Khrapitchev *et al.* [11].)

9.3.2 SERPENT

A pulsed gradient spin-echo experiment very similar to VEXSY was proposed by Stapf *et al.* [12], in which three or more sequential gradient pulses are applied to the spin ensemble, each with an arbitrary amplitude, but with the requirement that the time integral of the gradient waveforms over the whole experiment be zero, the condition required for echo formation. This experiment was termed SERPENT, for SEquential Rephasing by Pulsed field gradients Encoding N Time intervals. The comparison of VEXSY and SERPENT is discussed in detail in reference [13]. In the simplest SERPENT manifestation of three successive gradient pulses, the echo constraint determines that two gradient pulses may be independently varied, thus implying a 2-D encoding space. The three-pulse SERPENT and the VEXSY experiments contain precisely the same information, but with the equivalent mixing time of VEXSY being zero, so that the time interval over which changes in velocity are observed is essentially the velocity encoding time Δ. Because of the flexibility of mixing time setting in the VEXSY experiment, the latter might be more explicitly interpreted as a classic exchange experiment.

9.3.3   POXSY

In a sense, the simplest of all exchange experiments in the Fourier domain of gradient encoding is the positional exchange (POXSY) of Fig. 9.9(a), proposed by Han *et al.* [14]. The pulse sequence has the appearance of a simple Stejskal–Tanner two-

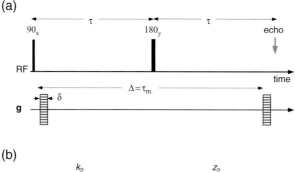

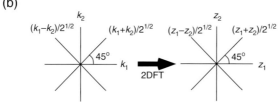

Fig. 9.9 (a) Pulse sequence for positional exchange spectroscopy (POXSY) along with (b) associated axis systems. Note that the diagonal and off-diagonal projections provide access to the mean position and mean displacement, respectively [14].

gradient pulse spin-echo experiment. But rather than having the two gradient pulses stepping in tandem, to ensure the echo condition $\int g^*(t)dt = 0$, two independently varied collinear gradient pulses of wavevector k_1 and k_2 are used to phase encode for position z along the gradient axes. The signal is then

$$S(k_1, k_2, \tau_m) = \iint \rho(z_1)P(z_1|z_2, \tau_m)\exp(ik_1 z_1 + ik_2 z_2)dz_1 dz_2 \tag{9.24}$$

so that 2-D Fourier transformation of $S(k_1, k_2, \tau_m)$ with respect to k_1 and k_2 returns the spectrum $\rho(z_1)P(z_1|z_2, \tau_m)$. This is indeed an exchange experiment in which the mixing time, $\tau_m = \Delta$ and where $\rho(z_1)P(z_1|z_2, \tau_m)$ is the probability that a spin-bearing particle at z_1 at time zero has moved to z_2 after time $\tau_m = \Delta$.

Han *et al.* point out that the diagonal and off-diagonal projections of the 2-D signal and its Fourier spectrum relate simply to the mean position $(z_1 + z_2)$ and mean displacement $(z_1 - z_2)$, as shown in the conjugate axes of Fig. 9.9(b).

9.3.4 Two-dimensional propagators

An entirely different approach to 2-D PGSE NMR has been taken by Stapf *et al.* [15] using a single effective gradient pair PGSE NMR sequence, but one in which the gradient pulses comprise q_X and q_Z components that are independently stepped, in effect traversing a 2-D Cartesian raster in q-space. Two dimensional Fourier transformation returns a 2-D average propagator $\bar{P}(X, Z, \Delta)$. The method is particularly interesting when applied to dispersive flow, in which Z refers to the flow direction while X is transverse to the flow, leading to intrinsic asymmetry in the distribution. Figure 9.10 shows an example of $\bar{P}(X, Z, \Delta)$ for water flowing at $\langle v \rangle = 2.5\,\mathrm{mm\ s^{-1}}$ through a pack of 0.6-mm diameter beads [15], for which $\tau_v = 240\,\mathrm{ms}$. Also shown are the projections of this distribution along each of the axes, in effect, the one-dimensional average propagators $\bar{P}(X, \Delta)$ and $\bar{P}(Z, \Delta)$, respectively, where $\Delta \approx \tau_v$. Note the similarity of the propagator $\bar{P}(Z, \Delta)$ to that shown in Fig. 9.7 under similar conditions.

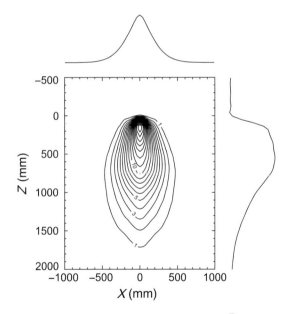

Fig. 9.10 Normalised, experimental 2-D average propagator $\bar{P}(X, Z, \Delta)$ in a 0.6-mm diameter bead pack at a flow rate of $42.0\,\mathrm{ml\ min^{-1}}$ and with $\Delta = 231\,\mathrm{ms}$. The projections, $\bar{P}(X, \Delta)$ and $\bar{P}(Z, \Delta)$, are drawn along the x- and z-axis, respectively. (Reproduced with permission from Stapf *et al.* [15].)

An intriguing aspect of the 2-D average propagator $\bar{P}(X, Z, \Delta)$ is the degree to which it differs from the joint probability formed by simply taking the product of $\bar{P}(X, \Delta)$ and $\bar{P}(Z, \Delta)$. Any difference must arise from correlations within the flow field. For example, the difference is zero if displacements X and Z are mutually independent, the case for Brownian motion in an isotropic system. Figure 9.11 shows both $\bar{P}(X, Z, \Delta)$ and $\bar{P}(X, Z, \Delta) - \bar{P}(X, \Delta)\bar{P}(Z, \Delta)$ for a range of Péclet numbers. The difference terms, which integrate to zero because of the normalisation employed, show quite remarkable positive and negative structure, clearly indicating the role of subtle correlations in the flow. However, the interpretation of these correlations is problematic, the authors quoting the statisticians' perspective [16] that 'the mathematical representation of the relationship, once demonstrated, is an art in itself.'

A more fundamental measure of flow correlations is perhaps given by the non-local dispersion tensor, the measurement of which was demonstrated in Chapter 8. For this tensor, arguably richer in detailed structure than the correlation distributions of Fig. 9.11(b), the statistical physics is explicit, and the patterns derived possibly more intuitive in their connection with the complexities of the flow field.

9.3.5   Inhomogeneous field exchange

Perhaps the simplest possible 2-D Fourier exchange method relevant to diffusion or flow in porous materials utilises spin phase shifts that arise from precession in the inhomogeneous local fields [17], rather than as a result of applied magnetic field gradients. Utilising the distribution of Larmor frequencies resulting from susceptibility inhomogeneity, such a 2-D exchange technique can be used to investigate molecular motion through a porous medium.

As imbibed liquid molecules diffuse through the inhomogeneous magnetic field of the pore space, they migrate between regions of different Larmor frequency, thus causing a broadening in the off-diagonal width of the 2-D spectra. As the mixing time is increased, the mean diffusion distance increases, resulting in an increasing off-diagonal intensity.

The pulse sequence used and shown in Fig. 9.12(a) is similar to a NOESY (Nuclear Overhauser Exchange SpectroscopY) [19], with a frequency encoding (t_1) and acquisition time (t_2) separated by a mixing time, τ_m. The method has been demonstrated with a bead pack comprising 100-micron diameter glass spheres saturated with water. At a proton Larmor frequency of 400 MHz, the resulting inhomogeneous linewidth was 5.2 kHz (FWHM). Using 256 data points at 20 kHz acquisition and evolution bandwidths, the longest evolution time is thus 12.8 ms, corresponding to a diffusion length of around 7 microns, much smaller than the pore dimensions, so that it is possible to approximate the pulse sequence of Fig. 9.12(a), resulting in an approximately 'instantaneous' frequency encoding so that all molecular motion may be considered to occur over the mixing period, τ_m. Clearly for such experiments, large linewidths or large length scales are advantageous in underpinning this assumption. These results are shown in Fig. 9.13, where each 2-D spectrum has been normalised to constant total intensity so as to remove T_1 relaxation effects. A clear growth of off-diagonal intensity is observed as the mixing time is increased. The line broadening can be quantified by taking an average of two intensities located at equidistant points in the off-diagonal direction from the centre peak, and a plot of intensity versus τ_m can be used to estimate the characteristic exchange time for molecular migration across pores.

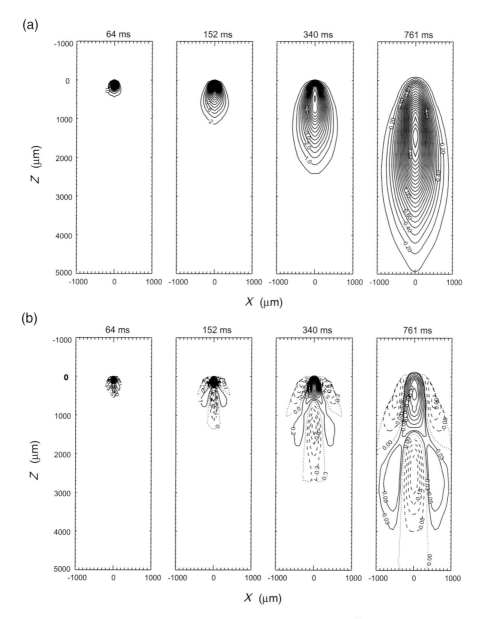

Fig. 9.11 (a) Normalised, experimental 2-D average propagator $\bar{P}(X, Z, \Delta)$ in a 0.6-mm diameter bead pack at a flow rate of $42.0\,\mathrm{ml\ min^{-1}}$. Evolution times Δ are as indicated. (b) Corresponding difference propagator $\bar{P}(X, Z, \Delta) - \bar{P}(X, \Delta)\bar{P}(Z, \Delta)$. Positive values are indicated by solid lines and negative values by dashed lines. (Reproduced with permission from Stapf *et al.* [15].)

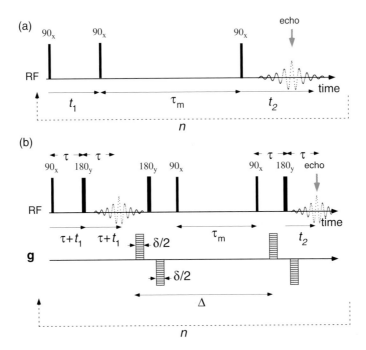

Fig. 9.12 (a) 2-D pulse sequence used for a local field exchange experiment. The evolution time, t_1, is incremented for each n. The mixing time, τ_m, is kept constant for each repetition, but varies for each experiment. (Adapted from reference [17].) (b) Sequence used for a propagator-resolved version of the local field exchange experiment in which both t_1 and the field gradient pulse amplitude g are independently stepped. Note the use of bipolar gradients to reduce susceptibility artifacts in the propagator dimension. (Adapted from reference [18].)

Propagator-resolved inhomogeneous field exchange

The inhomogeneous local field exchange experiment may be extended [18] to include a third Fourier space, as shown in Fig. 9.12(a), where independently stepped gradient pulses allow for a propagator dimension. By this means one may distinguish molecules that have diffused different distances along the field gradient axis. Hence, rather than measuring an off-diagonal intensity that is a function of mixing time alone, the additional spatial dimension allows a spatio-temporal analysis of the diffusion between sites of differing field.

9.4   Fourier–Laplace methods

In a pioneering 1981 paper, Stilbs suggested [20] that the molecular self-diffusion coefficient could be used to separate high-resolution NMR spectra according to molecular size, an effect he termed 'size-resolved NMR'. With the practical realisation of this idea through the advent of new data processing techniques, this method has come to be known as DOSY (for Diffusion Ordered SpectroscopY)—a term coined by Morris and Johnson [21]—or, in another variant, CORE-NMR (for COmponent REsolved NMR spectroscopy) [22]. From a broad perspective, these 2-D methods maintain the

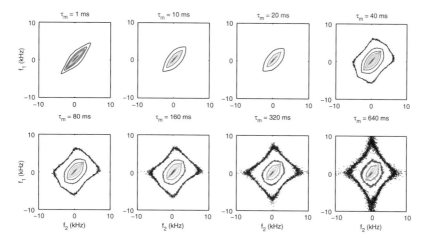

Fig. 9.13 2-D ^{1}H NMR exchange spectra obtained from water in 100-micron diameter bead pack at 400 MHz with a 20 kHz bandwidth. The eight examples show mixing times ranging from 1 to 640 ms. In each case the T_1 relaxation that occurs over the mixing time is compensated by normalising the total intensity in the spectrum, leading to noisier spectra at large τ_m. (Reproduced with permission from reference [17].)

full high resolution NMR spectrum as dimension in a Fourier domain, while distributing these spectra by diffusion coefficient in an inverse Laplace domain. They are not the only examples of possible 2-D Fourier–Laplace spectroscopies. Clearly one might correlate NMR spectra and relaxation rates, though the latter are more often chemical site-specific rather than molecule-specific, and hence do not present quite the same 'molecular separation' utility as CORE or DOSY. In the physics of porous media, the relaxation time may often be used as a marker for pore size, so that the Fourier–Laplace planes of inhomogeneous field-T_2 or displacement propagator-T_2 can provide useful insights. In this section we present examples of such 2-D Fourier–Laplace variants.

9.4.1 Diffusion-resolved spectroscopy

One simple approach to diffusion-resolved spectroscopy was demonstrated by Morris and Johnson [21]. Using the stimulated-echo PGSE NMR sequence shown in Fig. 9.14, they generated a 2-D data set by stepping the gradient amplitude in the first dimension and acquiring the free induction decay in the second. In order to ensure that eddy current effects from the gradient pulse switching had decayed before spectral sampling, they stored the stimulated echo along the z-axis for a sufficient period τ_e (around 10 ms) before recall and acquisition, a method referred to as longitudinal eddy current decay (LED) [23]. During τ_e homospoil, gradient pulses may be used to remove unwanted signals from the acquisition period.

The signal, normalised to the zero gradient case $q = 0$, could be written in continuous form as

$$S(t, q) = \iint \exp(-i\omega t) \exp(-q^2 D \Delta) f(\omega, D) d\omega dD \qquad (9.25)$$

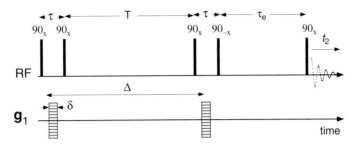

Fig. 9.14 2-D diffusion-ordered spectroscopy sequence of Morris and Johnson [21], in which a stimulated-echo sequence is employed with the gradients stepped to provide the first (inverse Laplace) dimension and the signal acquired in time in the second (Fourier) spectral dimension. The magnetisation arising at the echo after the third 90° RF pulse is stored along the z-axis for later recall, to allow all eddy current effects from the last gradient pulse to decay before recall of the signal for spectral analysis.

and, where necessary, with Δ replaced by $\Delta - \delta/3$ to allow for finite gradient pulse-width effects. Discretely encoded, we would represent these integrals by

$$S(t_i, q_j^2) = \sum_{k=1}^{M_1} \sum_{l=1}^{M_2} exp(-i\omega_k t_i) \exp(-q_j^2 D_l \Delta) f(\omega_k, D_l) + \epsilon_{i,j} \qquad (9.26)$$

where t_i ($i = 1..N_1$) are the sampling times of the acquisition dimension t_2, and q_j^2 ($j = 1..N_2$) are the squared-wavevectors of the evolution dimension associated with gradient stepping. An *a priori* domain of discrete frequencies ω_k and diffusion coefficients D_l is implied, the former being the usual Fourier domain of the NMR signal acquisition from which the spectrum arises. The second domain requires inverse Laplace transformation to reveal the diffusion coefficient distribution.

Morris and Johnson carried out this inversion both using regularised 1-D NNLS (CONTIN [6]) and by a programme that fitted to a predefined maximum number of exponential decays. To the extent that a known number, N, of molecular species is present in the sample, eqn 9.27 can be rewritten

$$S(t_i, q_j^2) = \sum_{n=1}^{N} \sum_{k=1}^{M_1} A_n(\omega_k) \exp(-i\omega_k t_i) \exp(-q_j^2 \Delta D_n) + \epsilon_{i,j} \qquad (9.27)$$

where $A_n(\omega_k)$ is the concentration-weighted NMR spectrum of the nth molecule. Of course the prior knowledge that each molecule has a unique diffusion coefficient argues against the use of a regularised NNLS inversion procedure in which broad diffusion distributions are favoured. Furthermore, in a large time-domain data set, perhaps 16k or 32k in size, the number of 1-D NNLS inversions to be performed is particularly large and therefore time-consuming. Where the number of molecular components is known, the use of simple multi-exponential fitting with a predetermined number of components is not only less time-consuming, but also produces a more physically realistic discrete diffusion spectrum. A powerful approach to inverting the q^2-domain data has been proposed by Stilbs *et al.* [22]. Their CORE-NMR uses a global procedure in which the

(a)

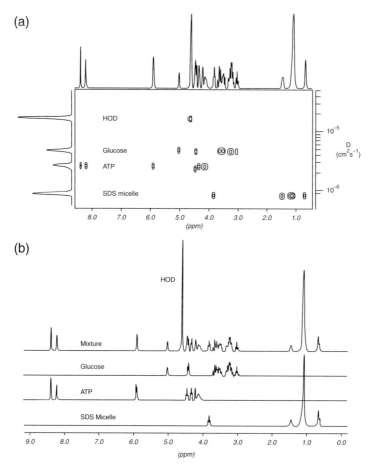

(b)

Fig. 9.15 Illustration of diffusion-ordered NMR: (a) 2-D frequency-diffusion coefficient spectrum for a mixture containing D_2O, glucose, ATP, and SDS micelles; (b) individual spectra of components in the mixture obtained by projection of intensities in the various diffusion ranges onto the chemical shift axis. (Reproduced with permission from Morris and Johnson [21].)

entire 2-D data set is fitted utilising the intrinsic common diffusive echo attenuations of individual components in the phased, absorption-mode spectra, the only fitting parameters in addition to the N diffusion coefficients being the N amplitudes for the contribution of each component. The method is efficient and ideally suited to large spectral data sets. The CORE approach relies on the 'prior knowledge' that a component bandshape (i.e. all frequencies) identically follow the same PGSE decay pattern, thus stabilising the processing and enhancing the signal-to-noise ratio.

9.4.2 Propagator-resolved T_2

Britton *et al.* [24] have used Fourier–Laplace spectroscopy to correlate relaxation times with molecular translational displacements in the case of water flowing through a packed bed of porous hydrogel particles ranging from 1 mm to 3 mm in diameter.

Their pulse sequence, shown in Fig. 9.16, incorporates a stimulated echo as the PGSE segment, with z-storage of magnetisation for later recall to a CPMG echo train of echo spacing τ_E. In order to accurately represent the propagators for displacement following Fourier transform with respect to q, a full complex signal must be acquired, necessitating storage of both x- and y-components of magnetisation during the PGSE segment, with appropriate recall and CPMG refocusing of both components through the remainder of the sequence. In consequence, the chosen phase cycle must reflect these requirements.

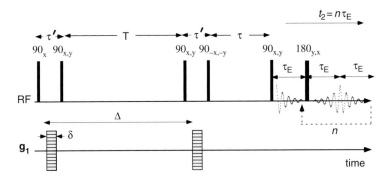

Fig. 9.16 RF and gradient pulse sequences for a T_2-resolved propagator experiment, comprising a pulsed gradient stimulated echo experiment followed by a CPMG train. As for the LED sequence, magnetisation is stored along the z-axis for later recall. Note that the phase cycle must store both x- and y-components of the magnetisation for correct reconstruction of the propagator.

The particles were formed by spraying droplets of alginate solution into 0.1 M calcium chloride. Water T_1 relaxation times and self-diffusion coefficients were the same inside the gel beads and in the outside pore fluid, at 1.8 s and $2 \times 10^{-9} \, \text{m}^2\text{s}^{-1}$, respectively, at 293 K, suggesting an open unrestricted gel structure. T_2 values are, however, widely distributed, in approximately three groups: short ($T_2 < 30 \, \text{ms}$), intermediate ($T_2 \sim 300 \, \text{ms}$), and long ($T_2 > 1 \, \text{s}$). These groups had been tentatively identified in an earlier T_2-only study [25] as water inside the beads and, for the medium and long T_2 values, as water in the pores between beads. Once flow is initiated, the propagator enables one to clearly recognise water resident within beads, water exchanging between the bead and the pore space, and water freely flowing along the interconnected pore space. Figure 9.17 shows the 2-D propagators that result from Fourier transformation with respect to q to produce a propagator for each echo in the CPMG train, followed by 1-D inverse Laplace transformation of the CPMG decays for each point (Z) in the propagator, by means of CONTIN.

The correlation apparent in Fig. 9.17 is quite powerful. The longest and shortest relaxation components are respectively associated with fluid remaining predominantly in the pore (the largest fluid displacements) and bead (the least fluid displacement). The finite displacement associated with the intermediate T_2 peak is consistent with

(a)

(b)

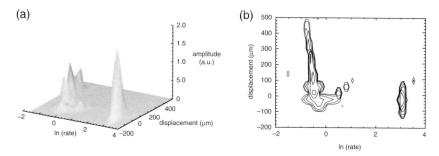

Fig. 9.17 (a) Surface and (b) contour plots of the T_2-resolved propagators, $\bar{P}(Z, \Delta, T_2)$, for water flowing through 3-mm diameter beads at an average flow velocity of $0.5\,\mathrm{mm\ s^{-1}}$. (Reproduced with permission from Britton *et al.* [24].)

exchange between the beads and the pore space. This example is instructive. Where a 1-D experiment provides the opportunity to reasonably speculate, such a 2-D correlation experiment allows a much more confident interpretation, one which enables the subsequent use of efficient 1-D techniques with greater information content.

9.5 Multidimensional Laplace–Laplace correlation methods

The idea that one could perform a 2-D NMR experiment in which Laplace inversions are performed in each dimension goes back to a relaxation exchange experiment proposed by Lee *et al.* in 1993 [26], in which independent 1-D transformations were applied to each dimension in succession, with the consequential disadvantages discussed in Section 9.2.3. The first integrated 2-D Laplace–Laplace NMR experiments were the $T_1 - T_2$ and $T_2 - D$ correlations carried out by Song, Hürlimann, and co-workers [27–29]. 2-D exchange measurements followed [30], and soon a plethora of possible dimensions were demonstrated, involving relaxation rates, diffusion coefficients, and internal magnetic field gradients.

Our summary begins with correlation methods. In each case these involve two or more encodings in sufficiently quick succession that the parameters to be extracted by ILT in each dimension correspond to molecular-borne spins at 'the same time'. Ideally, such encodings would be simultaneous and of infinitesimal duration. In practice, they will always require finite time, and almost invariably follow in succession. In consequence, interpretation of the measurements as true correlations requires prior knowledge that molecular behaviours are indeed consistent over the total encoding time.

9.5.1 Relaxation–relaxation

A $T_1 - T_2$ correlation pulse sequence is shown in Fig. 9.18 in which there is an evolution time, t_1, associated with spin-lattice relaxation, and a multi-echo acquisition time, t_2, associated with spin spin relaxation, where t_2 relates to echo maxima at time positions $t_2 = n\tau_E$. Assuming continuous time and relaxation-rate variables, and continuous relaxation-spectrum density, $f(R_1, R_2)$, we would write the signal, normalised to its value at $t_1 = t_2 = 0$, as

$$S(t_1, t_2) = \iint (1 - 2\exp(-R_1 t_1)) \exp(-R_2 t_2) f(R_1, R_2) dR_1 dR_2 \qquad (9.28)$$

where R_1 and R_2 refer to the relaxation rates T_1^{-1} and T_2^{-1}. Again, to make clear the discrete form,[5] and using explicit discretisation of the two independent encoding times, the signal can be written

$$S(t_{1i}, t_{2j}) = \sum_{k=1}^{M_1} \sum_{l=1}^{M_2} (1 - 2\exp(-R_{1k} t_{1i})) \exp(-R_{2l} t_{2j}) f(R_{1k}, R_{2l}) + \epsilon_{i,j} \qquad (9.29)$$

Note that the use of an inversion recovery stage implies that the kernel in the Fredholm integral problem, $(1 - 2\exp(-R_1 t_1)) \exp(-R_2 t_2)$, can have both positive and negative sign. The principles outlined in Section 9.2 remain the same, however. A more subtle issue concerns our assumptions regarding the sign of $f(R_1, R_2)$. The numerical ILT assumes positive amplitudes for $f(R_1, R_2)$. As we will see, there are circumstances in which they may be negative. For the moment, however, we will discuss an example where this problem does not arise.

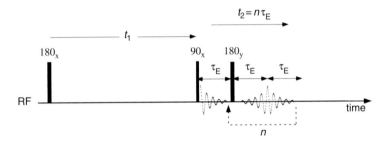

Fig. 9.18 RF pulse sequences for a $T_1 - T_2$ correlation experiment comprising an inversion-recovery segment followed by a CPMG train.

Figure 9.19 gives an example of the results of an experiment carried out on a sedimentary (limestone) rock sample saturated with brine, in which the distribution of pore sizes covers a wide range, and the T_1 and T_2 distributions are broad [27]. The time-domain data comprised 30 values of t_1 ($N_1 = 3$) and 4096 T_2 points ($N_2 = 4096$), where t_1 was varied logarithmically from $10\,\mu s$ to a few seconds, while the values of T_2 were equally spaced with a step size of $200\,\mu s$. Sixty-four scans were accumulated, at $5\,s$ repetition time, to improve the signal-to-noise ratio, with the total experiment lasting $2.7\,h$. The size of the matrix in the Laplace domain ($M_1 \times M_2$) was 100×100. The $T_1 - T_2$ spectrum reveals two separate peaks and a shoulder at $T_2 \approx 20\,ms$. The major peak at long T_1 and T_2 approaches the line of $T_1 = T_2$ expected for bulk water. By contrast the peak at shorter relaxation times is close to the line $T_1 = 4T_2$, typical for water in pores, where surface relaxation effects dominate. The behaviour

---

[5] From now on we shall use a continuous representation only, the discretisation process being apparent.

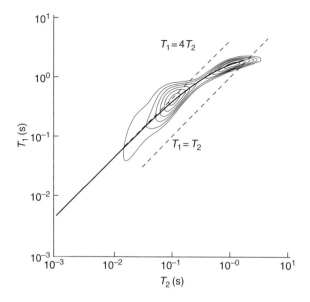

Fig. 9.19 Contour plots of $T_1 - T_2$ correlation spectrum for water in limestone. The dashed lines are for $T_1 = T_2$ and $T_1 = 4T_2$. The solid thick line is the theoretical behaviour of the sum of surface and bulk contributions to T_1 and T_2. (Reproduced with permission from Song *et al.* [27].)

is consistent with the rock structure indicated from optical microscopy, where grains containing small pores are close-packed with large inter-grain void spaces of around $100\,\mu$m in size. The solid line in Fig. 9.19 is obtained from a simple model in which the relaxation rate is assumed to be a sum of surface and bulk rates, under the condition that the ratio of the surface T_1 and T_2 relaxation times is 4.5.

9.5.2 Negative peaks and coupled eigenmodes

The T_1-T_2 correlation experiment is intriguing in the case of porous media where surface relaxation effects determine both relaxation modes. Under special circumstances, negative off-diagonal peaks may appear in the spectrum [31, 32]. Clearly the numerical inverse Laplace method cannot reveal such peaks and so great care is needed in this case.

The argument is simply followed using the normal modes description of the Brownstein–Tarr relaxation model, as described in Section 6.3.2 Chapter 6, starting with the magnetisation eqn 6.38 rewritten in terms of the pulse sequence of Fig. 9.18 (as [32]:

$$M(\mathbf{r}, t_1) = \sum_{p=0}^{\infty} a_p u_{1,p}(\mathbf{r}) \exp(-t_1/T_{1,p}) \tag{9.30}$$

and

$$M(\mathbf{r}, t_1, t_2) = \sum_{q=0}^{\infty} b_q u_{2,p}(\mathbf{r}) \exp(-t_2/T_{2,q}) \tag{9.31}$$

where the 1 and 2 subscripts refer to T_1 and T_2, respectively, and where $b_q = \int u_{2,q}^*(\mathbf{r})M(\mathbf{r},t_1)d\mathbf{r}$. Separate normal modes may be assigned to T_1 and T_2 relaxation because of their differing boundary conditions, i.e.,

$$D\hat{\mathbf{n}} \cdot \nabla u_{1,n}(\mathbf{r}) + \bar{\rho}_1 u_{1,n}(\mathbf{r})|_S = 0 \tag{9.32}$$

and

$$D\hat{\mathbf{n}} \cdot \nabla u_{2,n}(\mathbf{r}) + \bar{\rho}_2 u_{2,n}(\mathbf{r})|_S = 0 \tag{9.33}$$

Of course $M(\mathbf{r}, t_1, t_2)$ is the measured 2-D signal and ideal inverse Laplace transformation therefore leads to a 2-D spectrum $S(T_1, T_2)$ where each relaxation mode $\exp(-t/T_n)$ is replaced by $\delta(T - T_n)$. Song *et al.* [32] have pointed out that off-diagonal peaks in the T_1–T_2 spectrum indicate the degree of non-orthogonality of $u_{1,n}$ and $u_{2,n}$, since

$$S(T_1, T_2) = \sum_{p,q} \langle f|u_{2,q}\rangle \langle u_{2,q}|u_{1,p}\rangle \langle u_{1,p}|i\rangle \delta(T_1 - T_{1,p})\delta(T_2 - T_{2,q}) \tag{9.34}$$

where the shorthand $\langle \phi|\varphi \rangle$ represents $\int \phi^*(\mathbf{r})\varphi(\mathbf{r})d\mathbf{r}$, $\langle u_{1,p}|i\rangle = a_p$, $|i\rangle$ being the initial magnetisation distribution, $M(\mathbf{r}, 0)$. Similarly $|f\rangle$ is the final magnetisation distribution and can be assumed uniform for the purpose of calculation.

Song *et al.* [32] have pointed out that off-diagonal peaks in the T_1-T_2 spectrum indicate the degree of non-orthogonality of $u_{1,n}$ and $u_{2,n}$, as expressed in $\langle u_{2,q}|u_{1,p}\rangle$. For example, if the T_1 and T_2 surface relaxivities are identical, then so are the modes, and orthogonality ensures $\langle u_{2,q}|u_{1,p}\rangle = 0$, so that no off-diagonal contributions appear. Similarly, if both T_2 and T_1 are in a fast diffusion limit, i.e. $\bar{\rho}_{1,2}a/D \ll 1$, the T_2 and T_1 are in a common zeroth-order mode, u_0, and no off-diagonal features will be apparent. However, these authors have shown that in the intermediate case $\bar{\rho}_{1,2}a/D \sim 1$, clear off-diagonal structure is apparent and, remarkably, that those off-diagonal peaks are antisymmetric [31], with negative peaks seen in the 'non-physical' region of the spectrum where $T_2 > T_1$. Song *et al.* [32] have also demonstrated the existence of negative off-diagonal T_1-T_2 peaks in the case where spins diffuse from pore to pore.

Clearly the matter of negative peaks is problematic. Numerical inverse Laplace transformations are based on an amplitude positivity constraint, and so such peaks will not only be invisible, but their presence may distort the apparent spectrum. Ideally, they should not appear in $T_1 - T_1$ or $T_2 - T_2$ exchange experiments (discussed in Section 9.6), where the modes are orthogonal, resulting in off-diagonal peaks that are positive and symmetric. However, relaxation during the intervening mixing time has the potential to mix modes, thus making negative off-diagonal peaks possible [33]. The problem may be less likely in other correlation experiments, such as $D - T_2$, where quite independent physics governs the different relaxation parameters. Nonetheless, where a physical model is proposed in order to generate a 2-D spectrum, it should be tested for amplitude sign prediction in the manner outlined above.

9.5.3 Diffusion–relaxation

The diffusion T_2 relaxation correlation experiment can be performed using exactly the same sequence shown in Fig. 9.16 for the propagator-T_2 experiment of Section 9.4.2. The only difference is that the relevant gradient-pulse-induced kernel is that for diffusion, with the 2-D ILT being used to generate the $D-T_2$ spectrum. Rewriting eqn 9.20 using the explicit Stejskal–Tanner expression involving the gradient amplitude g_1, we have the signal (normalised to $g_1 = 0$)

$$S(g_1, t_2) = \iint \exp(-\gamma^2 g_1^2 \delta^2 D(\Delta - \delta/3)) \exp(-t_2 R_2) f(D, R_2) dD dR_2 \qquad (9.35)$$

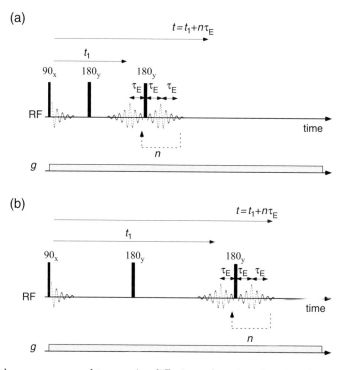

Fig. 9.20 Pulse sequence used to acquire diffusion-relaxation data in inhomogeneous fields. After the initial nominal 90° pulse, a series of echoes are generated with 180° pulses. The first two echo spacings, $t_1/2$, are longer than the later echo spacings, τ_E, such that the first echo amplitude is sensitive to diffusion while the echo train is sensitive only to T_2, because τ_E is so short. (a) and (b) show repeats, with different values of t_1 and the other independent time variable, $t = t_1 + n\tau_E$. [34].

Of course there are other means by which $f(D, R_2)$ may be measured. In particular, the diffusion-encoding can be performed using steady gradients and variable diffusion times, especially convenient in 'down bore hole' applications where pulsed gradient capabilities may be absent. Figure 9.20 shows just such a sequence [28] employed to take advantage of a constant background gradient, g. For down bore hole applications, the pulse sequence shown in Fig. 9.20 can be made more robust with respect to motion effects by using two initial echoes rather than a single echo [28].

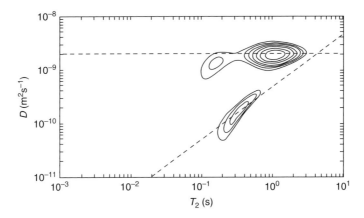

Fig. 9.21 Diffusion-relaxation correlation maps for a dolomite sample saturated with water and oil. The contour lines are at multiples of 12.5% of the maximum value in each graph. (Reproduced with permission from Hürlimann *et al.* [34].)

For such a sequence, the signal is given by

$$S(t_1, t) = \iint \exp(-\tfrac{1}{12}\gamma^2 g^2 D_k t_1^3) \exp(-tR_2) f(D, R_2) dDdR_2 \qquad (9.36)$$

Figure 9.21 shows the $f(D, R_2)$ distribution for a sample of Dolomite containing both water and oil. The horizontal dashed line indicates the diffusion coefficient of free water, while the sloping dashed line corresponds to the relaxation-diffusion correlation data for alkanes [35].

It is important to note that, for simple pore structures, where restricted diffusion and surface relaxation effects for imbibed fluid molecules play an important role in determining the NMR spin-echo signal behaviour, the dependence of the echo amplitude on gradient pulse strength will be complex because of q-space diffraction effects, and the decay of the echo amplitude with time will be multi-exponential because of the Brownstein–Tarr modes. Consequently, $D - T_2$ plots obtained from simple encoding sequences, such as that shown in Fig. 9.16, can lead to a quite complex response on 2-D inverse Laplace transformation [36]. An example of such a 2-D map calculated for spherical pores is shown in Fig. 9.22, over a wide range of diffusive observation times Δ and surface relaxivities $\bar{\rho} = M$, where these are expressed in appropriate dimensionless units relating to the spherical pore diameter a and the molecular self-diffusion coefficient D_0. In these plots there exists a dominant primary relaxation-diffusion mode, marked with a diagonal arrow. In order to more clearly reveal the weaker amplitude modes in these 2-D inverse Laplace plots, the analysis was carried out using a lower cutoff value for qa. The choice of cutoff only affects the relative amplitudes of the modes, but not their corresponding $D - T_2$ coordinates.

$D_0\Delta/\alpha^2$

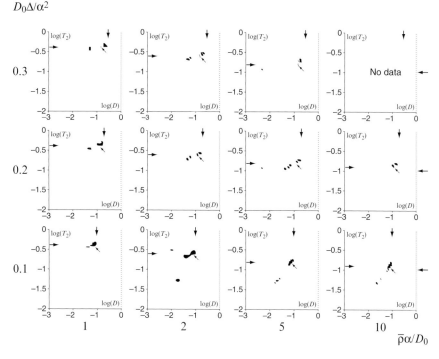

$\bar{\rho}\alpha/D_0$

Fig. 9.22 2-D $D-T_2$ maps for spherical pores, as a function of $\bar{\rho}a/D_0$ and $D_0\Delta/a^2$. D values are expressed in units of a^2/Δ and T_2 values in units of a^2/D_0. These maps were obtained by suppressing the amplitude of the primary relaxation-diffusion mode by using a lower cutoff for the q value. The diagonal arrow indicates the position of the primary relaxation-diffusion mode obtained from the low-q data. The vertical arrow indicates D_0, while the horizontal arrows on the left indicate the positions of the primary relaxation modes $T_2 = a^2/D_0\xi_0^2$. Horizontal arrows on the right indicate the Brownstein–Tarr limit for $\bar{\rho}a/D_0 \gg 1$ of $T_2 = a^2/D_0\pi^2$. (Reproduced with permission from reference [36].)

9.5.4   Diffusion-local field and relaxation-local field

As molecules diffuse in the local internal magnetic field gradients associated with field inhomogeneity, there is a dephasing akin to that experienced by spins in a steady applied magnetic field gradient. In a multispin-echo CPMG sequence, the sensitivity to that dephasing will be determined by the duration of the inter-echo period, the shorter the duration, the less the echo attenuation, according to the principles laid out in Chapter 5. Hence, it is possible to design a pulse sequence that has one of the encoding dimensions sensitive to this effect. Multi-dimensional examples incorporating local gradient dephasing include T_2-local field [37], T_1-local field [38], diffusion-local field [39], and a 3-D example incorporating diffusion, local field, and relaxation [39, 40].

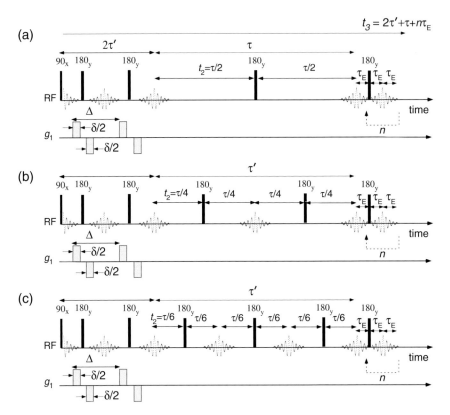

Fig. 9.23 3-D pulse sequence for measurement of correlations between diffusivity, internal gradients, and transverse relaxation. [37, 39]. Three segments operate. In the first dimension the pulsed gradient g_1 is stepped (note the use of a 13-interval sequence to reduce the effect of background gradients), in the second (t_2 dimension), over a fixed time τ, the numbers of $180°$ RF pulses is stepped—see (a) to (b) to (c)—thus changing the sensitivity of the sequence to diffusion in internal gradients. Finally the numbers of $180°$ RF pulses at fixed separation interval, τ_E, is incremented to provide the t_3 dimension sensitive to transverse relaxation.

For efficiency, we show a pulse sequence for such 3-D encoding in Fig. 9.23. Any 2-D subset of that sequence can be derived by eliminating the unwanted encoding element. Sequentially, the pulse train begins with a bipolar PGSE encoding (dimension g_1) designed to be sensitive to diffusion. Next comes a constant interval of time (τ) over which the number of $180°$ RF pulses in a CPMG train is varied so that the variable (second dimension) echo spacing interval becomes $t_2 = \tau/2n$, where n is an integer. As one increases the number of pulses in the fixed time period τ, the spin-bearing molecules have less time to diffuse through the internal gradients, lessening the dephasing effect. Below some value of n, the intensity of the measured echo will generally plateau, the echo spacing being sufficiently short that the influence of internal gradients upon the signal has become negligible. Finally, we have a CPMG train with a fixed (short) echo spacing but increasing total time as more RF pulses are added to the train. Hence we have the relaxation dimension, t_3. The signal attenuation is thus given by

$$S(g_1, t_2, t_3) = \iiint \exp(-\gamma^2 g_1^2 \delta^2 \Delta_r) \exp(-\tfrac{1}{3}\gamma^2 (Dg^2) t_2^2 \tau) \exp(-t_3 R_2)$$
$$\times f(D, (Dg^2), R_2) dDd(Dg^2) dR_2 \tag{9.37}$$

where Δ_r is the effective diffusion time.

Notice the importance of the various relevant length scales in our analysis. The dephasing length scale $l_g = (D_0/\gamma g)^{1/3}$ was introduced in Chapter 6. In a porous medium there exists a structural feature length scale, l_s, over which the magnitude of the internal gradient arising from local fields will change. There is also the distance, $l_E = (D_0 \tau_E)^{1/2}$ or $(D_0 t_2)^{1/2}$, diffused over the relevant echo time, τ_E or t_2. Provided $l_E \ll l_g$, the diffusion-induced echo attenuation between successive pairs of echoes is small, and the cumulative attenuation over many echoes is the result of successively small dephasing. For T_2 encoding through the multi-echo CPMG train, τ_E needs to be sufficiently short that T_2 effects are dominant. And, in order that we might sensibly use a local gradient description such that eqn 9.37 applies, we will require that $l_E \ll l_s$.

As an example of a relaxation local field experiment, Fig. 9.24 shows results obtained for water in a porous rock, where the pulse sequence has two encoding dimensions, inversion recovery for T_1, and a multi-echo train of variable echo spacing and total fixed time, to encode for Dg^2. In the interpretation of the data, D is taken to be constant and equal to the molecular self-diffusion coefficient of water. In consequence, a value for g may be derived in the inversion process. The examples shown in Fig. 9.24 feature increasing polarising field B_0, such that scaling laws for the dependence of T_1 and g on B_0 may be tested.

As B_0 is increased, the effect of susceptibility inhomogeneity is to cause the inhomogeneous local internal fields, $\Delta \chi B_0$, to proportionately increase, along with their local gradients, $\Delta \chi B_0 / l_s$. However, there is an upper limit to the effective gradient that can be measured, and this limit is reached when structural features have length scales on the order of l_g. Setting $g_{max} = \Delta \chi B_0 / l_s$, where $l_s = l_g$, one finds [41]

$$g_{max} = \left(\frac{\gamma}{D_0}\right)^{1/2} (\Delta \chi B_0)^{3/2} \tag{9.38}$$

with the critical structural length scale being

$$l^* = \left(\frac{D_0}{\gamma \Delta \chi B_0}\right)^{1/2} \tag{9.39}$$

For pores smaller than l^*, the fluctuations in gradient as molecules diffuse mean that the local gradients are averaged over the dephasing length. For larger pores, as l_s begins to approach l^*, the measured effective gradient approaches g_{max}, while for pores larger than l^*, the dephasing of the CPMG signal over the echo time, such that

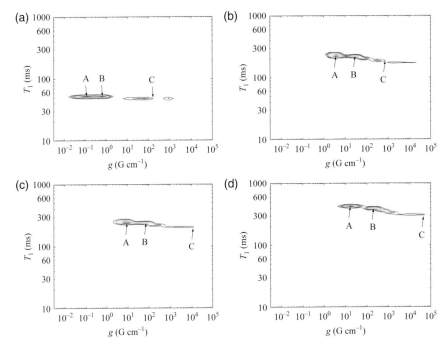

Fig. 9.24 $T_1 - g$ plots for water imbibed in a Mt. Gambier limestone at (a) 12 MHz, (b) 200 MHz, (c) 400 MHz, and (d) 900 MHz proton NMR frequency. A, B, and C arrows indicate the tracked intensity for regions of the 2-D inverse Laplace spectrum corresponding to pores of successively smaller size. Note the increasing local field and lengthening T_1 relaxation time as the Zeeman field applied to the sample is increased. (Reproduced with permission from Washburn *et al.* [38].)

$l_E \ll l_s$, arises from a local gradient that is effectively constant. This is the limit, applicable in eqn 9.37, where the overall echo train attenuation may be calculated by summing the contributions from each part of the distribution of gradients in the ensemble of spins.

9.5.5 DDCOSY

In Chapter 8, Section 8.5.3 introduced the concept of structural domains exhibiting local anisotropy, but where the the overall domain distribution exhibits global isotropy. The use of double wavevector encoding as a means of revealing such structural asymmetry was also outlined. In that treatment, the small wavevector limit was presented, under the condition that successive pairs of gradient pulses were applied along common or along different axes. In fact, this type of measurement can be multiplexed, using independent encoding of the gradient pairs to probe orthogonal domains of displacement in a 2-D correlation experiment, a method known as Diffusion–Diffusion COrrelation SpectroscopY or DDCOSY. This technique, where one compares 2-D maps obtained with either collinear or orthogonal gradient pulse pairs, has been applied to the study

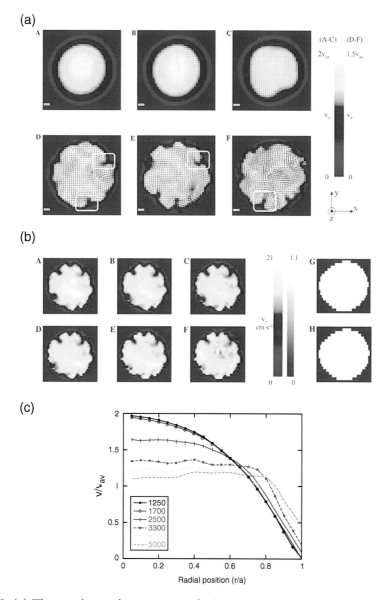

Fig. 10.9 (a) Three orthogonal component velocity images acquired at increasing Re of (A) 1250, (B) 1700, (C) 2500, (D) 3300, (E) 4200, and (F) 5000. The colour scale identifies the magnitude of the z-velocity. The flow velocity in the x–y plane of the image is shown by the vectors. The vector scale bar on each image corresponds to $1\,\mathrm{cm\,s^{-1}}$. (b) Six consecutive z-velocity images acquired at 20 ms time intervals are shown (A–F) for flow at $Re = 5000$. The average VACF maps calculated from pairs of images separated by time intervals of (G) 40 ms and (H) 80 ms are also shown. (c) Radial velocity profile of the z-velocity for Re in the range 1250–5000. (Reproduced with permission from Sederman *et al* [20].)

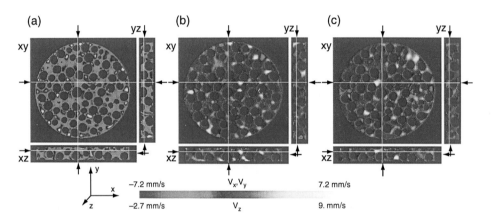

Fig. 10.16 Velocity fields within a packing of 5-mm diameter spheres within a column of internal diameter 46 mm. Fluid velocities in the (a) z-, (b) x-, and (c) y-directions are shown with slices taken in the xy, yz, and xz planes for each of the velocity components. For each image the positions at which the slices in the other two directions have been taken are highlighted. (Reproduced with permission from Sederman and Gladden [35].)

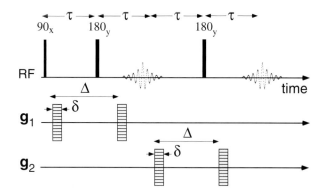

Fig. 9.25 DDCOSY pulsed gradient spin-echo NMR pulse sequence in which two successive pairs of diffusion encoding gradient pulses, with directions either collinear or orthogonal, are stepped independently.

of local diffusional asymmetry in lyotropic liquid crystals. The pulse sequence is given in Fig. 9.25.

In general, we may write the relationship between the signal and the diffusion coefficient distribution as

$$S(g_1, g_2) = \iint \exp(-\gamma^2 g_1^2 \delta^2 D_1 \Delta_r) \exp(-\gamma^2 g_2^2 \delta^2 D_2 \Delta_r) f(D_1, D_2) dD_1 dD_2 \quad (9.40)$$

where, as usual, $\Delta_r = \Delta - \delta/3$. Suppose we take a locally anisotropic diffusion tensor describable by the ellipsoid of revolution of Fig. 8.14. The relevant echo-attenuation expressions for E_{zz} and E_{zx} in the case of the double wavevector encoding, with equal magnitude gradient pulses, are those of eqn 8.45. Here these equations are modified to allow for independent q-vectors $\mathbf{q}_1$ and $\mathbf{q}_2$, such that

$$E(q_{1z}, q_{2z}) = \int_0^1 d\cos\theta \exp\left(-(q_{1z}^2 + q_{2z}^2)\Delta[D_\| \cos^2\theta + D_\perp(1 - \cos^2\theta)]\right)$$

$$E(q_{1z}, q_{2x}) = (2\pi)^{-1} \int_0^1 d\cos\theta \exp\left(-q_{1z}^2 \Delta[D_\| \cos^2\theta + D_\perp(1 - \cos^2\theta)]\right)$$

$$\times \int_0^{2\pi} d\phi \exp\left(-q_{2x}^2 \Delta[D_\|(1 - \cos^2\theta)\cos^2\phi + D_\perp \sin^2\phi + D_\perp \cos^2\theta \cos^2\phi]\right)$$

$$(9.41)$$

Equations 9.41 for E_{zz} and E_{zx} provide quite different responses, depending on the successive orientation of the gradient directions. Despite the isotropic distribution of directors, the chance of capturing a fast diffusion direction in either of the two encoding pairs is enhanced when the gradient direction is switched in the second pair. Of course, for a known, axially symmetric, diffusion anisotropy, the diffusion distribution, $f(D_1, D_2)$, can be deduced directly from eqn 9.41. Figure 9.26 shows the result of such a calculation carried out in the case of a prolate local diffusion tensor ($D_\perp = 0.1D_\|$), perhaps describing a 'random array of pipes', as well as the oblate case ($D_\parallel = 0.1D_\perp$) relevant to a polydomain lamellar phase liquid crystal in which solvent molecules diffuse freely along the lamellae. These ideal distributions become distorted by the pearling process in the inverse Laplace transformation of the echo attenuation data generated using eqn 9.41. Nonetheless, the patterns are reproducible and consistent and, most importantly, highly characteristic of either oblate or prolate symmetry [42].

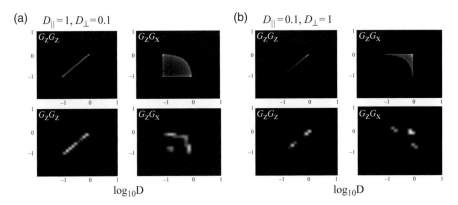

Fig. 9.26 (a) Calculated DDCOSY spectra for isotropic distribution of locally anisotropic diffusion elements in the case $D_\perp = 0.1D_\parallel$. The upper panels are calculated directly from the analytic distribution of diffusion coefficients apparent in eqn 9.41, while the lower panels are calculated by using eqn 9.41 to generate echo attenuations, which are then subject to 2-D inverse Laplace transformation. Note the distinctive off-diagonal features when the gradient directions are switched. (b) as for (a) but for the case $D_\parallel = 0.1D_\perp$. (Reproduced with permission from reference [42].)

An experimental demonstration of DDCOSY is shown in Fig. 9.27 in which theory and experiment are compared for water diffusing in a polydomain lamellar phase of the lyotropic liquid crystal, 40 wt % non-ionic surfactant C10E3 ($C_{10}H_{21}O(CH_2CH_2O)_6H$) in H_2O [43]. Strong off-diagonal features are apparent in the case where the successive gradients are applied in orthogonal directions, while a diagonal spectrum is apparent for collinear gradients. The agreement between theory and experiment provides nice confirmation for the existence of a multi-domain lamellar phase. Note that the 'slow' and 'fast' diffusion limit $D_\parallel$ ad $D_\perp$ can be read directly from these maps.

Crucially the domain size must be smaller than the diffusive observation time, Δ, if the molecules are to reside in a common domain over the DDCOSY encoding. In the present example $\Delta = 30$ ms corresponding to a water molecule diffusion distance of abut 10 μm. The absence of off-diagonal intensity in the collinear gradient plot provides confirmation that common domain behaviour is observed, and hence the domain size must be larger than this.

The question arises of whether the domain size can be measured by increasing the observation time using the collinear gradient pulse experiment. The answer, of course,

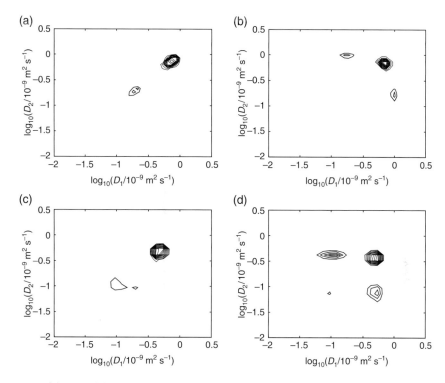

Fig. 9.27 (a) and (b) Calculated DDCOSY spectra for isotropic distribution of locally anisotropic diffusion elements in the case $D_\parallel = 0.1 D_\perp$, with $D_\perp = 1 \times 10^{-9}\,\mathrm{m^2\,s^{-1}}$. In (a) the gradient pulse pairs are collinear and in (b) they are orthogonal. The maps are calculated by using eqns 9.41 to generate echo attenuations, which are then subjected to 2-D inverse Laplace transformation. The contours represent a linear intensity scale with ten equal steps. (c) and (d) Equivalent maps but for an actual experiment on water molecules in the polydomain lamellar phase of a 40 wt % C10E3/water sample. (Reproduced with permission from reference [43].)

is 'yes', and the ideal way to do this is to maintain Δ short, but to introduce a well-defined mixing time, τ_m, as a delay between the encoding pulse pairs. Thus we move from a correlation experiment to an exchange experiment. Exchange methods are the subject of the next section.

9.6 Multidimensional Laplace–Laplace exchange methods

The measurement of exchange involves encoding for the same property at two different times, in order to ascertain any changes due to system dynamics. Multi-dimensional exchange methods rely on the fact that unchanging parameters will lie along the plot diagonal, while those that change between encodings will move off-diagonal. In order to ensure a common molecular identity for the parameter to be compared, each unique molecular magnetisation participating in the experiment must prevail over the mixing time, τ_m, the time over which the system is allowed to evolve between encodings.

This limits us to mixing times on the order of the slowest relaxation process from which we can protect the spins, typically T_1, and hence with protection imparted by storing magnetisation along the longitudinal axis. z-storage will be the norm in the experiments to follow.

Measuring the intensity of off-diagonal features as a function of the variable mixing time gives access to the dynamical rate constants. Here, consistent definitions of the region of intensity to be integrated are essential. In multi-dimensional exchange experiments involving the ILT, the effect of pearling is often to gather intensity into identifiable 'islands' in the plot. These can provide just the definition we seek when tracking τ_m-dependence of intensity, but, of course, caution is needed to ensure that sufficient stability is present in the inversion process to justify this type of analysis.

9.6.1 DEXSY

Diffusion Exchange SpectroscopY (DEXSY), compares molecular self-diffusion coefficients after a mixing interval. It is particularly useful for examining systems in which molecules are partitioned according to their diffusive behaviour, but exchange between these partitions over NMR-accessible timescales of milliseconds to seconds. The pulse sequence for DEXSY is essentially the same as that used for the VEXSY experiment and shown in Fig. 9.5. The variant shown here in Fig. 9.28 uses z-storage using the mixing interval, so that the spins experience only the slower T_1 relaxation. For our

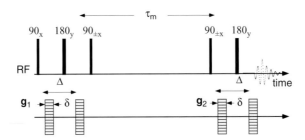

Fig. 9.28 DEXSY PGSE experiment in which two collinear gradient pulses pairs separated by a mixing time τ_m are stepped independently. z-storage of magnetisation is used during the mixing period.

first example of the use of DEXSY, we return to the polydomain lamellar lyotropic 40 wt % C10E3/water phase discussed in the previous section. Water molecules will diffuse over the mixing time and, in doing so, some will reach a new domain, while some will remain at the end of the mixing period in their original domain. For the latter, their signal contribution takes the form of the first of eqns 9.41, which we denote $E(q_{1z}^2, q_{2z}^2)_{orig}$. This signal contributes a completely diagonal DEXSY spectrum when subject to 2-D inverse Laplace transformation. By contrast, molecules that have diffused to a new domain will have a local director completely uncorrelated with their starting director so that their contribution to the echo attenuation is given by

$$E(q_{1z}^2, q_{2z}^2)_{new} = \int_0^1 d\cos\theta \exp\left(-q_{1z}^2\Delta[D_\| \cos^2\theta + D_\perp(1-\cos^2\theta)]\right)$$

$$\times \int_0^1 d\cos\theta \exp\left(-q_{2z}^2\Delta[D_\| \cos^2\theta + D_\perp(1-\cos^2\theta)]\right)$$

$$(9.42)$$

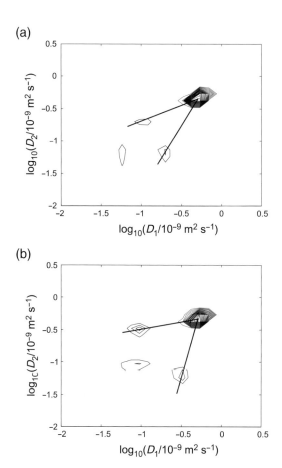

Fig. 9.29 DEXSY (D_1, D_2) exchange maps obtained from water molecules in the polydomain lamellar phase of the 40 wt % C10E3/water sample, using (q_{1z}^2, q_{2z}^2) encoding and in which the total effective mixing time τ_m' is (a) 230 ms and (b) 830 ms (Reproduced with permission from reference [43].)

The relative proportions of each population are determined by diffusion between the lamellar domains. For diffusion coefficient $D_\perp$ and given a rms domain size d, the resulting echo-attenuation expression can be written

$$E(q_{1z}^2, q_{2z}^2) = E(q_{1z}, q_{2z})_{orig} \left(1 - \int_d^\infty 2\pi r dr (\pi D_\perp \tau_m)^{-1} \exp(-r^2/4D_\perp \tau_m) \right)$$

$$+ E(q_{1z}^2, q_{2z}^2)_{new} \int_d^\infty 2\pi r dr (\pi D_\perp \tau_m)^{-1} \exp\left(-r^2/4D_\perp \tau_m \right)$$

$$(9.43)$$

in which r is a 2-D displacement along the lamellae (in the $\perp$ direction). Strictly, τ_m should be replaced with the effective mixing time $\tau'_m = \tau_m + \Delta$. When used to generate 2-D echo-attenuation functions that are subsequently subjected to the ILT, eqn 9.43 yields a map like that shown in Fig. 9.29, on which the off-diagonal peaks subtend an angle that depends on the mixing time [43]. A measurement of that angle enables one to estimate the domain size d. For the data shown in Fig. 9.29, $d \approx 43\,\mu$m.

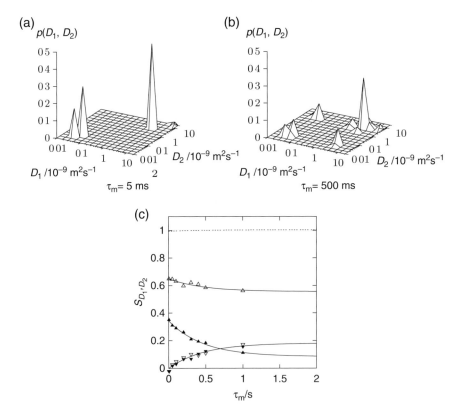

Fig. 9.30 (a) and (b) 2-D inverse Laplace transformed NMR signal maps of DEXSY experiments for pentane in zeolite, at different mixing times τ_m. (c) T_1 and T_2 corrected intensities of both (off-diagonal) exchange peaks (inverted triangles) and the (diagonal) free (open triangles), and adsorbed peaks (filled triangles) acquired by DEXSY, as well as their corresponding fits (—). For the fit of both exchange peaks their mean value was used. The sum of all four fitted intensities is also shown (- - -). Experimental uncertainties are on the order of the symbol size. (Reproduced with permission from Gratz *et al.* [44].)

The DEXSY method is particularly effective in analysing exchange of diffusing molecules between compartments in a partitioned system. It has been used to study diffusion of dextran with molecular weights 4.4 and 77 kDa through polyelectrolyte multilayer hollow capsules consisting of four bilayers of polystyrene sulfonate/polydiallyl dimethylammonium chloride. Intra-capsule and free dextran diffusion coefficients were identified, with off-diagonal DEXSY peaks arising from dextran migrating between the capsule and solvent spaces, exchange times of around 1 s being found [45].

Another example concerns diffusion of n-pentane between the interior and the exterior of zeolite crystals [44]. These crystals had diameters of around 15 μm, and were loaded with 500 mg g^{-1} n-pentane (2.5 times the maximum loading) such that there exists both an adsorbed phase (molecules in the zeolite) and a free liquid phase (molecules in the inter-crystalline space). Figure 9.30(a) shows the DEXSY map at short (5 ms) mixing time, where few molecules have time to diffuse between the adsorbed and free liquid phases. Two regions of intensity are seen on the diagonal: one corresponding to free pentane and the other, with much slower diffusion coefficient, to pentane within the zeolite crystals. As the mixing time is increased, as seen in Fig. 9.30(b), off-diagonal peaks arise due to molecules switching phases, and the plot in (c) shows the growth in intensity of these peaks, as well as the corresponding decay of the on-diagonal peaks as the mixing time is increased. Note that the intensities of these peaks has been corrected for both T_1 and T_2 relaxation. Fitting to a simple exponential behaviour yields an exchange time of 460 ± 50 ms, slower than that expected due to diffusion across the known crystal size, and therefore indicative of surface barriers [44].

9.6.2   Relaxation exchange spectroscopy

The very first 2-D ILT experiment carried out by Lee *et al.* [26] concerned T_2-T_2 exchange. In porous media, spin relaxation rates depend on pore surface-to-volume ratios, and hence can be used to derive pore size distributions [46]. In consequence, the change of spin-relaxation rates as molecules 'exchange' from pore to pore provides a window on inter-pore transport. However, the method is equally applicable to other materials, for example in soft matter, with exchange between liquid crystal domains, or in biophysics, where metabolites exchange between the interior and exterior of cells. In each of these it is useful to be able to quantify the details of exchange between specific sub-populations.

As for DEXSY, Relaxation EXchange SpectroscopY (REXSY) correlates molecular properties at two different times—in the pulse sequence shown in Fig. 9.31, their spin–spin (T_2) relaxation rates, preserving the magnetisation of individual molecules over a variable 'mixing time', separate the two encoding intervals when relaxation behaviour influences the spin magnetisation. The pulse sequence begins with a 90° RF excitation pulse, followed by the first T_2 encoding, in the form of a CPMG echo train in which the echo time is fixed while the number (n) of refocusing 180° pulses is varied. The use of closely spaced 180° RF pulses avoids any additional dephasing due to diffusion in the local gradients associated with internal diamagnetic susceptibility differences. A second 90° RF stores the magnetisation along the z-axis for the mixing time, τ_m, after which a third 90° RF pulse returns the magnetisation to the transverse plane, where the second T_2 encoding is efficiently performed in conjunction with multi-echo data acquisition.

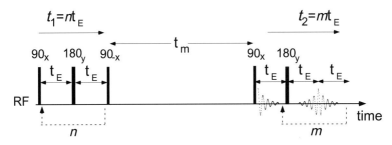

Fig. 9.31 Pulse sequence for the measurement of transverse relaxation exchange. At any fixed time interval τ_E, the numbers (n, m) of 180° RF pulses in the first and second T_2 encoding periods is independently varied. A succession of measurements are then made as a function of the mixing time τ_m.

The first application of the method was by McDonald and Korb [47] and Lee et al. [26], who observed T_2 exchange at a fixed time separation. Figure 9.32 shows the results of a variable mixing-time experiment carried out for water imbibed in a Castlegate sandstone [48]. The first T_2 encoding involves 42 logarithmically spaced echo train durations over the range 0.05–51.2 ms, while the second uses single-point acquisition at the top of each of 1024 CPMG echoes, giving a (42 × 1024) 2-D data set, from which second domain sampling points are chosen to produce a 42 × 42 matrix for subsequent inverse Laplace transformation. Castlegate sandstone exhibits T_2 relaxation times much shorter than T_1. However, a T_1-T_2 correlation experiment was used to assess the T_1 value for each T_2 component so that a correction could be made for longitudinal relaxation during the mixing time z-storage.

Figure 9.32 shows the 2-D T_2 exchange distribution at a mixing time of 160 ms. Along the diagonal, pearling results in discrete peaks seen at approximately 0.2 ms (labelled A), 1.6 ms (labelled B), 8 ms (labelled C), and 32.6 ms (labelled D). For the 0.2 ms, 1.6 ms, and 8 ms peaks, exchange between T_2 environments can be clearly seen, though no exchange appears between the 32.6 ms peak and those associated with shorter T_2 values. The longest relaxation peak is interpreted as belonging to a 'bulk' water phase, from which exchange is very slow.

The data shown in Fig. 9.32 can be analysed to reveal exchange times using the Zimmerman–Brittin method of Section 6.3.1. A somewhat simplistic first-order approach is to overlook the three sites involved and use a binary site model for each exchange peak. For the A–B exchange, the numbers of molecules, $N_{AB}(t)$, starting in A but residing in B after a mixing time t, corresponds to the A–B off-diagonal peak intensity. The rate equation will be $dN_{AB}/dt = N_{AA}/\tau_{AB} - N_{AB}/\tau_{BA}$, where N_{AA} is the number remaining in A after time t, and τ_{AB}^{-1} and τ_{BA}^{-1} are determined by the rates at which molecules in A migrate to B and vice versa. Such rates will be governed by the molecular self diffusion coefficient, the tortuosity of the pore throats, and the inter-pore distance. With initial condition $N_{AB}(0) = 0$, one may write [48]

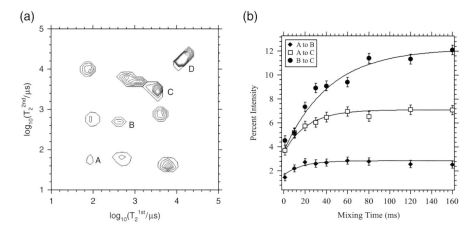

Fig. 9.32 (a) 2-D T_2 exchange distribution for water in Castlegate sandstone at a 160 ms mixing time. (b) Off-diagonal intensities, $N_{\alpha\beta}(t)$, as a function of mixing time.

$$N_{AB}(t) = \frac{N_B \tau_{AB}}{\tau_{AB} + \tau_{BA}} (1 - \exp(-\lambda t)) \tag{9.44}$$

with $\lambda = \tau_{AB}^{-1} + \tau_{BA}^{-1}$ and $\tau_{ex}^{AB} = \lambda^{-1}$ This exponential growth relationship is evident in Fig. 9.32, and using eqn 9.44 the exchange times are found to be $\tau_{ex}^{AB} = 27 \pm 3$ ms, $\tau_{ex}^{AC} = 31 \pm 3$ ms, and $\tau_{ex}^{BC} = 90 \pm 10$ ms. Similar relations can be derived for the exponential decay of the diagonal peaks where a self-consistent set of exchange times are observed.

A nice quantitative validation of REXSY has been carried out by Mitchell *et al.* [49] using a model porous medium comprising mixtures of nonporous borosilicate and soda lime glass spheres (diameter 100 μm) in water, thus providing an approximately constant characteristic pore dimension throughout the structures, but where water in different pore space regions had significantly different T_2 relaxation rates because of the two glass types. The packed beds were constructed with controlled glass-type domain sizes so as to test a model for region-to-region exchange of water. Using REXSY-determined exchange times along with the known self-diffusion coefficient of water, these authors calculated corresponding length scales that closely corresponded with the known domain dimensions.

9.6.3 Symmetry and peak amplitude sign in exchange and correlation spectroscopy

Single pore multimodes

The multi-eigenmodes of the Brownstein–Tarr relaxation-diffusion problem for a single pore imply the existence of multiple T_1 and T_2 values, except in the fast diffusion limit, $\bar{\rho}a/D \ll 1$, where a is the pore size. Hence, in slow and intermediate diffusion a single pore may return a multi-component $T_2 - T_2$ or $T_1 - T_2$ spectrum. As we have seen in Section 9.5.2, for the $T_1 - T_2$ correlation experiment, where the T_1 and T_2 relaxivities $\bar{\rho}_1$ and $\bar{\rho}_2$ are different, the eigenmodes of the T_1 and T_2 relaxation processes may be non-orthogonal and off-diagonal peaks may occur with negative amplitude [32], and hence not be amenable to an ILT process based on the non-negative amplitude assumption. Such curious negative peaks correspond to coupling between modes i and j in which $T_2^{(i)} > T_1^{(j)}$, that is, 'above the diagonal' of the correlation map if T_2 is the vertical axis. Single-pore $T_2 - T_2$ correlation may also exhibit off-diagonal intensity, provided the pulse sequence includes an intervening T_1 relaxation process between the two T_2 encodings, but for such $T_2 - T_2$ correlation the off-diagonal intensities are always positive and the correlation map symmetric [32].

Diffusion between interconnected pores

For a sample comprising pores of differing sizes, multi-component T_2 or T_1 spectra are found, even for fast exchange in each pore and local single modal relaxation. In this case, the observed relaxation rate provides a pore label based on the local surface-to-volume ratio, as discussed in Section 9.6.2. Here diffusive exchange between interconnected pores results in off-diagonal peaks, but in the case of pure $T_2 - T_2$ exchange, without intervening relaxation, the spectra are necessarily symmetric, and all off-diagonal peaks are positive in amplitude. But if T_1 relaxation occurs during the mixing time, mode admixture can, in principle, lead to negative, albeit weak, off-diagonal peaks [33].

There does exist the possibility of an unusual exchange experiment in which T_1 and T_2 are successively encoded, separated by a mixing time. In the case where diffusion between pores is rapid compared with intrapore diffusion, a common T_1 and T_2 value is found and the $T_1 - T_2$ spectrum exhibits a single peak. Where diffusion between pores is much slower than intrapore diffusion, separate peaks labelling the (T_1, T_2) coordinates of each pore will be found. However, at intermediate exchange rates, cross-peaks will appear. Here again, symmetry is no longer implied for $T_1 - T_2$, and negative off-diagonal peaks are possible [31, 32], but predominantly where $T_2^{(i)} > T_1^{(j)}$ [33]. It is clear that (T_1, T_2) experiments require considerable care in interpretation and that the role of negative amplitude contributions can significantly complicate the use of the standard non-negativity constraint approach to inverse Laplace transformation.

9.7 Diffusion tensor measurement

In Chapter 1, Section 1.3.5, the diffusion tensor was introduced, and its measurement using pulsed magnetic field gradients was discussed briefly in Chapter 5, Section 5.5.5. Acquiring the signals needed to calculate the diffusion tensor involves a multidimensional encoding [50]. Here we expand a little on the discussion in Section 5.5.5 to describe the data processing and methodology in a little more detail.

9.7.1 Frame transformation

Because the random displacements associated with Brownian motion are uncorrelated along orthogonal Cartesian axes, the diffusion tensor, which describes velocity autocorrelation, is intrinsically diagonal. However, in a material medium in which molecules diffuse under conditions of local structural anisotropy, there will exist a principal axis frame associated with that structure, in which the diagonal components of the diffusion tensor are expressed. These numbers D_{11}, D_{22}, and D_{33} give the self-diffusion rates along each of the local structural axes. For example, if our system comprised water molecules diffusing in a fibrous structure, we might expect that there would be rapid diffusion along the fibre axis, perhaps labelled 1, and slower diffusion transverse to the fibres, along axes 2 and 3. In this case we would have $D_{11} > D_{22} = D_{33}$, the equality representing axial symmetry.

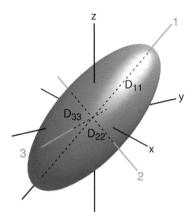

Fig. 9.33 Ellipsoid representation of diffusion tensor showing principal axis frame $(1, 2, 3)$, rotated with respect to the laboratory-frame coordinates (x, y, z). The eigenvalues of the tensor are the diagonal elements, D_{11}, D_{22}, and D_{33}, of $\underline{\underline{D}}_{local}$ in the principal axis frame, and are represented as the semi-axes of the ellipsoid.

In any attempt to measure local diffusion anisotropy using a laboratory-fixed axis frame, in which magnetic field gradients are applied along orthogonal directions within that frame, there is no *a priori* reason why this laboratory frame and the local structural principal axis frame should coincide. Hence the diffusion tensor measured in the lab frame, $\underline{\underline{D}}_{lab}$, will be related to the local diagonal tensor, $\underline{\underline{D}}_{local}$, using the Cartesian rotation matrices, $\underline{\underline{R}}$, by

$$
\begin{aligned}
\underline{\underline{D}}_{lab} &= \underline{\underline{R}}^T \underline{\underline{D}}_{local} \underline{\underline{R}} \\
&= \underline{\underline{R}}^T \begin{bmatrix} D_{11} & 0 & 0 \\ 0 & D_{22} & 0 \\ 0 & 0 & D_{33} \end{bmatrix} \underline{\underline{R}} \\
&= \begin{bmatrix} D_{xx} & D_{xy} & D_{xz} \\ D_{yx} & D_{yy} & D_{yz} \\ D_{zx} & D_{zy} & D_{zz} \end{bmatrix}
\end{aligned}
\tag{9.45}
$$

where $\underline{\underline{R}}$ transforms a vector in the laboratory frame to the local frame. Given eqn 9.45, it is obvious that $\underline{\underline{D}}_{lab}^T = \underline{\underline{D}}_{lab}$ or, in other words, $D_{\alpha\beta} = D_{\beta\alpha}$. Clearly only six elements of $\underline{\underline{D}}_{lab}$ are independent, but that represents a considerable increase over the three diffusion coefficients of the local frame, a consequence of not knowing the orientation of the principal axis frame in advance. The extra information we acquire in measuring these extra elements allows us to determine the local frame orientation.

Hence the process for finding D_{11}, D_{22}, and D_{33} involves the following steps. First, PGSE NMR experiments are performed using six linearly independent gradient directions in the (x, y, z) frame. By analysing those experiments, the independent coefficients D_{xx}, D_{xy}, D_{xz}, D_{yy}, D_{yz}, and D_{zz} are obtained, and $\underline{\underline{D}}_{lab}$ constructed. This laboratory-frame matrix is then diagonalised by solving the eigenvalue problem, the eigenvalues being D_{11}, D_{22}, and D_{33}, and the eigenvector transformations being given by the rotation $\underline{\underline{R}}$. At this point, the local diffusion anisotropy is known, along with the laboratory-frame directions of the local principal axes.

9.7.2   Echo experiment for diffusion tensor

Following eqns 5.73 and 5.74, allowing that for echo formation at time t the effective gradient $\mathbf{g}^*(t')$ obeys $\int_0^t \mathbf{g}^*(t')dt' = 0$, and neglecting flow, we have for the transverse magnetisation at the echo maximum,

$$M_+(t) = M_+(0)\exp\left(-\gamma^2 \int_0^t \left[\int_0^{t'} \mathbf{g}^*(t'')dt''\right]^T \underline{\underline{D}}\left[\int_0^{t'} \mathbf{g}^*(t'')dt''\right]dt'\right)exp(-t/T_2)$$

(9.46)

where the effective gradient vector is $\mathbf{g}^*(t) = \left(g_x^*(t), g_y^*(t), g_z^*(t)\right)^T$.

The standard approach to measuring $\underline{\underline{D}}$ is to employ a Stejskal–Tanner PGSE sequence where the effective gradient consists of two identical pulses of opposite sign. Suppose we consider PGSE gradient pulses of duration δ, separation Δ, and amplitude $\mathbf{g} = g\hat{\mathbf{g}}$, the latter represented by a magnitude g and with a direction determined by unit vector $\hat{\mathbf{g}}$. Following a convention used in diffusion measurement, we may write $b = \gamma^2\delta^2 g^2(\Delta - \delta/3)$, and so the normalised (to $g = 0$) echo amplitude may be written

$$E(\mathbf{g}) = \exp\left(-b\hat{\mathbf{g}}^T \underline{\underline{D}}\hat{\mathbf{g}}\right)$$
$$= \exp\left(-b\left[\hat{g}_x^2 D_{xx} + \hat{g}_y^2 D_{yy} + \hat{g}_z^2 D_{zz} + 2\hat{g}_x\hat{g}_y D_{xy} + 2\hat{g}_x\hat{g}_z D_{xz} + 2\hat{g}_y\hat{g}_z D_{yz}\right]\right)$$

(9.47)

where we have taken advantage of the symmetry of the diffusion tensor, $D_{\alpha\beta} = D_{\beta\alpha}$.

Alternatively we may define a 'b-matrix', given by

$$\underline{\underline{b}} = b\begin{pmatrix}\hat{g}_x \\ \hat{g}_y \\ \hat{g}_z\end{pmatrix}\begin{pmatrix}\hat{g}_x & \hat{g}_y & \hat{g}_z\end{pmatrix}$$

$$= b\begin{pmatrix}\hat{g}_x^2 & \hat{g}_x\hat{g}_y & \hat{g}_x\hat{g}_z \\ \hat{g}_y\hat{g}_x & \hat{g}_y^2 & \hat{g}_y\hat{g}_z \\ \hat{g}_z\hat{g}_x & \hat{g}_z\hat{g}_y & \hat{g}_z^2\end{pmatrix}$$

(9.48)

with

$$E(\underline{\underline{b}}) = \exp\left(-\underline{\underline{b}} : \underline{\underline{D}}\right)$$

(9.49)

and, taking advantage of the symmetry of the b-matrix elements,[6]

$$\underline{\underline{b}} : \underline{\underline{D}} = b_{xx}D_{xx} + b_{yy}D_{yy} + b_{zz}D_{zz} + 2b_{xy}D_{xy} + 2b_{xz}D_{xz} + 2b_{yz}D_{yz}$$

(9.50)

[6]$b_{\alpha\beta} = b\hat{g}_\alpha\hat{g}_\beta$.

9.7.3 Choice of diffusion gradient directions

A least six different gradient directions $\hat{\mathbf{g}}_i = (\hat{g}_{ix}, \hat{g}_{iy}, \hat{g}_{iz})$ will be needed to find the six independent elements of $\underline{\underline{D}}$. For any given direction, b may be varied, and the local image element signal analysed, in the standard manner of the Stejskal–Tanner experiment, through least squares fitting of $\ln(E)$ *vs* b, to find the effective diffusion coefficient D_i, where

$$D_i = \hat{g}_{ix}^2 D_{xx} + \hat{g}_{iy}^2 D_{yy} + \hat{g}_{iz}^2 D_{zz} + 2\hat{g}_{ix}\hat{g}_{iy}D_{xy} + 2\hat{g}_{ix}\hat{g}_{iz}D_{xz} + 2\hat{g}_{iy}\hat{g}_{iz}D_{yz} \quad (9.51)$$

From the $i = 1...6$ set of simultaneous linear equations for each of the six directions $\hat{\mathbf{g}}_i$, the six independent tensor elements, $D_{\alpha\beta}$, may be simply found. In a series of nice review papers on diffusion tensor imaging, Kingsley [51–53] has provided some direct matrix inversion routes to least squares and weighted least squares calculations of $D_{\alpha\beta}$, in the case where six or more directions are chosen for $\hat{\mathbf{g}}_i$.

The choice of diffusion gradient vectors will be determined by the need for linear independence of the vectors and, where slow diffusion is to be observed, a desire to optimise gradient strength, combinations of x, y, and z gradients providing a larger available gradient magnitude than can be obtained from any single direction. A detailed discussion of the various possible choices, and their respective advantages, is given in reference [52]. One commonly used choice is the set shown in Table 9.1, along with the corresponding matrix elements for b.

$u = 0$ cannot work, as it provides only three different gradient directions. $u = 1$ provides an efficient scheme, corresponding to six pairs of equal-magnitude gradients. $u = \frac{1}{2}(\sqrt{5}+1)$ or $u = \frac{1}{2}(\sqrt{5}-1)$ produces gradients pointing to the vertices of a regular icosahedron [54]. A nice feature of the set shown in Table 9.1 is that, having found the

Table 9.1 Calculation of the b-matrix for a common six-direction diffusion tensor gradient-encoding scheme, showing gradient unit vectors and associated b-matrix elements.

Direction index	$\hat{g}_x$	$\hat{g}_y$	$\hat{g}_z$	b_{xx}	b_{yy}	b_{zz}	b_{xy}	b_{xz}	b_{yz}
1	$\frac{1}{\sqrt{1+u^2}}$	$\frac{u}{\sqrt{1+u^2}}$	0	$\frac{b}{1+u^2}$	$\frac{bu^2}{1+u^2}$	0	$\frac{bu}{1+u^2}$	0	0
2	$\frac{1}{\sqrt{1+u^2}}$	$\frac{-u}{\sqrt{1+u^2}}$	0	$\frac{b}{1+u^2}$	$\frac{bu^2}{1+u^2}$	0	$\frac{-bu}{1+u^2}$	0	0
3	0	$\frac{1}{\sqrt{1+u^2}}$	$\frac{u}{\sqrt{1+u^2}}$	0	$\frac{b}{1+u^2}$	$\frac{bu^2}{1+u^2}$	0	0	$\frac{bu}{1+u^2}$
4	0	$\frac{1}{\sqrt{1+u^2}}$	$\frac{-u}{\sqrt{1+u^2}}$	0	$\frac{b}{1+u^2}$	$\frac{bu^2}{1+u^2}$	0	0	$\frac{-bu}{1+u^2}$
5	$\frac{u}{\sqrt{1+u^2}}$	0	$\frac{1}{\sqrt{1+u^2}}$	$\frac{bu^2}{1+u^2}$	0	$\frac{b}{1+u^2}$	0	$\frac{bu}{1+u^2}$	0
6	$\frac{-u}{\sqrt{1+u^2}}$	0	$\frac{1}{\sqrt{1+u^2}}$	$\frac{bu^2}{1+u^2}$	0	$\frac{b}{1+u^2}$	0	$\frac{-bu}{1+u^2}$	0

effective diffusion coefficients, D_i, corresponding to each of the $i = 1...6$ experiments, one may write down the diffusion matrix directly as [53].

$$
\begin{aligned}
D_{xx} &= h\left[D_1 + D_2 - u^2(D_3 + D_4) + u^4(D_5 + D_6)\right] \\
D_{yy} &= h\left[D_3 + D_4 - u^2(D_5 + D_6) + u^4(D_1 + D_2)\right] \\
D_{zz} &= h\left[D_5 + D_6 - u^2(D_1 + D_2) + u^4(D_3 + D_4)\right] \\
D_{xy} &= (D_1 - D_2)(1 + u^2)/4u \\
D_{xz} &= (D_3 - D_4)(1 + u^2)/4u \\
D_{yz} &= (D_5 - D_6)(1 + u^2)/4u
\end{aligned}
$$

(9.52)

where $h = (1 + u^2)/[2(1 + u^6)]$.

9.7.4 Diagonalisation and diffusion matrix parameters

The diffusion tensor is diagonalised by solving for the eigenvalues, D_{ii}, thus yielding the diagonal diffusion matrix, $\underline{\underline{D}}_{local}$ in its natural principal axis frame. The transformation from the laboratory frame to the principal axis frame is via a rotation matrix $\underline{\underline{R}}$, as shown in eqn 9.45. A suitable $\underline{\underline{R}}$ corresponds to an Euler rotation of coordinates footnoteFollowing convention, as successively rotated axes. successively α about z, β about y, and ζ about z so that

$$
\underline{\underline{R}} = \begin{pmatrix} \cos\gamma & \sin\gamma & 0 \\ -\sin\gamma & \cos\gamma & 0 \\ 0 & 0 & 1 \end{pmatrix} \begin{pmatrix} \cos\beta & 0 & -\sin\beta \\ 0 & 1 & 0 \\ \sin\beta & 0 & \cos\beta \end{pmatrix} \begin{pmatrix} \cos\alpha & \sin\alpha & 0 \\ -\sin\alpha & \cos\alpha & 0 \\ 0 & 0 & 1 \end{pmatrix}
$$

(9.53)

Defining $D_{ave} = \frac{1}{3}[D_{11} + D_{22} + D_{33}]$, the fractional diffusion anisotropy (FA) is

$$
\begin{aligned}
FA &= \sqrt{\frac{3\left[(D_{11} - D_{ave})^2 + (D_{22} - D_{ave})^2 + (D_{33} - D_{ave})^2\right]}{2\left[D_{11}^2 + D_{22}^2 + D_{33}^2\right]}} \\
&= \sqrt{\frac{3\left[(D_{xx} - D_{ave})^2 + (D_{yy} - D_{ave})^2 + (D_{zz} - D_{ave})^2 + 2(D_{xy}^2 + D_{xz}^2 + D_{yz}^2)\right]}{2\left[D_{xx}^2 + D_{yy}^2 + D_{zz}^2 + 2(D_{xy}^2 + D_{xz}^2 + D_{yz}^2)\right]}}
\end{aligned}
$$

(9.54)

Because of the invariance of the trace under a similarity transformation, $D_{ave} = \frac{1}{3}[D_{xx} + D_{yy} + D_{zz}]$. Conveniently, for the gradient directions chosen in Table 9.1, the average diffusion coefficient is given by

$$
\begin{aligned}
D_{ave} &= \tfrac{1}{3}[D_{xx} + D_{yy} + D_{zz}] \\
&= \tfrac{1}{6}[D_1 + D_2 + D_3 + D_4 + D_5 + D_6]
\end{aligned}
$$

(9.55)

Note that FA ranges from 0 for isotropic diffusion ($D_{11} = D_{22} = D_{33}$) to 1 for completely anisotropic diffusion ($D_{11} \neq 0$, $D_{22} = D_{33} = 0$).

Many more physical properties can be extracted from the diffusion matrix [53]. However, in addition to the average diffusion coefficient, the most important of these is undoubtedly the FA, which informs on the 'directedness' of the diffusion process, and, through a knowledge of $\underline{R}$, the orientation of the principal axis system and hence the local structural orientation of the material with respect to the laboratory frame.

9.7.5   Diffusion tensor imaging

Because diffusional anisotropy provides such a powerful insight regarding local structure in materials or biological tissue, measurement of the diffusion tensor is almost invariably combined with MRI, and commonly known as diffusion tensor imaging (DTI). In medical applications, DTI can be used to following the pathways of nerve fibres [55] to ascertain muscle orientation, and to determine the structural integrity of brain matter or indicate brain disease [56]. An example of a simple pulse sequence used in DTI is shown in Fig. 9.34, where the pulsed gradients used for diffusion encoding are labelled g_x, g_y, and g_z.

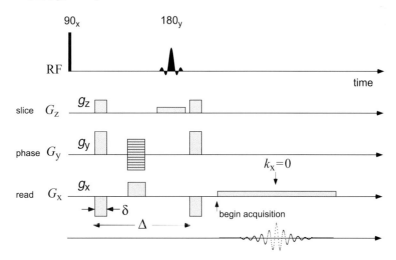

Fig. 9.34 DTI pulse sequence, based on spin warp with a soft $180°$ RF pulse for slice selection, in which the imaging gradients G_x, G_y, and G_z are supplemented by pulsed gradients g_x, g_y, and g_z, which are stepped from zero to provide the attenuation response arising from diffusion over time Δ. The direction of these diffusion gradients is determined by the vector $(g_x, g_y, g_z)^T$

The echo signal is, of course, encoded in both k- and q-space. The idealised Stejskal–Tanner DTI pulse sequence of Fig. 9.34 yields a multidimensional signal dependent on $\mathbf{k}$ and $\mathbf{g}$. Using the spin density function $\rho(\mathbf{r})$ to represent to local transverse magnetisation $M_+(\mathbf{r})$ in the absence of the PGSE gradients,

$$S(\mathbf{k}, \mathbf{g}) = \int \rho(\mathbf{r}) \exp(i\mathbf{k} \cdot \mathbf{r}) \exp\left(-b\hat{\mathbf{g}}^T \underline{\underline{D}}(\mathbf{r})\hat{\mathbf{g}}\right) d\mathbf{r}, \qquad (9.56)$$

$\underline{D}(\mathbf{r})$ being a spatially varying diffusion tensor, which may be used to report on local structural properties. Note that $\rho(\mathbf{r})$ will be influenced by T_2 relaxation over the spin warp sequence, as well as diffusive effects due to the imaging gradient $\mathbf{G}(t)$. These effects are discussed further in Section 9.7.6.

Subsequent to the image reconstruction (Fourier transformation with respect to $\mathbf{k}$) the local image element signals, $S(\mathbf{r}, \mathbf{g})$, for $b \neq 0$ and $S(\mathbf{r}, 0)$ when $b = 0$, are related by

$$S(\mathbf{r}, \mathbf{g}) = S(\mathbf{r}, 0) \exp \left(-b\hat{\mathbf{g}}^T \underline{D} \hat{\mathbf{g}} \right) \tag{9.57}$$

where $\underline{D}$ is implicitly $\underline{D}(\mathbf{r})$, the local diffusion tensor at $\mathbf{r}$.

9.7.6 Cancelling the effect of imaging gradients

If we analyse the response of the image to the stepping of the PGSE gradients $\mathbf{g}$, the relaxation attenuation and diffusive attenuation effects quadratic in $\mathbf{G}$ but independent of $\mathbf{g}$ will have no influence. They can, in effect, be normalised out by taking the ratio of the image with that obtained when $\mathbf{g} = 0$. However, the interaction of the imaging and PGSE gradients does involve sensitivity to diffusion due to bilinear terms, $g_\alpha G_\beta$. Clearly then, such artifacts in $E(\mathbf{k}, \mathbf{g})$ will depend (in the exponent) on the sign of $\mathbf{g}$, while the desired diffusion effect will depend only on $\mathbf{g}^2$. One simple means of removing these artifacts [58] is to carry out encodings with both positive and negative PGSE gradients, taking the corrected echo signal as

$$E_{corr}(\mathbf{k}, \mathbf{g}) = (E(\mathbf{k}, \mathbf{g}) E(\mathbf{k}, -\mathbf{g}))^{1/2} \tag{9.58}$$

In DTI applications it is customary to apply such a correction.

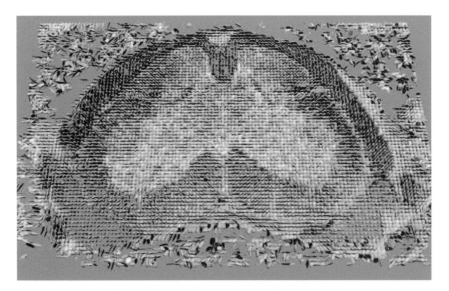

Fig. 9.35 Ellipsoid representation of mouse spinal cord. Note that right–left asymmetries in this image are due to disease. Significant changes are particularly noticeable in the ellipsoid shapes in the myelinated regions in the lower right quadrant of the image. (Reproduced by permission from Laidlaw *et al.* [57].)

9.7.7 Extracting and displaying the information content

One of the the biggest difficulties in DTI concerns how to display the spatially resolved tensor quantities. Pierpaoli and Basser [59] have proposed the use of ellipsoids, the semi-axes of which are proportional to the three diffusion eigenvalues, and which have an orientation set to the local principal axis frame. While these are quite informative individually, in a 3-D image the method does present problems. When visualising each voxel in the image dataset, only the outermost layer can be displayed on the screen, with internal data points being hidden. Second, ellipsoids will vary in size, according to D_{ave}, and so may overlap between voxels, making it hard to identify the continuity inherent in biological tissues [60]. One approach is to use normalised ellipsoids, where the largest dimensions are identical, irrespective of D_{ave}, so that they may be scaled to just fill each voxel without overlapping. An example is shown in the mouse spinal cord DTI image shown in Fig. 9.35.

References

[1] B. J. Davies. *Integral transforms and their Applications.* Springer-Verlag, Berlin, New York, 2002.

[2] E. I. Fredholm. Sur une classe d'equations fonctionnelles. *Acta Mathematica*, 27:365, 1903.

[3] I. S. Gradshteyn and I. M. Ryzhik. *Tables of Integrals, Series, and Products.* Academic Press, San Diego, 2000.

[4] S. J. Provencher. A constrained regularization method for inverting data represented by linear algebraic or integral equations. *Computer Physics Communications*, 27:213, 1982.

[5] A. N. Tychonoff and V. Y. Arsenin. *Solution of Ill-posed Problems.* Winston and Sons, Washington, 1977.

[6] S. W. Provencher. CONTIN: A general purpose constrained regularization program for inverting noisy linear algebraic and integral equations. *Comput. Phys. Commun.*, 7:229, 1982.

[7] K. P. Whittall and A. L. MacKay. Quantitative interpretation of NMR relaxation data. *J. Magn. Reson.*, 84:134, 1989.

[8] M. Prange and Y-Q. Song. Quantifying uncertainty in NMR T2 spectra using Monte Carlo inversion. *J. Magn. Reson.*, 196:54, 2009.

[9] L. Venkataramanan, Y-Q. Song, and M. D. Hüerlimann. Solving Fredholm integrals of the first kind with tensor product structure in 2 and 2.5 dimensions. *IEEE Transactions in Signal Processing*, 50:1017, 2002.

[10] P. T. Callaghan and B. Manz. Velocity exchange spectroscopy. *J. Magn. Reson. A*, 106:260, 1993.

[11] A. A. Khrapitchev, S. Stapf, and P. T. Callaghan. NMR visualization of displacement correlations for flow in porous media. *Phys. Rev. E*, 66:051203, 2002.

[12] S. Stapf, R. A. Damion, and K. J. Packer. Time correlations in fluid transport obtained by sequential rephasing gradient pulses. *J. Magn. Reson.*, 137:316, 1999.

[13] B. Bluemich, P. T. Callaghan, R. A. Damion, S. Han, A. A. Khrapitchev, K. J. Packer, and S. Stapf. Two-dimensional NMR of velocity exchange: VEXSY and SERPENT. *J. Magn. Reson.*, 152:162, 2001.

[14] S-I. Han, S. Stapf, and B. Bluemich. Two-dimensional PFG NMR for encoding correlations of position, velocity, and acceleration in fluid transport. *J.Magn. Reson.*, 146:169, 2000.

[15] S. Stapf, K. J. Packer, R. G. Graham, J-F. Thovert, and P. M. Adler. Spatial correlations and dispersion for fluid transport through packed glass beads studied by pulsed field-gradient NMR. *Phys. Rev. E*, 58:6206, 1998.

[16] G. P. Wadsworth and J. G. Bryan. *Introduction to Probability and Random Variables.* McGraw-Hill, New York, 1960.

[17] L. M. Burcaw and P. T. Callaghan. Observation of molecular migration in porous media using 2D exchange spectroscopy in the inhomogeneous magnetic field. *J. Magn. Reson.*, 198:167, 2009.

[18] L. M. Burcaw and P. T. Callaghan. Propagator resolved 2D exchange in porous media in the inhomogeneous magnetic field. *J. Magn. Reson.*, 205:209–215, 2010.

[19] J. Jeener, B. H. Meier, P. Bachmann, and R. R. Ernst. Investigation of exchange processes by two-dimensional NMR spectroscopy. *J. Chem. Phys.*, 71:4546, 1979.

[20] P. Stilbs. Molecular self-diffusion coefficients in Fourier transform nuclear magnetic resonance spectrometric analysis of complex mixtures. *Anal. Chem.*, 53:2135, 1981.

[21] K. F. Morris and C. S. Johnson. Resolution of discrete and continuous molecular size distributions by means of diffusion-ordered 2D NMR spectroscopy. *J. Am. Chem. Soc.*, 115:4291, 1993.

[22] P. Stilbs, K. Paulsen, and P. C. Griffiths. Global least-squares analysis of large, correlated spectral data sets: Application to component-resolved FT-PGSE NMR spectroscopy. *J. Phys. Chem.*, 100:8180, 1996.

[23] S. J. Gibbs and C. S. Johnson. A PFG NMR experiment for accurate diffusion and flow studies in the presence of eddy currents. *J. Magn. Reson.*, 93:395, 1991.

[24] M. M. Britton, R. G. Graham, and K. J. Packer. NMR relaxation and pulsed field gradient study of alginate bead porous media. *J. Magn. Reson.*, 169:203, 2004.

[25] A. D. Booth and K. J. Packer. Magnetic spin lattice relaxation in the presence of spin-diffusion. the one-dimensional, two-region system. *J. Magn. Reson.*, 62:811, 1987.

[26] J-H. Lee, C. Labadie, C. S. Springer, and G. S. Harbison. Two-dimensional inverse Laplace transform NMR: Altered relaxation times allow dectection of exchange correlation. *J. Am. Chem. Soc.*, 115:7761, 1993.

[27] Y-Q. Song, L. Venkataramanan, M. D. Hürlimann, M. Flaum, P. Frulla, and C. Straley. T1 T2 correlation spectra obtained using a fast two-dimensional Laplace inversion. *J. Magn. Reson.*, 154:261, 2002.

[28] M. D. Hürlimann, L. Venkataramanan, and C. Flaum. The diffusion-spin relaxation time distribution function as an experimental probe to characterize fluid mixtures in porous media. *J. Chem. Phys.*, 117:10223, 2002.

[29] M. D. Hürlimann and L. Venkataramanan. Quantitative measurement of two-dimensional distribution functions of diffusion and relaxation in grossly inhomogeneous fields. *J. Magn. Reson.*, 157:31, 2002.

[30] P. T. Callaghan and I. Furó. Diffusion-diffusion correlation and exchange as a signature for local order and dynamics. *J. Chem. Phys.*, 120:4032, 2004.

[31] L. Monteilhet, J-P. Korb, J. Mitchell, and P. J. McDonald. Observation of exchange of micropore water in cement pastes by two-dimensional T_2-T_2 nuclear magnetic resonance relaxometry. *Phys. Rev. E*, 74:061404, 2006.

[32] Y-Q. Song, L. Zielinski, and S. Ryu. Two-dimensional NMR of diffusion systems. *Phys. Rev. Lett.*, 100:248002, 2008.

[33] S. Rodts and D. Bytchenkoff. Structural properties of 2d NMR relaxation spectra of diffusive systems. *J. Magn. Reson.*, 205:315, 2010.

[34] M. D. Hürlimann, M. Flaum, L. Venkataramanan, C. Flaum, R. Freedman, and G. J. Hirasaki. Diffusion-relaxation distribution functions of sedimentary rocks in different saturation states. *Magn. Reson. Imaging.*, 21:305, 2003.

[35] S-W. Lo, G. J. Hirasaki, W. V. House, and R. Kobayashi. Mixing rules and correlations of NMR relaxation time with viscosity, diffusivity, and gas/oil ratio of methane/hydrocarbon mixtures. *Soc. Petroleum Eng. J.*, 7:24, 2002.

[36] P. T. Callaghan, S. Godefroy, and B. N. Ryland. Diffusion-relaxation correlation in simple pore structures. *J. Magn. Reson.*, 162:320, 2003.

[37] B. Sun and K.-J. Dunn. Probing the internal field gradients of porous media. *Phys. Rev. E*, 65:051309, 2004.

[38] K. E. Washburn, C. D. Eccles, and P. T. Callaghan. The dependence on magnetic field strength of correlated internal gradient-relaxation time distributions in heterogeneous materials. *J. Magn. Reson.*, 194:33, 2008.

[39] J. G. Seland, K. E. Washburn, H. W. Anthonsen, and J. Krane. Correlations between diffusion, internal magnetic field gradients, and transverse relaxation in porous systems containing oil and water. *Phys. Rev. E*, 70:051305, 2004.

[40] C. H. Arns, K. E. Washburn, and P. T. Callaghan. Multidimensional NMR inverse Laplace spectrsocopy in petrophysics. *Petrophysics*, 48:380, 2007.

[41] M. D. Hürlimann. Effective gradients in porous media due to susceptibility differences. *J. Magn. Reson.*, 131:232, 1998.

[42] P. T. Callaghan, S. Godefroy, and B. N. Ryland. Use of the second dimension in PGSE NMR studies of porous media. *Magn. Reson. Imaging*, 21:243, 2003.

[43] P. T. Callaghan and I. Furó. Diffusion-diffusion correlation and exchange as a signature for local order and dynamics. *J. Chem. Phys.*, 120:4032, 2004.

[44] M. Gratz, M. Wehring, P. Galvosas, and F. Stallmach. Multidimensional NMR diffusion studies in microporous materials. *Microporous and Mesoporous Materials*, 125:30, 2009.

[45] Y. Qiao, P. Galvosas, T. Adalsteinsson, M. Schoenhoff, and P. T. Callaghan. Multidimensional NMR diffusion studies in microporous materials. *J. Chem. Phys.*, 122:214912, 2005.

[46] R. L. Kleinberg and M. A. Horsfield. Transverse relaxation processes in porous sedimentary rock. *J. Magn. Reson.*, 88:9, 1990.

[47] P. J. McDonald, J-P. Korb, and L. Monteilhet. Surface relaxation and chemical exchange in hydrating cement pastes: A two-dimensional NMR relaxation study. *Phys. Rev. E*, 72:011409, 2005.

[48] K. E. Washburn and P. T. Callaghan. Tracking pore to pore exchange using relaxation exchange spectroscopy. *Phys. Rev. Lett.*, 97:175502, 2006.

[49] J. Mitchell, J. D. Griffith, J. H. P. Collins, A. J. Sederman, L. F. Gladden, and M. L. Johns. Validation of NMR relaxation exchange time measurements in porous media. *J. Chem. Phys.*, 127:234701, 2007.

[50] P. J. Basser, J.Mattiello, and D.LeBihan. MR diffusion tensor spectroscopy and imaging. *Biophysical Journal*, 66:259, 1994.

[51] P. B. Kingsley. Introduction to diffusion tensor imaging mathematics: Part I. Tensors, rotations, and eigenvectors. *Concepts Magn. Reson.*, 28A:101, 2005.

[52] P. B. Kingsley. Introduction to diffusion tensor imaging mathematics: Part II. Anisotropy, diffusion-weighting factors, and gradient encoding schemes. *Concepts Magn. Reson.*, 28A:123, 2005.

[53] P. B. Kingsley. Introduction to diffusion tensor imaging mathematics: Part III. Tensor calculation, noise, simulations, and optimization. *Concepts Magn. Reson.*, 28A:155, 2005.

[54] K. M. Hasab, D. L. Parker, and A. L Alexander. Comparison of gradient encoding schemes for diffusion-tensor MRI. *J. Magn. Reson. Imaging*, 13:769, 2001.

[55] P. J. Basser. *Magnetic resonance imaging of the brain and spine*, chapter Diffusion and diffusion tensor MR imaging. J.J. Attard ed., Lippincott, Williams and Wilkin, Philadelphia, 2002.

[56] M. Moseley. Diffusion tensor imaging and aging—A review. *NMR Biomed.*, 15:553, 2002.

[57] D. H. Laidlaw, E. T. Ahrens, D. Kremers, M. J. Avalos, R. E. Jacobs, and C. Readhead. Visualizing diffusion tensor images of the mouse spinal cord. In *Proceedings of the conference on Visualization*, Research Triangle Park, North Carolina, USA, 1998.

[58] M. Neeman, J. P. Freyer, and L. O. Sillerud. A simple method for obtaining crossterm free images for diffusion anisotropy studies in NMR microimaging. *Magn. Reson. Med.*, 21:138, 1991.

[59] C. Pierpaoli and P. Basser. Toward a quantitative assessment of diffusion anisotropy. *Magn. Reson. Medicine*, 36:893, 1996.

[60] S. Zhang, C. Demiral, and D. H. Laidlaw. Visualizing diffusion tensor MR images using streamtubes and streamsurfaces. *IEEE Trans. Visual Comput. Graphics*, 9:454, 2003.

10
Velocimetry

In Chapter 9 we saw how diffusion tensor imaging, based on PGSE contrast, can be used to provide a localised map of average diffusion coefficients, diffusion anisotropy, and the orientation of the diffusion tensor principal axis frame. Similarly, the PGSE method can be used in combination with imaging methods for the purpose of velocimetry, the mapping of the local velocity field in a fluid sample.

Fluid motion can result in bulk movement of elements within sample, leading to changes in local (fluid) spin density, $\rho(\mathbf{r})$. NMR imaging can follow such changes provided that the image capture is rapid enough that spin positions do not significantly change during that process, and successive images can reveal slow changes in fluid density. Alternatively, if the motion is periodic, the image capture process can be stroboscopically gated to allow a faithful representation of cyclic motion.

However, in this chapter, as elsewhere in this book, we will be concerned with a sample for which the spin density remains constant with time, even though the spin-bearing molecules are in motion. That motion may be steady-state, with an associated Eulerian velocity field that is constant, or in fluctuation, in which case any imaging of velocity will need to be sufficiently rapid to capture these changes. There exist numerous methods for flow contrast that can give a simple and direct visualisation of motion and, as discussed in reference [1], may be broadly categorised as 'steady state recovery', 'time of flight', or 'phase shift' determinations. Steady state recovery methods rely on a rapid repetition of RF and gradient pulses such that the saturation recovery or steady state free precession signal amplitude depends on the spin motion. In time of flight methods, the appearance of an image is compared in real space at two different times, or a 'bolus' of spins is tagged at one time and imaged at some later time. Here we focus solely on developments, post reference [1], in 'phase shift' PGSE methods, and in particular their use in combination with MRI, so that the Eulerian velocity field, $\mathbf{v}(\mathbf{r})$, may be measured. Generally, such experiments can be labelled NMR velocimetry or, more specifically, when spatially resolved, as MRI velocimetry.

There are some underlying physical constraints with which all NMR velocimetry methods must contend [1], and these warrant re-statement. First, the range of average velocities that can be accurately determined will be limited by the receiver coil dimension, l, at the maximum, and by molecular self-diffusion at the minimum. To have an observable signal the moving spins must remain in the receiver coil over the observational timescale, of order the echo time TE over which displacements are tracked, thus limiting velocities to less than or of order l/TE,[1] In non-imaging applications of flow measurement, where all spins within the coil are excited, there will always be a shift of spins located at the coil fringing field, and this effect will inevitably result in some signal degradation. In imaging, however, where slice selection is employed, such degradation can be avoided provided the excited spins remain within the receiver coil during image acquisition, although the slice selection gradient may itself introduce some additional velocity encoding.

The smallest average velocity that can be measured is limited simply by the superposed random motions of the molecules, although the degree to which random motion will mask the average flow is dependent on observational timescale. This is because flow displacements are linear with time, t, whereas rms diffusional displacements vary as $t^{1/2}$. The longest observational timescale, based on the use of stimulated echoes, is typically $t \sim T_1$. Thus the resolution of velocity measurements by NMR is of order $(D/T_1)^{1/2}$. For free water, this limit is around $50\,\mu m\ s^{-1}$. For macromolecules or large particles with substantially smaller self-diffusion coefficients, even allowing for shorter T_1 values, this limit may well be below $1\,\mu m\ s^{-1}$.

10.1 Imaging the propagator

Not long after the initial development of MRI techniques came the suggestion that bipolar (zero time integral) effective gradients could provide an image phase contrast proportional to velocity [5], and the imaging of mean flow rates by observing the dependence of pixel phase shifts on the first moment of $g^*(t)$ became well established [6, 7]. However, as we have seen earlier, the bipolar gradient pulse has the potential to provide a great deal more information than mean flow rates. In principle, it is possible to use narrow gradient pulse PGSE NMR to provide an average propagator for every pixel of the image, so that one is carrying out a measurement in both k- and q-space in the same experiment, a combination sometimes termed dynamic NMR imaging or dynamic NMR microscopy [8]. While costly in acquisition time, the combination of k-space and q-space imaging is a uniquely powerful method for determining local molecular ensemble behaviour in heterogeneous systems, providing in principle a resolution of order $10\,\mu m$ for the static dimension, $\mathbf{r}$, and $0.1\,\mu m$ for the dynamic dimension, $\mathbf{R}$.

10.1.1 Pulse sequence

The pulse sequence used for imaging the local propagator is essentially the same as that used for DTI. A spin warp version incorporating a single direction for q-encoding is shown in Fig. 10.1(a). With slice selection, imaging is performed in two dimensions, while the q-encoding represents the third dimension. In terms of total acquisition time, one k-dimension is a readout during acquisition, with the other a phase encoding requiring M separate acquisitions. M will typically be a multiple of 64, 128, or 256, depending on the spatial resolution and field of view required for the phase-encoding

---

[1]Some ingenious experiments by Pines and co-workers [2–4] have demonstrated that spins borne by flowing molecules can be spatially encoded in one coil, but remotely detected downstream, thus extending the limits to timing resolution.

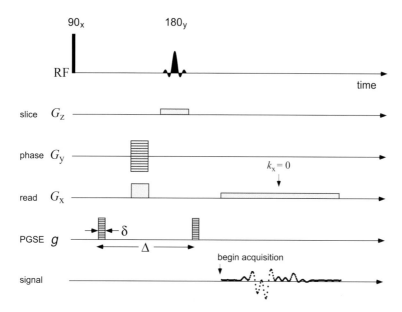

Fig. 10.1 Dynamic imaging pulse sequence incorporating both k- and q-space encoding. This particular spin warp imaging sequence, used in the experiments described in this chapter, employs a PGSE contrast period with, in this case, a spin echo. Long-Δ time interval experiments can be performed using a stimulated-echo version. Sampling of the echo as a function of k-space provides an image that has been further phase-encoded in q-space because of the PGSE gradients, which are stepped in amplitude between successive images.

dimension, the multiple being at least 2, for the most rudimentary 'add–subtract' phase cycle. Now include the q-steps, with the number again depending on the spatial resolution and field of view required for the displacement direction, and it is clear that on the order of 1000 or more signal acquisitions are needed.

10.1.2 Data processing

Data processing subsequent to the combined k- and q-encoding follows the flow chart illustrated in Fig. 10.2. First the inverse Fourier transform with respect to $\mathbf{k}$ is carried out, for each separate q-encoded image, generating separate real and imaginary image sets. The method is therefore akin to a multi-slice experiment where successive slices are obtained in q-space rather than real space. Generally the total number, n_D of 'q-slices' will be limited by the total available imaging time. However, a sufficient number of data points for accurate Fourier transformation, an analysis can be obtained by zero-filling the data, provided that the signal has sufficiently attenuated by the n_Dth slice. One commonly encountered ensemble motion comprises a combination of mean flow and molecular self diffusion. A velocity and diffusion map for each pixel in an image can be obtained by computing the width and offset of the propagator in Z space. In practice the dynamic profiles are computed by stepping the PGSE gradient in n_D steps to some maximum value g_m. Given that the q-space interval is $\gamma\delta(g_m/n_D)$, the

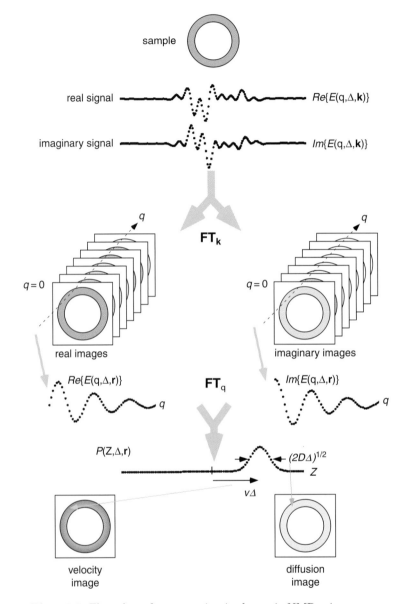

Fig. 10.2 Flow chart for processing in dynamic NMR microscopy.

complex signal, at pixel **r** of the nth q slice following the k-space FT, has separate oscillatory and damping components arising from flow, v, and diffusion, D, as

$$E(q, \Delta, \mathbf{r}) = E(0, \Delta, \mathbf{r}) \exp\left[iqv\Delta\right] \exp\left[-q^2 D\Delta\right]$$
$$= \rho(\mathbf{r}) \exp\left[i(\gamma\delta v\Delta)(g_m/n_D)n\right] \exp\left[-\gamma^2\delta^2(g_m/n_D)^2 Dn^2\Delta\right]$$

$$(10.1)$$

where the density image is recovered for $q = 0$ so that $E(0, \Delta, \mathbf{r})$ may be rewritten as the local NMR image density, $\rho(\mathbf{r})$, corresponding to that pixel. By implication v and D will depend on the chosen pixel and are therefore functions of $\mathbf{r}$. Further Fourier transformation with respect to q along the n dimension results in a one-dimensional spectrum for each pixel,[2] corresponding to the dynamic displacement profile (averaged propagator) at $\mathbf{r}$.

Suppose that an N-point digital FT is performed, resulting in N displacement spectrum points between $\pm Z_{max}$, where the field of displacement is $2Z_{max} = (2\pi n_D)/(\gamma\delta g_m)$. Given the product shown in eqn 10.1, the propagator is the convolution

$$\bar{P}(k/N, \Delta) = (\pi n_D^2/\gamma^2\delta^2 g_m^2 D\Delta)^{1/2} \exp\left[-\pi^2 k^2 n_D^2/\gamma^2\delta^2 g_m^2 N^2 D\Delta\right]$$
$$\otimes \delta(k/N - \gamma\delta v\Delta g_m/2\pi n_D) \tag{10.2}$$

where k is the digital value along the displacement direction and ranges between $-N/2$ and $N/2 - 1$. The peak centre occurs at the digital value

$$k_v = N\gamma\delta v\Delta g_m/2\pi n_D \tag{10.3}$$

and so the value of the mean molecular velocity in the pixel corresponding to the profile is

$$v = 2\pi n_D k_v/N\gamma\delta v\Delta g_m \tag{10.4}$$

The full-width-half-maximum (FWHM) of $\bar{P}$ in digital units is given by

$$k_{FWHM} = (2/\pi)[\ln(2)]^{1/2} N\gamma\delta(g_m/n_D)(D\Delta)^{1/2} \tag{10.5}$$

and the value of the mean molecular self-diffusion coefficient in the pixel corresponding to this profile is

$$D = (n_D k_{FWHM})^2/[(4\ln(2)/\pi^2)\gamma^2\delta^2 g_m^2 N^2\Delta]$$
$$= 3.56(n_D k_{FWHM})^2/[\gamma^2\delta^2 g_m^2 N^2\Delta]] \tag{10.6}$$

The location of peak centre position, k_v, and peak FWHM, k_{FWHM}, can be achieved by a simple computer algorithm [9].

10.1.3 Velocity resolution limit

The lower limit to velocity resolution, gradient strength permitting, is determined by the competitive stochastic motion due to self-diffusion. The precision in the velocity measurement will be limited by the width of the Gaussian propagator, the standard deviation of which is $(2D\Delta)^{1/2}$, and, at first sight, one might assume that the error in the velocity is $\sqrt{2D/\Delta}$. However, in practice the determination is made by locating the peak of a distribution, a process the precision of which is limited by the rms noise in the propagator, δP. Taking the derivative of the propagator near its peak with

[2]Note that at this point, correct 'phasing' of the spectrum can be applied.

respect to small deviations in velocity, δv, it is clear that the velocity determination error is given by [10]

$$\delta v = \sqrt{\frac{2D}{\Delta}\frac{\delta P}{P}} \tag{10.7}$$

where $\delta P/P$ is the noise-to-signal ratio in the propagator measurement. Clearly the best velocity resolution is achieved using the longest possible encoding time Δ, but at the same time the smallest flow displacement that can be measured results from using the smallest possible Δ consistent with the requirement that the random displacements due to self-diffusion do not dominate.

10.1.4 Velocity null experiments

With typical repetition times on the order of seconds, the pulse sequence of Fig. 10.1 implies a 30-min experiment at least, fine for steady-state flow, but of limited use when dealing with transient phenomena. Any speeding up requires trade-offs, and these are discussed in later sections. For the moment, we note that the fully multiplexed pulse sequence of Fig. 10.1 can be simply modified to allow for investigations of fluctuations, as shown in Fig. 10.3.

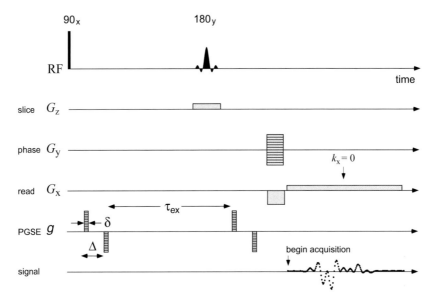

Fig. 10.3 Double PGSE version of the dynamic imaging pulse sequence of Fig. 10.1 used to investigate flow fluctuations.

By use of double PGSE encoding, a null displacement is returned where the dynamics are steady, the width of the average propagator being determined by self-diffusion alone. Any fluctuations over the time τ_{ex} between the PGSE pairs results in finite phase shifts, thus significantly broadening the propagator. The method does not directly visualise fluctuations, but it does detect their existence and their characteristic times. An example of this use is shown in Fig. 10.4, involving shear banded flow of

(a)

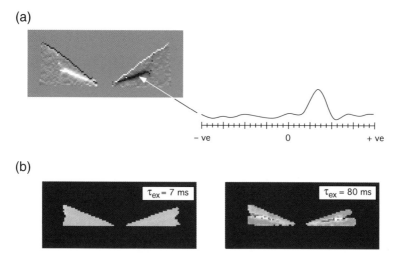

(b)

Fig. 10.4 (a) Velocity image for cetyl pyridinium chloride/sodium salicylate wormlike micelle solution in a 16-mm diameter 7-degree cone-and-plate system with a ×6 vertical gain (pixel size $156\,\mu$m×$26\,\mu$m), at an apparent shear rate of $10.7\,\mathrm{s}^{-1}$. Note the shear band indicative of a lower viscosity fluid state, near the centre of the gap. The average propagator shown is taken from a pixel inside the shear band, as indicated. (b) FWHM image for solution (a) at an apparent shear rate of $14.3\,\mathrm{s}^{-1}$ in the 25 mm, 7-degree cone-and-plate system with ×3 vertical gain, obtained using the double PGSE pulse sequence with exchange times of 7 ms and 80 ms. (Adapted from reference [11].)

wormlike micelles in a cone-and-plate geometry.[3] Figure 10.4(a) shows the gap of the 7-degree cone-and-plate cell with the vertical (y) gain expanded by a factor of 6. The grey scale intensity is a measure of the local shear rate $\partial v_z/\partial y$, where z is the flow direction normal to the page. A band of high shear rate is visible at gap centre, a consequence of the micellar solution undergoing a non-equilibrium phase transition to co-existing states of differing viscosity. A sample propagator is shown taken from a pixel within the low viscosity shear band.

Figure 10.4(a) is the result of a double PGSE encoding experiment at two different exchange times of 7 ms and 80 ms [11]. Here the grey scale plots the FWHM of the propagator. At $\tau_{ex} = 7$ ms the FWHM is common to all pixels in the image, irrespective of proximity to any shear band, while at 80 ms, a distinct FWHM enhancement results, thus revealing the existence of banding fluctuations at intermediate timescale.

10.2   Single-step phase encoding for velocity

Imaging the complete propagator comes at the price of multiple encoding steps in q-space. While the consequent commitment of imaging time provides valuable information concerning details of the molecular translational dynamics in each pixel, where all we seek is the mean local flow rate, a simpler, more time-efficient procedure is to

[3]See Section 10.4.3.

use a single q step. Of course, the more values of q we use, the more precisely we can determined the local velocity, but if we are prepared to trade away some of this precision, on the understanding that our intrinsic image signal-to-noise ratio is sufficient, then the combination of a reference $q = 0$ image and a finite q image allows for a velocity map to be generated.

10.2.1 Reference phase processing

Consider a particular pixel of two corresponding images obtained with $q = 0$ and $q = q_1$. The resulting complex signals are $E(0, \Delta, \mathbf{r})$ and $E(q_1, \Delta, \mathbf{r})$. If the mean flow in that pixel is v, then we may write

$$\frac{E(q_1, \Delta, \mathbf{r})}{E(0, \Delta, \mathbf{r})} = A \exp(i\phi) \tag{10.8}$$

where $\phi = q_1 v \Delta$ and $A \lesssim 1$ is the attenuation factor due to diffusive effects. The phase angle ϕ may be directly calculated from the complex number represented by eqn 10.8, for example using the Matlab function reference, 'angle'.

For any sensible experiment, we will require that the diffusive attenuation be not so great that the signal-to-noise ratio of the finite q image is significantly deteriorated beyond that of the $q = 0$ image. In other words, if f_σ is the fractional rms noise in each pixel, then we require $f_\sigma \ll A$.

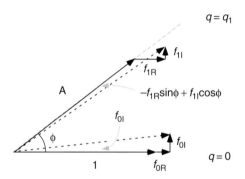

Fig. 10.5 Argand plane representation of $q = 0$ and $q = q_1$ signals for a particular image pixel, showing the influence of noise. $A \lesssim 1$ is the attenuation factor due to diffusive effects, where we require $A \gg f_\sigma$, f_σ being the rms value of the real and imaginary noise contributions.

Let us write the complex fractional noise in the cases $q = 0$ and $q = q_1$ for the pixel at $\mathbf{r}$ as $f_{0R} + if_{0I}$, and $f_{1R} + if_{1I}$, respectively, each noise term having the same rms value f_σ. Reference to Fig. 10.5 shows that the resulting error in the determination of the angle ϕ is

$$\delta\phi = -f_{0I} - f_{1R} \sin\phi + f_{1I} \cos\phi \tag{10.9}$$

By summing the noise power $f_{0I}^2 + f_{1R}^2 \sin^2\phi + f_{1I}^2 \cos^2\phi$, we see that the rms error in the angle determination is

$$\langle(\delta\phi)^2\rangle^{1/2} = \sqrt{2}f_\sigma \tag{10.10}$$

Calculating the mean flow by finding the argument of the ratio $E(q_1, \Delta, \mathbf{r})/E(0, \Delta, \mathbf{r})$ represents a quick and efficient data-processing scheme. An alternative is to use a Fourier analysis akin to that applied for multi-q experiments as shown in Fig. 10.2. In the next section, we show that these approaches are practically equivalent.

10.2.2 Fourier analysis

Having obtained $q = 0$ and $q = q_1$ images, we can generate a multi-q data set by adding zero-filled images. Suppose we make an N image set. Fourier transforming with respect to q will generate an average propagator for each pixel, but the dominant broadening for that propagator will arise not from diffusion, but from the effect of truncation at the second q step. In effect, the propagator is a convolution of the ideal propagator with a sinc function, the spectrum of the step function that truncates the data set. Labelling the digital steps in the q-domain by $0 < n < N - 1$ and in the propagator-domain by $-N/2 < k < N/2 - 1$, and using the definition of the digital Fourier transform given in eqn 4.44, we have the propagator given by

$$P(\frac{k}{N}) = 1 + \exp(i2\pi\frac{k}{N} - i\phi) \tag{10.11}$$

An example of this function, for $N = 128$ and $\phi = 0.2$ radians, is shown in Fig. 10.6. Our task is to find the k-coordinate of the maximum of this function, enabling us to directly determine ϕ. In the absence of noise, $\phi = 2\pi k_v/N$. Noise introduces uncertainty $\delta\phi$.

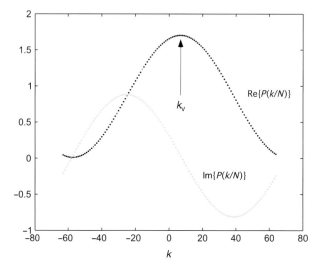

Fig. 10.6 Apparent propagator generated from Fourier transformation of a truncated q-space in which only $q = 0$ and $q = q_1$ steps are used. The real part of the phased propagator has a maximum at $k_v = N\phi/2\pi$. In this example $\phi = 0.2$ radians and $N = 128$.

Now we include the role of noise. The real and imaginary parts of eqn 10.11 become

$$Re\{P(\frac{k}{N})\} = 1 + \cos(2\pi\frac{k}{N} - \phi) + f_{0R} + f_{1R}\cos(2\pi\frac{k}{N}) - f_{1I}\sin(2\pi\frac{k}{N})$$

$$Im\{P(\frac{k}{N})\} = -\sin(2\pi\frac{k}{N} - \phi) + f_{0I} + f_{1I}\cos(2\pi\frac{k}{N}) + f_{1R}\sin(2\pi\frac{k}{N})$$

$$(10.12)$$

In fact, eqn 10.12 represents the correctly phased propagator. The unknown absolute phase of the NMR signal is unimportant when we calculate the flow-induced phase shift by taking the ratio in eqn 10.8. However, in using the Fourier method an additional autophasing step is needed. Here we assess the separate contributions to $\delta\phi$ from the maximum finding and autophasing procedures.

In autophasing, an arbitrary phase angle θ is adjusted in a multiplicative term, $\exp(i\theta)$, until the integral of the real part of $P(\frac{k}{N})$ is maximised, or until the integral of the imaginary part of $P(\frac{k}{N})$ is zero. Any error in θ due to the effect of noise becomes an error in ϕ. Since the integral of the f_{1I} and f_{1R} noise terms in eqn 10.12 are necessarily zero, the noise contribution to an error in the integral of $Im\{P(\frac{k}{N})\}$ is $iN f_{0I}$. The discrepancy in the signal integral between the correct phase θ and the shifted phase $\theta + \delta\phi$ is $N[\exp(i\delta\phi) - 1] \approx iN\delta\phi$. Whence, for autophasing, $\delta\phi = f_{0I}$.

In finding the maximum of $Re\{P(\frac{k}{N})\}$ the signal shape is unaffected by the constant noise term f_{0R}, but distorted by the sinusoidal variation of $f_{1R}\cos(2\pi\frac{k}{N}) - f_{1I}\sin(2\pi\frac{k}{N})$ in the vicinity of the maximum, $2\pi n/N = \phi$. Given a positional discrepancy of $\delta\phi$, the problem reduces to finding the null in the derivative of $Re\{P(\frac{k}{N})\}$, i.e. the function $\sin(\delta\phi) + f_{1R}\sin(\phi + \delta\phi) + f_{1I}\cos(\phi + \delta\phi)$. The small signal null offset, $\sin\delta\phi$, is counterbalanced by the local noise term contribution to the slope, to first order, $f_{1R}\sin\phi + f_{1I}\cos\phi$. Thus we have, for maximum finding,

$$\delta\phi = f_{1R}\sin\phi + f_{1I}\cos\phi \qquad (10.13)$$

Combining the autophasing and peak finding errors, we end up with a total error $\langle(\delta\phi)^2\rangle = \sqrt{2}f_\sigma$, independent of the number of points, N, used in the zero filling, and identical to that given in eqn 10.10 for the ratio method of phase determination.

10.2.3 Choice of method

Having established the equivalence of the ratio and Fourier spectrum methods in the case of a single phase-encoding q step, it is worth pointing out that both methods allow easy inclusion of extra phase-encoding steps, with consequent gains in the precision of velocity determination. Again, both appear equivalent in their precision, but perhaps the advantage of the Fourier method being an asymptotic behaviour towards an accurate representation of the pixel propagator as the number of q steps is increased sufficiently to significantly attenuate the image.

The result $\langle(\delta\phi)^2\rangle = \sqrt{2}f_\sigma$ means that some care is needed in experimental design. For example, a signal-to-noise ratio in each pixel of 10:1, a good result in any imaging experiment, implies that angular precision is on the order of 0.1 radians. The essential requirement in single-step phase-encoding is the need to ensure that the magnitude of

q_1 is large enough to give a phase shift significantly larger than the fractional signal-to-noise ratio.

Finally, it is worth noting that single-step encoding requires only one reference image at $q = 0$. Subsequent measurements need only single q_1 images to monitor a changing velocity field. Therein lies its real power.

10.3   Fast encoding and real-time velocimetry

The use of single-step phase encoding for flow certainly reduces the 'frame rate' at which a transient velocity field may be monitored. In this section are outlined some general principles concerning velocimetry speed. First we must clearly understand two questions. What are the characteristic times associated with the velocimetry pulse sequence? What are the characteristic times associated with the flow? So prepared, we are in a position to understand how these timescales intersect.

10.3.1   Flow timescale

The most important subdivision of fluid flow is between steady-state and transient. Steady-state refers to the time-independence of the Eulerian flow field, $\mathbf{v}(\mathbf{r})$. In transient flow, this field acquires a time dependence, $\mathbf{v}(\mathbf{r}, t)$. But NMR is essentially a Lagrangian flow method, in which the signal is derived from spins co-moving with the fluid. These velocities may vary with time, even under steady-state flow, as spins migrate to different positions in the flow field. And when we encode for position, as in MRI, the spatial coordinates acquired derive from the location of those spins at the time of that encoding.

Perhaps the best way to visualise the problem is by way of simple examples, as shown in Fig. 10.7. Here we see, in (a), (b), and (c), three examples of steady-state flow, namely laminar pipe flow, laminar converging flow, and laminar cylindrical Couette flow. Example (f) shows the flow within a falling droplet driven by surface stress due to a surrounding viscous fluid. Again the flow is steady state within the droplet, though stroboscopic methods might be needed to 'lock' the droplet frame to the velocimetry frame. Any periodic flow falls into the same category as example (f). By contrast, in examples (d) and (e), showing, respectively, trickle bed flow under fluctuating conditions and turbulent flow, we see the Eulerian flow field changing, with perhaps a characteristic time τ_c.

10.3.2   Velocimetry timescale

There are three timescales associated with velocimetry. These are the phase-encode time for velocity, Δ, the time to acquire the spatial locations of the spins, τ_{MRI}, and the image-refresh time, τ_R.

Δ is never a limiting timescale for fast flows. To ascertain the mean flow rate in an image pixel, we need only allow for a displacement that overwhelms the broadening due to diffusion. For convenience we may take the example of water molecules, where the diffusion length over $1\,\mathrm{ms}$ is on the order of a few microns. Flow rates of $\gtrsim 1$ m s^{-1} can be captured in $1\,\mathrm{ms}$. Slower rates take accordingly longer.

Now to position capture: in a 1-D spatial imaging experiment with single-step phase-encoding for velocity, and a readout gradient for position, $\tau_{MRI} = \tau_{read}$ can be

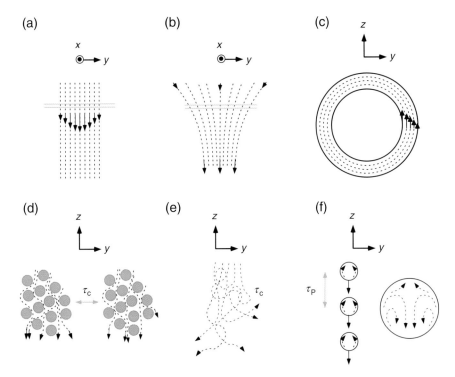

Fig. 10.7 Schematic representation of different flows in which characteristic times play a role. In each a a 2-D image acquisition is assumed and the imaging plane indicated by the axes. (a) Steady-state laminar pipe flow. (b) Laminar converging flow. (c) Laminar cylindrical Couette flow. (d) Trickle bed flow under fluctuating conditions. (e) Turbulent flow. (f) Periodic flow with stroboscopic imaging of a falling droplet.

as short as a few milliseconds. The refresh interval will be the sum of Δ, τ_{read}, and any recovery time associated with spin relaxation. In some fortuitous circumstances, for example where the velocity being measured is predominantly normal to the imaging strip in a 1-D image, the flow itself brings fresh spins to the image so that relaxation delays are no longer limiting. Alternatively, and provided one is prepared to pay the consequent signal-to-noise ratio price, small tip-angle excitation pulses may be used to allow further longitudinal magnetisation recall. In short, 1-D velocimetry is well suited to rapid flows, with potential refresh times on the order of a few milliseconds. The real challenge comes with 2-D or 3-D imaging of flow fields. However, as we shall see, for steady-state flow, the problem vanishes.

2-D and 3-D MRI images may be acquired in spin-warp mode by using readout (frequency encoding) for one dimension and phase encoding for the remaining dimensions. Thus the basic sub-unit of the imaging sequence comprises the repetitive 'excitation pulse and collect' acquisition under readout gradient, allowing a repeat interval determined by T_1 recovery alone, perhaps in the range 0.1 s to 1 s, during which the phase gradients are stepped to their next value. The total image may take minutes to acquire, but if the Eulerian velocity field is steady state, the only characteristic imaging times relevant to any artifact in determining the velocity field are the times Δ and the readout time, τ_{read}. For distortion-free imaging, we will require that the sum of these not exceed the time to move across one pixel in the 2-D or 3-D frame. Examples (b) and (c) in Fig. 10.7 are instructive in this regard. Here we have components of flow velocity 'within the imaging plane'. As a simple exercise, for pixel dimensions of $50\,\mu$m and Δ and τ_{read} on the order of millseconds, speeds of tens of metres per second are easily manageable. The same 'steady-state robustness' is available with repetitive flows, such as the falling droplet example of (f), providing the 'excitation pulse and collect' is stroboscopically locked to the repetitive phenomenon.

As a result, the real timescale challenge for MRI velocimetry in 2-D or 3-D concerns fluctuating flows. If we are to capture the essentials of the flow timescale, we must complete the entire velocity and position encoding with a total refresh time τ_R shorter than any flow correlation time τ_c.[4] The key to a rapid refresh rate is the ability to capture a 2-D image in the time interval associated with a single cycle involving a initial excitation RF pulse and the use of the consequent transverse magnetisation over a single relaxation interval. Three methods stand out, in order of decreasing speed and increasing robustness to image artifact. These are 'echo planar imaging' (EPI), 'rapid acquisition relaxation enhanced imaging' (RARE), and 'fast low angle shot imaging' or FLASH. Each of these is discussed in succession. Finally, we describe a slower, but highly robust velocimetry method based on pure phase encoding, known as 'single-point ramped imaging with T_1-enhancement' or SPRITE.

10.3.3   Echo planar imaging

The fastest possible velocimetry technique uses the EPI scheme, the fast imaging method proposed and demonstrated by Mansfield and co-workers [12–15]. Following a slice excitation, k-space is sampled in two dimensions using a sequence of readout gradients applied in opposite directions, so that, by means of 'blips' of intervening phase gradient pulses advancing along the orthogonal axis one line at a time, a Cartesian plane is mapped out. These gradients can be seen in the repeat elements of the repeated EPI scheme shown in Fig. 10.8. Because multiple gradient echoes are used in the 2-D sweep, rather than employing a multiple spin-echo sequence to subdivide the successive readouts, EPI minimises the number of RF pulses needed, allows the greatest time compression in the spatial encoding, and avoids the defocusing of PGSE-encoded spin phases associated with multiple RF pulse trains, a subtle point to which we shall return in the next section when discussing RARE.

However, EPI suffers two particular drawbacks. Because it relies on gradient rather than spin echoes, it does suffer from dephasing, and hence image distortion, caused by background-field inhomogeneity. This need not be a limiting factor provided that the readout gradients are much larger than those contributed by the sample. The second problem with EPI concerns the fundamental asymmetry inherent in the opposite direction readout sweeps, unproblematic in an ideal implementation, but leading in practice to slightly inequivalent acquisitions on negative and positive gradient pulses,

---

[4]Note, however, the trick outlined in Section 10.1.4, in which a velocity-null experiment can be used to reveal fluctuations on order of Δ.

and hence a k-space modulation that can lead to severe image distortion and 'ghost-ing' effects. Exquisite care is needed in the implementation of EPI to minimise such artifacts and proper setup requires considerable skill.

Th first use of PGSE-encoded EPI for rapid velocimetry was by Kose [16–18], although other researchers have reported the use of standard EPI-based techniques to investigate fluid flow from its influence on signal intensities in the image [19].

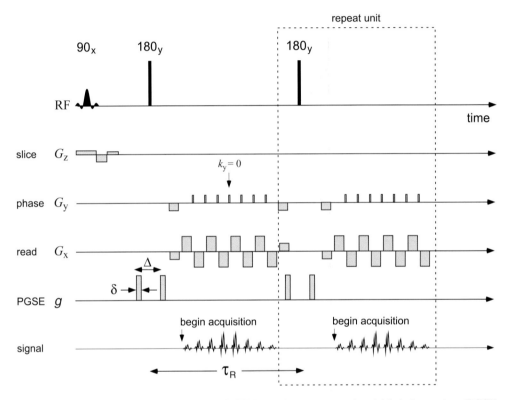

Fig. 10.8 GERVAIS version of a multiple EPI imaging sequence in which independent PGSE phase encoding for motion is performed in each successive repeat, thus allowing for rapid frame rate investigation of velocity fields, with a refresh time of τ_R.

However, the most comprehensive and powerful implementation of PGSE-encoded EPI is due to Sederman *et al.*, in a variant shown in Fig. 10.8, quixotically named GERVAIS for gradient echo rapid velocity and acceleration imaging sequence [20]. This group have been highly successful in implementing this EPI velocimetry method, achieving total image acquisition time as short as 10 ms, a repeat time τ_R of 20 ms, and a complete recycle time (duration to next slice excitation) of around 500 ms. Note the use of flow-compensated slice excitation, the positioning of that slice being made near the inlet end of the sensitive region within the RF coil, so to minimise outflow effects.

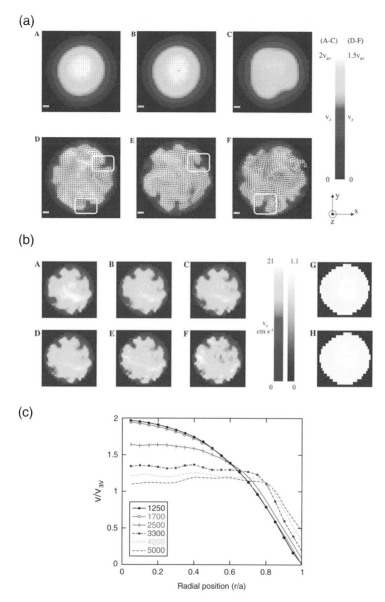

Fig. 10.9 (a) Three orthogonal component velocity images acquired at increasing Re of (A) 1250, (B) 1700, (C) 2500, (D) 3300, (E) 4200, and (F) 5000. The colour scale identifies the magnitude of the z-velocity. The flow velocity in the x–y plane of the image is shown by the vectors. The vector scale bar on each image corresponds to $1\,\mathrm{cm\,s^{-1}}$. (b) Six consecutive z-velocity images acquired at 20 ms time intervals are shown (A–F) for flow at $Re = 5000$. The average VACF maps calculated from pairs of images separated by time intervals of (G) 40 ms and (H) 80 ms are also shown. (c) Radial velocity profile of the z-velocity for Re in the range 1250–5000. (Reproduced with permission from Sederman *et al* [20].) This figure is reproduced in colour in the colour plate section.

Figure 10.9 shows a remarkable application of this rapid velocimetry technique in a study of the transition to turbulence for water in pipe flow as the Reynold's number is increased from 1000 to 5000. Figure 10.9(a) shows sets of images obtained over a 60 ms period, in which the flow-encoding gradient q is successively switched between the x, y, and z directions in successive 20 ms repeat intervals. With increasing Re a transition to turbulence is seen, in which significant transverse velocity components emerge. By tracking the longitudinal velocity at 20 ms intervals, a picture emerges regarding the characteristic fluctuation rate, from which a velocity autocorrelation function map can be calculated. Finally the v_z flow profile shows the classic transition from Poiseuille flow to plug flow as the turbulent regime is entered.

10.3.4 Rapid acquisition relaxation enhanced imaging

While RARE cannot match EPI for speed, it has the significant advantage of image robustness. The reason for this is its resilience to instrumental errors, due to the fundamental symmetry of the k-space sampling scheme, along with resilience to sample susceptibility heterogeneity by the use of a multiple spin-echo train, which refocuses unwanted phase shifts caused by background field gradients. In a pure RARE imaging experiment, the matter of symmetry under a CPMG RF pulse train is dealt with nicely. This can be seen by reference to the RARE part of the velocimetry sequence shown in Fig. 10.10. Each echo acquisition under the read gradient involves a single traverse (from left to right) across the k_x axis of the Cartesian grid. The precursor phase gradient has swept the spins to start the k-space traverse along a particular k_x line positioned as desired on the orthogonal k_y axis, but at the end of the sweep an unwind phase gradient returns the signal to $k_y = 0$. Thus, at the start of every echo interval, the spins are identically positioned in k-space.

However, as we shall see, when RARE is combined with a precursor PGSE gradient pulse pair to encode for motion, the CPMG RF pulse train introduces a new problem, namely the ability of multiple imperfect 180° pulses to preserve the in-phase component of magnetisation, while permitting phase decoherence to occur for the out of phase component. This problem needs to be specifically addressed with PGSE-RARE.

Sederman *et al.* [21] investigated pipe flow at 40 mm s^{-1} using a RARE sequence without PGSE encoding, while Scheenen *et al.* combined RARE with both PGSE and spin-echo encoding [22, 23] to study the water uptake in plant stems, e.g. liquids under very small flow rates ($\sim$ 0.5 mm s^{-1}). A modified version [24] of the PGSE-encoded RARE pulse sequence of Scheenen *et al.* is shown in Fig. 10.10. Here we look in detail at the way in which flow introduces new artifacts, and how these may be suitably countered in the pulse sequence design.

Soft-pulse-quadrature-cycled PGSE-RARE

The use of hard 180° RF pulses in the RARE part of the imaging sequence can lead to significant artifacts in which a part of the intensity in the central part of the image, where flow occurs, is transferred to the image border, the so-called 'ghosting' associated with $\pm$ signal modulation with respect to the successive phase steps. Such an amplitude alternation leads, by the shift theorem, to a 'ghost' image displaced by half the field of view. The intensity of this artifact depends on the velocity-encoding

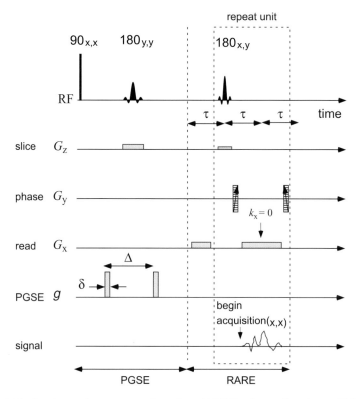

Fig. 10.10 Velocimetry pulse sequence based on RARE using soft pulses within the CPMG train. Note the use of shorter, more intense soft pulses in conjunction with a smaller slice gradient in the train to excite a large sample volume, but within the homogeneous region of the RF coil. The phase-encoding scheme involves wind and unwind pulses sandwiching each echo acquisition, so that the signal is return to the k_y (phase direction) origin before the next 180° RF pulse.

gradient strength applied, as well as on the velocity of the fluid itself. In other words, while these artifacts are a consequence of the imaging process, adding flow and q-gradients enhances them. The explanation is obvious and as old as the original paper by Carr and Purcell [25], who noted that odd and even echoes alternate in amplitude due to RF pulse imperfections, while flow induces even more severe odd–even alternation. 'Ghosting' artifacts can be mostly eliminated by the use of only the even echoes within the RARE part of the pulse sequence, the price paid being a doubled pulse sequence duration. Further, replacing the hard pulses in the CPMG train by soft pulses with the slice gradient chosen to include only spins within the uniform RF field region of the resonator causes a significant reduction in ghosting artifacts.

The second RARE artifact associated with PGSE phase encoding concerns the loss of the imaginary part of the NMR signal during the multi 180° RF pulse segment of the pulse sequence, a general feature of sequences based on the CPMG sequence [25]. This problem is not necessarily fatal, but is certainly inconvenient. The loss of the imaginary signal leads to a symmetrical positive and negative flow mirroring of the propagator in the subsequent Fourier transformation of the acquired data with respect to the q direction, as the input data are no longer modulated by an exponential function with a complex argument but instead experience a cosine modulation. While such mirroring may be acceptable at large propagator displacements, at low velocities it leads to overlap distortions. Further, with single q-step encoding, the mirroring effect is highly undesirable, since it impedes a proper phase angle calculation and renders the truncated propagator ambiguous.

The means by which this effect may be overcome is seen in the variant known as soft-pulse-quadrature-cycled PGSE-RARE [24], shown in Fig. 10.10. To avoid the cosine modulation effect and restore the full complex signal, one needs a two-step phase cycle in which the excitation phases are identical, while the phase of the 180° pulse train switches between x and y in two separate RARE acquisition trains, so that real and imaginary magnetisation is independently protected and stored. Thus the whole complex NMR signal is accumulated in the acquisition memory. One might expect that the acquisition phase should alternate between x and $-x$ because of the changing sign of the echo under 180_y° and 180_x° RF pulses, though in practice a fixed acquisition phase works equally well. The reason is that the 'non-protected' component dies away rapidly along the CPMG train and contributes little to the image, resulting in separate storage of real and imaginary signals in the two separate parts of the phase cycle. Hence (x, x) or $(x, -x)$ acquisition phases simply represent different sign PGSE-induced phase shifts.

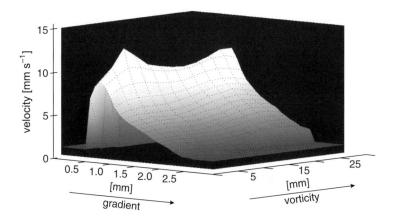

Fig. 10.11 Velocity of wormlike micellar solution under shear in the gap of a cylindrical Couette cell. The two spatial axes correspond to position across the cell gap (velocity gradient direction) and position along the cell axis (vorticity direction). (With permission from Kirk Feindel—see reference [26].)

Of course this phase cycling costs additional time, doubling the number of $q \neq 0$ encode images so that at least two RARE trains in succession are required. Given that a typical soft-pulse-quadrature-cycled PGSE-RARE train might take on the order of

500 ms, allowing for the finite duration of the 180° pulses, and the use of only every second echo, the refresh time, τ_R, is typically on the order of 1 to 2 s, T_1 recovery permitting.

An example of a 2-D velocity map taken at a refresh time of 2 s is shown in Fig. 10.11. It shows the velocity of a fluid comprising wormlike micelles being sheared in the gap of a cylindrical Couette cell, where the spatial axes correspond to the position across the cell gap (velocity gradient direction), and the position along the cell axis (vorticity direction). The image not only exhibits classic shear banding effects, in which the velocity gradient (strain rate) across the gap discontinuously changes, but also shows the degree to which the velocity profile is a function of the vorticity direction. This image is taken from one frame of a succession in which rapid fluctuations are apparent, image by image.

FLIESSEN

There is a brute force manner in which the PGSE phase shifts may be removed before each 180° RF pulse of the RARE train, thus avoiding the need to retain both the real and imaginary parts of the velocimetry phase along the entire duration of the CPMG echo train. In this variant, termed FLIESSEN, for FLow Imaging Employing a Single-Shot ENcoding [27], the PGSE gradient pulse pair is simply applied as a gradient echo of two oppositely-signed equal area pulses, within each echo cycle, and before each readout in the echo intervals, only to be unwound with an opposed pair, before the application of the next 180° pulse. The idea is simple and effective, but comes at a price of requiring that the PGSE flow observation time, Δ, be included in the echo interval for each readout, thus considerably lengthening the RARE sequence duration. The method is, however, highly suitable for rapid flows, where short Δ values are effective, and under steady-state conditions where the refresh time, τ_R, is permitted to be longer.

10.3.5   FLASH

Amongst the options which balance speed with robustness, the Fast Low Angle SHot imaging or FLASH [28] method provides a nice compromise. FLASH is based on the recall of small amounts of longitudinally stored magnetisation in a sequence of small tip-angle excitations, α_x, each of which contributes to a readout under different phase gradient positional encoding. At the penalty paid of a loss in signal-to-noise ratio due to the smaller $(\sin\alpha)$ component of magnetisation tipped to the transverse plane, the method does have the advantage of a completely fresh recall at each excitation pulse, and a consequent robustness, with no contribution from recycled magnetisation contributing to image artifacts. The duration of the FLASH train is comparable with RARE, with a complete image being acquired in a few 100 ms.

Just as a precursor PGSE gradient pulse pair can be used to store flow-encoded phase in the magnetisation prior to a RARE imaging sequence, so it can be similarly applied, as shown in Fig. 10.12. Like RARE, it suffers from the fact that the PGSE-encoded phase survives as a real component only, in the case of FLASH the reason being that the flow-related phase shifts are stored in longitudinal magnetisation used for later small tip-angle recall. Of course, like RARE, it is possible to phase cycle the

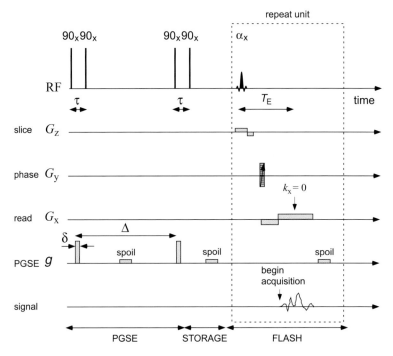

Fig. 10.12 Velocimetry pulse sequence based on FLASH using multiple small-tip-angle soft pulses, α_x, to generate each read sweep from longitudinal magnetisation that has been stored with a phase modulation due to prior PGSE-encoding for flow.

sequence so as to capture real and imaginary parts. The FLASH method has been used effectively to image xylem transport in plants [29], using a suitable phase cycle to suppress the stationary water signal.

10.3.6   SPRITE

The ultimate in imaging robustness is to be found by avoiding frequency encoding completely, so that k-space is sampled in a succession of pure phase-encoding steps. In a variant known as single point imaging [30], the spin system, initially in longitudinal Zeeman equilibrium, is subject to an applied magnetic field gradient and a broadband excitation RF pulse used to excite transverse magnetisation. Following a short time interval t_p, the NMR signal is acquired as a single point, such that the phase evolution of the spins over the period t_p provides a record of their positions under that gradient field. By repeating the process for a succession of different gradient fields, k-space is successively mapped out. The particular advantage of such single-point sampling is that all spin interactions unrelated to the changing gradient field provide a common amplitude or phase shift in the sampled signal, independent of k, and are therefore unable to contribute to image distortion. In particular, such unwanted interactions might include susceptibility-related field inhomogeneity, chemical shifts, dipolar interactions, or spin relaxation. By making t_p sufficiently short, the effect of rapid spin–spin relaxation can be overcome, thus making single point imaging suitable for application in solids, albeit with a concomitant requirement for magnetic field gradient amplitudes sufficiently large to induce the desired phase shift over t_p.

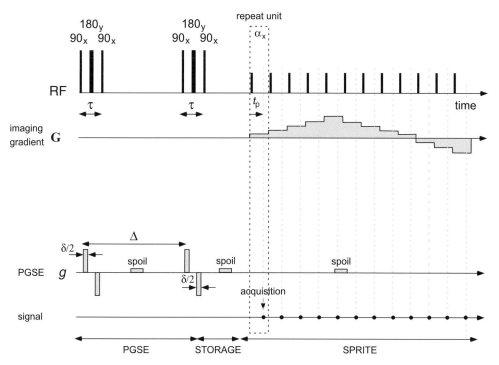

Fig. 10.13 Velocimetry pulse sequence suitable for porous media based on SPRITE using multiple small tip-angle hard pulses, α_x, to generate each single-point acquisition. A precursor 13-interval (Cotts) PGSE encoding is used to provide maximum resilience to background susceptibility related fields. The imaging part is based on a standard centric-scan SPRITE readout and the vector **G** is successively stepped to sweep out 2-D or 3-D k-space as desired. (Adapted from Li *et al.* [31].)

Of course the method is costly in time, a feature of all pure phase-encoding methods, each step in k requiring a fresh acquisition, allowing for longitudinal relaxation recovery and gradient settling following gradient pulse switching. However, a very efficient variant, known as single-point ramped imaging with T1 enhancement, or SPRITE [32], has been proposed by Balcom *et al.* SPRITE uses a smooth transition between gradients, in a gradual ramping process, to avoid the need for gradient settling delays, and a small-tip-angle excitation pulse to allow for rapid recall of fresh transverse magnetisation. Because the method involves a very large number of phase-encoding steps (especially for two or three dimensions) a steady-state transverse magnetisation is achieved, related to the T_2^* relaxation over t_p, and the equilibrium longitudinal magnetisation associated with a sampling repeat interval T_R, tip angle α, and longitudinal relaxation time, T_1 [33]. Thus, the signal at each point in the image density $\rho(\mathbf{r})$ is given by the relation [32]

$$S(\mathbf{r}) = \rho(\mathbf{r}) \exp(-t_p/T_2^*) \left[\frac{1 - \exp(-T_R/T_1)}{1 - \cos\alpha \exp(-T_R/T_1)} \sin\alpha \right] \qquad (10.14)$$

The use of larger than optimal turn angle[5] has the effect of introducing a strong T_1 contrast, a useful characteristic in some applications. Because of the robustness

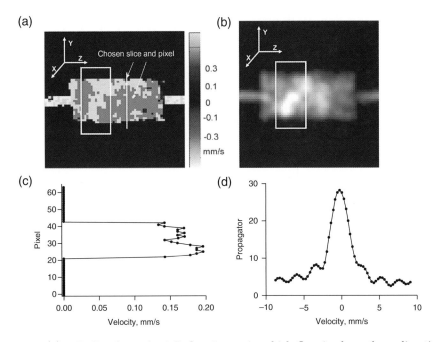

Fig. 10.14 (a) 2-D slice from the 3-D flow image in which flow is along the z-direction. A region of higher velocity, marked by the white square, indicates a high permeability channel in the rock. (b) The same 2-D slice but extracted from a 3-D T2-weighted image. The high intensity area, marked by the white square, indicates a high permeability channel. (c) 1-D profile of the flow map extracted from the line in (a). (d) Propagator extracted from the marked pixel in (a). (Reproduced with permission from Li *et al.* [31].)

of SPRITE, it does lend itself to challenging velocimetry imaging, for example in porous media, where strong susceptibility inhomogeneity can cause image distortion under frequency encoding, or in rapid turbulent flows where the short characteristic time interval, t_p, can be turned to significant advantage. We deal first with the porous medium case. Here the distortion-free aspect of SPRITE imaging is paramount, and the PGSE flow encoding is the standard precursor, as seen earlier in RARE or FLASH. In the present example, however, a 13-interval PGSE sequence is shown, so as to minimise the influence of background gradients due to susceptibility inhomogeneity [31].

---

[5]The optimum is the 'Ernst' angle such that $\cos\alpha_E = \exp(-T_R/T_1)$.

An example of the application of this method is shown in Fig. 10.14, where flow is imaged in a mapping of water flow through a 2.5-cm diameter, 5.24-cm-long carbonate limestone core plug coated with epoxy, at a flow rate of $2\,\mathrm{ml}\ \mathrm{min}^{-1}$. The core has a porosity of 27.6% and permeability of $9\,\mathrm{mD}$. The experiment was carried out at $300\,\mathrm{MHz}$ proton frequency in a field of $7\,\mathrm{T}$, giving a sample T_2^* of $500\,\mu s$. Despite the short T_2^* good velocity maps are obtained and the method is able to obtain velocity distributions in a localised pixel.

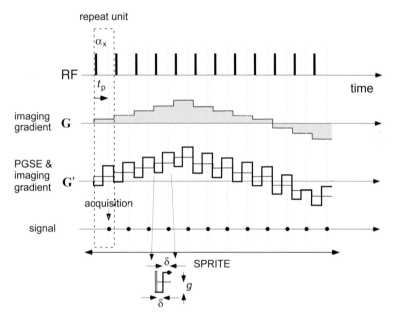

Fig. 10.15 SPRITE velocimetry pulse sequence suitable for fast flow. **G** shows the first few gradient values of an N-point spiral trajectory, while **G'** shows a motion sensitised version of that trajectory. One phase-encoded data point is acquired at each circle, at a time t_p after each RF pulse. **G'** differs from **G** only in a symmetrical excursion ($\pm\mathbf{g}$) either side of **G**. Such a superposition of $\pm\mathbf{g}$ excursions upon any k-space sampling scheme is possible. (Adapted from Newling *et al.* [34].)

Because of the large number of phase-encoding steps needed to acquire a 2-D image, SPRITE is not a fast refresh-time method, and is not well suited to the study of transient or fluctuating flows. However, it is suited to exceedingly fast flow under steady-state conditions, provided the motion-encoding time is sufficiently short. Here the inherently short characteristic sampling time of SPRITE can be turned to advantage. Figure 10.15 shows a variant of the SPRITE sequence in which the phase-encoding gradients are modulated within each sampling interval t_p, so as to provided an effective gradient echo phase encoding for spin displacement. This method has been successfully applied to the measurement of average velocity and eddy diffusivity in turbulent flow [34].

10.4 Velocimetry applications in materials science, biology, and medicine

MRI velocimetry has made a significant impact in a diverse range of fields in which complex fluid flows are significant. These include chemical engineering applications, involving catalysis and filtration, multi-phase flow in porous media, and the study of granular flow in particulate systems. In soft condensed matter physics, the method has been used to reveal deformational flow fields, both extensional and shear, as well as slip, shear banding, and flow instability. In chemical physics, MRI velocimetry has also proven useful in elucidating electro-phoretic and electro-osmotic flow, while in biological applications in-vivo flow of xylem and phloem transport in plants has bean measured.

Some of these applications are briefly reviewed here. The list is intended to be illustrative rather than comprehensive.

10.4.1 Porous media flow

Amongst the advantages offered by MRI velocimetry is its inherently 3-D capacity, allowing, in principle, the measurement of the complete vector flow field. One of the more remarkable examples of such an application is shown in Fig. 10.16, where the full 3-D velocity field for water flowing in a glass bead pack has been obtained using standard 3-D spin warp with PGSE encoding for flow [35]. The selected field of view is $50.0\,\mathrm{mm} \times 50.0\,\mathrm{mm} \times 6.25\,\mathrm{mm}$, and the the isotropic spatial resolution is $(195\,\mu\mathrm{m})^3$. The flow is steady-state and the total data acquisition time was 3 h. Furthermore, such 3-D flow fields obtained using MRI methods have been verified using lattice-Boltzmann simulations on identical flow geometries [36–38].

In porous media applications, it is not always necessary to encode for velocity in order to gain powerful insight. In some cases, merely obtaining a succession of MRI density images at high refresh rate can reveal subtle phenomena. This is particularly the case in two-phase (gas/liquid) flow, for example in 'trickle-bed' reactors, where the streamlines occupied by the fluid and gas phases may fluctuate with time, as in the hydrodynamic transition from trickle-to-pulsing flow [39]. Using a rapid FLASH MRI sequence, in which $16 \times 16 \times 32$ data arrays were acquired in 280 ms, Anadon *et al.* [40] achieved a refresh rate of 3.6 images per second, at a spatial resolution comparable with that of an individual packing element within the bed. These MRI movies are able to reveal the transition from trickle-to-pulsing flow, confirming that the transition is initiated by local pulsing events on a size-scale typical of the packing elements contained within the column [40].

Finally, it is important to remember that spatial resolution of the Eulerian flow field does not offer the only potential for powerful insight. Measurements of the distribution of velocities represented in the Lagrangian ensemble can be performed with great precision, trading away spatial encoding for a greater number of PGSE-encoding steps, allowing the generation of displacement distributions, $\bar{P}(\mathbf{R}, \Delta)$, for a range of flow times, Δ. Figure 10.17 shows an example of such propagators, for both longitudinal and transverse displacements in a bead pack, along with lattice-Boltzmann simulations. The agreement is excellent.

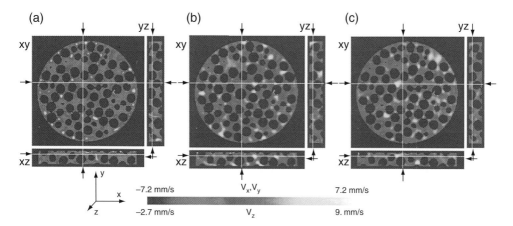

Fig. 10.16 Velocity fields within a packing of 5-mm diameter spheres within a column of internal diameter 46 mm. Fluid velocities in the (a) z-, (b) x-, and (c) y-directions are shown with slices taken in the xy, yz, and xz planes for each of the velocity components. For each image the positions at which the slices in the other two directions have been taken are highlighted. (Reproduced with permission from Sederman and Gladden [35].) This figure is reproduced in colour in the colour plate section.

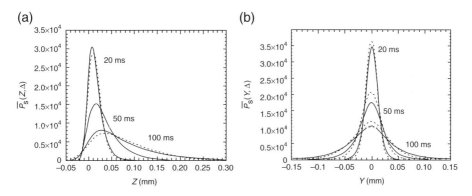

Fig. 10.17 Experimental (solid line) vs simulated (dashed line) displacement propagators along the (a) axial and (b) transverse directions for varying values of Δ for water flow in a 10-mm diameter bead pack comprising 1-mm diameter glass spheres at an average flow rate of $0.77 \, \text{mm s}^{-1}$. (Adapted from Manz *et al.* [36].)

10.4.2 Inertial flow and turbulence

One of the more challenging uses of NMR velocimetry is in the study of inertial effects in fluid flow and the transition to turbulence. In cylindrical Couette flow at high Reynolds number, centrifugal inertial effects can lead to a complex 3-D vortex structure known as Taylor–Couette flow. The dimensionless parameter that governs this phenomenon is the so-called Taylor–Couette number [41],

$$T = \frac{2\eta^2}{1-\eta^2} \frac{\Omega^2 d^4}{\nu^2} \tag{10.15}$$

where Ω is the inner cylinder rotation rate, ν the kinematic viscosity, d the gap width, and η the inner-to-outer radius ratio, R_i/R_o. The critical Taylor number for Taylor–Couette flow is 2510 [41], but due to the rotating bottom boundary and column aspect ratio, the onset of the instability can occur at values significantly below this. Figure 10.18 shows an example of stable 3-D velocity structure in Taylor vortices,

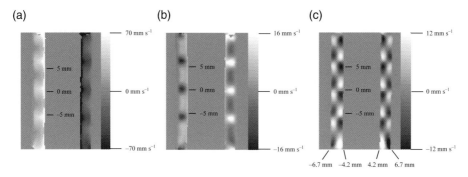

Fig. 10.18 Velocity fields for Taylor vortex motion of pentane in the gap between two concentric cylinders, at a Taylor number of 246. The cylinder is rotating such that the left-hand side is moving out of the page. (a) Radial velocity $v_r(r, z)$, (b) tangential velocity $v_\theta(r, z)$, and (c) axial velocity $v_z(r, z)$. (Reproduced with permission from Seymour *et al.* [42].)

obtained using a spin warp velocimetry sequence on pentane in a flow cell consisting of a precision-milled glass cylinder of radius $R_i = 4.2$ mm, residing in a glass tube of internal radius $R_o = 6.7$ mm, with a gap width of 2.5 mm [42]. Taylor vortex motion has also been investigated by MRI velocimetry using FLASH [43].

Applications of NMR velocimetry to turbulent flow are relatively recent [34, 44, 45], presenting significant challenges in terms of temporal resolution, but allowing for analysis of eddy diffusivity through the effective dispersion coefficient measured [34] in an experiment where PGSE gradients are stepped and a propagator analysis is undertaken.

10.4.3 Rheo-NMR

While rheology is primarily concerned with the characterisation of complex fluids and soft matter by mechanical means, flow visualisation provides a vital adjunct. In many flows there exists a need to obtain detailed information about the flow field generated by the device used to induce deformational flow, and in particular to be able to visualise, in real time, strain-rate heterogeneity and discontinuity, for example wall slip, sample fracture, and shear-banding. MRI velocimetry offers particular advantages over competing optical or ultrasound methods, both of which involve scattering from refractive index or density heterogeneity. MRI velocimetry is not subject to optical transparency constraints and, unlike these other methods, full three-dimensional velocity fields may be determined in any geometry whatsoever. The real advantage of optical or other scattering techniques lies in their inherent speed, and hence an ability to visualise fluctuations at quite short times. However, recent advances in fast MRI velocimetry have made available refresh rates below 1 s so that many systems exhibiting transient flow behaviour have become accessible.

The idea of 'rheo-NMR' was first suggested by Nakatani *et al.* [46] in the context of NMR spectroscopy carried out under deformational flow. The effects of shear or extension on molecular orientation, organisation, and dynamics can potentially affect spin relaxation rates or even alter terms in the nuclear spin Hamiltonian, such as the dipolar or quadrupolar interaction. Hence, the use of MRI velocimetry to provide a map of the complex flow field represents only a part of the contribution that NMR can make to rheology. The real challenge is to be able to perform spectroscopic studies at different locations in heterogeneous flow fields, using spectroscopic MRI or spatially selective NMR spectroscopy.

In this section, we briefly traverse some examples of what is possible with rheo-NMR in the context of MRI velocimetry, noting that the subject is reviewed in greater detail elsewhere [47–50]. Furthermore, there are numerous examples involving spectroscopic applications without imaging or velocimetry, of which references [51–54] provide but a sample.

Slip and power-law flow

Figure 10.19 provides an example of how MRI velocimetry may be used to measure slip, and to characterise shear-thinning flow by power-law behaviour. The sample consists of a 46% volume fraction of 370-nm diameter core-shell latex particles suspended in a 2% (w/w) aqueous dispersion of polyvinyl alcohol [55]. The core of these particles is liquid hexadecane, making them NMR-visible. The sample has been inserted between 5-mm outer diameter (OD) inner glass cylinder and a 9.2-mm inner diameter (ID) outer cylinder and centred by two Teflon disks, this comprising the annular region of a cylindrical Couette cell. However, the same sample material also fills the inner cylinder, rotating in rigid-body motion, and thus providing a nice 'marker' from which the inner cylinder wall velocity may be deduced by extrapolation.

The velocity profiles, obtained at 300 MHz using a spin warp sequence, are shown for rotational frequencies $f = 8 \times 10^{-4}$ Hz, 3.2×10^{-3} Hz and 0.16 Hz, corresponding to gap-averaged shear rates $\dot{\gamma}$ of $7 \times 10^{-3}\,\mathrm{s}^{-1}$, $0.028\,\mathrm{s}^{-1}$ and $1.4\,\mathrm{s}^{-1}$, respectively. At the lowest frequency, the velocity profile across the gap is completely flat and the whole sample moves like a solid plug. As the rotational frequency is increased, the shear stress across the gap increases, and in the inner regions of the gap the local shear stress exceeds the apparent yield stress σ_c of the sample, such that the outer parts of the sample at lower stress are still in the solid-like state, while the sample contained in the higher stress region near the inner wall now flows with a liquid-like, shear-thinning behaviour. Similar two-phase behaviour has been seen using MRI velocimetry in a clay and silica suspension, a concentrated oil-in-water emulsion, and a cement paste [50, 56].

For the core-shell suspension, a power-law exponent[6] n of 0.2 can be obtained from fitting the inner part of the velocity profile to the analytical expression for the angular velocity at radius r of a power-law fluid in a wide-gap Couette geometry

---

[6]In the power-law model $\sigma \sim \dot{gamma}^n$.

$$v(r) = \omega r \frac{1 - (r/r_o)^{2/n}}{1 - (r_i/r_o)^{2/n}} \qquad (10.16)$$

where r_i and r_o are the inner and outer radii bounding the annulus, and $\omega = 2\pi f$ is the inner-wall rotation frequency. At the highest rotational frequencies, see Fig. 10.19(c), the whole sample is liquid-like and the shear-thinning somewhat less severe, with $n \approx 0.4$.

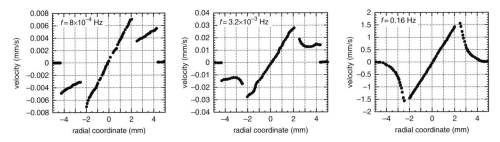

Fig. 10.19 Velocity profiles for core-shell latex particles flowing in a Couette geometry at different rotation speeds. (a) For the lowest rotational frequency $f = 8 \times 10^{-4}$ Hz, the sample behaves like a solid plug. Note the presence of slip at both at the inner and at outer walls. (b) The annulus of sample near the inner cylinder, and at higher stress has undergone a solid-to-liquid transition and flows with power-law behaviour ($n = 0.2$), but the outer parts are still in a solid-like state. (c) At higher rotational speeds ($f = 0.16$ Hz), the whole sample is liquid-like and shows a power-law behaviour with $n = 0.4$. (Reproduced with permission from reference [10].)

The self-diffusion coefficient of these core-shell particles is around 10^{-13} m^2 s^{-1}, corresponding to a diffusion distance of 450 nm over the PGSE encoding time of 0.5 s, around seven times the distance diffused by intra-particle molecular self-diffusion alone. Enlargement of the middle part of the velocity profile in Fig. 10.19 for the lowest rotational frequency of $f = 8 \times 10^{-4}$ Hz shows that smallest velocity measured is around 200 nm s^{-1} [10], with a displacement of 100 nm over the encoding time Δ, a factor of four smaller than the particle size. Such measurements show the potential of NMR velocimetry to measure ballistic particle collision events at very small particle velocities.

Shear banding

One of the more active areas of complex fluid research in recent years concerns the phenomenon of shear banding, in which a fluid undergoing shear flow sub-divides into two fluids of differing viscosity experiencing differing strain rates at similar stress. The effect is usually associated with the existence of a plateau in the flow curve, a phenomenon exhibited by entangled semi-dilute wormlike micelles, which undergo dissociation–recombination reactions in conjunction with reptational dynamics [57–59]. The inflected constitutive relation of Fig. 10.20(a) leads just such a plateau. In this simple picture [60], the fluid experiencing an averaged shear rate applied in the unstable (declining stress branch) region would subdivide into two coexisting 'phases' of high and low viscosity represented by the strain rate states $\dot{\gamma}_1$ and $\dot{\gamma}_2$, the relative proportions being determined by a simple first-order lever rule.

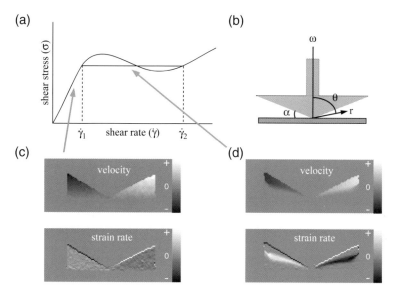

Fig. 10.20 (a) Schematic flow curve for a fluid exhibiting double-valued stress vs rate-of-s-train behaviour. In the phase-separation model for shear banding, $\dot{\gamma}_1$ and $\dot{\gamma}_2$ correspond to coexisting shear rates at a single stress value. (b) Cone and plate system used in NMR microimaging experiments, showing the gap angle α. (c) and (d) show velocity and shear rate images for 100 mM/60 mM cetylpyridinium chloride/NaSal wormlike surfactant solution. The grey scales indicate the velocities and shear rates in arbitrary units. Note the opposite sign shear for the receding and advancing segments of fluid on opposite sides of the gap. (c) In a 4° gap (vertical gain ×6) with free exterior fluid surface and at an apparent shear rate, $\omega/\tan\alpha$ of $1.5\,\mathrm{s}^{-1}$, a uniform strain rate is apparent. (d) At $16\,\mathrm{s}^{-1}$ distinct shear bands are observed. (Adapted from references [61, 62].)

While optical birefringence measurements indicated the existence of banded structures in wormlike micelles [63, 64], the first direct evidence of shear-rate banding came from MRI velocimetry [61] using the wormlike micelle system 100 mM/60 mM cetylpyridinium chloride/NaSal in a cone-and-plate geometry. The device comprises a 24-mm diameter cone and plate made from the machinable glass, MACOR, with two available gap angles of 4° and 7°. Figure 10.20(c) and (d) show spin warp velocity maps, along with their derived shear rate maps, obtained from the 4° device in a 2-mm thick planar slice of spins, is excited normal to the velocity-encoding direction, such that a two-dimensional image of that plane is obtained. Note that the vertical scale is multiplied by six to provide a better visualisation of the 4° gap. At a shear rate $\omega/\tan\alpha$ of $1.5\,\mathrm{s}^{-1}$, below that required to reach the stress plateau, a uniform strain-rate fluid is apparent. By contrast, at $16\,\mathrm{s}^{-1}$ shear banding is clearly observed. A feature of shear-banded flow is the existence of complex fluctuation dynamics, first seen in MRI velocimetry measurements [11] and shown in Fig. 10.4. These can arise from coupling

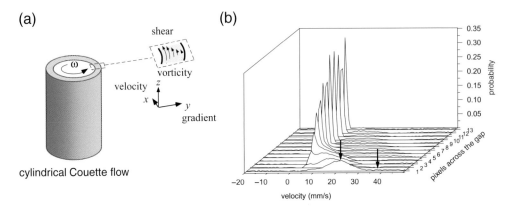

(a)

shear

vorticity

velocity z

x y

gradient

cylindrical Couette flow

(b)

0.35
0.30
0.25
0.20
0.15
0.10
0.05

probability

pixels across the gap

-20 -10 0 10 20 30 40

velocity (mm/s)

Fig. 10.21 (a) Schematic cylindrical Couette cell showing velocity, gradient, and vorticity axes. (b) Velocity distributions for the wormlike micelle system, 10% equimolar cetylpyridinium chloride/sodium salicylate in brine, in each pixel across the 1-mm gap of a 17-mm/19-mm diameter cylindrical Couette cell at a gap-averaged shear rate of $37\,\mathrm{s}^{-1}$. The lower velocity arrows show the positions of the mode velocities at the inner wall while the other shows the wall velocity, thus indicating slip. The broadening of distributions in the high shear band is indicative of fluctuations. (Adapted from López-González *et al.* [65].)

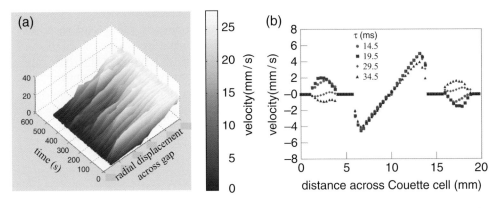

(a)

40
20
0
600
500
400
300
200
100
0

time (s)

radial displacement across gap

25
20
15
10
5
0

velocity(mm/s)

(b)

8
6
4
2
-0
-2
-4
-6
-8

velocity(mm/s)

τ (ms)
• 14.5
■ 19.5
• 29.5
▲ 34.5

0 5 10 15 20

distance across Couette cell (mm)

Fig. 10.22 (a) Velocity profiles for the wormlike micelle system of Fig. 10.21 across the 1-mm gap, at successive 1 s intervals, in the same 17-mm/19-mm diameter cylindrical Couette cell, also at $37\,\mathrm{s}^{-1}$. Here, fluctuations are directly visible. (b) Velocity profiles following the cessation of steady shear flow in a Couette cell, acquired using GERVAIS with delays of 14.5, 19.5, 29.5, and 34.5 ms. The inner cavity contains water and the outer annulus contains CPyCl/NaSal solution. (Adapted from López-González *et al.* [65] and Davis *et al.* [72].)

between flow and microstructure, as well as being the result of an intrinsic instability of planar shear-banded flow with respect to perturbations with wave vector in the plane of the banding interface [66–69], with both spatio-temporal oscillations and chaotic banded flows possible. Such fluctuations have been found in ultrasound velocimetry measurements [70] and in rapid MRI velocimetry using RARE [26, 71].

Figures 10.21 and 10.22 show two ways in which these fluctuations may be observed. In Fig. 10.21(b), spin warp MRI velocimetry, with full propagator acquisition over a 30-min experiment, reveals significantly broader velocity distributions in the vicinity of the high shear rate band, indicative of fluctuations. In Fig. 10.22(a), a single phase-encoded RARE sequence is used to obtain velocity profiles with 1 s refresh time. Here the fluctuations may be directly observed. Finally, in a most impressive demonstration of velocity image refresh speed, Fig. 10.22(b) shows the result of a cylindrical Couette cell shear-cessation experiment [72] on a wormlike micellar system using GERVAIS. Transient oscillations are seen in the wormlike micelle system occupying the annular gap, while water in the central cavity continues to rotate in the initial direction.

Molecular alignment

A fundamental driver for imaging in rheology is the role of heterogeneous deformational flow in a range of interesting rheological phenomena, including slip, shear banding, jamming and fracture, yield stress behaviour, and flow fluctuations. In consequence, a major goal for rheo-NMR is spatially resolved spectroscopic measurements. Over recent years there has been significant progress, including measuring deuterium and proton NMR line-broadening associated with shear-induced nematic order in a wormlike micelle solution, across the gap of a Couette cell [73, 74], observation of the the sudden director transition that occurs in extensional flow in a four-roll mill of a magnetically aligned liquid crystalline polymer [75], and the measurement of the full alignment tensor for an entangled polymer melt [76–78]. Figure 10.23 shows elements of the alignment tensor $S_{\alpha\beta}$ for polydimethylsiloxane obtained from deuterium NMR using a 10% deuteration benzene probe molecule. The method uses selective excitation from a region of the Couette cell where the magnetic field is parallel to the relevant tensor element of interest and the resulting strain-rate-dependent tensor elements provide a powerful test of the Doi–Edwards theory [80, 81], as well as second-order variants, such as convected constraint release [82].

10.4.4 Electro-osmotic and electrophoretic flow

One potentially useful application of MRI velocimetry concerns the imaging of velocity fields in electrophoretic [83–85] or electro-osmotic flow [86, 87]. Figure 10.24 shows how a conventional spin warp velocimetry sequence can be modified to allow for the application of an electric field within the sample. Note that to avoid charge buildup effects at electrodes, such fields are generally applied with opposite sign in successive pulse cycles.

Ions undergoing electrophoretic migration collide with neutral solvent molecules, which transfer momentum in the direction of ion flow. For an ionic solution in which the charge distribution is homogeneous, the oppositely directed flow of ions and cations causes local cancellation of such effects resulting in a stationary solvent. But any inhomogeneity in charge distribution can drastically alter the local balance in electrophoretic drag forces, such that local solvent migration effects become visible, a transport phenomenon known as electro-osmosis. A nice example is provided when an ionic solution is bounded by charged surfaces, for example in a glass capillary, where the surface screening results in a separation of ions of opposite charge. Here the relocation

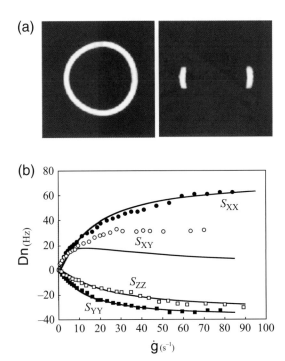

Fig. 10.23 (a) ^{1}H NMR images of polydimethylsiloxane polymer in the annular gap of the horizontal Couette cell, along with that obtained using a selective pulse sequence used to localise a part of the fluid. (b) Deuterium quadrupole splittings, $\Delta\nu$, as a function of shear rate obtained using localised spectroscopy from selected regions of the horizontal Couette cell in which the velocity direction (solid circles) and gradient direction (solid squares) are parallel to the magnetic field and at a 45° angle giving S_{XX}, S_{YY}, and S_{XY} (open circles), respectively. Similar measurements were done in a vertical Couette cell in which the vorticity axis is parallel to the B_0 field (open squares), giving S_{ZZ}. The solid lines are fits using Doi–Edwards theory and a polymer tube disengagement time of $\tau_d = 210$ ms. (Adapted from Cormier *et al.* [79].)

of counterions close to the surface results in an inhomogeneous distribution of mobile charges, so that application of an electric field causes inhomogeneous electrophoretic currents. An interesting consequence is that a significant slip develops within the Debye screening layer where mobile counterions predominate and fluid experiences significant drag in the direction of electrophoretic migration. Outside this layer the flow profile is determined by the hydrodynamics of a neutral fluid experiencing the electro-osmotic flow as a boundary condition [87].

Figure 10.25 shows a set of velocity images obtained for a transverse section of a water-filled 3.4-mm ID glass capillary at successively increasing values of delay time T_l using a 100 V EFP using platinum electrodes 70 mm apart. A symmetric, well-behaved velocity distribution is found below $T_l = 500$ ms, but subsequently asymmetric flow

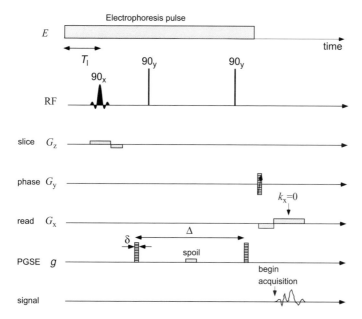

Fig. 10.24 Spin warp pulse sequence used to phase-encode the image for velocity in which an electrophoretic pulse (EFP) is included, this pulse being switched off subsequent to velocity encoding but prior to signal acquisition. Note that T_l defines the time delay between the start of the EFP and the start of the soft RF pulse. A further delay occurs before the start of velocity encoding. (Adapted from Manz *et al.* [87].)

results. The insets show a comparison of theory and experiment. Note the net zero flow in these profiles, along with the boundary slip effect [87].

10.4.5 Biology

In medical MRI, the use of velocimetry to measure blood flow is so widespread that no brief review could here suffice. Indeed, the key words 'magnetic resonance' and 'angiography' result in 11,000 references, whilst restrictions to 'rat' give over 100. There have been some quite spectacular uses of MRI velocimetry in plant science, including the measurement of xylem flow in a wheat grain *in vivo* [88], as well as in the xylem and phloem of castor bean *in vivo* using both spin warp [89] and FLASH velocimetry [29]. Recently MRI velocimetry has been used to measure cytoplasmic streaming velocities in single living cells, the spatial variation of the longitudinal velocity field in cross-sections of internodal cells of *Chara corallina* being imaged with a spatial resolution of 16 μm [90].

10.4.6 Granular flow

By using solid particles with liquid interior, it is also possible to use MRI velocimetry to study granular flow [91–93]. A particularly effective 'MR-visible' solid particle is provided by mustard seeds, and Fukushima and co-workers have used this device to

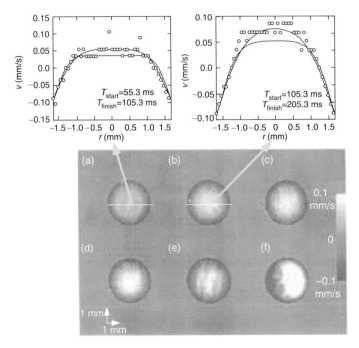

Fig. 10.25 Velocity images obtained for a transverse section of a 3.4-mm ID glass capillary at successively increasing values of delay time T_l, following application of an electric field. These delays are, respectively, (a) 50 (b) 100, (c) 200, (d) 300, (e) 500, and (f) 1000 ms. Note the asymmetric flow distribution in the image obtained at the longest time delay. Also shown are diametral profiles of velocity obtained from the images along with two theoretical profiles calculated using delay times corresponding to the start and end of velocity phase-encoding. These times are shown in the graphs for (a) $T_1 = 50$ ms and (b) $T_l = 100$ ms. (Adapted from Manz *et al.* [87].)

investigate the velocity depth profile of the flowing layer near the axial centre of a half-filled 3-D drum, confirming that it follows a quadratic form, except very close to the free surface, this deviation due to particles reaching the surface with large components of their velocity in the azimuthal direction.

Examples of their work are shown in Fig. 10.26, where in (a) a DANTE tagging sequence has been used to provide a simple flow visualisation.

10.5   Potential artifacts

Note that any use of PGSE gradient encoding is subject, in principle, to a number of systematic errors and artifacts. In particular, the measurement of self-diffusion, which relies on phase spreading, is inherently more susceptible to systematic artifacts than the measurement of velocity. This is because any effect that introduces phase incoherence will enhance the apparent diffusion rate. Such effects include sample container movement, spectrometer RF or field instability, and imperfect gradient pulse matching due to induced eddy current effects or to noise or ripple in the gradient (a)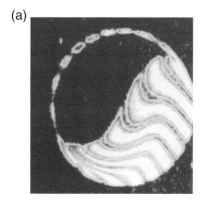

(b)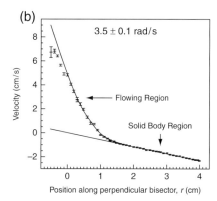

Fig. 10.26 (a) MRI tagging experiment for 2-mm diameter mustard seeds half-filling a rotating drum, in which the grains clearly separate between solid-body rotation and a flowing region. (Reproduced with permission from Nakagawa *et al.* [91].) (b) Velocity depth profiles for 2-mm mustard seeds in a half-filled cylinder along the perpendicular bisector of the free surface, averaged over five experimental runs. (Reproduced with permission from Sanfratello *et al.* [92].)

power supply. These are very standard but nonetheless troublesome problems that must be addressed by the experimenter before embarking on either standard PGSE or dynamic imaging/velocimetry experiments. Suggestions concerning counteracting these problems are given in Chapter 9 of reference [1] and Chapter 12 of this book. There are, in addition, effects peculiar to dynamic imaging and to the measurement of self-diffusion in the presence of flow, including gradient-dependent phase shifts, inflow–outflow effects, the influence of the slice selection gradient, the influence of gradient non-uniformity, and diffusion across streamlines. Again, a discussion of these is found in Chapter 8 of reference [1].

References

[1] P. T. Callaghan. *Principles of Nuclear Magnetic Resonance Microscopy.* Oxford University Press, New York, 1991.

[2] J. Granwehr, E. Harel, S. Han, S. Garcia, A. Pinesand, P. N. Sen, and Y-Q. Song. Time-of-flight flow imaging using NMR remote detection. *Phys. Rev. Lett.*, 95:075503, 2005.

[3] E. Harel, J. Granwehr, J.A. Seeley, and A. Pines. Multiphase imaging of gas flow in a nanoporous material using remote-detection NMR. *Nature Materials*, 5:321, 2006.

[4] E. Harel, C. Hilty, K. Koen, E. E. McDonnell, and A. Pines. Time-of-flight flow imaging of two-component flow inside a microfluidic chip. *Phys. Rev. Lett.*, 98:017601, 2007.

[5] P. R. Moran. A flow velocity zeugmatographic interlace for NMR imaging in humans. *Magn. Reson. Imaging.*, 1:197, 1982.

[6] T. W. Redpath, D. G. Norris, R. A. Jones, and J. M. S. Hutchison. A new method of NMR flow imaging. *Phys. Med. Biol.*, 29:891, 1984.

[7] D. G. Taylor and M. C. Bushell. The spatial mapping of translational diffusion coefficients by the NMR imaging technique. *Phys. Med. Biol.*, 30:345, 1985.

[8] P. T. Callaghan, C. D. Eccles, and Y. Xia. NMR microscopy of dynamic displacements: k-space and q-space imaging. *J. Phys. E*, 21:820, 1988.

[9] Y. Xia. *MSc Thesis*. Massey University (unpublished), 1988.

[10] H. Wassenius and P. T. Callaghan. Nanoscale NMR velocimetry. *J. Magn. Reson.*, 169:250, 2004.

[11] M. M. Britton and P. T. Callaghan. Shear banding instability inwormlike micellar solutions.. *Eur. Phys. J. B*, 7:237, 1999.

[12] P. Mansfield. Multi-planar image formation using NMR spin echoes. *J. Phys. C: Solid State Physics*, 10:L55, 1977.

[13] P. Mansfield and I. L. Pykett. Biological and medical imaging by NMR. *J. Magn. Reson.*, 29:355, 1978.

[14] M. Doyle, B. Chapman, R. Turner, R. J. Ordidge, M. Cawley, R. Coxon, P. Glover, R. E. Coupland, G. K. Morris, B. S. Worthington, and P. Mansfield. Real-time cardiac imaging of adults at video frame rates by magnetic resonance imaging. *Lancet*, 328:682, 1986.

[15] R.J. Ordidge, A. Howseman, R. Coxon, R. Turner, B. Chapman, P. Glover, M. Stehling, and P. Mansfield. Microscopic imaging of slow flow and diffusion: a pulsed field gradient stimulated echo sequence combined with turbo spin echo imaging. *Mag. Reson. Med.*, 10:227, 1989.

[16] K. Kose. Instantaneous flow-distribution measurements of the equilibrium turbulent region in a circular pipe using ultrafast NMR imaging. *Phys. Rev. A*, 44:2495, 1991.

[17] K. Kose. One-shot velocity mapping using multiple spin-echo EPI and its application to turbulent flow. *J. Magn. Reson.*, 92:631, 1991.

[18] K. Kose. Visualization of local shearing motion in turbulent fluids using echo-planar imaging. *J. Magn. Reson.*, 96:596, 1992.

[19] J. C. Gatenby and J. C. Gore. Echo-planar-imaging studies of turbulent flow. *J. Magn. Reson. A*, 121:193, 1996.

[20] A. J. Sederman, M. D. Mantle, C. Buckley, and L. F. Gladden. MRI technique for measurement of velocity vectors, acceleration, and autocorrelation functions in turbulent flow. *J. Magn. Reson.*, 166:182, 2004.

[21] A. J. Sederman, M. D. Mantle, and L. F. Gladden. Single excitation multiple image rare (semi-rare): ultra-fast imaging of static and flowing systems. *J. Magn. Reson.*, 161:15, 2003.

[22] T. W. J. Scheenen, D. van Dusschoten, P. A. de Jager, and H. van As. Microscopic displacement imaging with pulsed field gradient turbo spin-echo NMR. *J. Magn. Reson.*, 142:207, 2000.

[23] T. W. J. Scheenen, F. J. Vergeldt, C. W. Windt, P. A. de Jager, and H. van As. Microscopic imaging of slow flow and diffusion: a pulsed field gradient stimulated echo sequence combined with turbo spin echo imaging. *J. Magn. Reson.*, 151:94, 2001.

[24] P. Galvosas and P. T. Callaghan. Fast magnetic resonance imaging and velocimetry for liquids under high flow rates. *J. Magn. Reson.*, 181:119, 2006.

[25] H. Y. Carr and E. M. Purcell. Effects of diffusion on free precession in nuclear magnetic resonance experiments. *Phys. Rev.*, 94:630, 1954.

[26] K. W. Feindel and P. T. Callaghan. Anomalous shear banding: multidimensional dynamics under fluctuating slip conditions. *Rheo. Acta.*, 49:1003, 2010.

[27] A. Amar, F. Casanova, and B. Bluemich. Rapid multiphase flow dynamics mapped by single-shot MRI velocimetry. *Chem. Phys. Chem.*, 11:2630, 2010.

[28] A. Haase, J. Frahm, D. Matthei, and K. D. Merbold. FLASH imaging: rapid NMR imaging using low flip angle pulses,. *J. Magn. Reson.*, 67:258, 1986.

[29] M. Rokitta, U. Zimmermann, and A. Haase. Fast NMR flow measurements in plants using flash imaging. *J. Magn. Reson.*, 137:29, 1999.

[30] S. Emid and J. H. N. Creyghton. High resolution NMR imaging in solids. *Physica B*, 128:81, 1985.

[31] L. Li, Q. Chen, A. E. Marble, L. Romero-Zerón, B. Newling, and B. J. Balcom. Flow imaging of fluids in porous media by magnetization prepared centric-scan SPRITE. *J. Magn. Reson.*, 197:1, 2009.

[32] B. J. Balcom, R. P. Macgregor, S. D. Beyea, D. P. Green, R. L. Armstrong, and T. W. Bremner. Single-point ramped imaging with T_1 enhancement (SPRITE). *J. Magn. Reson. A*, 123:131, 1996.

[33] R. R. Ernst, G. Bodenhausen, and A. Wokaun. *Principles of Nuclear Magnetic Resonance in One and Two Dimensions*. Clarendon Press, Oxford, 1987.

[34] B. Newling, C. C. Poirier, Y. Zhi, J. A. Rioux, A. J. Coristine, D. Roach, and B. J. Balcom. Velocity imaging of highly turbulent gas flow. *Phys. Rev. Lett.*, 93:154503–1, 2004.

[35] A. J. Sederman and L. F. Gladden. Magnetic resonance visualisation of single- and two-phase flow in porous media. *Magnetic Resonance Imaging*, 19:339, 2001.

[36] B. Manz, L. F. Gladden, and P. B. Warren. Flow and dispersion in porous media: Lattice-Boltzmann and NMR studies. *AIChE J.*, 45:1845, 1999.

[37] M. D. Mantle, B. Bijeljic, A. J. Sederman, and L. F. Gladden. MRI velocimetry and Lattice-Boltzmann simulations of viscous flow of a Newtonian liquid through a dual porosity fibre array. *Magnetic Resonance Imaging*, 19:527, 2001.

[38] M. D. Mantle, A. J. Sederman, and L. F. Gladden. Single- and two-phase flow in fixed-bed reactors: MRI flow visualisation and Lattice-Boltzmann simulations. *Chemical Engineering Science*, 58:523, 2001.

[39] I. Iliuta, A. Ortiz-Arroyo, F. Larachi, B. P. A. Grandjean, and G. Wild. Hydrodynamics and mass transfer in trickle-bed reactors: an overview. *Chemical Engineering Science*, 54:5329, 2001.

[40] L. D. Anadon, A. J. Sederman, and L. F. Gladden. Mechanism of the trickle-to-pulse flow transition in fixed-bed reactors. *AIChE J.*, 52:1522, 2006.

[41] E. L. Koschmeider. *Benard Cells and Taylor Vortices*. Cambridge University Press, Cambridge, 1993.

[42] J. D. Seymour, B. Manz, and P. T. Callaghan. Pulsed gradient spin echo nuclear magnetic resonance measurements of hydrodynamic instabilities with coherent structure: Taylor vortices. *Phys. Fluids*, 11:1104, 1999.

[43] K. W. Mosera, G. Raguin, A. Harris, D. Morris, J. Georgiadis, M. Shannon, and M. Philpott. Visualization of taylor-couette and spiral poiseuille flows using a

snapshot flash spatial tagging sequence. *Magnetic Resonance Imaging*, 18:199, 2000.

[44] M. Sankey, Y. Yang, L. F. Gladden, M. L. Johns, D. Lister, and B. Newling. SPRITE MRI of bubbly flow in a horizontal pipe. *J. Magn. Reson.*, 199:126, 2009.

[45] L. Saetran C.J. Elkins, M.T. Alley and J.K. Eaton. Three-dimensional magnetic resonance velocimetry measurements of turbulence quantities in complex flow. *Experiments in Fluids*, 46:285, 2009.

[46] A. I. Nakatani, M. D. Poliks, and E. T. Samulski. NMR investigation of chain deformation in sheared polymer fluids. *Macromolecules*, 23:2686, 1990.

[47] P. T. Callaghan. Rheo-NMR: nuclear magnetic resonance and the rheology of complex fluids. *Rep. Prog. Phys.*, 62:599, 1999.

[48] P. T. Callaghan. Rheo-NMR and velocity imaging. *Current Opinion in Colloid and Interface Science*, 11:13, 2006.

[49] P. T. Callaghan. Rheo NMR and shear banding. *Rheol. Acta*, 47:243, 2008.

[50] P. Coussot, L. Tocquer, C. Lanos, and G. Ovarlez. Macroscopic vs. local rheology of yield stress fluids. *J. Non Newton. Fluid Mech.*, 158:85, 2009.

[51] A. Veron, A. E. Gomes, C. R. Leal, J. Van der Klink, and A. F. Martins. NMR study of flow and viscoelastic properties of PBLG/m-cresol lyotropic liquid crystal. *Mol. Cryst. Liquid. Cryst.*, 331:499, 1999.

[52] H. Siebert, P. Becker, I. Quijada-Garrido, D. A. Grabowski, and C. Schmidt. In-situ deuteron NMR investigations of sheared liquid crystalline polymers. *Solid State NMR*, 22:311, 2002.

[53] L. A. Madsen, T. J. Dingemans, M. Nakata, and E. T. Samulski. Thermotropic biaxial nematic liquid crystals. *Phys. Rev. Lett.*, 92:145505, 2004.

[54] P. Becker, H. Siebert, L. Noirez, and C. Schmidt. Shear-induced order in nematic polymer. *Macromolecules*, 220:111, 2005.

[55] H. Wassenius and P.T. Callaghan. NMR velocimetry and the steady-shear rheology of a concentrated hard-sphere colloidal system. *Eur. Phys. J. E.*, 18:69, 2005.

[56] P. Coussot, J. S. Raynaud, F. Bertrand, P. Moucheront, J. P. Guilbaud, H. T. Huynh, S. Jarny, and D. Lesueur. Coexistence of liquid and solid phases in flowing soft-glassy materials. *Phys. Rev. Lett.*, 88:218301, 2002.

[57] M. E. Cates. Nonlinear viscoelasticity of wormlike micelles (and other reversibly breakable polymers). *J. Phys. Chem.*, 94:371, 1990.

[58] N. A. Spenley, M. E. Cates, and T. C. B. McLeish. Nonlinear rheology of wormlike micelles. *Phys. Rev. Lett.*, 71:939, 1993.

[59] N. A. Spenley, X. F. Yuan, and M. E. Cates. Nonmonotonic constitutive laws and the formation of shear-banded flows. *J. Phys. II (France)*, 6:551, 1996.

[60] M. E. Cates, G. Marrucci, and T. C. B. McLeish. The rheology of entangled polymers at very high shear rates. *Europhys. Lett.*, 21:451, 1993.

[61] M. M. Britton and P. T. Callaghan. Two-phase shear band structures at uniform stress. *Phys. Rev. Lett.*, 78:4930, 1997.

[62] M. M. Britton and P. T. Callaghan. NMR visualisation of anomalous flow in cone-and-plate rheometry. *J. Rheol.*, 41:1365, 1997.

[63] J. P. Decruppe, R. Cressel, R. Makhloufi, and E. Cappelaere. Flow birefringence experiments showing a shear banding structure in a ctab solution. *Colloid and Polym. Sci.*, 275:346, 1995.

[64] J-F. Berret, D. C. Roux, and G. Porte. Isotropic-to-nematic transition in wormlike micelles under shear. *J. Phys. II (France)*, 4:1261, 1994.

[65] M.R. López-González, W. M. Holmes, and P. T. Callaghan. Rheo-NMR phenomena of wormlike micelles. *Soft Matter*, 2:855, 2006.

[66] S. M. Fielding and P. D. Omsted. Spatiotemporal oscillations and rheochaos in a simple model of shear banding. *Phys. Rev. Lett.*, 92:084502, 2004.

[67] S. M. Fielding and P. D. Omsted. Nonlinear dynamics of an interface between shear bands. *Phys. Rev. Lett.*, 96:104502, 2006.

[68] S. M. Fielding. Complex dynamics of shear banded flows. *Soft Matter*, 2:1262, 2007.

[69] S. M. Fielding and H. J. Wilson. Shear banding and interfacial instability in planar poiseuille flow. *Journal of Non-Newtonian Fluid Mechanics*, 165:196, 2010.

[70] S. Manneville, L. Becu, and A. Colin. High-frequency ultrasonic speckle velocimetry in sheared complex fluids. *Eur. Phys. J. Appl. Phys.*, 28:361, 2004.

[71] M. R. López-González, P. Photinos, W. M. Holmes, and P. T. Callaghan. Fluctuations and order for wormlike micelles under shear. *Phys. Rev. Lett.*, 93:268302, 2004.

[72] C. J. Davies, A. J. Sederman, C. J. Pipe, G. H. McKinley, L. F. Gladden, and M. L. Johns. Rapid measurement of transient velocity evolution using GERVAIS. *J. Magn. Reson.*, 202:93, 2010.

[73] E. Fischer and P. T. Callaghan. Shear banding and the isotropic to nematic transition in wormlike micelles. *Phys. Rev. E*, 6401:1501, 2001.

[74] W. M. Holmes, M. R. López-González, and P. T. Callaghan. Shear-induced constraint to amphiphile chain dynamics in wormlike micelles. *Europhys. Lett.*, 66:132, 2004.

[75] R. J. Cormier, C. Schmidt, and P. T. Callaghan. Director reorientation of a side-chain liquid crystalline polymer under extensional flow. *J. Rheol.*, 48:881, 2004.

[76] M. L. Kilfoil and P. T. Callaghan. Selective storage of magnetization in strongly relaxing spin systems. *Journal of Magnetic Resonance B*, 150:110, 2001.

[77] M. L. Kilfoil and P. T. Callaghan. NMR measurement of the alignment tensor for a polymer melt under strong shearing flow. *Macromolecules*, 33:6828, 2000.

[78] R. J. Cormier and P. T. Callaghan. Molecular weight dependence of segmental alignment in a sheared polymer melt: A deuterium NMR investigation. *J. Chem. Phys.*, 116:10020, 2002.

[79] R. J. Cormier, M. L. Kilfoil, and P. T. Callaghan. Bi-axial deformation of a polymer under shear: NMR test of Doi-Edwards model with convected constraint release. *Physical Review E*, 6405:1809, 2001.

[80] M. Doi and S.F. Edwards. Dynamics of concentrated polymer solutions-ii: molecular motion under flow. *J. Chem. Soc., Faraday Trans. 2.*, 74:1802, 1978.

[81] M. Doi and S.F. Edwards. *The Theory of Polymer Dynamics*. Clarendon Press, Oxford, 1986.

[82] G. Marrucci and G. Ianniruberto. Stress tensor and stress-optical law in entangled polymers. *J. Non-Newt. Fluid Mech.*, 79:225, 1998.

[83] S. J. Gibbs and C. S. Johnson. Pulsed field gradient NMR study of probe motion in polyacrylamide gels. *Macromolecules*, 24:6110, 1991.

[84] P. Stilbs and I. Furó. Electrophoretic NMR. *Current Opinion in Colloid and Interface Science*, 11:3, 2006.

[85] E. Pettersson, I I. Furó, and P. Stilbs. On experimental aspects of electrophoretic NMR. *Concepts in Magnetic Resonance*, 22A:61, 2004.

[86] X. F. Zhang and A. G. Webb. Magnetic resonance microimaging and numerical simulations of velocity fields inside enlarged flow cells used for coupled NMR microseparations. *Analytical Chemistry*, 77:1338, 1998.

[87] B. Manz, P. Stilbs, B. Joensson, O. Söderman, and P. T. Callaghan. NMR imaging of the time evolution of electroosmotic flow in a capillary. *J. Phys. Chem.*, 99:11297, 1995.

[88] C. F. Jenner, Y. Xia, C. D. Eccles, and P. T. Callaghan. Circulation of water within the wheat grain revealed by NMR micro-imaging methods. *Nature*, 336:399, 1988.

[89] W. Koeckenberger, J. M. Pope, Y. Xia, K. R. Jeffrey, E. Komor, and P. T. Callaghan. A non-invasive measurement of phloem and xylem water flow in castor bean seedlings by NMR microimaging. *Planta*, 201:53, 1997.

[90] J. W. van de Meent, A. J. Sederman, L. F. Gladden, and R. E. Goldstein. Measurement of cytoplasmic streaming in single plant cells by magnetic resonance velocimetry. *J. Fluid. Mech.*, 642:5, 2010.

[91] M. Nakagawa, S. A. Altobelli, A. Caprihan, and E. Fukushima. *Powders and Grains*, chapter NMR measurement and approximate derivation of the velocity depth-profile of granular flow in a rotating, partially filled, horizontal cylinder, page 449. R.P. Behringer and J.Y. Jenkins, eds., Balkema, Rotterdam,, 1997.

[92] L. Sanfratello, A. Caprihan, and E. Fukushima. Velocity depth profile of granular matter in a horizontal rotating drum. *Granular Matter*, 9:1, 2007.

[93] D. Bonn, S. Rodts, M. Groenink, S. Rafai, N. Shahidzadeh-Bonn, and P. Coussot. Some applications of magnetic resonance imaging in fluid mechanics: Complex flows and complex fluids. *Ann. Rev. Fluid. Mech.*, 40:209, 2008.

11
Translational dynamics and quantum coherence

In all that we have encountered so far, the evolution of spin phase under the effect of magnetic field gradients has been considered to arise from single-quantum coherence, that simple classical transverse magnetisation generated when the ensemble of nuclear spins is disturbed from thermal equilibrium by a resonant RF pulse. But, as we learned in Chapters 3 and 4, terms in the spin Hamiltonian bilinear in the spin operator can act to transform such magnetisation into various states of quantum coherence. Such terms include homo- and heteronuclear scalar couplings, dipolar interactions and quadrupolar interactions. All are capable of generating double-quantum coherence or, for two-spin interactions, zero-quantum coherence, and—given sufficient spin couplings or in the case of quadrupole interactions, sufficient single-spin quantum number, I— states of n-quantum coherence, where $n > 2$. Once created, these will evolve in phase under the influence of the Zeeman interaction at n times that of the single-quantum state. By such means, sensitivity to magnetic field gradients may be enhanced.

Moving from single to n-quantum coherence changes none of the physics encountered so far. The use of multiple-quantum coherence as a vehicle for imprinting phase information arising from translational dynamics merely represents a tool for amplifying the apparent magnetogyric ratio or gradient amplitude. And of course, as with all enticing tricks, there may be a price to be paid. With what efficiency can we transform our initial magnetisation to a state of multiple-quantum coherence, and then reconvert to the single-quantum state required by our detector? What are the relaxation rates associated with higher quantum coherence, and if faster, will we lose too much signal during coherence preparation, evolution, and reconversion? In short, are the signal-to-noise ratio costs worth the heightened phase-shift sensitivity? While each case must be evaluated on its own merits, we will see that there are systems where the gains are indeed real.

There is another sense in which differing quantum states of the spin system can be enlisted to enhance sensitivity to translational dynamics. A fundamental limit to PGSE NMR concerns the time over which a ensemble phase coherence can be preserved: the maximum value of Δ over which it is possible to observe molecular translation. Conventional wisdom suggests that the best way to maximise the time is to use a stimulated echo, so that the limiting timescale is determined by T_1 relaxation during z-storage. But in fact longer-lived states of spin coherence are possible, in particular certain specially prepared singlet states. In this chapter we will see how such states may be generated and utilised to stretch our time window.

Finally, we will encounter a curious aspect of quantum coherence in which diffusion plays an inherent role in determining the signal observed. In liquids, intra-molecular dipolar interactions between spins are motionally averaged to zero by the rapid rotational tumbling of the host molecule. What then of inter-molecular dipolar interactions between spins on different molecules? Here we might hope that translational diffusion acts to cause pairs of molecules to undergo a migrational dance, in which the internuclear vectors between their respective spins will reorient sufficiently rapidly to average these interactions as well. And indeed, if the molecules are sufficiently close, the dance suffices. But as molecular separation increases, the time taken for translational diffusion to permit an orbit in which the vectors sample all possible directions will diverge.

Of course, the strength of the dipolar interactions drops with separation distance r as r^{-3}, so that we might expect these more distant molecular neighbours to impose unaveraged dipolar interactions too weak to matter. But now we face the vastly increasing number of that outer pool of neighbours as we integrate their contributions out to the limits of our sample. A logarithmic divergence presents itself, and what we find is that a residual inter-molecular dipolar interaction persists, with a size determined by two primary factors. The first is the least separation distance, where translational diffusion is no longer capable of dancing away the dipolar interaction. The slower the diffusion, the tighter that inner radius, and the stronger the dipolar contribution. The second factor concerns symmetry. The inner volume determined by the diffusion limit is essentially a sphere, and as we integrate contributions out to the limits of our sample, we do so in spherical shells, so that whatever our sample shape, the effects die away before such shape is capable of breaking spherical symmetry, a symmetry that sums all dipolar contributions to zero. To observe the intermolecular dipolar sum, we need to externally break that spherical symmetry by manipulating the bulk magnetisation phase, and this we may do by imposing magnetic field gradient pulses.

These remarkable intermolecular dipolar interactions were first observed in helium-3 NMR [1], then rediscovered in proton NMR, fully explained and brought to the attention of the wider magnetic resonance community by Warren and co-workers [2–7]. The cumulative interaction strengths are small, on the order of hertz or tens of hertz, but they increase with increasing polarising field strength and decreasing diffusion coefficient, and they particularly appear when magnetic field gradients are applied. For that reason alone we need to understand them. Furthermore, these effects have potential applications in NMR microscopy, in which a subtle interplay of long-range dipolar correlations allow inferences regarding sample internal structure.

11.1   Diffusion measurement using multiple-quantum coherences

11.1.1   Use of dipolar couplings

The first demonstrations of diffusion measurement using multiple-quantum coherences employed the dipolar interaction to convert from the the single-quantum state [8, 9]. Martin *et al.* [8] used double-quantum coherence to measure the diffusion coefficient of dichloromethane (CH_2Cl_2) dissolved at $< 10\%$ concentration in thermotropic liquid crystals in which the nematic order of the host induced a dipolar splitting of several kilohertz, while Zax and Pines measured the diffusion of benzene dissolved in liquid crystals at higher concentration, using various quantum orders.

As we have seen in Chapter 4, the dipolar interaction between a pair of spins, i and j, involves a bilinear spin operator $\omega_{ij}[3I_{iz}I_{jz} - \mathbf{I_1} \cdot \mathbf{I_2}]$, where ω_{ij} is the 'dipolar precession frequency'. Suppose we start with an equilibrium longitudinal magnetisation represented by a density matrix proportional to $I_{iz} + I_{jz}$. After a 90° RF pulse, the evolution after time t can be described by

$$I_{iz} + I_{jz}$$
$$\xrightarrow{-(\pi/2)(I_{ix}+I_{jx})} I_{iy} + I_{jy}$$
$$\xrightarrow{3\omega_{ij}I_{iz}I_{jz}} (I_{iy} + I_{jy})\cos\phi - (2I_{ix}I_{jz} + 2I_{iz}I_{jx})\sin\phi$$

$$(11.1)$$

where $\phi = \frac{3}{2}\omega_{ij}t$. Only the density matrix term $(I_{iy} + I_{jy})$ is directly observable using Faraday detection, and so the transverse magnetisation detected in our receiver coil is modulated by $\cos(\frac{3}{2}\omega_{ij}t)$, corresponding to two spectral components at frequencies $\pm\frac{3}{2}\omega_{ij}$. This doublet is split by angular frequency $3\omega_{ij}$. Note that the term $\mathbf{I_1} \cdot \mathbf{I_2}$ commutes with the operator $I_{iy} + I_{jy}$, and as for the scalar coupling when the chemical shift is zero, produces no observable interaction.

RF and gradient pulse scheme
To see how RF pulses may be used to create a state of double-quantum coherence from the two-spin dipolar interaction, consider the following sequence of two identical 90° excitations.

$$I_{iz} + I_{jz}$$
$$\xrightarrow{-(\pi/2)(I_{ix}+I_{jx})} I_{iy} + I_{jy}$$
$$\xrightarrow{3\omega_{ij}I_{iz}I_{jz}} (I_{iy} + I_{jy})\cos\phi - (2I_{ix}I_{jz} + 2I_{iz}I_{jx})\sin\phi$$
$$\xrightarrow{-(\pi/2)(I_{ix}+I_{jx})} - (I_{iz} + I_{jz})\cos\phi - (2I_{ix}I_{jy} + 2I_{iy}I_{jx})\sin\phi$$

$$(11.2)$$

For $\phi = (2p+1)\frac{\pi}{2}$ where $p = 0, 1, 2...$, we have a transfer to $(2I_{ix}I_{jy}+2I_{iy}I_{jx})$, a state of double-quantum coherence. For a spectral splitting $\Delta\nu$ in hertz, the simple case $p = 0$ corresponds to a time delay between the two RF pulses of $(2\Delta\nu)^{-1}$. Note that higher-order spin couplings (3-spin, 4-spin) can lead to higher orders of coherence. These will arise from different coupling strengths and hence may be generated over longer times with lower efficiency. For the moment, however, we will consider a sequence optimised for double-quantum generation and focus solely on $n = 2$.

Using the spherical tensor language of Table 3.8 and Section 3.4.4, $I_{iy} + I_{jy}$ or $T_{11}(10, s) + T_{11}(01, s)$ evolves under the influence of $T_{20}(11)$ to generate $T_{22}(11) - T_{2-2}(11)$, a superposition of +2QC and -2QC. Note that $[T_{2\pm2}(11), T_{20}(11)] = 0$. Once created, the state of double-quantum coherence is invariant under the effect of the dipolar interaction. This means that it is not subject to dephasing as a result of an inhomogeneous distribution of dipolar terms. However, it does precess about any Zeeman term in the spin Hamiltonian, such as is imposed by a magnetic field gradient pulse. And it does so at twice the Larmor frequency, since the commutator yields

$$[\omega_0 (T_{10}(10) + T_{10}(01)), T_{2\pm2}(11)] = \pm 2\omega_0 T_{2\pm2}(11) \tag{11.3}$$

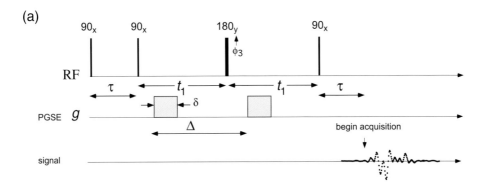

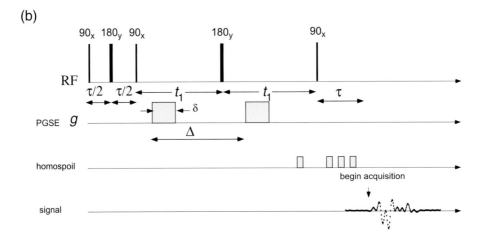

Fig. 11.1 (a) double-quantum spin-echo sequence for diffusion measurement. The final 90° RF pulse at the double-quantum echo transfers coherence from $(2I_{ix}I_{jy} + 2I_{iy}I_{jx})$ to $(2I_{ix}I_{jz} + 2I_{iz}I_{jx})$, which subsequently evolves under the dipolar interaction to observable transverse magnetisation at time τ later. The RF pulses are phase cycled to optimise for coherence selection. (Adapted from reference [8].) (b) Another version of the pulse sequence in which a $90° - 180° - 90°$ echo is used in the initial evolution, refocusing chemical shift or field inhomogeneity, but leaving the dipolar evolution unchanged, and in which homospoil gradient pulses are used to select only those pathways involving n-quantum coherence during the echo periods, T. (Adapted from Zax and Pines [9].)

The double-quantum superposition state, $T_{22}(11) - T_{2-2}(11)$, evolves under the Zeeman Hamiltonian, $\omega_0(I_{iz} + I_{jz}) = \omega_0 (T_{10}(10) + T_{10}(01))$, as

$$T_{22}(11) - T_{2-2}(11)$$

$$\xrightarrow{\omega_0(I_{iz}+I_{jz})} (T_{22}(11) - T_{2-2}(11)) \cos 2\omega_0 t$$
$$+ i(T_{22}(11) + T_{2-2}(11)) \sin 2\omega_0 t$$
$$= (2I_{ix}I_{jy} + 2I_{iy}I_{jx}) \cos 2\omega_0 t$$
$$+ (2I_{iy}I_{jy} - 2I_{ix}I_{jx}) \sin 2\omega_0 t$$

$$(11.4)$$

These two tensors generated in the dephasing process possess the required property for refocusing using $180°$ RF pulses, namely, one being negated while the other being invariant.[1] Just as for transverse magnetisation, the phase shifts experienced by the double-quantum state may be refocused using $180°$ RF pulses, exactly as needed for echo formation and, more importantly, allowing phase offsets to result from the translational motion of spins in the presence of a magnetic field gradient.

Of course, the double-quantum state, having acquired the phase shifts imprinted by the spin motion, must be subsequently reconverted back to single-quantum coherence for Faraday detection. This is achieved by means of a final $90°$ mixing pulse, as shown in Fig. 11.1.

The doubled phase shifts experienced by the two-quantum state in a spin-echo experiment are a direct consequence of the factor of 2, on the right of eqn 11.3, effectively doubling the apparent gradient strength, or the apparent magnetogyric ratio. Later, when we come to discuss the effects of heteronuclear j-couplings, we will see that the latter perspective is the right one. In general, for an n-quantum coherence, the effective γ is multiplied by n, for example, leading, in the case of simple diffusion in a two-pulse PGSE NMR experiment, to a modified Stejskal–Tanner equation,

$$E(g) = -\exp\left(-(n\gamma g)^2\delta^2 D(\Delta - \delta/3)\right) \tag{11.5}$$

The phase cycle

The density matrix transformations outlined all depend on perfect RF pulses. In practise RF inhomogeneity means that many spins will experience pulses that deviate from $90°$ or $180°$, thus causing a mixing of unwanted coherences into the evolution pathway. The means by which such effects may be eliminated is outlined in Section 4.6.4. By choice of an appropriate phase cycle for the RF pulses, only desired coherences are retained. Crucial in the two-quantum PGSE experiment is that phase shifts generated under the influence of the magnetic field gradient pulses arise from evolution of a 2-QC state, with no contribution from single (or indeed any $n \neq 2$ quantum coherences), for if such unwanted states did contribute, the status of the factor n^2 in eqn 11.5 would be ambiguous. The change, Δp, in quantum order under the refocusing (third) RF pulse in Fig. 11.1(a) gives us the means by which to make our selection.

---

[1] In the standard spin echo, transverse magnetisation $I_x \cos \phi + I_y \sin \phi$ has one term negated and one invariant, whether a $180°_x$ or $180°_y$ RF pulse is used.

As can be seen from eqn 11.4,[2] this $180°$ RF pulse turns $T_{2\pm2}(11)$ into $T_{2\mp2}(11)$, and so $\Delta p = 4$. Consider then an eight-step phase cycle in which the first, second, and fourth RF pulse phases remain fixed, while the phase ϕ_3 of the third pulse is incremented in $45°$ steps as $\phi_3 = 0°, 45°, 90°, 135°\ldots$ From eqn 4.102 we can see that the introduced phase factor $\exp(i4\phi_3)$ will cause the final signal to alternate in sign as $1, -1, 1 - 1\ldots$ All other coherence orders will cause a quite different pattern of shifts in their contribution to the final signal. Hence by toggling the acquisition phase between $0°$ and $180°$, and summing the successive phase cycle contributions, unwanted signals decohere and cancel as we select signal arising purely from the double-quantum coherence pathway.

The eight-step cycle can be further improved by adding a second eight steps in which the phase of the first two pulses are changed from $\phi_1 = \phi_2 = 0$ to $\phi_1 = \phi_2 = 180°$. Such a 16-step procedure has been termed a 'hexadecacycle' [10]. All phase cycles depend on such an analysis of the effects of each RF pulse. Having established the basic principles with this simple example, further unravelling of the mechanics of such phase cycles will be left as an exercise for the reader.

Gradient selection of quantum order

In Fig. 11.1(b) a different approach to the selection of desired quantum order is shown. Here a small homospoil gradient pulse of wavevector $\mathbf{q}$ is inserted into the second period during which the desired n quantum state evolves, so that $T_{2\pm2}$ evolves to $T_{2\pm2}\,\overline{\exp(\pm i2\mathbf{q}\cdot\mathbf{r})}$. Once converted back to single-quantum coherence by RF pulse 4, any refocusing of this phase-spread will occur at only the single-quantum rate, and so require twice as many homospoil pulses to generate a non-zero signal. Similarly, for selection of n-quantum coherence during the echo periods T, n homospoil pulses are needed after RF pulse 4. This method has been very effectively used by Zax and Pines to select quantum orders from 1 to 6 in a diffusion measurement on benzene dissolved in a liquid crystal, where multiple spin dipolar couplings permit higher-order coherences to be generated.

11.1.2   Use of the quadrupole interaction

Quadrupolar nuclei ($I > 1/2$) provide another opportunity for the generation of multiple-quantum states and an enhanced sensitivity to magnetic field gradients. Given that all stable quadrupolar nuclei have a magnetogyric ratio somewhat smaller that the proton, this opportunity is worth using. With quadrupolar nuclei, it is really helpful to will look at the evolution process using the spherical tensor formalism introduced in Chapter 3. Using Table 3.6, we may write, in the case of a uniaxial quadrupolar interaction, the rotating frame Hamiltonian under resonance offset $\Delta\omega$ as

$$\mathcal{H} = \mathcal{H}_{Zeeman} + \mathcal{H}_{Q_0}$$
$$= \Delta\omega T_{10} + PT_{20} \tag{11.6}$$

---

[2] *i.e.* $T_{2\pm2}(11) = \pm(I_{ix}I_{jy} + I_{iy}I_{jx}) - i(I_{iy}I_{jy} - I_{ix}I_{jx})$ and a $180°$ RF pulse inverts the first term only.

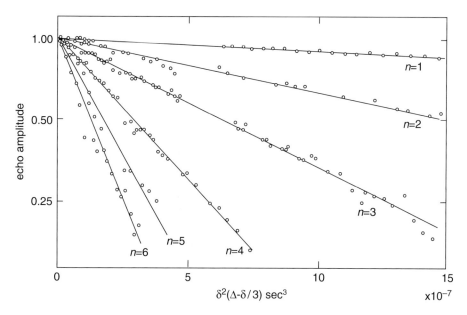

Fig. 11.2 NMR diffusion experiment for all n-quantum orders of benzene dissolved at 25% concentration in Eastman liquid crystal No. 15320. The plots show the normalised echo amplitude vs the gradient pulse timing parameters. The straight lines are linear least-squares fits to the accumulated data the slopes of which vary as n^2, indicating the increasing sensitivity to diffusion as the multiple-quantum order is increased. (Adapted from Zax and Pines [9].)

where only the secular part, $\mathcal{H}_{Q_0}$, of $\mathcal{H}_Q$ is retained, as a first-order perturbation with respect to the dominant laboratory-frame Zeeman term. P represents the strength of the quadrupolar term, and is given by

$$P = \frac{3eV_{zz}Q}{4I(2I-1)\hbar}\sqrt{6}P_2(\cos\theta) \qquad (11.7)$$

where θ is the angle between the principal axis of the electric field gradient and the polarising magnetic field. For a system in which the molecule bearing the quadrupolar nuclear fluctuates rapidly between different local electric field gradient directions, but with average polar angle α, for example where the bond directions fluctuate during tumbling motion, $P_2(\cos\theta)$ may be replaced by the product of an order parameter and $P_2(\cos\alpha)$, as discussed in Section 4.4.4.

The quadrupole frequency

Suppose the spin system is excited with a 90° RF pulse to generate the density-matrix state $I_y = iT_{11}(s)$, corresponding to transverse magnetisation.[3] How will this evolve under the quadrupole interaction? For simplicity, let us start with the case $I = 1$. Figure 3.11(a) shows the relevant precession diagram, while Table 3.7 gives the

---

[3]Note that $T_{11}(s) = \frac{1}{\sqrt{2}}[T_{11} + T_{1-1}]$.

commutator algebra on which it is based. From the latter, it may easily be shown that, on resonance,

$$T_{11}(s) \xrightarrow{PT_{20}} T_{11}(s)\cos(\omega_Q t)$$
$$+i(T_{21} - T_{2-1})\sin(\omega_Q t)$$

$$(11.8)$$

where $\omega_Q = \sqrt{\frac{3}{2}}P$. The second term in the evolved state is unobservable by Faraday detection. Hence the observable, I_y, is modulated by $\cos(\omega_Q t)$, corresponding to spectral contributions at $\pm\omega_Q$. In consequence a doublet is observed with splitting $2\omega_Q$. On this basis we could replace P in eqn 11.6 by $2\omega_Q/\sqrt{6}$.

For $I = 1$, evolution of I_y under the quadrupole interaction generated a state involving a superposition of $T_{2\pm1}$. But for higher-spin quantum numbers, while the quadrupole beat frequency remains the same, the density matrix evolution is more complex, with tensor states up to rank $k = 2I$ being generated [11]. The amplitude of these maximum rank states, $A^{1,2I}(\omega_Q t)$ are given in reference [11], up to $I = 5/2$. Under the quadrupolar interaction T_{20}, these higher-rank states, with correspondingly higher orders, q, will precess at multiples of ω_Q, so that for a spin quantum number I there are $2I$ spectral terms centred on the resonant frequency, all separated by $2\omega_Q$.

Order of quantum coherence

To understand the effect of any subsequent RF pulse, we need to return to eqn 3.65. Hard RF pulses result in simple rotations. A pulse of flip angle θ and phase ϕ with respect to the y-axis of the rotating frame transforms a spherical tensor as

$$T_{kq}(s) \xrightarrow{(\theta)_\phi} \sum_{q'=-k}^{k} d_{qq'}^{(k)}(\cos\theta)T_{kq'}\exp(-i\Delta q\phi)$$

$$(11.9)$$

$d_{qq'}^{(k)}(\cos\theta)$ being an element of the Wigner rotation matrix, and $\Delta q = q' - q$. Hence the rotation induced by an RF pulse can change the order, but not the rank, of the tensor. However, the orders generated range up to the value of the rank, each one a p-quantum coherence with p given by the q' value for that term. In the same manner that the simple two-spin dipolar interaction was used to generate double-quantum coherence, the combination of evolution under the quadrupole interaction followed by a subsequent RF pulse can generate states of multiple quantum coherence ranging up to $2I$. And of course, in accordance with eqn 3.71, such a state of coherence will precess under the Zeeman interaction of a magnetic field gradient pulse at a rate $2I$ times faster than for simple magnetisation.

The pulse sequences used to generate multiple-quantum coherence in quadrupolar systems are the same as those shown in Fig. 11.1 for the dipolar interaction. Van Dam *et al.* [12] have analysed the transfer efficiency to various coherence orders for the case where the evolution time τ following the first RF pulse is set to $\omega_Q\tau = \pi/2$. Using the relevant Wigner rotation matrix elements, they calculate the transfer efficiencies, $A^{1,2I}(\omega_Q t)$, and $A^{2I,1}(\omega_Q t)$, to generate the maximum rank states, and to return, following the final mixing pulse to transverse magnetisation, and find, for $I = 1, \frac{3}{2}, 2, \frac{5}{2}, 3$, and $\frac{7}{2}$, efficiencies of $1, \frac{9}{10}, \frac{4}{5}, \frac{5}{7}, \frac{9}{14}$, and $\frac{7}{12}$.

Example for $I = \frac{3}{2}$

Van Dam *et al.* demonstrated the measurement of diffusion using a three-quantum coherence in a spin-$\frac{3}{2}$ system. Their sample comprised macroscopically ordered fibres of DNA in which lithium ions diffused. The single-quantum single-pulse spectrum of this system gives rise to a ^{7}Li quadrupolar spectrum consisting of three peaks with a quadrupolar line-splitting of $105\,\mathrm{Hz}$. Figure 11.3 compares the result of a single-

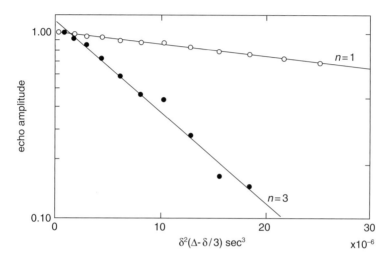

Fig. 11.3 NMR diffusion experiment for all $n = 1$- and $n = 3$-quantum states for ^{7}Li ions diffusing in a sample comprising oriented DNA fibres. The straight lines are linear least-squares fits to the data. (Adapted from van Dam *et al.* [12].)

quantum PGSE NMR experiment (open circles) with that obtained using the maximum available quantum coherence, $n = 3$ (closed circles), an experiment they refer to as total quantum PGSE or TQPGSE. Using the same diffusion coefficient for both experiments, a ratio of 2.7 is found between the square root of the slopes, close to the theoretical ratio of 3.

11.1.3   Scalar couplings and heteronuclear states

The limitations of both dipolar and quadrupole coupling approaches to the generation of multiple-quantum states are obvious. Many nuclei of interest are spin-$\frac{1}{2}$, ruling out the quadrupole route, and many systems of interest have no orientational order, making them unamenable to both quadrupolar and dipolar pathways. However, the the remaining Hamiltonian term that is also bilinear in the spin operator is the scalar coupling $2\pi J\mathbf{I}_i \cdot \mathbf{I}_j$. This interaction, which is effective in most molecules of more than rudimentary structure (the latter, sadly, including water), provides possible routes to quantum coherences and, because it acts in both heteronuclear and homonuclear mode, and like the dipolar interaction permits the formation of zero-quantum states, offers a greater range of possibilities. One might imagine that zero-quantum coherence would be unlikely to offer amplification of the effective precession rate under the influence of a gradient pulse. However, as we shall see, in heteronuclear mode, such a gain is indeed possible.

Homonuclei

For the weak-coupling case, we may use the secular form, $2\pi J I_{iz} I_{jz}$, of the two-spin homonuclear scalar coupling and our analysis precisely mirrors the discussion in Section 11.1.1. However, greater coupling complexity exists in the case of intramolecular J-couplings, a nice example being the AMX spin system of ^{19}F nuclei in the molecule 1,1,3-trichloro-2,2,3-trifluorocyclobutane, as studied by Kay and Prestegard [13]. In this system the geminal pair has a large splitting of 200 Hz, while a weaker coupling exists between the geminal and vicinal ^{19}F.

When homonuclear couplings are used to generate two-quantum coherence, the variant of the multi-quantum PGSE sequence shown in Fig. 11.1(b) is desirable, the 180° RF pulses in the evolution period (along with that in the multiquantum echo) acting to refocus heteronuclear couplings and chemical shifts. For the molecule used in reference [13], and given suitable phase cycling, the two J-couplings lead, at the end of the preparation period τ, to a double-quantum state with amplitudes determined by J_{12} as well as J_{13}, and J_{23}, i.e.

$$\rho_{2QT} = (2I_{1x}I_{2y} + 2I_{1y}I_{2x})\cos(\pi J_{12}\tau)\left(\cos(\pi J_{13}\tau) + \cos(\pi J_{23}\tau)\right) \tag{11.10}$$

where $i = 1, 2$ refers to the geminal fluorines, while $i = 3$ refers to the vicinal ^{19}F in this molecule. Given $J_{12} \gg J_{13}, J_{23}$, maximum amplitude is achieved for $\tau = 1/2J_{12}$, leading to a double-quantum state with amplitude $(\cos(\pi J_{13}/2J_{12}) + \cos(\pi J_{23}/2J_{12}))$. Such subtle modulation effects will generally be a feature of multiple-quantum coherence generated by homonuclear couplings, each molecule requiring its own density matrix evolution analysis.

Heteronuclei

With heteronuclear couplings used to generate multi-quantum states, an interesting questions arises. Just what is the multiplication factor for γ given that more than one magnetogyric ratio plays a role? This question was first addressed by Kuchel and Chapman [14], who demonstrated measurement of diffusion under double-quantum coherence in the heteronuclear coupled system ^{1}H-^{31}P in neutralised phosphoric acid, and later by Dingley *et al.* [15] for both double and zero-quantum coherence using the ^{1}H-^{15}N coupling for a protein in aqueous solution.

Suppose we consider a state of two-quantum coherence generated between a nucleus I and a heteronucleus S, namely $I_x S_y + I_y S_x$, and allow each nucleus to to evolve under its imposed Zeeman interactions, $\gamma_I B I_z$, and $\gamma_S B S_z$, to be acted on by 180° RF pulses for each nucleus, and to evolve under a differing Zeeman field B', such as might result from the fields resulting from applied gradient, with intervening translational motion. We could write

$$I_x S_y + I_y S_x$$

$$\xrightarrow{-(\gamma_I B I_z + \gamma_S B S_z)} \quad (I_x \cos\phi_I - I_y \sin\phi_I)(S_y \cos\phi_S + S_x \sin\phi_S)$$
$$+(I_y \cos\phi_I + I_x \sin\phi_I)(S_x \cos\phi_S - S_y \sin\phi_S)$$

$$\xrightarrow{-(\pi/2)(I_y + S_y)} \quad (-I_x \cos\phi_I - I_y \sin\phi_I)(S_y \cos\phi_S - S_x \sin\phi_S)$$
$$+(I_y \cos\phi_I - I_x \sin\phi_I)(-S_x \cos\phi_S - S_y \sin\phi_S)$$

$$\xrightarrow{-(\gamma_I B' I_z + \gamma_S B' S_z)} \quad (-I_x(\cos\phi_I \cos\phi_I' + \sin\phi_I \sin\phi_I')$$
$$-I_y(\sin\phi_I \cos\phi_I' - \cos\phi_I \sin\phi_I'))$$
$$\times(S_y(\cos\phi_S \cos\phi_S' + \sin\phi_S \sin\phi_S')$$
$$-S_x(\sin\phi_S \cos\phi_S' - \cos\phi_S \sin\phi_S'))$$
$$+(I_y(\cos\phi_I \cos\phi_I' + \sin\phi_I \sin\phi_I')$$
$$-I_x(\sin\phi_I \cos\phi_I' - \cos\phi_I \sin\phi_I'))$$
$$\times(-S_x(\cos\phi_S \cos\phi_S' + \sin\phi_S \sin\phi_S')$$
$$-S_y(\sin\phi_S \cos\phi_S' - \cos\phi_S \sin\phi_S'))$$

$$= \quad -(I_x S_y + I_y S_x)\cos(\Delta\phi_I + \Delta\phi_S)$$
$$+(I_x S_x - I_y S_y)\sin(\Delta\phi_I + \Delta\phi_S) \tag{11.11}$$

where $\Delta\phi_I + \Delta\phi_S = \phi_I - \phi_I' + \phi_S - \phi_S'$. We see that the state of double-quantum coherence precesses at a rate given by the sum of the I and S spin Larmor frequencies, and that in the PGSE experiment the phase offset due to translational motion is the sum of that experienced by the I and S spins. It is as though the effective magnetogyric ratio of the heteronuclear double-quantum coherence state is $\gamma_S + \gamma_I$.

As an exercise, the reader could follow the same process for the heteronuclear zero-quantum coherence state $I_x S_y - I_y S_x$. Now we find that the effective magnetogyric ratio is $\gamma_S - \gamma_I$. By the way, there is no doubt that for heteronuclear scalar couplings, product operator algebra is more transparent than is possible using spherical tensors

The pulse sequence used by Kuchel and Chapman is shown in Fig. 11.4(a). It is based on the inverse DEPT sequence [16].[4] The experiment starts with an excitation of the heteronucleus, ^{31}P, labelled S, but with final detection on the more sensitive proton, here labelled I. In this example, a simple two-spin heteronuclear coupling, $2\pi J I_z S_z$, applies. Because of the markedly different Larmor frequencies in the heteronuclear case, this secular approximation is effectively exact. It is instructive to follow the density matrix through the time points, shown in Fig. 11.4(a), using the initial step of the phase cycle [14], $\phi_1 = x$, $\phi_2 = x$, $\phi_3 = y$, $\phi_4 = x$, $\phi_5 = x$, $\phi_6 = x$, $\phi_7 = -y$, acquisition phase $= y$. These states are as follows:

$$\rho_1 = I_z + S_z$$
$$\rho_2 = I_z + S_y$$
$$\rho_3 = I_z + S_y \cos(\pi J \tau_1) - 2I_z S_x \sin(\pi J \tau_1)$$
$$\rho_4 = I_y - S_y \cos(\pi J \tau_1) - 2I_y S_x \sin(\pi J \tau_1) \tag{11.12}$$

---

[4] Also known as heteronuclear multiple-quantum coherence (HMQC).

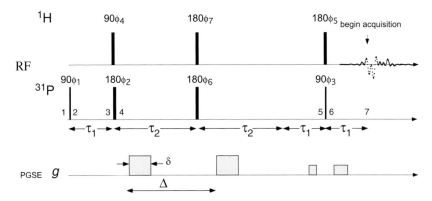

Fig. 11.4 Pulse sequence based on inverse DEPT used to measure diffusion via the double-quantum state generated by the heteronuclear ^{31}P-^{1}H J coupling. The phase labels follow those used in reference [14], and specific evolution points of the density matrix are marked with numerals. Note the selection of coherence order by careful choice of homospoil pulses. (Adapted from Kuchel and Chapman [14].)

By the choice $\tau_1 = 1/2J$, or any odd multiple of $1/2J$, a partial state of two-quantum coherence is generated,[5] so that $\rho_4 = I_y - 2I_y S_x$. The subsequent PGSE echo over the time period $2\tau_1$, involving $180°$ RF pulses for both I and S spins, serves to refocus any chemical shift dephasing over that period. Meanwhile the J coupling acts relentlessly on I_y over the total time period, while, of course, the two-quantum coherence state, $I_y S_x$ is invariant under the scalar coupling. Neglecting for the moment the Zeeman precession under the PGSE gradient pulses, this brings us to density matrix state 5 where

$$\rho_5 = I_y \cos(\pi J(2\tau_2 + \tau_1)) - 2I_x S_z \sin(\pi J(2\tau_2 + \tau_1)) - 2I_y S_x$$
$$\rho_6 = -I_y \cos(\pi J(2\tau_2 + \tau_1)) + 2I_x S_x \sin(\pi J(2\tau_2 + \tau_1)) + 2I_y S_z$$
$$\rho_7 = \text{``unobservables''} - I_x, \qquad (11.13)$$

The first term in ρ_6 is rendered unobservable by the unequal area homospoil pulses which sandwich the final RF pulse. The second term clearly remains unobservable while the final $I_y S_z$ term precisely evolves to the observable $-I_x$ over the duration τ_1.

Note that the density matrix term $I_y S_x$ existing over the PGSE echo period is only part of the full two-quantum coherence state. The remaining $I_x S_y$ component could be generated in the latter steps of a phase cycle, but this is unnecessary. We can instead regard $I_x S_y$ as a superposition of the double quantum state $I_x S_y + I_y S_x$ and the single quantum state $I_x S_y - I_y S_x$, and then ensure that remaining parts of the pulse sequence select for the former. The means by which this is achieved is via the use of a pair of homospoil gradient pulses before and after the final RF pulses. This results in precession of the $2I_y S_x + 2I_x S_y$ component, from which the final observable derives,

---

[5] $I_y S_x$ is a superposition of double- and zero-quantum coherence.

at rate given by an effective magnetogyric ratio, $\gamma_S + \gamma_I$, while the zero quantum part, $2I_yS_x - 2I_xS_y$ precesses at a rate determined by $\gamma_S - \gamma_I$. After the last RF pulse, the observable term precesses at γ_I. Thus for double quantum coherence selection, the ratio of the pulse areas before and after the last RF pulse needs to be $\gamma_I/(\gamma_S + \gamma_I)$.

Now we see the answer to our question. During the multiple-quantum coherence, and compared with a single-quantum PGSE experiment using spin I, the precession rate is enhanced, not by n^2 as in eqn 11.5, but by [14]

$$n^2 \rightarrow \left(\frac{\gamma_s + \gamma_I}{\gamma_I}\right)^2 \qquad (11.14)$$

Equation 11.14 poses an interesting problem. Suppose γ_s and γ_I have opposite sign? In that case we will be worse off by travelling via the double-quantum pathway. The solution, proposed by Dingley *et al.*, is to employ the state of zero-quantum coherence, where $\gamma_S - \gamma_I$ is the effective magnetogyric ratio. This opposite sign situation applies in ^{1}H-^{15}N experiments, highly relevant for isotopically labelled proteins.[6]

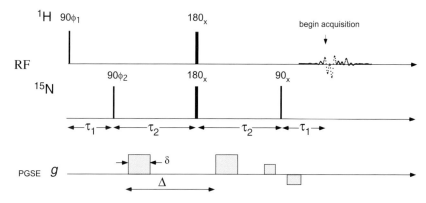

Fig. 11.5 Modified version of Fig. 11.4. (Adapted from Dingley *et al.* [15].) Note again the selection of coherence order by careful choice of homospoil pulses.

Their pulse sequence, shown in Fig. 11.5, is a simplified version of that used by Kuchel and Chapman, which also generates a mixture of double- and zero-quantum coherence. Using the methods outlined above, it is a straightforward process to derive the density matrix at various steps of the evolution pathway. The important point to note is that the final homospoil gradient selection filter is designed to select the zero-quantum state from the PGSE evolution period.

Remarks on multiple-quantum PGSE NMR
In the discussion so far, no reference has been made to that bugbear of PGSE NMR, the loss of signal due to transverse relaxation while the spin ensembles are being encoded for translational motion. In most of the examples discussed, where relaxation

[6]$\gamma_{1\,H} = 267.513$ rad s^{-1} T^{-1} and $\gamma_{15\,N} = -27.116$ rad s^{-1} T^{-1}.

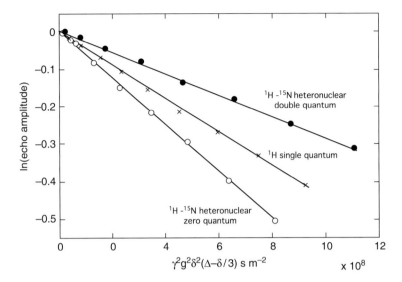

Fig. 11.6 Showing the sensitivity of different coherence orders to field-gradient pulses for ^{1}H-^{15}N heteronuclear states in selected ubiquitin resonances. Paradoxically, the zero-quantum coherence is most sensitive to diffusion because of the opposite signs of the ^{1}H and ^{15}N magnetogyric ratios. The ratios of the slopes obtained from these fits are 0.65 and 1.36 for DQ/SQ and ZQ/SQ, respectively, which are in reasonable agreement with the theoretical values of 0.81 and 1.21 (see eqn 11.14). (Adapted from Dingley *et al.* [15].)

rates are quoted, these are not significantly worse in double-quantum states, perhaps up to a factor of two, and in some cases longer [8]. However, the generation of these higher-order quantum states does require additional evolution time in the transverse plane and hence greater T_2 relaxation loss.

Before leaving this topic, it is worth remarking that uses of multiple-quantum coherence to measure diffusion, despite increase sensitivity in the precession, are, in addition to those discussed above, remarkably few [17, 18], probably reflecting the greater complexity of the pulse sequences needed, as well as the loss of signal due to additional relaxation and inefficiency in coherence transfer and selection.

11.2   Singlet states and time extension

One factor that limits PGSE NMR is the upper limit to available time window, Δ, for molecular migration, T_2 and the longer T_1 for spin and stimulated echoes respectively. For most liquids, T_1 values range from 100 ms for macromolecules in solution to around 10 s in de-oxygenated water. In nearly every case, T_1 relaxation arises from dipolar interactions, primarily intramolecular, since the internuclear proximity r affects the relaxation rate as r^{-6}, though more distant intermolecular dipolar interactions will also contribute. In what follows we see how to suppress the effect of intramolecular dipole–dipole interactions, given suitable conditions, and how this suppression can lend itself to the production of nuclear spin coherences sufficiently long-lived to allow much longer molecular migration times, Δ.

11.2.1 Product states, singlet–triplet states, and symmetry

Consider a pair of nuclear spins with scalar coupling J and chemical shift difference $\Delta\omega$. Let us assume they are homonuclei, calling them I_1 and I_2. Provided the chemical shift difference is much larger than the coupling, i.e. $\Delta\omega \gg 2\pi J\hbar$, then the product operator basis $\{|\uparrow\uparrow\rangle, |\uparrow\downarrow\rangle, |\downarrow\uparrow\rangle, |\downarrow\downarrow\rangle\}$ is approximately diagonal,[7] and the scalar coupling may be treated as a first-order perturbation, $2\pi J I_{1z} I_{2z}$. If on the other hand $\Delta\omega \ll 2\pi J\hbar$, the near-diagonal basis is the singlet–triplet set $\{|S_0\rangle, |T_{-1}\rangle, |T_0\rangle, |T_1\rangle\}$ where

$$|S_0\rangle = \sqrt{\tfrac{1}{2}} \, [|\uparrow\downarrow\rangle - |\downarrow\uparrow\rangle]$$
$$|T_{-1}\rangle = |\uparrow\uparrow\rangle$$
$$|T_0\rangle = \sqrt{\tfrac{1}{2}} \, [|\uparrow\downarrow\rangle + |\downarrow\uparrow\rangle]$$
$$|T_1\rangle = |\downarrow\downarrow\rangle$$

$$(11.15)$$

and the full scalar coupling $2\pi J \mathbf{I_1} \cdot \mathbf{I_2}$ is diagonal, with the good quantum numbers provided by the total angular momentum I, and its azimuthal projection m. $I = 0$ and $m = 0$ for the singlet and $I = 1$ and $m = 1, 0, -1$ for the triplet. Note that the singlet state is antisymmetric under exchange of spins 1 and 2, while the triplet states are symmetric. Further details concerning operator algebra for the product and singlet–triplet bases, and transformations between these bases can be found in reference [19].

Now consider the through-space intramolecular dipole–dipole interaction between spins 1 and 2. As pointed out by Caravetto and Levitt [20–22], this interaction is symmetric under exchange of the spins and therefore cannot couple singlet and triplet states. As a result, a non-equilibrium state containing a population difference between the singlet and triplet manifolds cannot be brought into equilibrium by intramolecular dipole–dipole relaxation alone, while relaxation between the various sub-manifolds of the triplet state is indeed possible, and is indeed the basis of T_1 relaxation for the two-spin system. When the chemical shift $\Delta\omega$ is non-zero, the singlet and triplet states are admixed, so that T_1 relaxation applies to the whole spin system. If $\Delta\omega = 0$ then the singlet state may be isolated, and its T_1 relaxation suppressed. What will remain will be a much slower relaxation process, T_S, perhaps involving the much weaker dipolar interactions from more distant spins.

Why then do we not normally see this slower relaxing component in coupled spin systems? First, the thermal equilibrium density matrix state, $I_{1z} + I_{2z}$, is a superposition of T_1 and T_{-1} with no S_0 component. Even if we could transfer coherence to S_0, the singlet state generates no net magnetisation and cannot be observed directly in a conventional NMR experiment.

---

[7]Some authors use α to represent spin-up $(\tfrac{1}{2})$ and β for spin-down $(-\tfrac{1}{2})$.

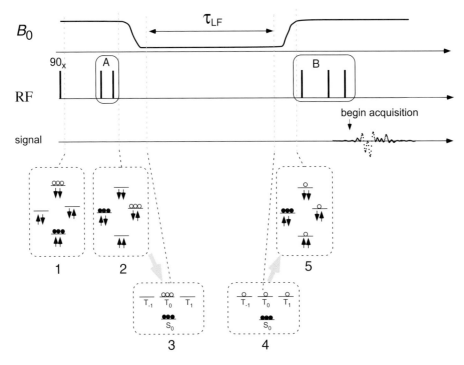

Fig. 11.7 Schematic idea of transfer of magnetisation to singlet state for-long term storage over time τ_{LF}, during which relaxation occurs at a rate T_S^{-1}, much slower than T_1^{-1}. In this version, the singlet state is isolated from the triplet by reducing the magnetic field strength. A suitable RF pulse train can achieve the same effect. Pulse segment A performs conversions from magnetisation states $I_{1x} + I_{2x}$ and $2I_{1x}I_{2z} + 2I_{1z}I_{2x}$ to the superposition of singlet and triplet states, while B returns the coherence to detectable magnetisation, $I_{1x}+I_{2x}$ or $I_{1y}+I_{2y}$. During τ_{LF}, relaxation occurs amongst the triplet sub-levels while the singlet is protected. (Adapted from Carravetta *et al.* [20].)

Levitt and co-workers discuss the potential for coherence transfer in and out of the singlet state. In systems where the coupled spins are in equivalent chemical sites ($\Delta\omega = 0$), there is no obvious mechanism to convert $I_{1z}+I_{2z}$ to the long-lived singlet state by population transfer to or from other spin states, since most experimental manipulations such as RF pulses, or evolution under to the spin Hamiltonian, act symmetrically on the two spins [22]. And so we turn our attention back to to systems where the coupled spins are in non-equivalent chemical sites ($\Delta\omega \neq 0$). Here the manipulation tools are available, but once the transfer to the singlet state has taken place, the chemical shift interaction breaks the spin exchange symmetry, thereby destroying any dynamic isolation.

The trick then is to be able to turn the chemical shift on and off at will. As Levitt *et al.* point out, this can be done by lowering the polarising field B_0 or by the use of refocusing the RF pulse train. The latter is clearly the easier route! Therein lies the potential for creating a slowly relaxing state.

11.2.2   Sequence for translation measurement

The first demonstration of diffusion measurement using a long-lived singlet state was by Bodenhausen, Cavadini, and co-workers [23], using the simple scalar two-spin system [21], 2-chloroacrylonitrile (i.e., $H_{I_1}H_{I_2}C{=}CRR'$ with $R = Cl$ and $R = CN$. The coherence evolution pathway chosen is to first excite a zero-quantum coherence, $ZQC_y = \frac{1}{2}(2I_{1y}I_{2x} - 2I_{1y}I_{2x})$, and allow it to evolve under the effects of the chemical shifts into $ZQC_x = \frac{1}{2}(2I_{1x}I_{2x} + 2I_{1y}I_{2y})$. RF irradiation at the midpoint between the I_1 and I_2 resonances leads to the suppression of the chemical shifts, while the $2\pi J\mathbf{I_1 \cdot I_2}$ interaction remains, converting the two-spin system into one where the two nuclei have, in effect, become magnetically equivalent. In the process, the two eigen-states $|\uparrow\downarrow\rangle$ and $|\downarrow\uparrow\rangle$ become degenerate and can be expressed as a superposition of a triplet state $|T_0\rangle\langle T_0|$ and a singlet state $|S_0\rangle\langle S_0|$, the zero-quantum coherence ZQC_x becoming $\frac{1}{2}(|T_0\rangle\langle T_0| - |S_0\rangle\langle S_0|)$, and then, over the irradiation period T, the triplet component relaxing as T_1 and the singlet relaxing much more slowly as T_S.

Figure 11.8 shows the pulse sequence used to generate the singlet state for the purpose of diffusion measurement. For the moment we will ignore the effect of the gradient pulses and focus solely on the coherence transfers, using a rotating frame with reference frequency ω_0 centred between the two resonant frequencies, ω_1 and ω_2. After the first $90^\circ_x - \tau_1 - 180^\circ_x - \tau_1$ RF pulse sandwich, with $2\tau = 1/2J$, chemical shift offsets have been refocused and the initial longitudinal magnetisation $I_{1z} + I_{2z}$ has been converted to $-2I_{1x}I_{2z} - 2I_{1z}I_{2x}$. Suppose the next interval, τ_2, is set to the 'difference precession interval' $\pi/(2|\omega_0 - \omega_{I_1}|) = \pi/(2|\omega_0 - \omega_{I_2}|) = \pi/(|\omega_{I_1} - \omega_{I_2}|)$. Then the transverse components undergo a precession in opposite directions through one-quarter of a full cycle and at the end of τ_2 one is left with the density matrix state $2I_{1y}I_{2z} - 2I_{1z}2I_{2y}$ and the 90°_y pulse at the end of the τ_2 interval converts this into $2I_{1y}I_{2x} - 2I_{1x}2I_{2y}$, the state of zero-quantum coherence, ZQC_y.

The next interval, τ_3 is half τ_2, i.e. $\tau_3 = \pi/(2|\omega_{I_1} - \omega_{I_2}|)$, allowing ZQC_y to precess into $ZQC_x = 2I_{1x}I_{2x} + 2I_{1y}2I_{2y}$, and, at the onset of RF irradiation, when magnetic equivalence is established, ZQC_x may be rewritten $\frac{1}{2}(|T_0\rangle\langle T_0| - |S_0\rangle\langle S_0|)$. During the irradiation for period T, the triplet state relaxes at rate T_1^{-1}, leaving only $-\frac{1}{2}|S_0\rangle\langle S_0|$ surviving. When the RF is switched off at the end of the period T, the chemical shift once more breaks the symmetry and $-\frac{1}{2}|S_0\rangle\langle S_0|$ may be rewritten in the product basis as $-\frac{1}{2}(I_{1x}I_{2x} + I_{1y}I_{2y}) - I_{1z}I_{2z} + \frac{1}{2}(I_{1x}I_{2y} - I_{1y}I_{2x})$. Note that this superposition is invariant under a Zeeman Hamiltonian and is therefore unaffected by any homospoil gradient pulse. By this means, unwanted coherences may be removed by a homospoil pulse, while the desired superposition state is preserved.

As an exercise to make the process explicit for the reader, we can follow the density matrix through the time points, shown in Fig. 11.8. These states are as follows:

$$\rho_1 = I_{1z} + I_{2z}$$
$$\rho_2 = I_{1y} + I_{2y}$$
$$\rho_3 = I_{1y}\cos(\pi J\tau_1) - 2I_{1x}I_{2z}\sin(\pi J\tau_1) + I_{2y}\cos(\pi J\tau_1) - 2I_{1z}I_{2x}\sin(\pi J\tau_1)$$
$$= -2I_{1x}I_{2z} - 2I_{1z}I_{2x}$$
$$\rho_4 = 2I_{1y}I_{2z} - 2I_{1z}I_{2y}$$

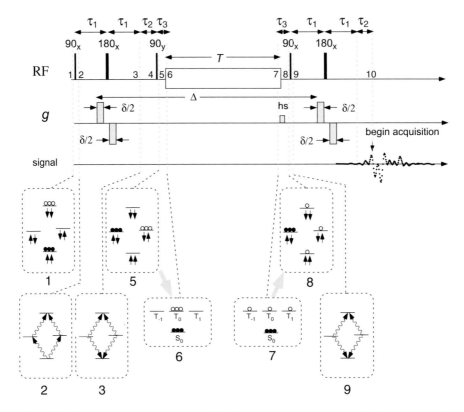

Fig. 11.8 Singlet-state single-quantum diffusion pulse sequence applicable to a molecule with a two-spin scalar coupling. The information about spatial localisation is stored in the form of singlet state populations with a relaxation time T_s during the interval T, while RF irradiation is applied at the midpoint of the Larmor frequencies of the spins. The intervals τ_1, τ_2, and τ_3 are chosen, as described in the text, to cause specific evolutions of the density matrix that generate the desired pathway. (Adapted from Cavadini *et al.* [23].)

$$\rho_5 = 2I_{1y}I_{2x} - 2I_{1x}I_{2y}$$
$$\rho_5 = 2I_{1x}I_{2x} + 2I_{1x}I_{2x} \tag{11.16}$$

RF irradiation during the period T, with the transmitter set at the midpoint between the resonances at ω_1 and ω_2, and using an RF field amplitude such that the nutation frequency is well in excess of the difference $\omega_1 - \omega_2$, essentially spin-locks the density matrix in the new Zeeman frame of the RF field, where now I_1 and I_2 have a common Larmor frequency.[8] The scalar coupling, $2\pi\mathbf{I_1} \cdot \mathbf{I_2}$ is unaffected by the RF field and remains to act on the spin system, with eigenstates of the pure singlet–triplet basis. Thus the ZQC_x state of ρ_6 must be transformed to the new basis:

---

[8]Yes, the chemical shift still remains, but since it is proportional to the RF field amplitude rather than the B_0 field amplitude, its absolute size is insignificant compared with the unmodified scalar coupling. Hence strong coupling conditions are established.

$$\rho_6 = \tfrac{1}{2}(|T_0\rangle\langle T_0| - |S_0\rangle\langle S_0|)$$
$$\rho_7 = -\tfrac{1}{2}|S_0\rangle\langle S_0|$$
$$\rho_7 = -\tfrac{1}{2}(I_{1x}I_{2x} + I_{1y}I_{2y}) - I_{1z}I_{2z} + \tfrac{1}{2}(I_{1x}I_{2y} - I_{1y}I_{2x})$$
$$\rho_8 = \tfrac{1}{2}(I_{1y}I_{2y} - I_{1x}I_{2x}) - I_{1z}I_{2z} + \tfrac{1}{2}(I_{1y}I_{2x} - I_{1x}I_{2y})$$
$$\rho_9 = -\tfrac{1}{2}(I_{1z}I_{2z} + I_{1x}I_{2x}) - I_{1y}I_{2y} - \tfrac{1}{2}(I_{1x}I_{2z} + I_{1z}I_{2x}) \tag{11.17}$$

With no more RF pulses to follow, only the last term of ρ_9 can contribute to observable magnetisation and so

$$\rho_9 = -\tfrac{1}{2}(I_{1x}I_{2z} + I_{1z}I_{2x})$$
$$\rho_{10} = \tfrac{1}{2}(I_{1y} + I_{2y}) \tag{11.18}$$

Finally, we consider the effect of the bipolar gradient pairs placed around the 180° RF pulses, thus giving them the same effective sign, with total pulse q-vector amplitude $\gamma g \delta$. The two pairs act to induce phase shifts $\phi = \mathbf{q} \cdot \mathbf{r}$ and $\phi' = \mathbf{q} \cdot \mathbf{r}'$, depending on the locations $\mathbf{r}$ and $\mathbf{r}'$ of the spin-bearing molecules at the respective times of encoding separated by time interval Δ. The final signal therefore is modulated by a ensemble average factor $\overline{\cos \phi \cos \phi'} = \tfrac{1}{2}\left(\overline{\cos(\phi - \phi')} + \overline{\cos(\phi + \phi')}\right)$, the latter term averaging to zero over the sample, while the first, $\tfrac{1}{2}\overline{\cos(\phi - \phi')}$, comprises the desired encoding for motion.

There many ways in which the singlet state may be generated and used for diffusion measurement. The pulse sequence of reference [23], which is analysed above, is one example. However, by understanding in detail the mechanism of one such example, the reader can easily invent other schemes [19, 24]. The crucial factor is that the process is relatively uncomplicated and easy to implement. The trick is to find the molecule with the desired dominant two-spin scalar coupling.

Finally, it is worth noting that the prerequisite for creating the singlet state, namely the existence of the intramolecular two-spin scalar coupling, also enables the generation of two-quantum coherence. By this means the enhanced diffusion sensitivity discussed in Section **??** and the enhanced diffusion timescale enabled by singlet state creation can be combined in the one experiment [23].

11.2.3 Measurement of diffusion via the singlet state

The first point to establish in a practical implementation of singlet-state time extension is the degree to which T_S exceeds T_1. In the $H_{I_1} H_{I_2} C{=}CRR'$ example of Cavadini *et al.* [23], the result is really quite spectacular, as can be seen in Fig. 11.9, with well over an order of magnitude achieved.

The singlet state method has been used effectively to measure the diffusion of the 2-chloroacrylonitrile at diffusive observation times of up to 20 s, an example of such a measurement obtained at $\Delta = 12$ s being shown in Fig. 11.10. Even with the loss of $\tfrac{1}{2}$ signal amplitude during the storage of the singlet state, the time extension available with decay at the rate T_S^{-1} offers a very significant advantage over traditional z-storage, where T_1 relaxation would normally result in a severe loss of signal amplitude over a storage period of a few seconds.

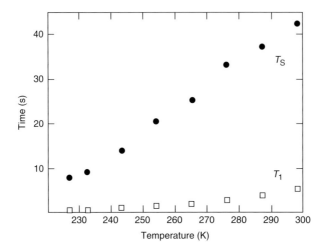

Fig. 11.9 Temperature dependence in the range from −46 to +25°C (227 to 298 K) of the spin-lattice relaxation time T_1 and the singlet state lifetime T_S of the protons I_1 and I_2 in 10 mM 2-chloroacrylonitrile, $H_{I_1}H_{I_2}C{=}CClCN$, dissolved in a mixture of deuterated DMSO-d6/D$_2$O. (Adapted from Cavadini *et al.* [23].)

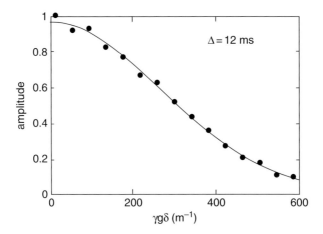

Fig. 11.10 Echo amplitude versus q-vector amplitude from the protons I_1 and I_2 in the 10 mM 2-chloroacrylonitrile $H_{I_1}H_{I_2}C{=}CClCN$/deuterated DMSO-d6/D$_2$O system of reference [23] obtained at 245 K and with diffusion observation time $\Delta{=}12$ ms. The Gaussian decay is fitted using the Stejskal–Tanner relation to yield a diffusion coefficient of $(5.7 \pm 0.2) \times 10^{-11}$ m^{2}s^2. (Adapted from Cavadini and Vasos [19].)

11.3 Intermolecular quantum coherence

It has long been known that in samples with large magnetisation density, non-linearities play a role in the Bloch equations, resulting in complex evolution of the spin magnetisation. A classic example of such a phenomenon is the effect of radiation damping [25], where induced emfs in the receiver coil are sufficiently large to interact back on the precessing magnetisation, thus causing a free-induction decay rate faster than that due to relaxation alone. However, in 1979 Deville *et al.* [1] discovered multiple echoes in ^{3}He NMR experiments which appeared to arise from intermolecular dipolar interactions. Even more remarkably, Warren and co-workers in 1993 demonstrated, in two-dimensional proton NMR experiments, the appearance of cross peaks between protons on molecules of completely different molecular species, indicative of intermolecular multiple-quantum transitions [3, 4]. Warren *et al.* argued that the origin of these intermolecular coherences was the same long-range dipolar field responsible for the appearance of multiple echoes following a simple two RF pulse spin-echo sequence, presenting in the process a convenient quantum description based on density matrix formalism.

The 1993 experiment [3, 4] that caused such surprise to the NMR community was dubbed by its authors 'CRAZED'[9], a suitable title given the level of consternation that resulted. A schematic example of the experiment is shown in Fig. 11.11, involving proton cross-peaks from benzene and chloroform molecules in a liquid mixture, found in a two dimensional correlation experiment in which gradient pulse filters are used to select for two-quantum coherence during the t_1 evolution period. The results astound for three specific reasons. First, there are no J-couplings that can cause magnetisation transfer between different molecular species; indeed for these molecules, there are no *intramolecular* J-couplings, because of the simple chemically equivalent nature of the hydrogens in each molecular species. Second, we are dealing with a liquid, in which rapid molecular tumbling is more than adequate at removing spin-pairwise intermolecular dipolar couplings at all length scales, a matter we shall address in more detail in Section 11.3.1. Third, we have to ask, how can there exist a state of two-quantum coherence after the first 90_x RF pulse, when no obvious bilinear spin term in the Hamiltonian is present to help create such a coherence? In any case, our experience suggests the need for at least two RF pulses sandwiching evolution under a bilinear secular term, if two-quantum coherence is to be created. CRAZED not only challenges some of our deepest assumptions but it provides a warning that we need to be alert to unexpected signals in NMR experiments in simple liquids, especially where magnetic field gradients are used. For this reason alone, these effects are of interest to experimenters carrying out PGSE NMR or NMR imaging experiments.

While the original multi-echo experiments were explained through an understanding of the role of non-linearities in the Bloch equations arising from the distant dipolar field, CRAZED, with its remarkable resemblance to quantum coherence effects, called out for fresh insight. As a consequence, a significant debate took place regarding the precise physical origin of all these experimental phenomena. While there was agreement that the additional demagnetising field arising from long-range dipolar interactions was central to any understanding of these effects, the dispute concerned whether these arose solely from classical Bloch equation non-linearities, or whether the effects were inherently quantum mechanical in origin. Thanks to a cooperative effort by several interested authors [5, 26, 27], we now understand that the effects

---

[9]Correlated 2-D spectroscopy Revamped by Asymmetric Z-gradient Echo Detection.

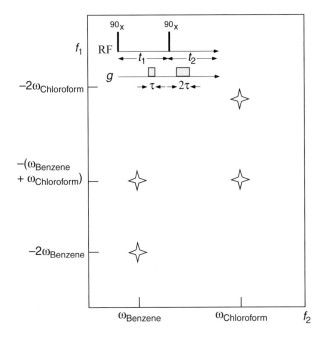

Fig. 11.11 CRAZED spectrum of a mixture of benzene and chloroform. The inset shows the simple two RF pulse sequence along with coherence-selecting homospoil gradient pulses. By conventional theory, this spectrum should be blank. Instead, there are peaks with all properties of intermolecular double-quantum peaks. (Adapted from Richter and Warren [7].)

of classical non-linearities and a description based on quantum coherence phenomena really are one and the same.

11.3.1   The quantum description

What made the quantum description of Warren *et al.* so appealing was its inherent simplicity, its connection with well-understood NMR density matrix operator algebra, and its ability to explain previously unobserved phenomena. In particular, the experimental demonstration of echo order selection by means of homospoiling gradient pulses of differing duration could be explained—exactly as is found in multiple-quantum coherence phenomena in high resolution NMR experiments in which bilinear terms in the spin Hamiltonian provide the means to explore different coherence pathways. At the heart of the physics is the existence of high-order terms in the equilibrium density matrix. Indeed, the breakdown of the usual high temperature approximation (HTA) is central to Warren's description. Furthermore the quantum operator formalism requires that we represent ensemble-averaged spin behaviour, namely the collective sum of pairwise dipolar interactions, in the same manner as we would represent a simple two-spin interaction through a single term in the spin Hamiltonian. These unusual requirements, which can be formally justified, allow for very familiar and NMR-relevant treatment of the intermolecular dipolar phenomena.

Examination of the high-temperature approximation
We start by considering the density matrix in thermal equilibrium,

$$\rho_{eq} = \frac{\exp(-\beta\mathcal{H})}{Tr\left(\exp(-\beta\mathcal{H})\right)} \tag{11.19}$$

where $\beta = 1/k_B T$. In the case of an N-spin system of coupled spin-$\frac{1}{2}$ particles, $\mathcal{H}$ is a $2N \times 2N$ matrix and is given by

$$\mathcal{H} = \sum_i^N \omega_i I_{iz} + \sum_{i,j}^N D_{ij} I_{iz} I_{jz} \tag{11.20}$$

where D_{ij} is the internuclear dipolar interaction, as defined in eqns 3.76 and 4.53, and given by

$$D_{ij} = \frac{\mu_0 \gamma^2 \hbar}{4\pi} \frac{1}{r_{ij}^3} \frac{1}{2} (1 - 3\cos^2\theta_{ij}) 3 I_{iz} I_{jz}$$
$$= 3\omega_{ij} I_{iz} I_{jz} \tag{11.21}$$

Equation 11.21 does not include all secular terms from the dipolar interaction Hamiltonian, but only the $I_{iz} I_{jz}$ associated with the formation of multiple echoes. Notice that conventionally we neglect these terms. In solids the sum of dipolar coupling terms for the i-th spin is generally 10^4 smaller than the Zeeman term, and in liquids some further 10^4 lower due to motional averaging. Thus dipolar interactions are usually ignored as far as the composition of the equilibrium density matrix is concerned, and it is generally sufficient to use a single-spin picture when calculating the equilibrium magnetisation. But when considering the sum of all possible intermolecular dipolar interactions, and especially at high polarising fields, this assumption requires closer examination.

In a single-spin picture, we would write the thermal equilibrium density matrix in the HTA as

$$\sigma_{eq} \approx \tfrac{1}{2}\left(1 + \beta\omega_i I_{iz}\right) \tag{11.22}$$

Hence for an N spin system, the density matrix becomes

$$\rho_{eq} = \sigma_{1eq} \otimes \sigma_{2eq} \otimes \sigma_{3eq} \ldots\ldots \otimes \sigma_{Neq}$$
$$= 2^{-N} \Pi_i \left(1 + \beta\omega_i I_{iz}\right) \tag{11.23}$$

This product involves N one-spin operators on the order of size $\gamma B_0/k_B T$, $N^2/2$ two-spin operators (on the order of size $(\gamma B_0/k_B T)^2$), and higher-order spin operators in succession. Despite the fact that $\gamma B_0/k_B T \sim 10^{-4}$ for protons at 10-T field strengths, the increasing multiplicity of higher-order terms by successive powers of N means that higher-order terms in the density matrix cannot be ignored without some further justification.

Conventionally we justify ignoring bilinear terms in the density matrix by the fact that all subsequent Zeeman interactions produce density matrix terms that remain bilinear and so cannot result in the NMR observables I_x or I_y. But the dipolar field introduces bilinear Hamiltonian terms that are capable of generating observable coherences from higher-order density matrix elements. Here we might argue that molecular self diffusion will cause such interactions to fluctuate and therefore average to zero, but this argument can only hold for spins that are in close proximity. Beyond some characteristic distance, such Brownian motion will not suffice, as we will see. Finally, we might hope that the $1/r^3$ dependence of D_{ij} would render the effect of distant spins insignificant. In fact this is not the case, especially when asymmetry of the surrounding dipolar magnetisation allows for a significant integral effect. Such asymmetry will always play a significant role when magnetic field gradients are used, and so these dipolar effects become of particular importance in PGSE NMR and NMR imaging.

Beyond the 'diffusion sphere'

Consider two particles, labelled 1 and 2, which migrate via Brownian motion over some fixed time, as illustrated in Fig. 11.12. If the particles are proximate, their internuclear vectors can undergo a significant reorientation due to the relative Brownian motion as seen in (a). In (b), where the spins are more distant, the same relative translational motion leads to a minor change in interparticle vector orientation, thus making motional averaging of the dipolar interaction much less likely. As pointed out by Warren *et al.*, molecular self-diffusion in liquids causes dipolar interactions to be averaged to zero for spins on nearby molecules, but beyond a diffusion length corresponding to

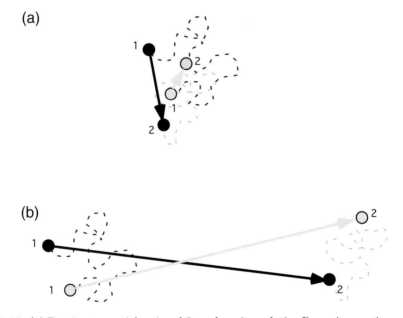

Fig. 11.12 (a) Proximate particles, 1 and 2, undergoing relative Brownian motion over some fixed time interval. (b) The same relative Brownian motion but for distant particles. Here the interparticle vectors reorient much less significantly.

the characteristic NMR timescale such averaging does not apply. That length is easily calculated by finding the distance such that the dipolar interaction strength in hertz at that spin separation equates to the inverse time required to diffuse that distance. For pairs of protons in water, the dipolar interaction strength at about 0.5 nm separation is on the order of 500 Hz, while the diffusion time for 0.5 nm is around 6×10^{-11} s, so that there is no doubt that the dipolar interaction is motionally averaged to zero by diffusion. At greater distances, as the pairwise dipolar interaction decreases as r^{-3}, while the diffusion time grows as r^2, strict motionally averaging requirements are even more strongly observed. But what makes the long-range intermolecular dipolar contribution behave differently than the simple two-spin picture is the sum of contribution from N spins. What counts therefore is the characteristic time associated with the dipolar field generated by the sample nuclear magnetisation, M_{eq}. This magnetic field strength is $\mu_0 M_{eq}$, and so the characteristic precession time for spin i in this field is $\tau_d = (\gamma \mu_0 M_{eq})^{-1}$. The equilibrium nuclear magnetisation is given in eqn 3.89 and for water protons at 300 K and 14.1 T (600 MHz) $M_{eq} = 0.045$ JT^{-1}m^{-3} and $\tau_d = 0.066$ s. The characteristic water molecule diffusion length for this time is thus on the order of 10 microns and within this length motional averaging caused by translational diffusion is effective.

Beyond that diffusion sphere radius, dipolar contributions from remote spins do contribute to the field experienced by any spin within that sphere, and the volume integral of $r^2 dr$ r over the r^{-3} dependence of the dipolar interaction ensures a logarithmic contribution, the magnitude of which depends on the ratio of the sample dimension to the diffusion length.

For spherical samples the angular dependence of the dipolar interaction in the integral contribution from all spins ensures that the sum of the dipolar field vanishes. However, if the spherical symmetry is broken, for example by application of a magnetic field gradient, then the effects of the dipolar field will be felt by the spins and the conversion of higher-order coherences into observable magnetisation becomes possible.

11.3.2 Multi-echo and CRAZED phenomena

Figure 11.13 shows two versions of a simple two 90° RF pulse sequence. The first involves a steady background gradient and the second is a two-dimensional multiple-quantum experiment, with independent time domains t_1 and t_2, using pulsed gradients of differing time interval to select for higher-order coherence in the evolution period t_1. This latter sequence we met earlier in the CRAZED experiment.

The generation of the observables

Let us begin by trying to understand the multi-echoes using the quantum mechanical density matrix approach. We can begin by stating the obvious: thermal equilibrium density matrix terms linear in I_z contribute to a primary echo at $t_2 = t_1$. So let us see how the gradient and dipolar interactions combine with density matrix terms bilinear in I_z to play a role in making multiple echoes possible. Ignoring the effects of spin–spin relaxation, we start by considering the evolution of an element of the second-order term $(\gamma B_0 / k_B T)^2 I_{iz} I_{jz}$, arising in the equilibrium density matrix, ρ_{eq}.

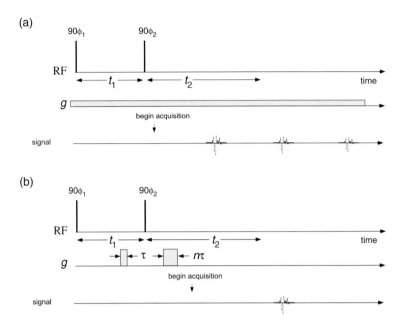

Fig. 11.13 (a) Pulse sequence for demonstrating the multiple spin-echo effect due to intermolecular dipolar interactions experiment. ϕ_1 and ϕ_2 denote the phase of the RF pulses. g is a steady background gradient in the z direction unless otherwise stated. The multiple spin-echo signals occur at $t_2 = t_1$, $t_2 = 2t_1$, $t_2 = 3t_1 \ldots$ (b) The nth-order CRAZED experiment [3], in which homospoil gradient pulses of different area can be used to select specific echoes according to their n quantum origin during the period t_1. τ is the period of the first gradient pulse, n is an integer multiple, and the the CRAZED signal occurs at $t_2 = nt_1$. (Adapted from Minot *et al.* [28].)

To clearly distinguish the Larmor and dipolar precessions, we use the symbols w_i and D_{ij}, respectively, noting that the dipolar interaction causes I_{iy} to evolve to $\cos(\frac{1}{2}D_{ij}t)I_{iy} + \sin(\frac{1}{2}D_{ij}t)I_{ix}I_{jz}$, while the Larmor precession takes I_{iy} to $\cos(w_i t)I_{iy} + \sin(w_i t)I_{ix}$. For the moment, we will allow that the Larmor precession may be associated with the intrinsic chemical shift of the spin where a molecular mixture exists, as well as from the local position in a magnetic field gradient.

The first 90°_x pulse converts $I_{iz}I_{jz}$ to $I_{iy}I_{jy}$, then via the Zeeman and dipolar interactions to

$$\begin{aligned}
\rho(t_{1-}) = \cos^2(\tfrac{1}{2}D_{ij}t_1)&[I_{iy}I_{jy}\cos(w_i t_1)\cos(w_j t_1) + I_{ix}I_{jy}\sin(w_i t_1)\cos(w_j t_1) \\
&+ I_{iy}I_{jx}\cos(w_i t_1)\sin(w_j t_1) + I_{ix}I_{jx}\sin(w_i t_1)\sin(w_j t_1)] \\
+ \sin(\tfrac{1}{2}D_{ij}t_1)&\cos(\tfrac{1}{2}D_{ij}t_1)[2I_{ix}I_{jz}I_{jy}\cos(w_i t_1)\cos(w_j t_1) \\
&- 2I_{iy}I_{jz}I_{jy}\sin(w_i t_1)\cos(w_j t_1) \\
&+ 2I_{ix}I_{jz}I_{jx}\cos(w_i t_1)\sin(w_j t_1) - 2I_{iy}I_{jz}I_{jx}\sin(w_i t_1)\sin(w_j t_1)]\ldots
\end{aligned}$$

$$(11.24)$$

Since the intermolecular dipolar interaction is weak, and $D_{ij}t_1, D_{ij}t_2 \ll 1$, only terms involving one $\sin(\frac{1}{2}D_{ij}t_2)$ factor need be considered as making any significant contribution to the signal.

Following the second 90_x° RF pulse the density matrix becomes

$$
\begin{aligned}
\rho(t_{1+}) = {}& \cos^2(\tfrac{1}{2}D_{ij}t_1)[I_{iz}I_{jz}\cos(\omega_i t_1)\cos(\omega_j t_1) + I_{ix}I_{jz}\sin(\omega_i t_1)\cos(\omega_j t_1) \\
&+ I_{iz}I_{jx}\cos(\omega_i t_1)\sin(\omega_j t_1) + I_{ix}I_{jx}\sin(\omega_i t_1)\sin(\omega_j t_1)] \\
&- \sin(\tfrac{1}{2}D_{ij}t_1)\cos(\tfrac{1}{2}D_{ij}t_1)[-2I_{ix}I_{jy}I_{jz}\cos(\omega_i t_1)\cos(\omega_j t_1) \\
&+ 2I_{iz}I_{jy}I_{jz}\sin(\omega_i t_1)\cos(\omega_j t_1) \\
&- 2I_{ix}I_{jy}I_{jx}\cos(\omega_i t_1)\sin(\omega_j t_1) + 2I_{iz}I_{jy}I_{jx}\sin(\omega_i t_1)\sin(\omega_j t_1)]... \quad (11.25)
\end{aligned}
$$

Subsequently, over the period t_2, the density matrix evolves to over 100 different terms. However, the only terms capable of generating observable magnetisation under the influence of the dipolar interaction are those involving products such as $I_{ix}[I_{jz}I_{jz}...]$ or $I_{iy}[I_{jz}I_{jz}...]$. Terms involving $I_{ix}I_{jz}I_{jz}$ will generate $n = 3$ echoes, $I_{ix}I_{jz}I_{jz}I_{jz}$, $n = 4$ and so on. Let us focus on $n = 2$. For example, consider the term $\cos^2(\frac{1}{2}D_{ij}t_1)I_{ix}I_{jz}\sin(\omega_i t_1)\cos(\omega_j t_1)$ in eqn 11.25. Under the dipolar interaction and the Larmor precessions it evolves during t_2 to

$$
\begin{aligned}
\rho_{visible}(t_1, t_2) = \sum_{i=1}^{N}\sum_{j=1}^{N} \sin(\tfrac{1}{2}D_{ij}t_2)\cos^2(\tfrac{1}{2}D_{ij}t_1)[I_{iy}\sin(\omega_i t_1)\cos(\omega_j t_1)\cos(\omega_i t_2) \\
+ I_{ix}\sin(\omega_i t_1)\cos(\omega_j t_1)\sin(\omega_i t_2)] \quad\quad (11.26)
\end{aligned}
$$

To identify the relevant precession frequencies of contributing terms, the products of sines and cosines need to be factorised in terms of sum and difference frequencies $\omega_i + \omega_j$ and $\omega_i - \omega_j$.

The magnetic field gradient and nuclei with common chemical shift

Suppose we have a sample of a common molecular species—for simplicity, one with a single chemical shift, such as water or polydimethylsiloxane, such that Larmor frequency differences are entirely due to applied magnetic field gradients. What we seek is an understanding of how the $n = 2$ echo arises (and by implication all higher-order echoes) and what will govern its amplitude. To appreciate the role of the applied magnetic field gradient, we allow that the Larmor frequencies ω_i and ω_j arise from the different locations of spins i and j. For a gradient g directed along an axis s, then spin i at position s_i will have a Larmor frequency $\omega_i = \gamma(B_0 + gs_i)$. Signal from transverse magnetisation terms in eqn 11.26, modulated by sinusoidal functions of absolute position, $\sin(\omega_i + \omega_j)t$ or $\cos(\omega_i + \omega_j)t$, will be averaged to zero. Also averaged to zero will be odd functions of relative position $\sin(\omega_i - \omega_j)t$. By contrast, signal from terms modulated by even functions of relative position, $\cos(\omega_i - \omega_j)t) = \cos \gamma g(s_i - s_j)t$, will not average to zero because the coefficient $D_{ij}t_2$ introduces the factor r_{ij}^{-3}, thus giving the sinusoids a decay envelope that ensures that even functions of relative position have a non-zero integral over the sample space.

The subtlety of the multiple-quantum coherences

At first sight, the precession rate during the t_1 period of the signal-contributing term in eqn 11.26 appears to involve a double-quantum coherence $I_{ix}I_{jy}$, whereas during the t_2 period the term precesses as a single-quantum coherence. This alone would suggest the need for a factor of two difference in gradient time integrals in the t_1 and t_2 periods if an echo is to be formed. In truth, the matter is a little more subtle. Unlike the case for scalar coupled spins, the intermolecular dipolar interaction means that spins i and j are at very different locations and have very different Larmor frequencies in the presence of the gradient. It is in the decomposition of the $\sin(\omega_i t_1)\cos(\omega_j t_1)\sin(\omega_i t_2)$ factor, and the requirement that the signal arises only from terms modulated by even functions of relative position, that the relationship between t_2 and t_1 is set. For the constant gradient experiment, as in Fig. 11.13(a), this means that a $\frac{1}{4}\cos(\omega_i t_1 + \omega_j t_1 - \omega_j t_2)$ term reduces to the required 'slow'-oscillatory $\cos(\omega_i - \omega_j)t_1$ form when $t_2 = 2t_1$.

CRAZED

CRAZED is a two-dimensional experiment where t_1 and t_2 are independently varied, and we need to allow that the separate effect of different chemical shifts (say ω_{0i} and ω_{0j}) in the molecular mixture is accounted for, as well as the effect of the homospoil gradient pulses. Allowing for both gradients and chemical shift, now our $\frac{1}{4}\cos(\omega_i t_1 + \omega_j t_1 - \omega_j t_2)$ term decomposes and factorises (keeping only the terms modulated by even functions of relative position), as

$$
\begin{aligned}
\cos(\omega_i t_1 + \omega_j t_1 - \omega_j t_2) &= \cos(\omega_{0i} t_1 + \omega_{0j} t_1 + \gamma g s_i \tau + \gamma g s_j \tau - \omega_{0j} t_2 - \gamma g s_j 2\tau) \\
&= \cos(\omega_{0i} t_1 + \omega_{0j} t_1 - \omega_{0j} t_2 + \gamma g (s_i - s_j)\tau) \\
&\rightarrow \cos(\omega_{0i} t_1 + \omega_{0j} t_1) \cos(\omega_{0j} t_2) \cos(\gamma g (s_i - s_j)\tau)
\end{aligned}
$$

$$(11.27)$$

And, of course, similar terms arise with ω_{0i} as the frequency in the t_2 dimension. The final factor governs the signal amplitude in the t_2 domain, and has required the factor of two difference in gradient pulse area. It is just as the condition for a double-quantum coherence experiment, but with the subtlety of a maximum formed by integration of a slowly oscillatory phase factor over macroscopic sample dimensions, rather than a simple phase cancellation. For the chemical shift locations in the f_1 dimension, the sum of molecular chemical shifts arises, exactly as seen in Fig. 11.11, while in the acquisition, f_2 dimension, single frequency oscillation occurs. The resemblance to a double-quantum COSY experiment is compelling. But underlying this effect are not single-spin operators, but pseudo-density matrix operators made up of macroscopic ensemble averages.

The intermolecular integral and gradient symmetry breaking

Now we are in a position to calculate the total contribution to the higher-order signal components represented by the visible terms from eqn 11.26 that emerge with gradient phase modulation and dipolar amplitude as $I_{xi}\cos(\gamma g t_1 (s_i - s_j)D_{ij}t_2$. Suppose we integrate these contributions over the entire sample outside the diffusion sphere around spin i. Then the contribution will be determined by the factor

$$\int_{volume} \cos(\gamma g t_1 \mathbf{r} \cdot \hat{\mathbf{s}}) \frac{3\cos^2\theta - 1}{r^3} r^2 \sin\theta d\theta d\phi dr \qquad (11.28)$$

where $\hat{\mathbf{s}}$ is the unit vector describing the gradient direction, and $\mathbf{r}$ defines the position of the remote spin in polar coordinates, where θ is taken with respect to the polarising field direction, $\hat{\mathbf{z}}$. The term $(3\cos^2\theta - 1)/r^3$ introduces the spatial dependence of D_{ij}. It can be shown [1] that eqn 11.28 reduces to

$$[3(\hat{\mathbf{s}} \cdot \hat{\mathbf{z}})^2 - 1] \int_{r_{min}} F(\gamma g t_1 r) dr \qquad (11.29)$$

where the function F contains all r dependence, and has an integral that is greatest when $\gamma g t_1 r$ ranges between two and four in a large fraction of the sample, i.e. when the gradient imparts to the spins a helical phase twist with wavelength smaller than the sample dimensions but larger than the diffusion radius, r_{min}.

11.3.3   Experimental verification of multiple echo effect

The smaller the diffusion radius, r_{min}, the larger the integral of F. Hence larger multi-order echoes from intermolecular dipolar effects are observed for more slowly diffusing liquids. Figure 11.14 shows the results obtained using a sample of 5000-Da polydimethylsiloxane at 7 T polarising field strength [28]. Echoes out to order four are clearly visible. The phases of the higher-order echoes vary, in a predictable manner, according to the density matrix terms from which they originate.

Figure 11.15 shows the result of the CRAZED 'multi-quantum' experiment where homospoil gradient pulses are used to select for echo order, according to the value n chosen for the ratio of the gradient pulse integral during t_2 to that of the period t_1. Note that specific echo orders can also be selected by appropriate phase cycling, and examples of these cycles up to $n = 3$ are given in reference [28].

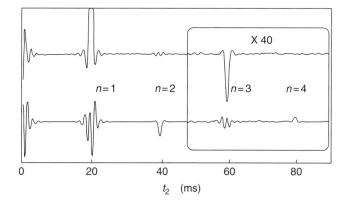

Fig. 11.14 Time-domain signals from the multi-echo experiment with steady gradient (Fig. 11.13(a)) where $\phi_1 = \phi_2 = x$ and $g = 0.04\,\text{mT m}^{-1}$. The upper line is the imaginary channel and the lower line is the real channel. Data within the box are magnified 40 times. (Adapted from Minot *et al.* [28].)

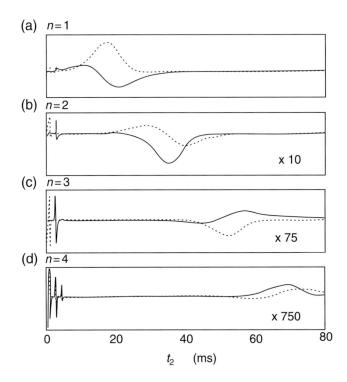

Fig. 11.15 Time-domain signals from the nth order multi-echo (CRAZED) experiment, with pulsed gradient (Fig. 11.13(b)). The solid lines show the real channel data and the dashed lines show the imaginary channel. Magnification relative to (a) is shown in (b) to (d). Here $\phi_1 = \phi_2 = x$, $t_1 = 20\,\mathrm{ms}$ and the gradient duration τ is $2\,\mathrm{ms}$. (Adapted from Minot *et al.* [28].)

Note the role of the factor $[3(\hat{\mathbf{s}}\cdot\hat{\mathbf{z}})^2-1]$. Changing the applied gradient direction from $\hat{\mathbf{z}}$ to one transverse to $\hat{\mathbf{z}}$ changes the sign of the echo. And of course, when $\hat{\mathbf{s}}\cdot\hat{\mathbf{z}}=1/3$, in other words when the gradient direction is set to the magic angle $\theta=54.4°$, the symmetry-breaking effect of the gradient disappears and any intermolecular dipolar echoes must arise from sample shape alone. Here the damping effect in the integral of F ensures that sample shape will not contribute to the formation of multi-echoes, provided the sample size is much larger than the gradient-induced wavelength. By this means, it is possible to avoid the appearance of unwanted intermolecular dipolar echos in PGSE NMR experiments by using a sufficiently large homospoil gradient oriented at the magic angle.

11.3.4   Probing structure via intermolecular coherences

The role of the function F in determining the amplitude of the higher-order echo signal suggests a possible use in structural determination [29–32]. Returning to eqn 11.28, we can see that it assumes a uniform sample magnetisation modulated in phase by the applied gradient. Suppose instead we allowed for a more general contribution of magnetisation phase and amplitude which depended on sample structure. In that case, we need to rewrite the integral allowing for the sample density $\rho(\mathbf{r})$ at the origin of the i spin and $\rho(\mathbf{r}')$ at the remote j spin, separately integrating each set of coordinates. Hence we could rewrite eqn 11.28 as

$$\int_{volume} d\mathbf{r}d\mathbf{r}' \rho(\mathbf{r})\rho(\mathbf{r}') \cos(\gamma g t_1 (\mathbf{r} - \mathbf{r}') \cdot \hat{\mathbf{s}}) \frac{3\cos^2\theta_{rr'} - 1}{|\mathbf{r} - \mathbf{r}'|^3} \qquad (11.30)$$

By adjusting the spatial period of the gradient-induced phase shift, the higher-order echo signal can be used to probe correlations in structure $\rho(\mathbf{r})$ and $\rho(\mathbf{r}')$ at differing spatial separations, $\mathbf{r} - \mathbf{r}'$. The inverse problem to be solved is not trivial, but the potential advantage of the method is that, unlike conventional NMR microscopy, the entire sample contributes to the signal, as the resolution is improved by increasing the gradient wavevector, thus maintaining high signal-to-noise ratio. However, the lower limit to resolution is still constrained by the diffusion sphere, and so in aqueous systems, is around 10 microns.

The theoretical description and mathematical details of the problem are complex, though somewhat assisted by working in the Fourier space of $\mathbf{r}$, in which case expression 11.30 can be written in terms of overlap integrals. Complementary treatments of the problem are described in detail in references [31] and [29]. While the method clearly has some potential, the connection between sample structure and the gradient wavevector-dependent multi-echo signal is not particularly transparent. However, it can be useful in demonstrating structural anisotropy in a porous material [31].

References

[1] G. Deville, M. Bernier, and J. M. Delrieux. NMR multiple echoes observed in solid 3-He. *Phys. Rev. B*, 19:5666, 1979.

[2] M. A. McCoy and W. S. Warren. 3-quantum nuclear magnetic resonance spectroscopy of liquid water-intermolecular multiple quantum coherence generated by spin cavity coupling. *J. Chem Phys.*, 93:858, 1990.

[3] Q. H. He, W. Richter, S. Vathyam, and W. S. Warren. Intermolecular multiple-quantum coherences and cross correlations in solution nuclear magnetic resonance. *J. Chem. Phys.*, 98:6779, 1993.

[4] W. S. Warren, W. Richter, A. H. Andreotti, and B. T. Farmer. Generation of impossible cross-peaks between bulk water and biomlecules in solution NMR. *Science*, 262:2005, 1993.

[5] W. S. Warren, S. Lee, W. Richter, and S. Vathyam. Correcting the classical dipolar demagnetizing field in solution NMR. *Chemical Physics Letters*, 247:207, 1995.

[6] S. Lee, W. Richter, S. Vathyam, and W. S. Warren. Quantum treatment of the effects of dipole-dipole interactions in liquid nuclear magnetic resonance. *J. Chem. Phys.*, 105:874, 1996.

[7] W. Richter and W. S. Warren. Intermolecular multiple quantum coherences in liquids. *Concepts in Magnetic Resonance*, 12:396, 2000.

[8] J. F. Martin, L. S. Selwyn, R. R. Void, and R. L. Void. The determination of translational diffusion constants in liquid crystals from pulsed field gradient double quantum spin echo decays. *J. Chem. Phys.*, 76:2632, 1982.

[9] D. Zax and A. Pines. Study of anisotropic diffusion of oriented molecules by multiple quantum spin echoes. *J. Chem. Phys.*, 78:6333, 1983.

[10] G. Bodenhausen, R. R. Void, and R. L. Void. Multiple quantum spin-echo spectroscopy. *J. Magn. Reson.*, 37:93, 1980.

[11] G. J. Bowden, W. D. Hutchison, and J. Kachan. Tensor operator formalism for multiple quantum NMR. I. Spin-1 nuclei. *Molecular Physics*, 67:415, 1986.

[12] L. van Dam, B. Andreasson, and L. Nordenskiold. Multiple-quantum pulsed gradient NMR diffusion experiments on quadrupolar ($i > 1/2$) spins. *Chemical Physics Letters*, 262:737, 1996.

[13] L. E. Kay and J. H. Prestegard. An application of pulse-gradient double-quantum spin echoes to diffusion measurements on molecules with scalar-coupled spins. *J. Magn. Reson.*, 67:103, 1986.

[14] P. W. Kuchel and B. E. Chapman. Heteronucelar double quantum coherence selection with magnetic field gradients in diffusion experiments. *J. Magn. Reson.*, A101:53, 1993.

[15] A. J. Dingley, J. P. Mackay, G. L. Shaw, B. D. Hambly, and G. F. King. Measuring macromolecular diffusion using heteronuclear multiple-quantum pulsed-field-gradient NMR. *Journal of Biomolecular NMR*, 10:1, 1997.

[16] M. R. Bendall, D. T. Pegg, D.M. Doddrell, and J. Field. Inverse DEPT sequence. Polarization transfer from a spin-1/2 nucleus to n spin-1/2 heteronuclei via correlated motion in the doubly rotating reference frame. *J Magn. Reson.*, 51:520, 1983.

[17] K. I. Momot and P. W. Kuchel. Convection-compensating diffusion experiments with phase-sensitive double-quantum filtering. *J. Magn. Reson.*, 574:229, 2005.

[18] G. Zheng, A. M. Torres, and W. S. Price. MQ-PGSTE: A new multi-quantum STE-based PGSE NMR sequence. *J. Magn. Reson.*, 198:271, 2010.

[19] S. Cavadini and P. R. Vasos. Singlet states open the way to longer time-scales in the measurement of diffusion by NMR spectroscopy. *Concepts in Magnetic Resonance*, 32A:68, 2008.

[20] M. Carravetta, O. G. Johannessen, and M. H. Levitt. Beyond the T_1 limit: singlet nuclear spin states in low magnetic fields. *Phys. Rev. Lett.*, 92:153003–1, 2004.

[21] M. Carravetta and M. H. Levitt. Long-lived nuclear spin states in high-field solution NMR. *J. Am. Chem. Soc.*, 126:6228, 2005.

[22] M. Carravetta and M. H. Levitt. Theory of long-lived nuclear spin states in solution nuclear magnetic resonance. *J. Chem. Phys.*, 122:214505, 2005.

[23] S. Cavadini, J. Dittmer, S. Antonijevic, and G. Bodenhausen. Slow diffusion by singlet state NMR spectroscopy. *J. Am. Chem. Soc.*, 127:15744, 2005.

[24] N. N. Yadav, A. M. Torres, and W. S. Price. q-space imaging of macroscopic pores using singlet spin states. *J. Magn. Reson.*, 234:306, 2010.

[25] A. Abragam. *Principles of Nuclear Magnetism.* Clarendon Press, Oxford, 1961.

[26] J. Jeener, A. Vlassenbroek, and P. Broekaert. Unified derivation of the dipolar field and relaxation terms in the bloch-redfield equations of liquid NMR. *J. Chem. Phys.*, 103:1309, 1995.

[27] M. H. Levitt. Demagnetization field effects in two-dimensional solution NMR. *Concepts Magn. Reson.*, 8:77, 1996.

[28] E. D. Minot, P. T. Callaghan, and N. Kaplan. Multiple echoes, multiple quantum coherence, and the dipolar field: Demonstrating the significance of higher order terms in the equilibrium density matrix. *J. Magn. Reson.*, 140:200, 1999.

[29] R. Bowtell and P. Robyr. Structural investigations with the dipolar demagnetizing field in solution NMR. *Phys. Rev. Lett..*, 76:4971, 1996.

[30] R. Bowtell and P. Robyr. Nuclear magnetic resonance microscopy in liquids using the dipolar field. *J. Chem. Phys.*, 10:467, 1997.

[31] L. S. Bouchard and W. S. Warren. Reconstruction of porous material geometry by stochastic optimization based on bulk NMR measurements of the dipolar field. *J. Magn. Reson.*, 170:299, 2005.

[32] L.S. Bouchard, F. W. Wehrli, C.L. Chin, and W.S. Warren. Structural anisotropy and internal magnetic fields in trabecular bone: Coupling solution and solid dipolar interactions. *J. Magn. Reson.*, 176:27, 2005.

12
Tricks of the trade

This book has largely concerned a description and explanation of the physical principles which underpin PGSE NMR. In the numerous examples presented, time-dependent magnetic field gradients are included in pulse sequences as though their role is routine, yet another facility provided by the NMR spectrometer. But in truth, the practical implementation of these sequences requires some care, an awareness of the potential role for experimental artifacts.

Consider for the moment the measurement of self-diffusion, using a typical NMR sample size of a few millimetres. The magnitude of q required to measure dynamic displacements n orders of magnitude smaller than the sample dimensions, will result in a spin dephasing of order 10^n cycles across the sample, with final rephasing in the echo required to be within a few degrees. Consequently, for a typical rms diffusion distance $\langle Z^2 \rangle^{1/2}$ on the order of 0.1 to 10 microns, the time integral of the two gradient pulses in the simple Stejskal–Tanner experiment, must be matched to better than 1 in 10^5, a demanding requirement and one that is not easily monitored by electronic means alone. In addition, if we are to effectively measure molecular displacements over micron or sub-micron distances, we had better be sure that any bulk sample movement over the echo period is significantly smaller. Yet switching magnetic fields on and off within the high polarising field results in Lorentz forces, which in turn induce mechanical motion in the gradient coils, manifest most obviously as an acoustic wave, a 'click' from the gradient coils. Can we ensure that these do not result in sample displacements that mask the underlying molecular dynamics we seek to measure?

In summary then, we need an awareness of the role of sample vibration or convective flow within our sample, of gradient pulse area matching, the role of Lorentz forces, and of eddy currents induced in surrounding conductors by time-varying magnetic fields. But more importantly we need to know how to mitigate these effects, or to avoid them completely. This chapter gives some pointers to the experimentalist that may assist in this regard. We do not seek to provide here a description of gradient coil or RF probe design, matters dealt with in the author's earlier book [1], but instead to give a short description of how to make these devices work effectively.

12.1 Instrumental limits

12.1.1 The diffusion baseline

In the measurement of diffusion, all instrumental artifacts have the effect of enhancing the diffusion coefficient. All PGSE NMR systems have a lower limit of mean-squared displacement below which artefactual attenuation exceeds diffusive attenuation. The challenge in diffusion measurement is to ensure that instrumental factors that contribute to spin phase-spreading are kept below the spread due to diffusion alone. For example, phase instability of the spectrometer, mismatch of the gradient pulse time integrals due to current instability, or vibrations of the sample; these all contribute to additional phase spreading artifacts and so cause a measured diffusion coefficient to appear larger than its true value. In consequence, the better the instrument, the lower the value of D that may be measured. More particularly, a phase variance $\langle (\Delta\phi)^2 \rangle$ gives an apparent mean-squared displacement $\langle Z^2 \rangle = 2q^{-2}\langle (\Delta\phi)^2 \rangle$, and an apparent diffusion coefficient $D = q^{-2}\Delta^{-1}\langle (\Delta\phi)^2 \rangle$, where q is the applied wavevector and Δ the displacement observation time.

Although instrumental artifacts enhance the apparent diffusion coefficient, there does exist a sample-related artifact where one measures too small a diffusion coefficient. Such a condition can arise where internal magnetic field gradients arising from susceptibility inhomogeneity have the effect of opposing the applied gradient pulse [2, 3].

12.1.2   Test samples

An ideal sample with which to test the apparatus is one with small D and long T_2 and T_1, in other words with inhibited translational mobility and relatively free rotational mobility. Polydimethylsiloxane melts meet such requirements, but these materials have considerable chain-length polydispersity and so exhibit a wide distribution of diffusion coefficients within the same sample, leading to multi-exponential decay in the PGSE NMR experiment. For mono-disperse behaviour, semi-dilute high molecular weight polystyrene solutions are ideal, with their relatively free local segmental motion, but with entanglements highly restricting self-diffusion. For example, a 5% w/v solution of 10 MDa polystyrene in (per-deutero) toluene solvent has (for aromatic protons) $T_2 = 70$ ms and $T_1 = 800$ ms, and a self-diffusion coefficient of 6.8×10^{-16} m^{2}s^{-1} [4], at the lower limits of what can be measured using PGSE NMR. Indeed, in such a polymer the spin phase spreading can be dominated not by Brownian motion of the molecules, but by spin-diffusion, the process of magnetisation diffusion along the polymer chain caused by energy-conserving mutual spin flip-flops. The same 5% 10 MDa polymer solution yields a spin-diffusion coefficient of 3.5×10^{-15} m^{2}s^{-1} [5], and this effect must be allowed for in order to obtain the underlying polymer segmental diffusion rate.

Of course, what is directly measured in PGSE NMR is the mean-squared spin displacement $\langle Z^2 \rangle$. The smallest rms distance so far measured is on the order of 25 nm, over a time Δ of around 10 ms [5]. It is this rms displacement limit with which we must compare instrumental phase errors. Note that at 10 ms such a mean-squared displacement, incorporating spin-diffusion effects, corresponds to an effective diffusion coefficient of 3×10^{-14} m^2 s^{-1} for the 5% 10 MDa polystyrene in d-toluene. The same sample returns 3.5×10^{-15} m^2 s^{-1} at $\Delta = 1$ s, (i.e. $\langle Z^2 \rangle^{1/2} = 265$ nm), the diffusion coefficient discrepancy arising from polymer reptation, a dynamics which exhibits a non-Fickian, fractal $\langle Z^2(t) \rangle$ time-dependence. But for the purposes of instrument testing, the sample provides a good reference with which to probe the g, δ, and Δ parameter space of the instrument to ensure that the measured $\langle Z^2 \rangle$ or D values do not exceed the known calibration values. Where these values are exceeded then the result is indicative of instrument artifact, setting a new lower limit for reliable measurement.

12.1.3   Non-Gaussian displacements

A further test of instrument reliability is to be found in the shape of the echo-attenuation data. In samples for which molecules undergo unrestricted Brownian motion, $E(q) = \exp(-\gamma^2\delta^2 g^2 D\Delta)$. A semi-log plot of echo-attenuation data allows us to test for a linear response of the exponent to $\delta^2 g^2 \Delta$ as g, δ, and Δ are independently varied. Most importantly, while deviation from linearity in the form of positive curvature can be associated with a distribution of diffusion coefficients, for example arising from molecular polydispersity, negative curvature is almost invariably another sign of instrumental artifact. Such negative curvature is often associated with increasingly incoherent signal averaging as the time integral of the pulse gradient is increased.

12.2   Conquering artifacts

Having suggested a means by which instrument artifact might be detected, we now lay out the various lines of attack whereby these unwanted phase fluctuations may be minimised. The experimenter needs to iterate amongst these avenues for improvement, using the slow diffusion sample, and the 'lowest $\langle Z^2(t)\rangle$' or 'lowest D' measure, as the indicator of success. But by far the most efficient route to a solution is by real-time monitoring of individual echoes, so that the experimenter has available an immediate visual response to any adjustment within the apparatus.

12.2.1   Real-time monitoring

Using the semi-dilute polymer sample, the spectrometer acquisition window should be set up so that a complete echo is sampled and displayed as acquired in real time. That echo should ideally be stable. Figure 12.1 shows three types of typical echo instability associated with pulsed gradient artifacts. The consequence of co-adding such unstable signals in the signal-acquisition process is a reduced signal amplitude, contributing to unphysical downward curvature in the semi-logarithmic echo-attenuation plot. By monitoring echo instability in real time, successive experimental modifications can be made and tested, the aim being to produce a consistent echo envelope, phase, and frequency.

Experimenters be warned! Modern NMR instruments permit the user to prepare many hours of automated experiments, traversing a wide range of parameter space. Failure to 'converse with the instrument', ensuring all such experiments experience stable echo conditions, can lead to the collection of worthless data.

12.2.2   Sample movement

Vibration

The most common source of echo instability is sample movement due to vibration. In instruments where the gradient coils produce a noticeable acoustic pulse on current switching, the associated vibration can easily cause sample movement relative to the gradient field. The matter of acoustic pulses and their minimisation is dealt with in the section 'small is beautiful'. But for a given coil set-up, the best remedy is to ensure that the sample is firmly inserted in the gradient/RF assembly, most easily with some Teflon tape wrapped round the glass sample tube to provide a firm fit. In systems

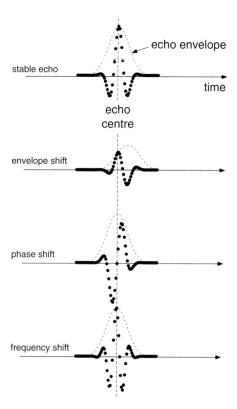

Fig. 12.1 Sequence of echoes with ideal echo at top and successively instability in the echo envelope, the echo phase, and the echo frequency. Sample movement, gradient pulse integral mismatch, and time-dependent fields due to eddy currents induced by gradient pulse switching can all contribute to a combination of these effects.

where the RF coil is free-standing with respect to the gradient coils, this may require that the tape or fixing collar is applied at a position where the sample tube is able to be directly registered with the gradient coil assembly.

This tight coupling works well for a viscous liquid sample where the relative motion of liquid and holder is strongly damped under the influence of the gradient coil acoustic pulse. However, for some samples, such as powders, the acoustic shock can cause particle jumping. Here mechanical decoupling of the sample from the gradient coil assembly can prove the most effective solution [6].

Note that many modern NMR spectrometers allow for the superconducting magnet to be raised on air-damped bearings. This can prove helpful in mitigating some vibration effects.

Convection

Whenever a liquid sample is used in the presence of a temperature gradient, convective flow is possible, an effect that plays havoc with diffusion measurements. The onset of convection is determined by a critical balance between hydrostatic and viscous forces, as reflected by the Rayleigh–Bénard number [7, 8]

$$Ra = \frac{\alpha \Delta T g L^3}{\nu \kappa} \tag{12.1}$$

where g is the acceleration due to gravity, and α, ν, and κ are the coefficients of thermal expansion, kinematic viscosity, and thermal diffusivity of the fluid, respectively. ΔT is the vertical temperature difference across the sample and L is the thickness of the fluid layer. For a fluid of infinite horizontal extent, the critical Rayleigh–Bénard number for the onset of convection is $Ra_c \sim 1700$. However, an NMR sample is typically a vertical cylindrical tube, characterised by a radius to depth ratio, r/l, much less than one. For such a geometry, the critical value of Ra for the onset of convection is given by [9]

$$Ra_c \sim 200 \frac{l^4}{r^4} \tag{12.2}$$

For example, a 4-mm ID NMR tube with sample length 20 mm, has $r/l = 10$, which suggests a critical Ra for the onset of convection of $\sim 2 \times 10^6$. Note the dependence on the aspect ratio as the fourth power. This tells us us that an effective means of suppressing convection is to use a narrow diameter capillary. Of course many slowly diffusing systems are often highly viscous and as such, Rayleigh–Bénard convection presents no problem. But as an example, consider the polymer system 5% w/w $M_w = 127,000$ Da polystyrene/cyclohexane solution, studied by Manz *et al.* [10], for which the polymer solution viscosity is $\eta = 1.43 \times 10^{-3}$ Pa s, close to that of water. For a 25-mm length sample, with 3.0-mm ID, Ra_c for the onset of convection $\sim 10^7$. Using the literature values of α and κ, this value of Ra occurs at a temperature difference from top to bottom of 4 K.

12.2.3   Eddy currents

Throughout this book we have represented gradient pulses as schematic rectangles, implying in the process an infinite rate of rise and fall of gradient currents and associated field. Of course, such a simplistic representation is impossible in practice. The maximum current switching speed, di/dt, is partly limited by the power-supply voltage, which must equal $Ri + Ldi/dt$ where L is the load inductance and R the load resistance. The other limitation is the power-supply bandwidth, expressed in the time domain as a 'slew rate' or maximum rate of change of current. And even if our power supply were able to deliver high slew rates, there is another reason why we would wish to limit the rate of rise and fall of our gradient current pulses. The rapidly changing magnetic fields arising from the gradient pulse switching interact with surrounding metal to induce eddy currents proportional to the current switching rate, di/dt, and these currents in turn have associated magnetic fields, which not only distort the gradient profiles around the sample but also can persist for tens of milliseconds after the gradient pulse has been turned off. While these 'eddy gradients' are not such a severe problem in the electromagnet, where the pole pieces present an unfavourable geometry for current flow, in the superconducting solenoid with its cylindrical tubes of surrounding metal these currents can be devastating in their effect. For this reason PGSE NMR systems in superconducting magnets benefit from the largest possible diameter bore space.

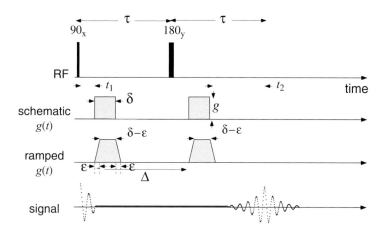

Fig. 12.2 Simple Stejskal–Tanner PGSE NMR sequence showing idealised rectangular gradient current pulses, along with a practical ramped alternative, designed to reduce the effect of eddy-current-induced field distortions.

Unless the gradient coils are much smaller than the magnet bore, some manner of eddy gradient compensation or suppression is essential. The simplest remedy is to ramp the rise and fall of the current pulses as shown in Fig. 12.2, with typical rise times, ϵ, of a few tens of microseconds often being sufficient to mitigate any noticeable eddy-current problems. Again, the best diagnostic is to monitor the real-time echo stability while the pulse rise time is adjusted. Note that the shift to a trapezoidal pulse shape implies a slight (but often insignificant) change to the Stejskal–Tanner expression for the echo attenuation, as outlined in Section 5.5.1.

There does exist another approach to eddy-current-induced field compensation, through what is known as 'pre-emphasis' and 'de-emphasis' of the coil current, a process that relies on Lenz's law requirement that the sign of fields associated with eddy currents will be opposed to the change that produced them. By deliberately overdriving the current at the leading and falling edges, the coils themselves produce fields that compensate the unwanted induced gradients. The optimisation of pre-emphasis and de-emphasis currents is a complex process, requiring adjustment of a multitude of time constants and of amplitudes for exponential currents, which are added to the desired waveform. This compensation can never be perfect since the spatial distribution of the additional fields produced by the gradient coils will never match those produced by current in the surrounding metal.

The best philosophy is to ensure eddy gradient suppression by using gradient sets the fields of which are zero outside the coil boundaries. Active shielding [1, 11] involves adding a second layer of current density outside the primary coil surface, in a design intended to compensate the primary coil field at all points in space outside the screen.

Because this approach is so successful at removing eddy current effects without the need for pre-emphasis or de-emphasis pulse shaping, it provides a means of generating rapidly switched gradient pulses in which ease of use is traded against complexity in coil design.

12.2.4   Pulse mismatch

The stability of the gradient-current supply sets the ultimate capability of the PGSE NMR experiment. k-space encoding merely requires that we provide a suitable integral of current through the gradient coils, to a precision and accuracy consistent with the desired spatial resolution, measured as a fraction of sample dimension and possibly on the order of 0.5%. By contrast, in q-space applications the need to control current pulse integral is considerably more stringent. In these experiments we may be seeking to measure sub-micron displacements with sample dimensions of several millimetres, thus demanding gradient pulse balance better than 10^{-5}. Ripple or noise on a current supply at these low levels is extremely difficult to observe directly via an oscilloscope. Indeed, the very best sensor for current quality will be the real-time PGSE NMR measurement.

Mains hum

First and foremost, the matter of mains 'hum' needs to be attacked. Any ripple due to residual AC components in the current supply will cause gradient pulse mismatch. This will be at a maximum when the separation Δ between the pulses is half a mains cycle, it will fluctuate when Δ is incommensurate, and will be minimised when Δ is exactly equal to a mains cycle. Mains frequencies are either 50 Hz or 60 Hz depending in which part of the world one is operating. Alternate Δ between an incommensurate value and 20 ms (50 Hz) or 16.67 ms (60 Hz), and look for an improvement in the echo stability. If such improvement exists, your power supply is inadequate to the task and unless you can reduce the ripple you may be condemned to only be able to use Δ values that are a multiple of the mains period. A feasible solution, which allows greater timing freedom, is to use a 'mains trigger" to set the pulse sequence timing [12]. This ensures a fixed gradient pulse integral mismatch at any particular separation Δ, a discrepancy which can be compensated by a small adjustment in the second gradient pulse duration or by means of a stabilising background gradient (see section 12.3.3).

Power-supply 'sagging'

Even the best power supply will have limits to its ability to supply transient power output at a consistent value. After any recovery period the first pulse will typically deliver a little more power than the succeeding. However, there will exist a steady-state output level in a repetitive train where the pulses deliver close to equal power. The trick, therefore, is to run the power supply using a train of 'dummy' pulses that have no influence on the spin system, selecting for the actual PGSE NMR measurement pulses later in the train when the power supply has settled. The idea is illustrated in Fig. 12.3.

Gradient pulse compatibility with power supply

The gradient $g(t)$ is commonly generated using a current $i(t)$ from a current source, a high-power operational amplifier driven in a current-controlled mode. This control

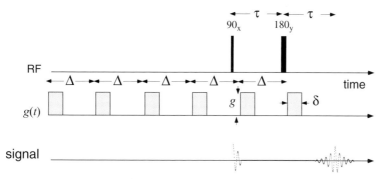

Fig. 12.3 Simple Stejskal–Tanner PGSE NMR sequence in which a train of 'dummy' gradient current pulses is applied prior to excitation by the 90_x° RF pulse. By ensuring equal spacing between all the gradient pulses, the power supply settles to a steady-state power output value before the final pulse pair comprising the actual measurement.

is achieved using a feedback loop involving a sensing resistor. The operational amplifier compares the voltage measured across this resistor with a voltage function $v(t)$, representing the the the desired time-dependence of the gradient pulse. Stallmach and Galvosas [13] have pointed out that if the rise or fall rate of the gradient current pulse is set too high, the voltage across the coil may exceed the maximum DC voltage of the power supply, v_B. When this happens, proper current control is lost and error currents caused by the ripple, hum, and noise of the DC power supply will be unsuppressed by the feedback loop to the gradient current. These authors point out that by optimally shaping the gradient current pulse using an appropriate setpoint function, $v(t)$, such overloads can be avoided. The best shape for maximum pulse-switching speed is an exponential growth controlled by the coil time constant L/R, where L and R are the coil inductance and resistance, respectively, and where the maximum slew rate is limited to v_B/L.

12.2.5 Small is beautiful

Given the opportunity to design and build one's own gradient coil, it is worth noting that most of the artifacts discussed in this chapter can be considerably alleviated by choosing to use the smallest practicable gradient coil assembly [14]. Downsizing leads to reduced coil inductance and hence reduced power-supply requirements. Most importantly, there will be the reduced Lorentz torque on the coil array and consequently less acoustic response and sample vibration. In addition, there will of course be significantly reduced induction of eddy currents in the surrounding magnet due to stray pulsed magnetic fields, in particular because a gradient coil comprises an opposed pair of dipoles and therefore represents an octopole the stray fields of which attenuate with distance r as $(r/a)^{-5}$, where a is the radius of the coil. Finally, a small gradient coil implies a smaller sample volume. While there is a signal-to-noise price to be paid, smaller sample volumes both inhibit convection and lead to a less stringent requirement for gradient pulse time integral matching, since the area mismatch results in a dephasing error proportional to the sample dimensions.

12.2.6 Fringe field diffusometry

A particularly effective means of avoiding most problems associated with gradient pulse mismatch, eddy currents, and sample movement is to avoid the use of current-switched pulsed gradients altogether. Kimmich and co-workers [15] suggested using the fringe field of a superconducting magnet to provide very high constant magnetic field gradients, on the order of $10\,\mathrm{T/m}$, which can be turned into pulses of the effective gradient by means of a stimulated-echo RF sequence. The idea is shown in Fig. 12.4.

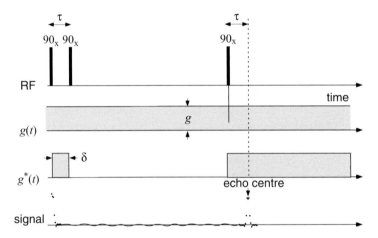

Fig. 12.4 Fringe field diffusometry pulse sequence in which a stimulated echo is used in the presence of a constant strong magnetic field gradient to create an effective gradient sequence, which appears to the spins as pulses. Note that the 90_x° RF pulses are slice selective and that the signal in the frequency domain is related to the RF pulse spectrum, so that the time-domain signal is reminiscent of the RF pulse shape.

The great advantage of the method is that, unlike gradient fields produced by pulsing current through coils, the highly stable superconducting fringe field of the magnet has practically no discernable temporal fluctuation and induced-eddy-current effects are absent because the formation of effective gradient pulses is via the 'framing' action of the RF pulses on the constant gradient. The disadvantages of the fringe field method are, first, that spectral resolution is lost by the need to detect the signal in the presence of the constant gradient and, second, that because of the spectral spread associated with the constant gradient, the signal is weak, the RF pulses being inevitably slice-selective, exciting spins residing only within their bandwidth. Even with high-power RF, and hence brief wide-bandwidth pulses, a typical sample slice contributing to the experiment might be on the order of 100 microns thick. These disadvantages are no particular handicap where one is working with a single dominant molecular species at high concentration, for example in studying the diffusion of polymer melts [16] or the restricted diffusion of a mono liquid in a porous matrix [17].

12.3 Pulse-sequence compensation

Having optimised the apparatus, there do exist pulse-sequence design strategies that can further compensate for remaining artifacts, including those caused by eddy-current-induced fields, convection, sample movement, and gradient pulse mismatch.

12.3.1 Eddy current fields

One of the problems associated with eddy currents in the magnet assembly induced by the switching of gradient pulses concerns the persistence of time-dependent fields which perturb the spectral resolution during signal acquisition. To assess the effect of eddy fields, an ideal test is to apply a gradient pulse followed by a variable delay time before an excitation 90° RF pulse and signal acquisition, as shown in Fig. 12.5. As the delay is increased, allowing for eddy-field decay, the experimenter gains insight regarding the longevity of these fields by examining the gradual improvement of the NMR spectrum quality.

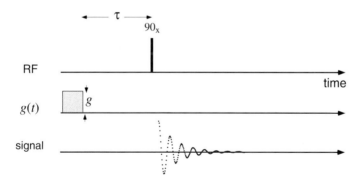

Fig. 12.5 Simple excitation and acquire RF pulse sequence, in which the 90° RF pulse is preceded by a gradient pulse and a variable delay. As the delay is increased, the decay of eddy fields results in decreasing spectral distortion. By this means decay time constants can be established.

A simple solution is the longitudinal eddy current decay (LED) method of Gibbs and Johnson [18], in which the magnetisation at the echo formation is stored along the longitudinal axis by means of 90° RF pulses, as shown in Fig. 9.14. During the storage period, τ_e, the unwanted fields can be allowed to attenuate before recall of the magnetisation for readout using a second 90° RF pulse. Clearly T_1 provides an upper limit for the time over which such storage is possible, and a suitable phase-cycling scheme is needed to ensure that only magnetisation arising from the initial excitation pulse contributes to the final signal.

12.3.2   Convection compensation

Inhibiting convection by restricting sample dimensions comes at the price of a loss of signal. An alternative strategy is to utilise flow compensation methods to remove unwanted phase shifts due to convective flow in the sample [19]. These are effective only if the convective flow is laminar and constant over the timescale of the flow-compensated PGSE gradient sequence. Where the convective flow fluctuates over the duration of the gradient pulse train, or where turbulence is present, the method cannot work effectively. However, many practical examples of convective flow driven by small temperature gradients across the sample do exhibit the necessary stability. Figure 5.11(b) shows an effective gradient sequence resulting in the echo condition $\int_0^t g^*(t')dt' = 0$, but with first moment $\int_0^t t'g^*(t')dt'$ zero, thus resulting in insensitivity to flow.

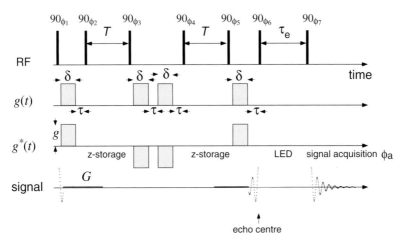

Fig. 12.6 Double stimulated-echo PGSE NMR sequence with zero first moment of the effective gradient, as suggested by Jerschow and Mueller, in which an LED segment (duration τ_e) is used to allow for the decay of fields associated with induced eddy currents. The phase cycle is $\phi_1 = x, y, -x, -y$; $\phi_2 = \phi_5 = \phi_6 = \phi_7 = x$; $\phi_3 = -x, -y$; $\phi_4 = 4(-x), 4(x)$; $\phi_a = x, x, -x, -x, -x, -x, x, x$. Homospoil gradients should be applied during the z-storage intervals. (Adapted from Jerschow and Müller [19].)

Figure 12.6 shows an example of a flow-compensated PGSE NMR sequence [19] based on a double stimulated echo, for which the echo-attenuation expression is given by

$$E(q) = \exp\left(-q^2(T + \frac{4\delta}{3} + 2\tau)\right) \qquad (12.3)$$

where $q = \gamma g \delta$. This is but one of a set of sequence options. Further examples are provided in references [20–22]. Note that one of the consequences of convection-compensating sequences is the lack of clear definition of the time over which diffusion occurs. This presents no disadvantage in the case of unrestricted free diffusion, where an exact expression for the echo attenuation is available. And, in the case of samples exhibiting restricted diffusion, where we seek a definitive timescale, the problem of convective perturbation is unlikely to arise.

However, there is a highly effective means of avoiding convection problems while retaining sensitivity to timescale-dependent diffusivity, and that is to shift to frequency-domain analysis using the CPMG methodology outlined in Section 5.7.2. Here flow effects are completely suppressed over the period T of the effective gradient cycle, while restricted diffusion can be studied by probing the diffusion spectrum through the gradient waveform frequency sweep.

12.3.3 Gradient pulse mismatch compensation

The effects of sample movement and gradient-pulse mismatch are exhibited in the echo phase of the PGSE NMR experiment. Movement results in a phase shift common to all spins, while mismatch results in position-dependent local phase shifts. If the entire sample moves by $\Delta\mathbf{r}$ between the first and second pulses, while the gradient mismatch is given by $\Delta\mathbf{q}$, then the narrow gradient pulse equation 5.85 must be rewritten

$$E(\mathbf{q}, \Delta) = \int \overline{P}(\mathbf{R}, \Delta) \exp(i\mathbf{q} \cdot \mathbf{R} + \Phi)d\mathbf{R}, \qquad (12.4)$$

where

$$\mathbf{q} \cdot \mathbf{R} + \Phi = (\mathbf{q} + \Delta\mathbf{q}) \cdot (\mathbf{r} + \mathbf{R} + \Delta\mathbf{r}) - \mathbf{q} \cdot \mathbf{r} \qquad (12.5)$$

We shall only be concerned with sample motion displacements Δz parallel to $\mathbf{q}$ and will presume that these are common to all spins in the sample. This 'rigid body' assumption is reasonable if the material being studied is sufficiently viscous that we wish to observe very slow motion. Noting further that Δq is parallel to $\mathbf{q}$,

$$E(q, \Delta) = \{ \int \overline{P}(Z, \Delta) \exp(iqZ)dZ \}$$
$$\times \{ \exp[i(q + \Delta q)\Delta z] \}$$
$$\times \{ \int \rho(z) \exp(i\Delta qz)dz \} \qquad (12.6)$$

The first bracketed term in eqn 12.6 is the Fourier transform of the average propagator, the quantity that contains the information about microscopic dynamic displacements. It is sensible to label this term $E_0(q, \Delta)$, representing the unperturbed echo attenuation that we seek to measure. The second term is a phase shift resulting from net motion of the sample, caused, for example, by vibration. The final term is the integral of position-dependent phase shifts and is clearly reminiscent of k-space encoding. While the second term could be removed by autophasing or modulus calculation, the third is only amenable to correction once the spatial dependence of the phase shifts is unravelled. This can be achieved by means of a read gradient, as shown in Fig. 12.7. Of course, the price paid is that NMR spectral resolution is lost. Hence the method is really only applicable when we are able to uniquely identify the deliberately broadened NMR spectrum with the desired molecular species.

By using the same gradient coil responsible for pulse mismatch in g to generate a much smaller read gradient G, the phase shifts can be resolved. Given pixels separated

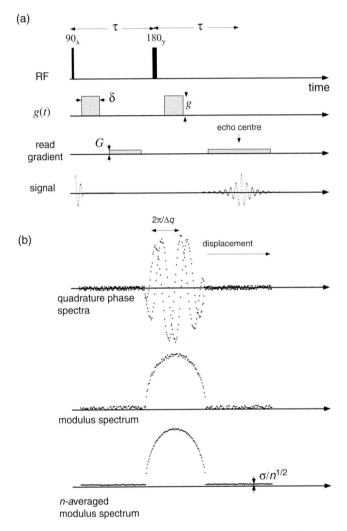

Fig. 12.7 (a) PGSE-MASSEY pulse sequence with q-gradient pulses (amplitude g) and k-gradient pulses (amplitude G). The read gradient enables the restoration of spatially dependent phase shifts at a time $-\Delta q/\gamma G$ with respect to the echo centre. (b) Simulated data for a PGSE experiment in which a q-pulse mismatch is present. The spatially dependent phase shift (fluctuating spatial period $2\pi/\Delta q$) is revealed by Fourier transformation of the signal acquired in the presence of a read gradient generated with the same coil. The phase artifacts are removed by taking the modulus of the spectrum.

by $1/NT$ it is clear that we require $G > \Delta q/\gamma NT$. Where Δq arises from a gradient fluctuation, Δg, G will need to be comparable with this difference. Note that the effect of the read gradient in the pulse sequence shown in Fig. 12.7 is to cause a coherent superposition at the instant $t = -\Delta q/\gamma G$, arising either before or after the expected echo centre depending on the sign of the mismatch. If G is made very large then the echo may be 'centred' by brute force, although this will result in a wide spectral spread and consequent signal-to-noise ratio reduction. The best result is obtained by Fourier transforming the echo signal with respect to k where $k = \gamma G t$. This yields

$$E(q, \Delta) = E_0(q, \Delta)\{\exp[i(q + \Delta q)\Delta z]\, \rho(z)\exp(i\Delta q z)\} \qquad (12.7)$$

The result of this transformation is shown in Fig. 12.7(a). Provided that the entire echo is sampled, $E_0(q, \Delta)$ can be recovered by computing the spectrum modulus, so that signal averaging can then proceed despite the fact that Δz and Δq may fluctuate from one acquisition to the next.

The process of signal averaging under power spectrum addition is easily represented, given a spectrum area A corresponding to the echo centre amplitude. For n acquisitions (labelled by i) and with m pixels (labelled by j), the power sum of the signal a_j and noise σ_{ij} in pixel j is

$$power(j) = \sum_{i=1}^{n}(a_j + \sigma_{ij})^2 = na_j^2 + n\sigma^2 + 2a_j\sum_{i=1}^{n}\sigma_{ij} \qquad (12.8)$$

where σ is the rms noise per pixel. The term $n\sigma^2$ is the noise power baseline and can be easily calculated (from data points outside the spectrum) and subtracted. Following this the net power is given by

$$power(j) = na_j^2(1 + 2\sigma_j/n^{1/2}a_j) \qquad (12.9)$$

σ_j being a noise amplitude with the same rms value, σ. The second term in the parentheses may be made arbitrarily small by averaging with n sufficiently large. Thus, using the binomial approximation, the square root of the power in pixel j is simply $n^{1/2}a_j + \sigma_j$ and the integral over the spectrum is

$$signal = n^{1/2}A + \sum_{j=1}^{m}o_j \qquad (12.10)$$

The noise sum is a random value centred about zero and with standard deviation $m^{1/2}\sigma$. In consequence the overall signal-to-noise ratio is $n^{1/2}A/m^{1/2}$, which represents a degradation by a factor $m^{1/2}$ compared with the case where no read gradient is employed. The use of a read gradient coupled with modulus-squared spectral addition has been dubbed PGSE-MASSEY [23] for Modulus Addition using Spatially Separated Echo spectroscopY.

The loss of signal-to-noise ratio by $m^{1/2}$ represents a very small price to be paid for the gain in PGSE resolution. For example, at the q-value where mismatch effects become important, an increase in q by 100 requires spectral-spatial spreading into 100 pixels, degrading the signal-to-noise ratio by 10. In many interesting applications, such as investigation of internal motion in high polymer melts, the PGSE experiment is frustrated by phase instability alone, and the facility to probe dynamic displacements on a distance scale of 1 to 10 nm, some two orders of magnitude lower than existing limits, is well worth the loss of a factor of ten in signal sensitivity.

12.3.4   Varying q

In varying the magnitude of the q-vector in PGSE NMR, either the gradient pulse amplitude g or the pulse duration δ may be swept. Clearly, g variation provides better time-definition of the pulse and allows one to fix the pulse duration to conditions where the 'narrow gradient pulse approximation' applies. Apart from these factors there is no reason why either g or δ might not be used as the control parameter for the experiment, and both approaches are used with equal success.

One particularly interesting aspect of g variation is the ease of traversing both negative and positive q-space, according to the sign of the current through the gradient coil. In the measurement of displacement propagators, the ability to sample the signal response to a sweep of both negative and positive q-domains confers particular advantage in that the propagator will have no dispersion component. This greatly simplifies the delicate task of 'phasing the spectrum', a process which can be a bit arbitrary when the propagator is broad, by simply seeking the imaginary spectrum null or, in a brute force manner, by calculating the modulus of the complex spectrum. Displacement propagators are invariably better represented when both signs of q-space are sampled.

12.4   Final thoughts

This chapter briefly summarises some of the better understood problems of practical implementation associated with PGSE NMR. The relative importance of each of these will depend on the particular apparatus. With awareness raised, the experimenter needs to explore the parameter space of the instrument and determine the boundaries beyond which dragons lie.

In the case of diffusion measurement, mere appearance of the echo signal as a Gaussian echo decay is no guarantee of a well-functioning experiment. Cruelly, many stochastic phase fluctuations unrelated to molecular translation will mimic just such an apparent adherence to the Stejskal–Tanner formula. Ultimately, there is no substitute for real understanding of the intrinsic physics and vigilant scepticism on the part of the experimenter, who should always verify the domain of measurement using a known diffusion coefficient for calibration.

By comparison with PGSE NMR, MRI is a safer method, plagued by fewer experimental difficulties. The reasons should be abundantly clear. In the measurement of translational dynamics by magnetic resonance, we traverse three to four orders of magnitude of translation time (milliseconds to tens of seconds), and, potentially, an even greater number of decades of translation distance (10 nm to 10 mm). There is nothing 'routine' about this remarkable experimental tool, nothing that permits the 'press of a button' and the delivery of a meaningful number, except where we confine the operator to a limited and well-proven domain. Yet, with care, calibration, and craft, good experimentalists have successfully used this method right to the edge of 'dragon lands', and done so with accuracy and precision. The key to success is physical insight and understanding. Hopefully, this book will contribute in some small measure to advancing those virtues.

References

[1] P. T. Callaghan. *Principles of Nuclear Magnetic Resonance Microscopy.* Oxford, New York, 1991.

[2] J. Zhong, R. P. Kennan, and J. C. Gore. Effects of susceptibility variations on NMR measurements of diffusion. *J. Magn. Reson.*, 95:267, 1991.

[3] S. Vasenkov, P. Galvosas, O. Geier, N. Nestle, F. Stallmach, and J. Kärger. Determination of genuine diffusivities in heterogeneous media using stimulated echo pulsed field gradient NMR. *J. Magn. Reson.*, 149:228, 2001.

[4] M. E. Komlosh and P. T. Callaghan. Segmental motion of entangled random coil polymers studied by pulsed gradient spin echo nuclear magnetic resonance. *J. Chem Phys*, 109:10053, 1998.

[5] M. E. Komlosh and P. T. Callaghan. Spin diffusion in semi-dilute random coil polymers studied by PGSE NMR. *Macromolecules*, 33:6824, 2000.

[6] N. K. Bar, J. Kärger, C. Krause, W. Schmitz, and G. Seiffert. Pitfalls in PFG NMR self-diffusion measurements with powder samples. *J. Magn. Reson. A*, 113:278, 1995.

[7] H. Bénard. Les tourbillons cellulaire dans une nappe liquide transportant de la chaleur par convection en regime permanent. *Annals Chim. Phys.*, 11:1261, 1900.

[8] Lord Rayleigh. On convective currents in a horizontal layer of fluid when the higher temperature is on the underside. *Phil. Mag.*, 32:529, 1916.

[9] G. S. Charlson and R. L. Sani. On convective instability in a bounded cylindrical fluid layer. *Int. J. Heat and Mass Transfer*, 14:2157, 1971.

[10] B. Manz, J. D. Seymour, and P. T. Callaghan. PGSE NMR observations of convection. *J. Magn. Reson.*, 125:153, 1997.

[11] P. Mansfield and B. Chapman. Active magnetic screening of coils for static and time-dependent magnetic field generation in NMR imaging. *J. Phys. E.: Scient. Instrum.*, 19:540, 1986.

[12] P. Galvosas, F. Stallmach, G. Seiffert, J. Kaerger, U. Kaess, and G. Majer. Generation and application of ultra-high-intensity magnetic field gradient pulses for NMR spectroscopy. *J. Magn. Reson.*, 151:260, 2001.

[13] F. Stallmach and P. Galvosas. Spin echo NMR diffusion studies. *Ann. Reports on NM Spectroscopy*, 61:52, 2007.

[14] P. T. Callaghan, M. E. Komlosh, and M. Nyden. High magnetic field gradient PGSE NMR in the presence of a large polarizing field. *J. Magn. Reson.*, 133:177, 1998.

[15] R. Kimmich, W. Unrath, G. Schnur, and E. Rommel. NMR measurement of small self-diffusion coefficients in the fringe field of superconducting magnets. *J. Magn. Reson.*, 91:136, 1991.

[16] E. Fischer, R. Kimmich, and N. Fatkullin. NMR field gradient diffusometry of segment displacements in melts of entangled polymers. *J. Chem. Phys.*, 104:9174, 1996.

[17] E. Farrhar, I. Ardelean, and R. Kimmich. Probing four orders of magnitude of the diffusion time in porous silica glass with unconventional NMR techniques. *J. Magn. Reson.*, 182:215, 2006.

[18] S. J. Gibbs and C. S. Johnson. A PFG NMR experiment for accurate diffusion and flow studies in the presence of eddy currents. *J. Magn. Reson.*, 93:395, 1991.

[19] A. Jerschow and N. Müller. Suppression of convection artifacts in stimulated-echo diffusion experiments. Double-stimulated-echo experiments. *J. Magn. Reson.*, 125:372, 1997.

[20] A. Jerschow and N. Müller. Convection compensation in gradient enhanced nuclear magnetic resonance spectroscopy. *J. Magn. Reson.*, 132:13, 1998.

[21] J. Krane J. G. H. Sorland, J. G. Seland and H. W. Anthonsen. Improved convection compensating pulsed field gradient spin-echo and stimulated-echo methods. *J. Magn. Reson.*, 142:323, 2000.

[22] K. I. Momot and P. W. Kuchel. Convection-compensating PGSE experiment incorporating excitation-sculpting water suppression (CONVEX). *J. Magn. Reson.*, 169:92, 2004.

[23] P.T. Callaghan. PGSE-MASSEY, a sequence for overcoming phase instability in very high gradient spin echo NMR. *J. Magn. Reson.*, 88:493, 1990.

Index